3O7

CW01432186

Circuits

Elektor Electronics (Publishing)

Elektor Electronics (Publishing)

P.O. Box 190 • Tunbridge Wells • England TN5 7WY

British Library Cataloguing in Publication Data
A catalogue record for this book is available from the British Library

ISBN 0 905705 62 9

Editor: Jan Buiting

Translation: Len Seymour, Jan Buiting, Kenneth Cox, Richard Hardy

Prepress production: Patrick Wielders

First published in the United Kingdom 2000

© Segment BV 2000

Printed in the Netherlands by WILCO, Amersfoort

Contents

Other books from Elektor Electronics

Safety

In all mains-operated equipment certain important safety requirements must be met. The relevant standard for most sound equipment is *Safety of Information Technology Equipment, including Electrical Business Equipment* (European Harmonized British Standard BS EN 60950:1992. Elecxtrical safety under this standard relates to protection from
* a hazardous voltage, that is, a voltage greater than 42.4 V peak or 60 V d.c.;
* a hazardous energy level, which is defined as a stored energy level of 20 Joules or more or an available continuous power level of 240 VA or more at a potential of 2 V or more;
a single insulation fault which would cause a conductive part to become hazardous;
* the source of a hazardous voltage or energy level from primary power;
* secondary power (derived from internal circuitry which is supplied and isolated from any power source, including d.c.)

Protection against electric shock is achieved by two classes of equipment.

Class I equipment uses basic insulation ; its conductive parts, which   may become hazardous if this insulation fails. must be connected to the supply protective earth.

Class II equipment uses double or reinforced insulation for use where there is no provision for supply protective earth (rare in sound engineering – mainly applicable to power tools).

If Class II equipment is totally enclosed by non-conductive durable insulating material , it is called 'insulation-encased Class II equipment'. If it is totally enclosed by a metallic enclosure in which double or reinforeced insulation is used throughout, it is called 'metal-encased Class II equipment'. Even so, where supply protective earth is available, it should be used if at all possible.

The use of a a Class II insulated transformer is preferred, but note that when this is fitted in a Class I equipment, this does not, by itself, confer Class II status on the equipment.

Electrically conductive enclosures that are used to isolate and protect a hazardous supply voltage or energy level from user access must be protectively earthed regardless of whether the mains transformer is Class I or Class II.

There is no requirement for a safety earth if the hazardous supply voltage or energy level area enclosure is non-conductive to Class II insulation requirements and the insulation in the mains transformer is Class II, although a mains earth may still need to be connected for functional purposes.

Always keep the distance between mains-carrying parts and other parts as large as possible, but never less than required.

If at all possible, use an approved mains entry with integrated fuse holder and on/off switch. If this is not available, use a strain relief on the mains cable at the point of entry. In this case, the mains fuse or circuit breaker should be placed after the double-pole on/off switch unless the fuse is a Touchproof® type or similar. Close to each and every fuse must be affixed a label stating the fuse rating and type. The rating of a slow fuse should be not greater than 1.25 times the normal operating current, whereas that of a fast fuse should be equal to the operating current. Fast fuses are used, for instance, in case multiple secondary windings, but if there is an electrolytic capacitor behind the secondary, a slow fuse must be used to allow for surges in the charging current.

The separate on/off switch, which is really a 'disconnect device', should be an approved double-pole type (to switch the phase and neutral conductors of a single-phase mains supply). In case of a three-phase supply, all phases and neutral (where used) must be switched simultaneously. A pluggable mains cable may be considered as a disconnect device. In an approved switch, the contact gap in the off position is not smaller than 3 mm.

The on/off switch must be fitted by as short a cable as possible to the mains entry point. All components in the primary transformer circuit, including a separate mains fuse and separate mains filtering components, must be placed in the switched section of the primary circuit. Placing them before the on/off switch will leave them at a hazardous voltage level when the equipment is switched off.

If the equipment uses an open-construction power supply which is not separately protected by an earthed metal screen or insulated enclosure or otherwise guarded, all the conductive parts of the enclosure must be protectively earthed using green/yellow wire (green with a narrow yellow stripe – do not use yellow wire with a green stripe). The earth wire must not be daisy-chained from one part of the enclosure to another. Each conductive part must be protectively earthed by direct and separate wiring to the primary earth point which should be as close as possible to the mains connector or mains cable entry. This ensures that removal of the protective earth from a conductive part does not also remove the protective earth from other conductive parts.

Pay particular attention to the metal spindles of switches and potentiometers: if touchable, these must be protectively earthed. Note, however, that such components fitted with metal spindles and/or levers constructed to the the relevant British Standard fully meet all insulation requirements.

The temperature of touchable parts must not be so high as to cause injury or to create a fire risk.

Most risks can be eliminated by the use of correct fuses or circuit breakers, a sufficiently firm construction, correct choice and use of insulating materials and adequate cooling through heat sinks and by extractor fans.

The equipment must be sturdy: repeatedly dropping it on to a hard surface from a height of 50 mm must not cause damage. Greater impacts must not loosen the mains transformer, electrolytic capacitors and other important components.

Do not use dubious or flammable materials that emit poisonous gases.

Shorten screws that come too close to other components.

Keep mains-carrying parts and wires well away from ventilation holes, so that an intruding screwdriver or inward falling metal object cannot touch such parts.

As soon as you open an equipment, there are many potential dangers. Most of these can be eliminated by disconnecting the equipment from the mains before the unit is opened. But, since testing requires that it is plugged in again, it is good practice (and safe) to fit a residual current device (RCD)*, rated at not more than 30 mA to the mains system (it may be fitted inside the outlet box or multiple socket. RCDs more sensitive than 30 mA need to be used only if the leakage current is expected to remain below 30 mA, which is rarely the case.

* Sometimes called residual current breaker – RCB – or residual circuit current breaker –RCCB)

1. Use a mains cable with moulded-on plug.
2. Use a strain relief on the mains cable.
3. Affix a label at the outside of the enclosure near the mains entry stating the equipment type, the mains voltage or voltage range, the frequency or frequency range, and the current drain or curent drain range.
4. Use an approved double-pole on/off switch, which is effectively the 'disconnect device'.
5. Push wires through eyelets before soldering them in place.
6. Use insulating sleeves for extra protection.
7. The distance between transformer terminals and core and other parts must be ≥ 6 mm.
8. Use the correct type, size and current-carrying capacity of cables and wires.
9. A printed-circuit board like all other parts should be well secured. All joints and connections should be well made and soldered neatly so that they are mechanically and electrically sound. Never solder mains-carrying wires directly to the board: use solder tags. The use of crimp-on tags is also good practice.
10. Even when a Class II transformer is used, it remains the on/off switch whose function it is to isolate a hazardous voltage (i.e., mains input) from the primary circuit in the equipment. The primary-to-secondary isolation of the transformer does not and can not perform this function.

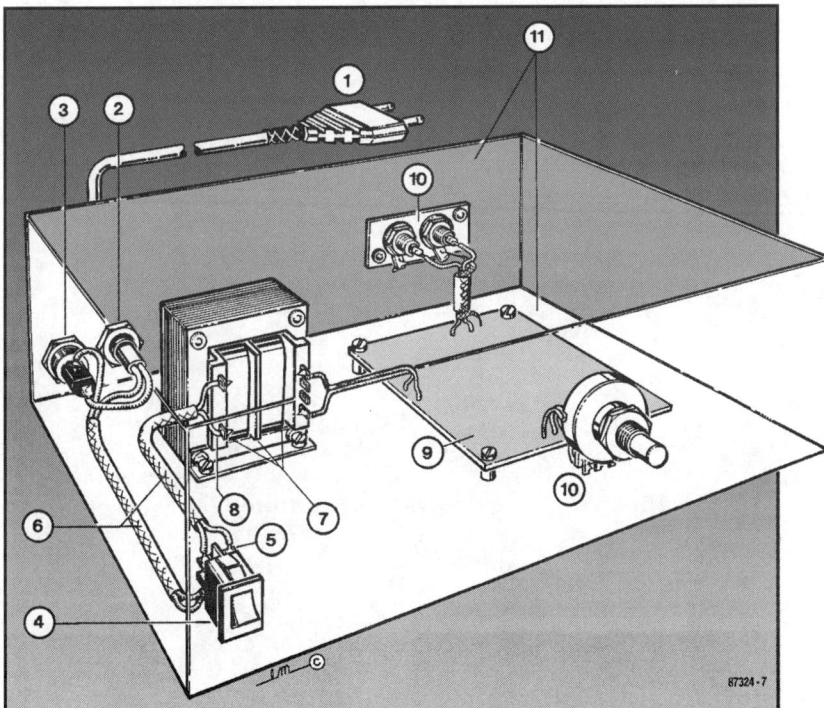

87324-7

1
Audio & hifi

AES/EBU-to-S/PDIF* converter

The converter is intended primarily for use with the sample rate converter published in the October 1996 issue of *Elektor Electronics*.

The conversion of a symmetrical signal to an asymmetrical signal requires no more than a small transformer. Amplification is not required since the core. The primary as well as the secondary are wound from enamelled copper wire of 0.5 mm dia. The transformer for Version A needs a primary winding of 18 turns and a secondary of 5 turns. That for Version B requires a primary of 20 turns and a secondary of 4 turns.

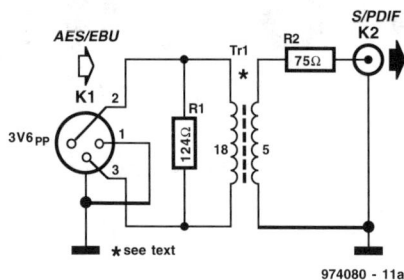

974080 - 11a

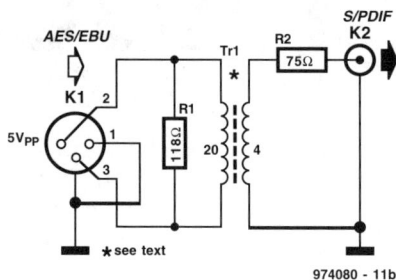

974080 - 11b

AES/EBU signal is strong enough to generate the S/PDIF signal (500 mV$_{pp}$ into 75 Ω). However, the quality of the conversion depends entirely on that of the home-made transformer.

The simplicity of the circuit means that the turns ratio depends on the level of the balanced input voltage. This is why the diagram shows two versions: that at the left is suitable for inputs of 3.6 V$_{pp}$ and that at the right for inputs of 5 V$_{pp}$.

The transformer is wound on a Type G2-3/FT12

The secondary impedance is transformed to the primary winding. Assuming that the system has a correctly terminated output of 75 Ω, the primary winding needs to be shunted by a resistor, R$_1$, of 124 Ω (left-hand version) or 118 Ω (righthand version) to give an input impedance of 110 Ω. This arrangement ensures a correct input impedance over a wide range of input frequencies. Only at 60 kHz (Version A) or 50 kHz (Version B) does the impedance drop by about 20%.

The bandwidth of the converter is ≥ 20 MHz.

* The S/PDIF – Sony/Philips Digital Interface Format – is the consumer version of the AES/EBU professional standard. It was devised by the AES and EBU to define the signal format, electrical characteristics, and connectors, to be used for digital interfaces between professional audio products. AES is the American Audio Engineering Society and EBU is the European Broadcasting Union.

AF input module

The use of a relay for input selection in a preamplifier is always better than a simple rotary switch – at least from a quality point of view. A relay obviates long signal paths to a common switch and may be controlled electronically. The module uses a bistable relay, because a standard relay needs a continuous energizing current. The slightly higher price of the bistable relay is more than offset by the high current requirement of the standard relay.

Although the module, which serves one stereo input, is intended primarily for use with the battery operated preamplifier published in the January 1997 issue of *Elektor Electronics*, it may also be used with most other preamplifiers. Note therefore that at least six of these modules are required to replace the existing input selector in the battery operated preamplifier.

The drive to inputs RST (reset) and ON is best obtained from the next article 'AF input selection'.

The two relays are energized via transistors T_1 and T_2. These provide a pulse of a few milliseconds, since the base drive is effected via a differentiating network. Current only flows through the relay coils when the charging current for C_3 or C_4 is sufficient to bring the base-emitter junction of the relevant transistor into conduction.

Transistor T_1 is switched on when the RST level changes from high to low and T_2 when ON goes from low to high. Resistors R_5 and R_7 ensure that when C_3 or C_4 is being discharged the maximum permissible reverse bias voltage of the relevant transistor is not exceeded, and that the switching pulse is well defined.

Diodes D_1 and D_2 short-circuit any voltage peaks caused by the relay coils when T_1 or T_2 switches off and thus protect the transistors.

Resistors R_1 and R_2 at the inputs are terminating resistors.

Networks R_3-C_1 and R_4-C_2 filter out any r.f. noise.

Each of the relay contacts is complemented by a jumper to earth. Use of these jumpers depends on the application, but only if the amplifier is a summing type. Non-selected inputs are then linked to earth and any crosstalk is suppressed effectively. Therefore, if the amplifier is a buffer, the jumpers should be left open.

Mind the correct polarity (shown on the relay) when connecting the relays, because incorrect polarity reverses the relay action. The marking on some relays

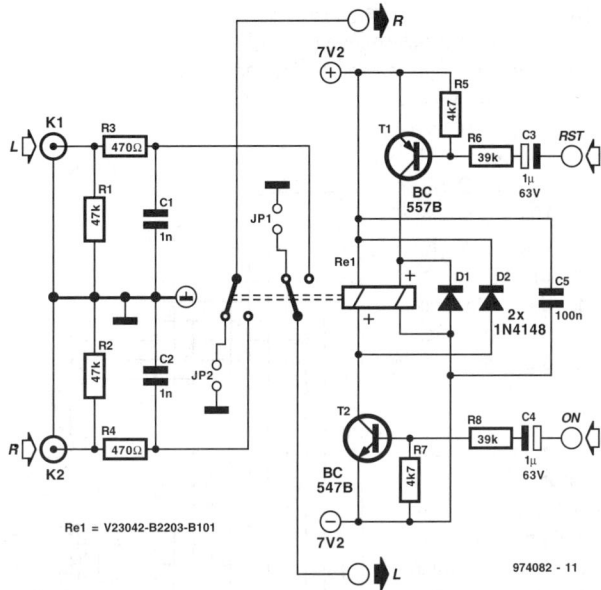

is by bullets and on others by plus signs. The bullets give the polarity for the same function (for instance, positive for the make contact), and the plus signs the polarity of the function of the relevant coil. There is, therefore, no difference in pinout (function).

The relays are 12 V types with 720 Ω coils.

AF input selection

An input selection in a battery operated preamplifier is likely to degrade the performance as far as crosstalk and channel separation are concerned. The present design is free from this drawback and also enables the use of up to 12 inputs. Each input source is linked to the circuit via a bistable relay.

The input source is selected with a single-pole, 12 position rotary switch, S_1. Each position of this switch is linked to an individual pull-down resistor. Since only one resistor is connected at any one time, the current drawn by the circuit is only 15 μA, which, in the case of a battery operated preamplifier, is a real benefit.

The 12 outputs of S_1 are linked to parity checker IC_3. The output of this device is high only if an odd number of inputs is high. When S_1 is turned, all inputs

go low briefly and so the output of S_3 also becomes low for an instant. The output of IC_3 then triggers monostable multivibrator (MMV) IC_{4a}. Since this is retriggerable, its output will be only one pulse even with contact bounce of S_1. As long as trigger pulses arrive during the period the output is active, the output pulse is stretched. To make absolutely sure, the time can be set between 0.1 s and 1 s with P_1.

The outputs of S_1 are also linked to D-type bistables IC_1 and IC_2, which ensure a stable change-over of the output levels. An advantage of the bistables is that they can be reset. This facility is made use of by resetting all relays before a change of input, so ensuring that only one input is linked to the circuit at any one time. This arrangement provides a dead time between the releasing of one relay and the tripping of another. Strictly speaking, the dead time is equal to the sum of the mono times of IC_{4a} and IC_{4b}. However, MMV IC_{4b} serves to clock the inputs of all the D-type bistable, so that, since this requires a pulse of only 10 μs, the dead time is determined primarily by IC_{4a}.

The Q output of IC_{4a} is used to reset the D-type bistables and also provides the reset pulse for all relays together. So, the new data from S_1 is accepted by the D-type bistables 10 μs after the reset pulse. To enable the position of S_1 to be assumed during the rise time of the supply lines, the bistables need an additional pulse and this is provided by R_5-C_3.

About 4 seconds after the supply has been switched on, the 13th input of the parity checker changes state, which results in the output of IC_3 changing from low to high and the triggering of IC_{4a}. This means that all relays are reset after switch-on, immediately followed by the enabling of the relevant input. This entire process is completed before the output of the preamplifier becomes active.

The circuit requires a power supply of about 15 V. The diagram shows how this may be derived.

The ICs are protected against overvoltage by zener diode D_1. Current source T_1 in series with D_1 ensures that the current through the diode is held within specified limits when the battery voltage is high.

When the supply voltage is lower than 15 V,

the drop across R_6 and T_1 may be ignored, but when it is higher, the current is limited to about 400 μA.

The value of capacitor C_8 is purposely large since this component provides the energy required for changing over the inputs.

Finally, diode D_2 prevents C_8 being discharged via T_1.

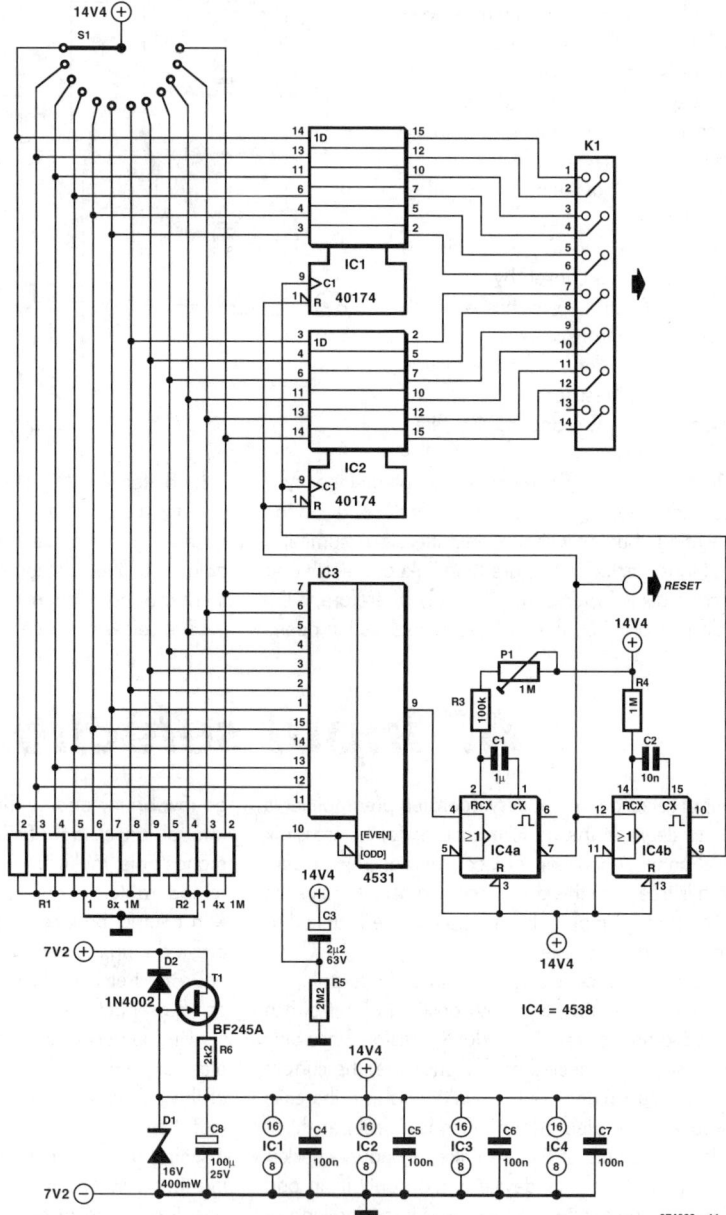

974083 - 11

4

Attenuator/limiter

The attenuator/limiter may be of interest to readers involved in producing sound tracks with (small-format) films (home movies, perhaps), or producing sound samples using 'performers' (including animals) whose dynamic range is too large.

The signal to be processed arrives on the input terminals shown at the left-hand side of the circuit diagram. Potentiometer P_1 allows the input signal to be adjusted before it reaches the input of IC_1, an MC3340P from Motorola. The level adjustment is necessary to prevent sudden variations in the input signal level from reaching the output of the circuit. The MC3340P offers an attenuation range of up to 80 dB for frequencies up to 1 MHz. The distortion introduced by it is smaller than 1% if the attenuation does not exceed 15 dB, or 3% for an attenuation of 40 dB.

The value of the RC networks at the input and the output is such that the lower cut-off frequency of the overall circuit does not exceed 12 Hz. Resistor R_8 defines the time constant of the rectifier formed by diodes D_1 and D_2. Its value prevents the distortion from increasing when the input signal frequency decreases. The current supplied by the MC3340 via pin 2 produces a voltage across R_8, which, without the action of buffer op amp IC_3, would cause excessive attenuation.

The two op amps in IC_2 form a kind of level adap-

tor that makes the conditioned signal available at the output terminals.

The circuit draws a current of about +18 mA and −4 mA. If it is decided to give it a dedicated power

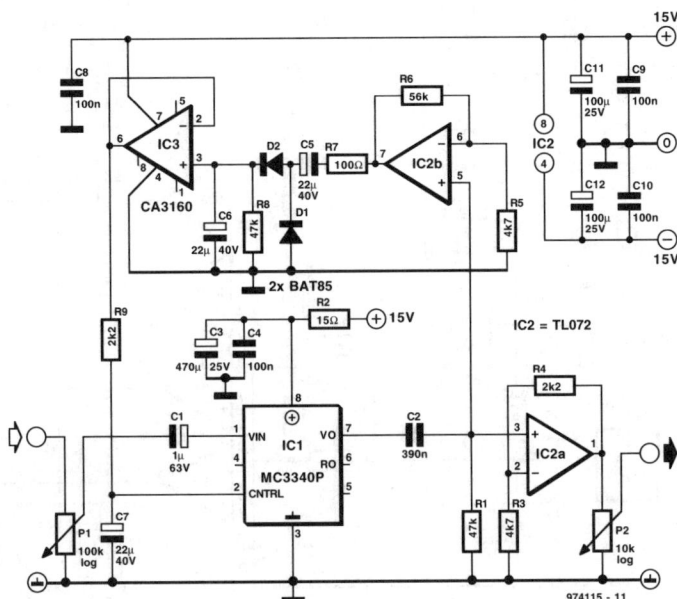

supply, a 'standard' design based on two integrated voltage regulators powered by a bridge rectifier, smoothing and filtering capacitors, and a mains transformer with a centre tap will do nicely.

The circuit has a small additional advantage: it helps to reduce the audio level of commercials and advertising on radio and TV, which are often broadcast at a higher sound level than the programmes they rudely interrupt. This attenuator circuit makes them fall into line again.

Auto volume control

The volume control is intended primarily for insertion between a car radio and its booster. It automatically adapts the volume to the amount of road and engine noise. This is done in four 5 dB steps based on the measured sound pressure in the interior of the car. This means that the volume can be increased by up to 20 dB with respect to the set volume level. Care

should, therefore, be taken to ensure that the booster and loudspeakers do not become overloaded.

In Figure 1, IC_{4a} and IC_{4b} operate as control amplifiers. The audio signal is input via K_1 and K_3 and applied to the booster via K_2 and K_4. The basis level is that registered with the electret microphone MIC_1.

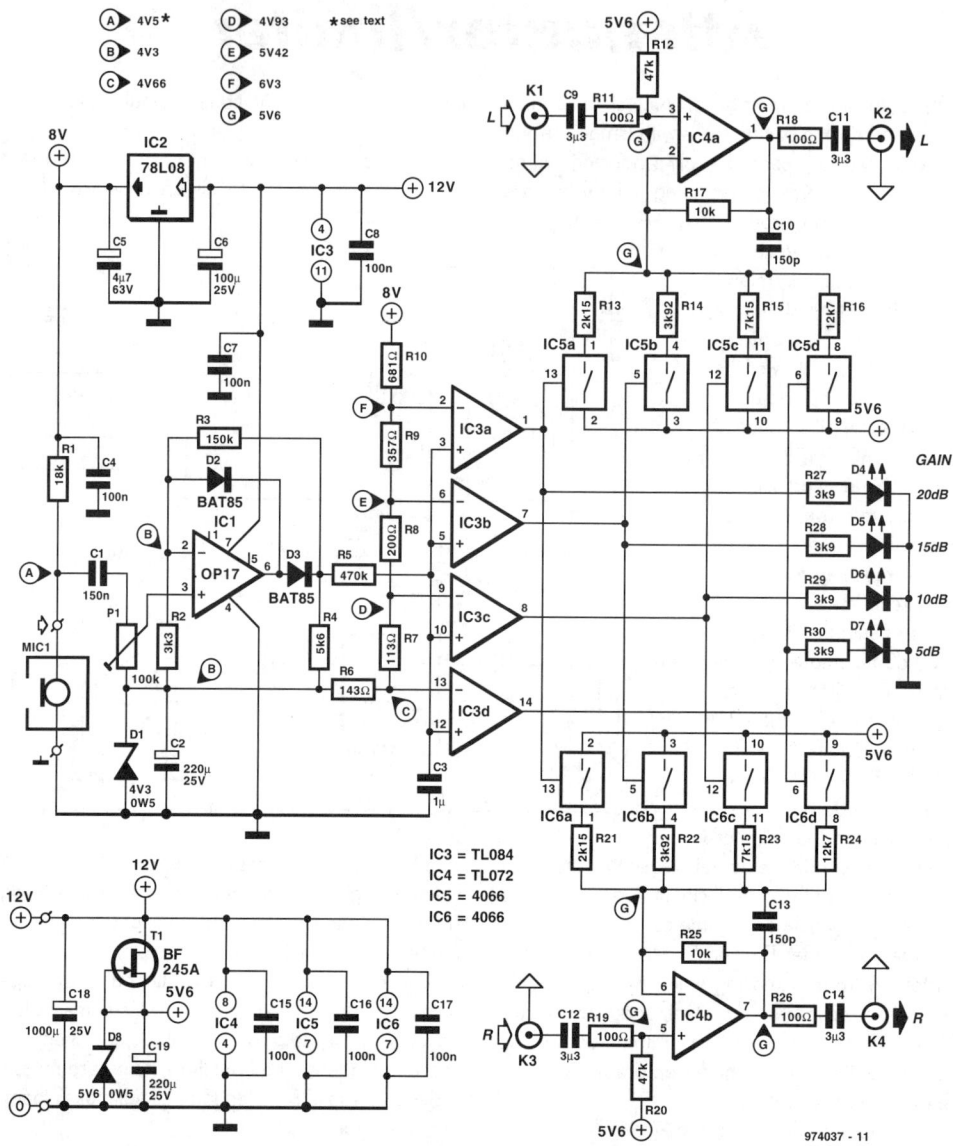

Schematic labels (selected):

A▷ 4V5 ★   D▷ 4V93   ★ see text
B▷ 4V3   E▷ 5V42
C▷ 4V66   F▷ 6V3
G▷ 5V6

IC2 78L08 · IC3 = TL084 · IC4 = TL072 · IC5 = 4066 · IC6 = 4066

GAIN: 20dB, 15dB, 10dB, 5dB

974037 - 11

The microphone should not be too sensitive to avoid overdrive and acoustic coupling between it and the loudspeakers. Its d.c. setting is arranged with resistor R_1 while its sensitivity is set with P_1.

The output of the microphone is applied to fast op amp IC_1 via the wiper of P_1. The op amp, arranged as a rectifier/amplifier, provides an amplification of x45. Its output is averaged by R_5-C_3 and then applied to comparators IC_{3a}–IC_{3d}. These compare the amplified signal and averaged signal, U_{AA}, with the potentials at the junctions of divider R_6-R_{10}.

Each of these potentials differs by 5 dB from the preceding or next one as the case may be.

The comparators control electronic switches IC_{5a}–IC_{5d} and IC_{6a}–IC_{6d}, which vary the degree of feedback of IC_{4a} and IC_{4b} on the basis of the control input. For instance, if none of the comparators in IC_3 has changed state, IC_{4a} operates as a voltage follower with unity gain. When U_{AA} exceeds the level at junction R_6-R_7, the gain of IC_{4a} is raised by 5 dB. When with increasing road and engine noise it exceeds the level at junction R_9-R_{10}, the switches are all closed so that R_{13}–R_{16} are in parallel, whereupon the gain of IC_{4a} is raised by 20 dB. The position of the automatic volume control is indicated by light-emitting diodes D_4–D_7.

The circuit is powered by the car battery. It is recommended that the battery voltage is well filtered.

The supply lines for the microphone and the voltage divider are held at 8 V by regulator IC_2. That for IC_4 is held at 5.6 V by T_1-D_8, irrespective of the battery voltage.

The circuit draws a current of 40 mA when the LEDs light.

The distortion of 0.0025% is well within the requirements for car hi-fi equipment.

The volume control is best built on the printed-circuit board in Figure 2, which is, however, not commercially available, but may be made with the aid of the relevant track layout in Appendix 1.

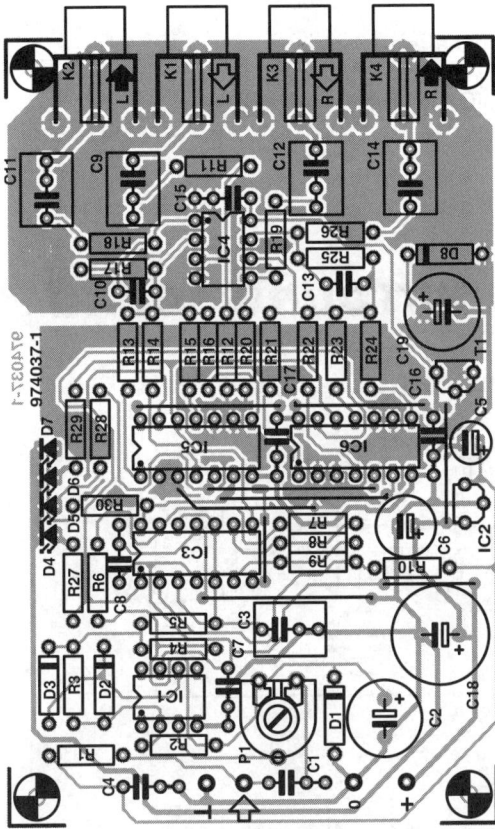

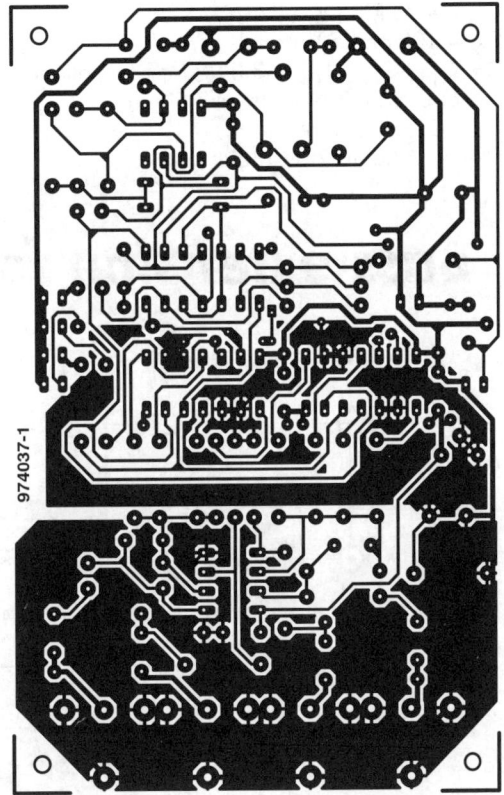

Parts list

Resistors:

R_1 = 18 kΩ
R_2 = 3.3 kΩ
R_3 = 150 kΩ
R_4 = 5.6 kΩ
R_5 = 470 kΩ
R_6 = 143 Ω
R_7 = 113 Ω
R_8 = 200 Ω
R_9 = 357 Ω
R_{10} = 681 Ω
R_{11}, R_{18}, R_{19}, R_{26} = 100 Ω
R_{12}, R_{20} = 47 kΩ

R13, R21 = 2.15 kΩ, 1%
R14, R22 = 3.92 kΩ, 1%
R15, R23 = 7.15 kΩ, 1%
R16, R24 = 12.7 kΩ, 1%
R17, R25 = 10 kΩ
R27–R30 = 3.9 kΩ
P1 = 100 kΩ preset

Capacitors:
C1 = 0.15 μF
C2, C19 = 220 μF, 25 V, radial
C3 = 1 μF, MKT (metallized polyester),
  pitch 5 mm or 7.5 mm
C4, C7, C8, C15–C17 = 0.1 μF
C5 = 4.7 μF, 63 V, radial
C6 = 100 μF, 25 V, radial
C9, C11, C12, C14 = 3.3 μF, MKT (metallized
  polyester), pitch 5 mm or 7.5 mm
C10, C13 = 150 pF
C18 = 1000 μF, 25 V, radial

Semiconductors:
D1 = zener, 4.3 V, 500 mW
D2, D3 = BAT85
D4–D7 = LED, high-efficiency
D8 = zener, 5.6. V, 500 mW
T1 = BF245A

Integrated circuits:
IC1 = OP17
IC2 = 78LO8
IC3 = TL084
IC4 = TL072
IC5, IC6 = 4066

Miscellaneous:
K1–K4 = audio socket for board mounting
MIC1 = electret microphone

Bass extension for surround sound

1

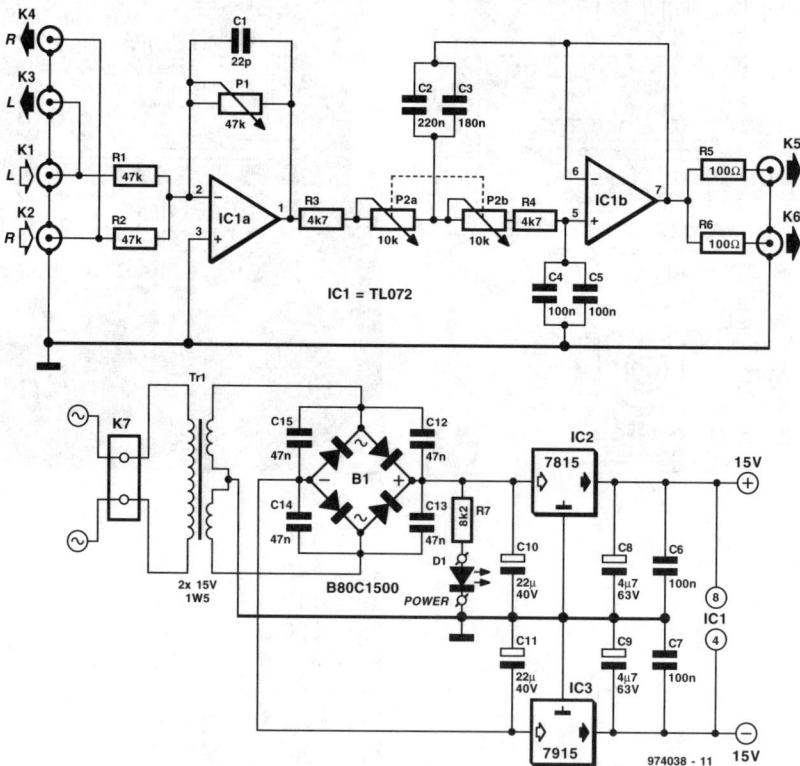

The extension is intended primarily for surround-sound installations that need some boosting of the bass frequencies but where an additional subwoofer cannot be afforded. It is based on a disused mono a.f.

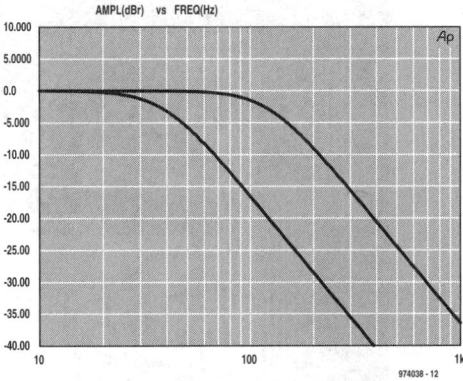

2

AMPL(dBr) vs FREQ(Hz)

amplifier and loudspeaker. If these provide reasonable bass performance, they can be converted into a fairly good subwoofer with the aid of an active low-pass filter—see Figure 1.

The input signals for the left-hand and right-hand channels are applied to audio sockets K_1 and K_2 respectively. They are output via audio sockets K_3 and K_4 to which the surround-sound decoder is connected.

The signals of the two channels are summed in IC_{1a}, which also functions as input amplifier. The amplification, and therefore the sensitivity of the 'sub-

woofer', can be adjusted with P_1.

The output of IC_{1a} is applied to a 2nd-order Butterworth low-pass filter. The cut-off frequency of this active filter can be set between 40 Hz and 120 Hz with stereo potentiometer P_2. The response characteristic of the filter at both these frequencies is shown in Figure 2. The actual cut-off point depends on individual taste.

The mono amplifier is connected to audio output sockets K_5 and K_6.

The power supply for the circuit is simple and consists of a small mains transformer, Tr_1, a bridge rectifier, B_1, antihunt capacitors C_{12}–C_{15}, a number of smoothing and decoupling capacitors, and two integrated voltage regulators, IC_2 and IC_3.

The filter circuit is best built on the printed-circuit board shown in Figure 3, which is, however, not commercially available, but may be made wit the aid of the relevant track layout in Appendix 1.

The filter should be housed in a metal enclosure. Also, P_1 and P_2 should preferably be types with a metal case. Hum is prevented by earthing the cases and the enclosure.

The harmonic distortion, with two input signals of 200 mV and a bandwidth of 22 kHz, is 0.0016% at 30 Hz.

Although not of prime importance at low frequencies, the polarity of the 'subwoofer' should be the reverse of that of the remainder of the system since the present circuit inverts the signals.

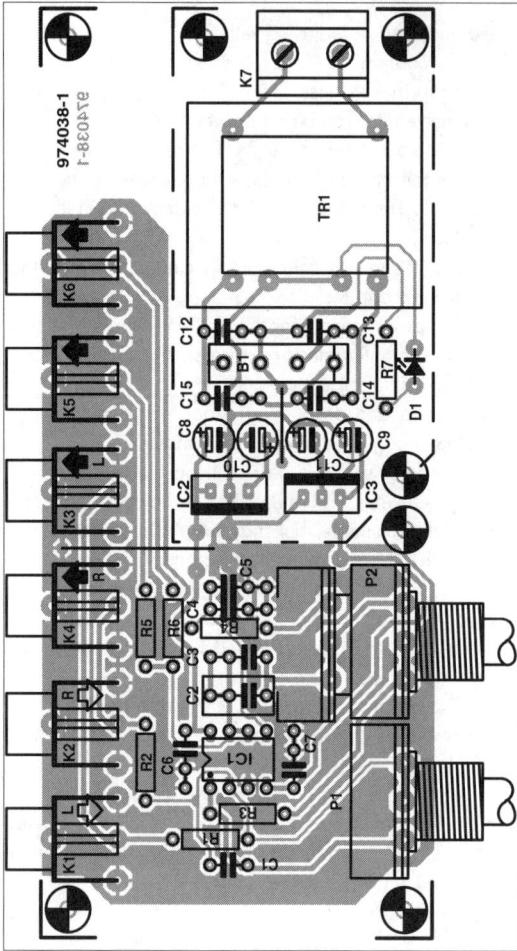

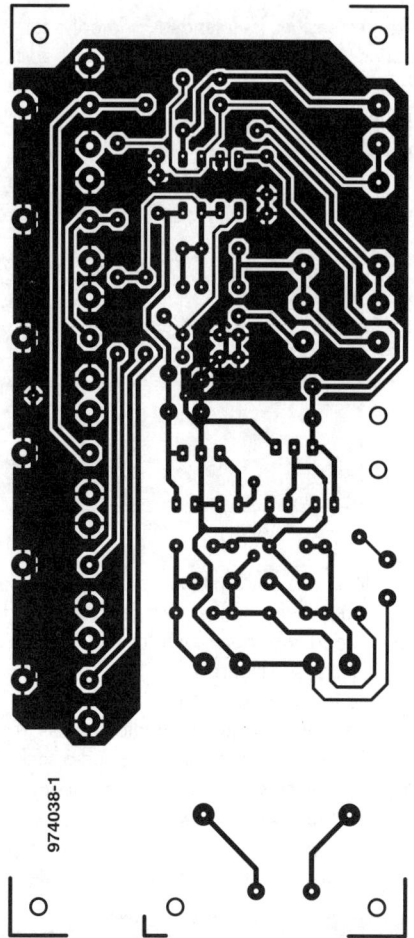

974038-1

974038-1

Parts list

Resistors:
R_1, R_2 = 47 kΩ
R_3, R_4 = 4.7 kΩ
R_5, R_6 = 100 Ω
R_7 = 8.2 kΩ
P_1 = 47 kΩ logarithmic potentiometer
P_2 = 10 kΩ, linear stereo potentiometer

Capacitors:
C_1 = 22 pF
C_2 = 0.22 μF
C_3 = 0.18 μF
C_4–C_7 = 0.1 μF
C_8, C_9 = 4.7 μF, 63 V, radial
C_{10}, C_{11} = 22 μF, 40 V, radial
C_{12}–C_{15} = 0.047 μF ceramic

Semiconductors:
D_1 = LED, high efficiency

Integrated circuits:
IC_1 = TL072CP
IC_2 = 7815
IC_3 = 7915

Miscellaneous:
K_1–K_6, K_8–K_9 = audio socket for board
    mounting
K_7 = 2-way terminal block, pitch 7.5 mm
B_1 = B80C1500
Tr_1 = mains transformer, 2x15 V secondaries,
    1.5 VA

Car booster adaptor

Judging by the cacaphony emanating from an increasing number of cars on the road, car radio boosters unfortunately remain popular with young people. Unfortunately, because deafness among these young people is becoming quite common.

From a technical point of view, the setup with a booster is often very in efficient, because these power monsters are normally connected simply to the loud-speaker terminals of the existing car radio installation via an attenuator. This puts the two output amplifiers in series, which is, as said, quite inefficient.

It is much better to take the signal from the wiper of the volume control in the car radio and use this as the input to the booster. This is normally not much of a job. The signal so obtained must, however, be buffered and sometimes also amplified.

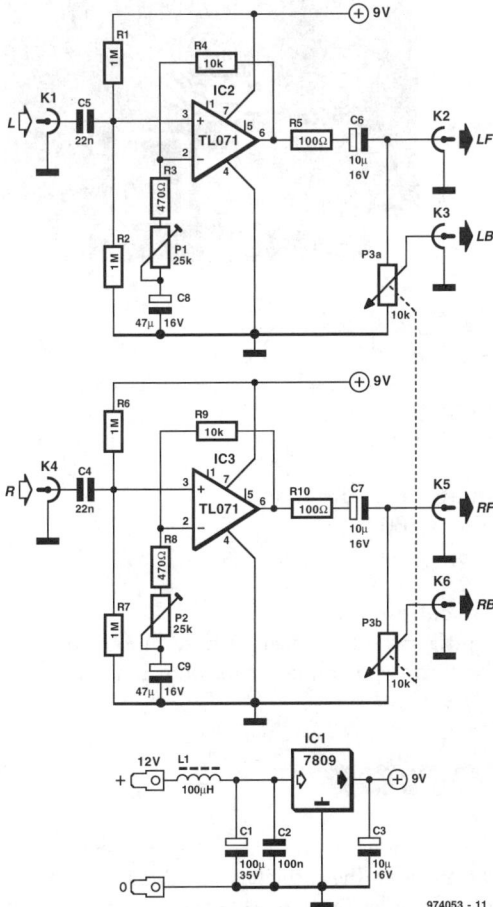

974053 - 11

The adaptor provides both these functions in a simple manner. The stereo signals are applied via K_1 and K_2 and buffered an amplified by an op amp in each channel. The amplification may be set between x1.5 and x22 with P_1 and P_2 respectively. These levels should be more than adequate for most situations. The peak output voltage is 2 V_{RMS}.

The output in each chan-nel is split into a front and a rear branch (left-hand front, LF, and left-hand back, LB, and RF and RB respectively). The volume of the rear speakers is set with P_3.

Regulator IC_1 provides a stable 9 V supply line for the op amps. The circuit draws a current of not more than 7 mA.

The adaptor is best built on the printed-circuit shown, which is, however, not commercially available, but may be made wit the aid of the relevant track lay-out in Appendix 1.

The input and output terminals are audio sockets for board mounting.

The battery voltage is applied to the circuit via two car-type connectors mounted on the board. When the adaptor is fitted in a small case, care must be taken that P_3 remains accessible.

11

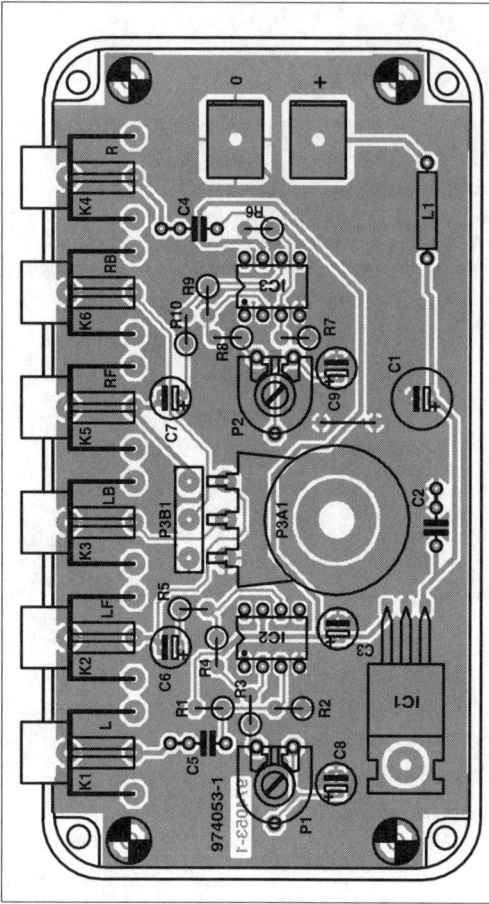

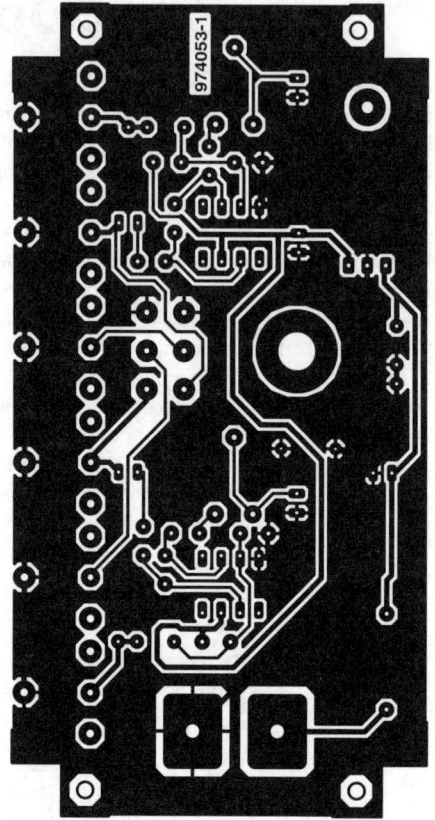

Parts list

Resistors:
$R_1, R_2, R_6, R_7 = 1\ M\Omega$
$R_3, R_8 = 470\ \Omega$
$R_4, R_9 = 10\ k\Omega$
$R_5, R_{10} = 100\ \Omega$
$P_1, P_2 = 25\ k\Omega$ preset
$P_3 = 10\ k\Omega$ log stereo pot

Capacitors:
$C_1 = 100\ \mu F$, 35 V, radial
$C_2 = 0.001\ \mu F$, high stability
$C_3, C_6, C_7 = 10\ \mu F$, 16 V, radial
$C_4, C_5 = 0.022\ \mu F$
$C_8, C_9 = 47\ \mu F$, 16 V, radial

Inductors:
$L_1 = 100\ \mu H$

Integrated circuits:
$IC_1 = 7809$
$IC_2, IC_3 = TL071CP$

Miscellaneous:
K_1-K_6 = audio socket for board mounting
2 off car-type connector for board mounting

DC detector

The detector is intended primarily to sense direct voltages at the output of power amplifiers. The signal so detected may be used to enable a protection circuit that, for instance, disconnects the loudspeakers from the amplifier. The circuit has the advantage of always reacting within75 ms at whatever level of direct voltage. It also reacts to signals >600 mV at very low frequencies below about 4 Hz, which are likely to damage the loudspeakers.

The circuit is configured symmetrically and may therefore be split into two. The upper part in the diagram processes positive input signals, and the lower part, negative signals.

The signal from the amplifier is applied to the sensor via R_{10}. Its level is limited by diodes D_2–D_5. The trip levels of comparators IC_{2a}–IC_{2b} are set to +600 mV and –600 mV by R_2-D_6 and R_3-D_1 respectively. This means that the output of IC_{2a} goes high when the input voltage is higher than +600 mV and that of IC_{2b} when the input voltage is lower than –600 mV.

It follows that the signals at the outputs of the comparators together form a square wave. This is used to charge C_3 and C_4 alternately to a potential that does

not exceed the trip level of the comparators. This situation changes, however, if, for instance because of a positive offset, the output of IC_{2a} remains high longer than usual. This causes C_3 to be charged to a higher potential, while at the same time T_1 is switched on via R_9 and C_4 is short-circuited. This causes T_2 to be blocked via R_6, so that the potential building up across C_3 cannot be removed via this transistor. This means that the trip level of IC_3 will be exceeded so that the output of the circuit changes from low to high.

occurs if owing to a negative offset the output of IC_{2b} remains high longer than usual. It is then C_4, however, that is charged, while IC_{1b} functions as the trigger.

Diodes D_7 and D_{10} protect T_1 and T_2 by preventing their base voltage dropping below –700 mV.

Clearly, the response time of the sensor depends not only on the trigger level of IC_{1a} and IC_{1b}, but also on the time constants R_4-C_4 and R_7-C_3. The HEF4093 used in the prototype triggered at 7.5 V (V_{DD} = 15 V), which resulted in a response time of 57 ms. However, the spread of trigger voltages in the 4093 series is appreciable and it may, therefore, be necessary to lower the values of R_4 and R_7.

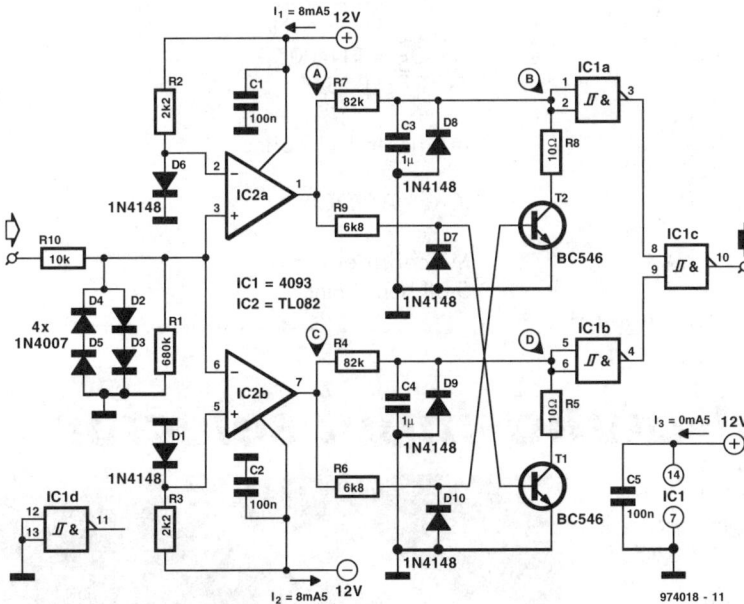

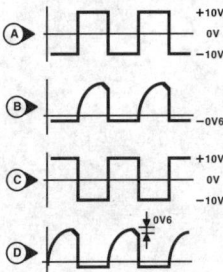

The detector is best built in the printed-circuit board shown, which is not commercially available, but may be made wit the aid of the relevant track layout .

The symmetrical power supply may have an output between ±10 V and ±18 V. The prototype draws a current not exceeding 10 mA.

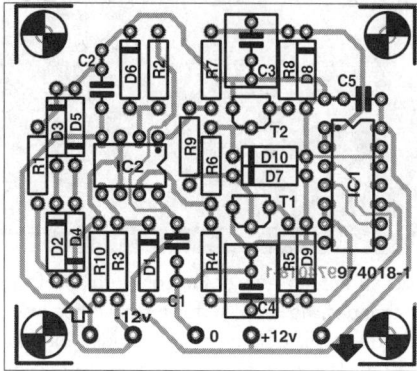

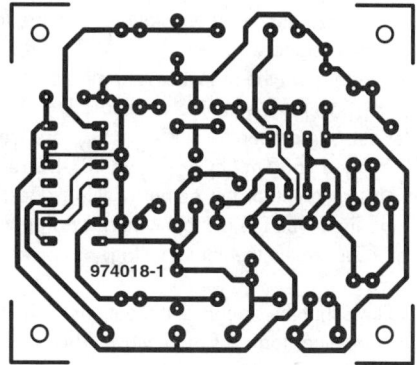

Parts list

Resistors:
R_1 = 680 kΩ
R_2, R_3 = 2.2 kΩ
R_4, R_7 = 82 kΩ
R_5, R_8 = 10 Ω
R_6, R_9 = 6.8 kΩ
R_{10} = 10 kΩ

Capacitors:
C_1, C_2, C_5 = 0.001 μF
C_3, C_4 = 1 μF, MKT (metallized polyester)

Semiconductors:
D_1, D_6–D_{10} = 1N4148
D_2–D_5 = 1N4007
T_1, T_2 = BC546

Integrated circuits:
IC_1 = 4093
IC_2 = TL082CP

Miscellaneous:
5 off board pins

Digital-audio-input selector

As the name indicates, the selector is intended to choose one of up to eight digital audio signal inputs, which it does with the aid of a multiplexer.

The multiplexer, IC_6, is controlled by preset up/down counter IC_2. The counter is set with DIP switch S_3 (note that the MSB switch is not used in this application).

The various inputs are selected with press-keys S_1 and S_2. Gates IC_{1d} and IC_{1e}, in conjunction with networks R_1-C_1 and R_3-C_2, provide effective debouncing of the keys. Resistor R_5 and capacitor C_3 ensure that the counter is set on power-up.

If fewer than eight inputs are needed, the

number can be reduced to four by resetting jumper J_1 so that pin 9 of IC_6 is linked to a fixed level. The non-used inputs of the multiplexer, pins 1, 2, 4, and 5, must be strapped to earth.

Which of the inputs is selected is indicated by one of four or eight LEDs that are controlled by 3-to-8 decoder IC_3 at the outputs of IC_2. If four inputs are used, D_5–D_8 must be omitted.

Since the digital-audio-input circuits are identical, only one is shown (in dashed lines at the top left-hand side of the diagram). Each has an optical input (IC_5) and a coaxial input (K_1). It needs only one inverter (here IC_{4a}); the others (IC_{4b}–IC_{4e}) are strapped to earth.

The selector has an optical output (IC_7) as well as a coaxial output (K_2).

The current drawn by the selector depends primarily on the number of optical modules (each of which draws 20–25 mA).

If standard LEDs instead of high-efficiency types are used, the value of R_{10} should be lowered to 220 Ω. The total current drain then rises by about 10 mA.

WARNING: Each input must have a separate IC_{4a}: do NOT use the remaining gates in the original IC_4 since this would increase the dissipation in the device to dangerously high levels.

974034 - 11

Drive indicator

The indicator shows the four important drive states of an output amplifier: Class A (low power), Class AB (normal operation), half power and full power. The indication is provided by a dual (red/green) LED, which also shows the colours orange and yellow. The circuit can be used with most types of output amplifier.

The input signal is derived from the output of the amplifier and is applied to a full-wave rectifier, IC_1, via attenuator network R_1-R_2-R_5. The symmetry of the signal is set with P_1. (This is done simply by first applying a positive direct voltage of, say, 15 V to the input and then a negative one at the same level. Turn P_1 until the output of IC_{1b} is the same in both cases: +5 V).

The rectified signal is compared in three-fold

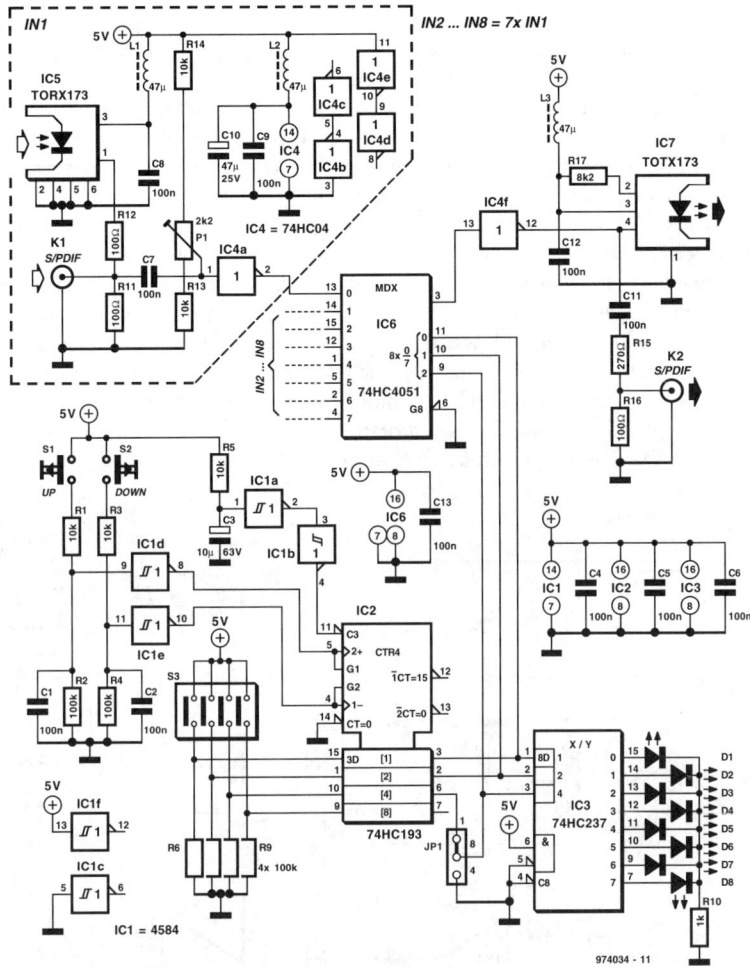

comparator IC_{2a}–IC_{2c} with reference voltages provided by potential divider P_2-R_7-R_8-R_9-R_{10}. They assume a normal load and represent powers of 2.5 W (limit of Class A operation); 25 W (half power) and 50 W (full power). The divider may be calibrated with the aid of P_2, which should be set to a position for which the potential at junction R_7-R_8 is 9.3 V. If this proves impossible, the value of R_7 should be altered as appropriate.

The outputs of the comparators are applied to MMVs (monostable multivibrators) IC_{3a}, IC_{4a}, and IC_{4b}. These retain the relevant levels for a short while, so that the appropriate LED lights clearly even with even with short power peaks. In all three cases, the time constant, R_{14}-C_1, R_{15}-C_2, and R_{16}-C_3, is 1 second.

When not one of the comparators is enabled (Class A), the Q output of IC_{3a} is high, whereupon the green section in D_3 is actuated via R_{19}.

When the first comparator has changed state (Class AB), the red section of D_3 is actuated via IC_{4b}. The value of R_{17} is such that the diode then emits a yellow(ish) colour light (since the green section has stayed on).

When the second comparator level is exceeded (half power), the red section of the LED gets extra drive, whereupon the colour becomes orange.

When IC_{2c} is enabled, the Q output of IC_{3a} changes from high to low, whereupon the green section goes out and the LED light is red to show that the amplifier has reached its drive level.

The supply is derived from that of the output amplifier via network R_{20}-R_{21}-D_4-D_5. The indicator draws a current of 25 mA from the +ve line and 50 mA from the –ve line. Note that the –ve line has to provide the current for the LED.

Since the zener diodes require some time to warm up, it is best to leave the circuit switched on for, say, 15 minutes before adjusting P_1 and P_2.

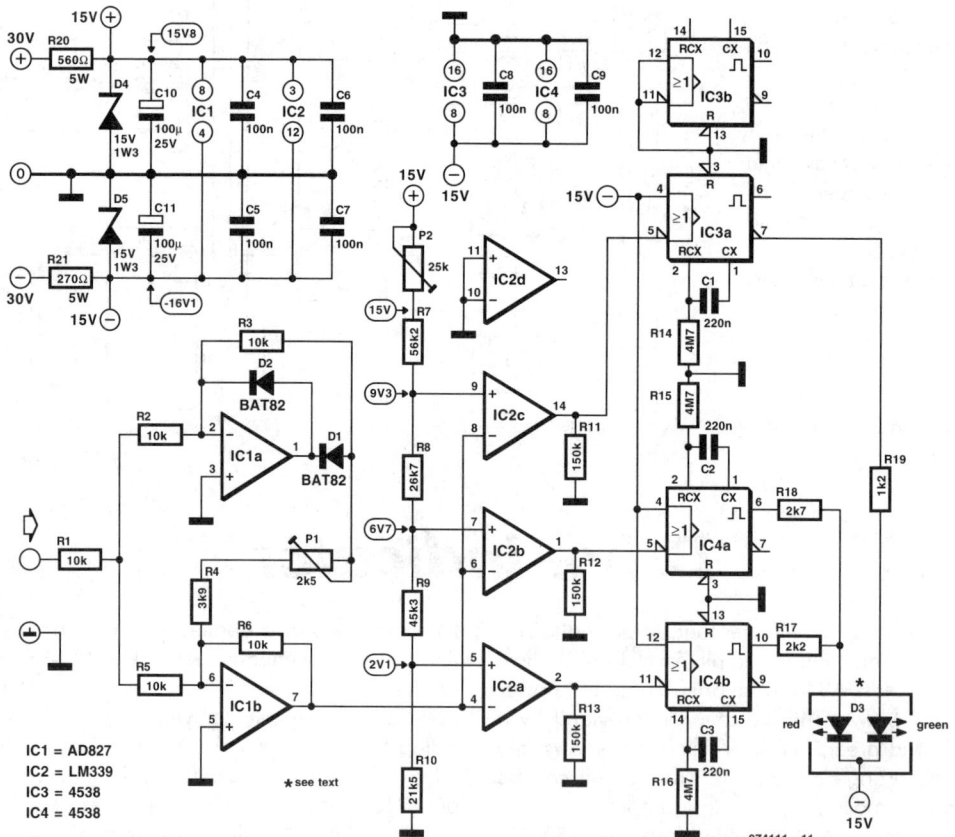

974111 - 11

High-end oscillator for digital audio

Jitter (phase noise) is a serious problem in the linking of two or more audio units. It is invariably caused by poorly designed oscillators in the recording equipment when this operates in the slave mode, that is, when it reproduces the system clock of the source equipment with the aid of a phase-locked loop (PLL).

The high-end oscillator may be used to replace such a poor reproduction or as a high-quality master oscillator. In the prototype no frequency shift was

sate for its parasitic parallel capacitance. The inductor also short-circuits any low-frequency noise. The value of the inductor is critical, but can be determined empirically.

• In the slave mode, the oscillator is detuned by varactor D_1, which is part of an external PLL. Since the capacitance of the varactor changes from 4–50 pF by an applied voltage of 1–25 V, the frequency can be shifted by about ±150 ppm. Since even small inter-

974106 - 11

detected under all kinds of operating conditions.

The high-end oscillator has these advantages over a usual design.

• The crystal is operated in the series mode instead of in the parallel mode as usual, since the resistance of the crystal at the resonance frequency is a minimum, while external resistances do not affect the Q factor to the same degree.

• The stability is enhanced by the use of an additional LC circuit (L_1-C_1-C_3) tuned to the fundamental frequency.

• The crystal is shunted by an inductor to compen-

ference signals cause fairly large changes in capacitance, the desired capacitance range should be kept as narrow as possible by using a different varactor, or by connecting a smaller capacitor in series with it. When used as a master oscillator, when the slightest jitter is noticeable, the varactor must be replaced by a fixed capacitor. The apparatus with which it is used must be equipped for genlock operation, that is, must have a separate clock input.

With reference to the complete circuit diagram of the high-end oscillator, some additional points should be noted.

Much attention has been paid to the decoupling of the supply lines. Also, the oscillator and buffer circuits have separate supply lines to ensure interference-free oscillator performance.

The clock is buffered by three stages of the non-buffered IC_1. The first stage, IC_{1a}, is arranged as a low-gain amplifier. Too much gain might cause feedback of harmonics into the oscillator. The clock is available at the output via R_{15}.

Diode D_2 ensures that the output of the final buffer is high when the oscillator is off. This arrangement enables several oscillators, providing various sampling frequencies, to be used over one clock line via an AND or NAND gate. The desired oscillator is enabled by applying 6.5 V to it.

Components R_7, R_8 and C_9 are part of the PLL, which determines their values. Surface mount devices (SMDs) T_1 and T_2 may be replaced by standard transistors Type BF494.

The relationship between the sampling frequency, f_s, the crystal frequency, f_c, and the value of C_2 in pF is

f_s (kHz)	f_c (kHz)	C_2 (pF)
32	12.288	47
38	14.592	27
44.1	16.9344	15
48	18.432	10

If the oscillator does not work owing to excessive tolerance of L_1, the parallel capacitance may be balanced by altering the value of C_3. It may also be necessary to alter the value of C_2.

• Adjust trimmer C_7 to give a voltage of $\frac{1}{2}U_{var}$ in PLL operation.
• Capacitor C_x extends the nominal frequency range of the oscillator downwards.
• The completed oscillator is best housed in a small tin-plate enclosure.
• The oscillator draws a current of about 40 mA.

S/PDIF-to-AES/EBU* converter

The converter translates the unbalanced S/PDIF format to the balanced (professional) AES/EBU format. Its timing and levels comply with the AES3-1992 Standard. This means that: (a) the output voltage must be 2–7 V_{pp} (transmitter load 100 Ω); (b) the rise and decay times must be 5–30 ns; (c) the output impedance must be 110 Ω ±20% (within the bandwidth of 0.1–6 MHz). These requirements are met in the design in the diagram (30 ns; 3.6 V_{pp}; 115 Ω respectively).

The circuit at the input, based on IC_1, converts the S/PDIF signal to HC levels. Op amp IC_{1a} is an analogue amplifier, while IC_{1b} raises the signal to the level of the supply lines. Resistor R_3 pulls IC_{1a} slightly from its centre of operation, so that the output buffer attains a logic level in the absence of an input signal.

The buffer to drive the output transformer is formed by a balanced circuit based on IC_{2a}–IC_{2d}. This arrangement ensures that the rise and decay times are equal and that the output voltage is large enough. The use of XOR gates ensures that the transfer times for inverting and non-inverting of the output of IC_{1b} are equal. Since the primary transformer voltage is 9.5 V, the secondary voltage could be decreased slightly. This is beneficial for the linearity of the impedance and the bandwidth of the converter.

The transformer is wound on a Type G2-3/FT12 core: the primary on one side and the secondary on the other. Both windings consist of enamelled copper wire of 0.5 mm dia. The core can accommodate a tin-plate screen for maximum common-mode suppression. Regulations require this to be $\geq$ –30 dB w.r.t. the nominal output level; in the present circuit it is –48 dB (with screen).

The output impedance, ignoring R_4 and R_5, is about 22 Ω. If a figure of exactly 110 Ω is wanted, R_4 and R_5 should have a value of 44.2 Ω.

Capacitors C_3–C_5 prevent any direct current flowing through the transformer in the absence of a signal as this would short-circuit IC_2. The use of three capacitors in parallel ensures that their total impedance and

---

* The S/PDIF – Sony/Philips Digital Interface Format – is the consumer version of the AES/EBU professional standard. It was devised by the AES and EBU to define the signal format, electrical characteristics, and connectors, to be used for digital interfaces between professional audio products. AES is the American Audio Engineering Society and EBU is the European Broadcasting Union.

loss resistance is (ceramic, high-stability types) low.

The AES/EBU signal is output via XLR connectors (to IEC268-12). Note that versions with male pins and female shells are used. Pin 1 is for the screen or the signal earth; pins 2 and 3 are for the signals – the phase is not important.

The circuit requires a 5 V power supply from which it draws a current of about 26 mA.

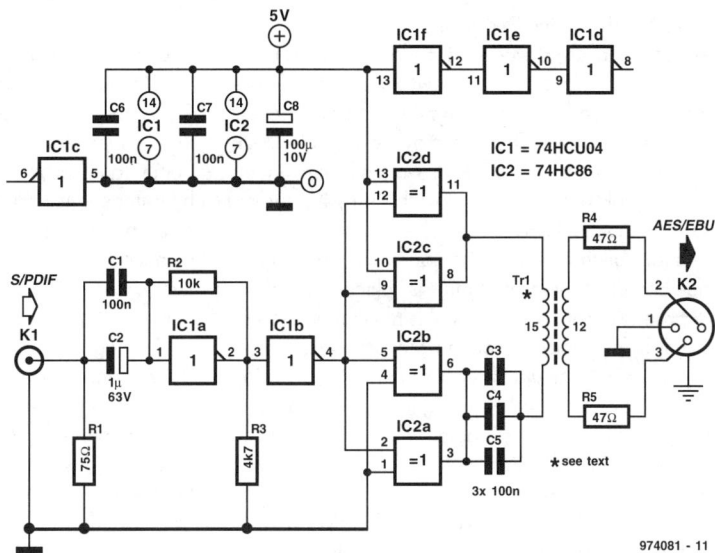

974081 - 11

XLR-to-audio-socket switch

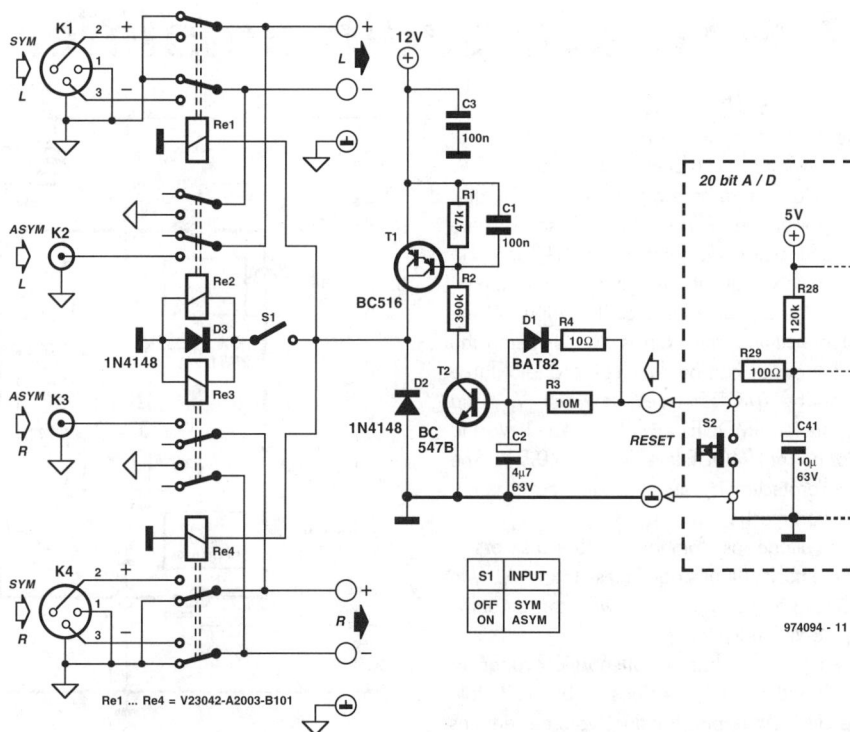

S1	INPUT
OFF	SYM
ON	ASYM

Re1 ... Re4 = V23042-A2003-B101

974094 - 11

The 20-bit ADC in the December 1996 issue of *Elektor Electronics* was fitted with balanced inputs (as is usual in a professional appliance). However, in view of non-professional readers, the unit was also able to process unbalanced signals. The unit may be extended with normal audio sockets to make working with unbalanced signals a little easier. How this may be done is shown in the diagram.

In fact, all that needs to be done to make the input stage accept unbalanced signals is linking pin 3 of the XLR buses to earth, whereupon this pin can be used as the earth for audio sockets. In the proposed circuit, this is done with the aid of relay Re_2 and Re_3. By using further relays, Re_1 and Re_4, for linking and passing the signal as relevant, a single switch (S_1) suffices to change over from XLR buses K_1–K_4 to audio sockets K_2 and K_3.

Transistors T_1 and T_2 have a dual function: (a) they provide level matching between the digital circuit in the ADC (5 V) and the relay circuit (12 V); (b) they provide a switch-on delay whereby after the power is switched on or after a reset (S_2) the inputs are briefly linked to earth via Re_1 and Re_4. This makes it possible for the input stage to be included in the offset calibration of the ADC.

When the reset switch (S_2) on the ADC is operated, C_2 is discharged rapidly via R_4 and D_1. Network R_3-C_2 then ensures that the relays remain reset for about four seconds. This network has been made high-impedance purpose, since it must not affect the reset network of the ADC itself (in dashed box). Capacitor C_1 provides further decoupling of any interference pulses. Diodes D_2 and D_3 are free-wheeling devices.

The signal earth must be kept separate for the relay earth.

It has been assumed that the relays are powered by a separate 12 V line. The 12 V line for the analogue input stage of the ADC must not be used, because it cannot provide the requisite current of some 80 mA.

It is inherent in the circuit that the unbalanced signals applied to K_2 and K_3 are passed on to pins 2 of K_1 and K_4, and that pins 3 of these buses are short-circuited, when S_1 is closed. This means that only one stereo source can be connected at any one time.

1-watt BTL audio amplifier

The TDA8581(T) from Philips Semiconductors is a 1-watt Bridge Tied Load (BTL) audio power amplifier capable of delivering 1 watt output power into an 8-W load at THD (total harmonic distortion) of 10% and using a 5-V power supply. The schematic shown here combines the functional diagram of the TDA8551 with its typical application circuit. The gain of the amplifier can be set by the digital volume control input. At the highest volume setting, the gain is 20 dB. Using the MODE pin the device can be switched to one of three modes: standby (MODE level between V_p and V_p–0.5 V), muted (MODE level between 1 V and V_p–1.4 V) or normal (MODE level less than 0.5 V). The TDA8551 is protected by an internal thermal shutdown protection mechanism.

The total voltage loss for both MOS transistors in the complementary output stage is less than 1 V. With a 5-V supply, an output power of 1 watt can be delivered into an 8-W loudspeaker.

The volume control has an attenuation range of between 0 dB and 80 dB in 64 steps set by the 3-state level at the UP/DOWN pin: floating: volume remains

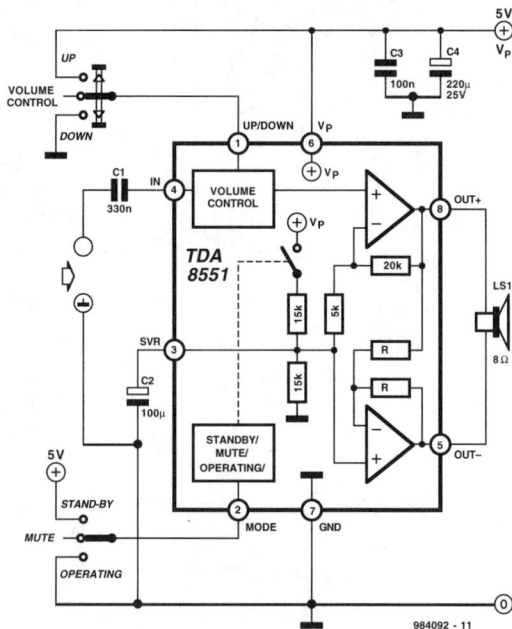

984092 - 11

unchanged; negative pulses: decrease volume; positive pulses: increase volume.

Each pulse at he UP/DOWN pin causes a change in gain of 80/64 = 1.25 dB (typical value). When the supply voltage is first connected, the attenuator is set to 40 dB (low volume), so the gain of the total amplifier is then –20 dB. Some positive pulses have to be applied to the UP/DOWN pin to achieve listening volume. The graph shows the THD as a function of output power. The specified maximum quiescent current drawn by the amplifier is 10 mA, to which should be added the current resulting from the output offset voltage divided by the load impedance.

10-band equalizer

The equalizer is suitable for use with hi-fi installations, public-address systems, mixers and electronic musical instruments.

The relay contacts at the inputs and outputs, in conjunction with S_2, enable the wanted channel to be selected. The input may be linked directly to the output, if desired. The input impedance and amplification of the equalizer are set with S_1 and S_3. The audio frequency spectrum of 31 Hz to 16 kHz is divided into ten bands.

active one to avoid a very large value of inductance. It is based in a traditional manner on op amp A_1.

The inductors used in the passive filters are readily available small chokes. The filter based on L_1 and L_2 operates at about the lowest frequency (62 Hz) that can be achieved with standard, passive components.

The Q(uality) factor of the filters can, in principle, be raised slightly by increasing the value of R_{19} and R_{23}, as well as that of P_1–P_{10}, but that would be at the expense of the noise level of op amp IC_1.

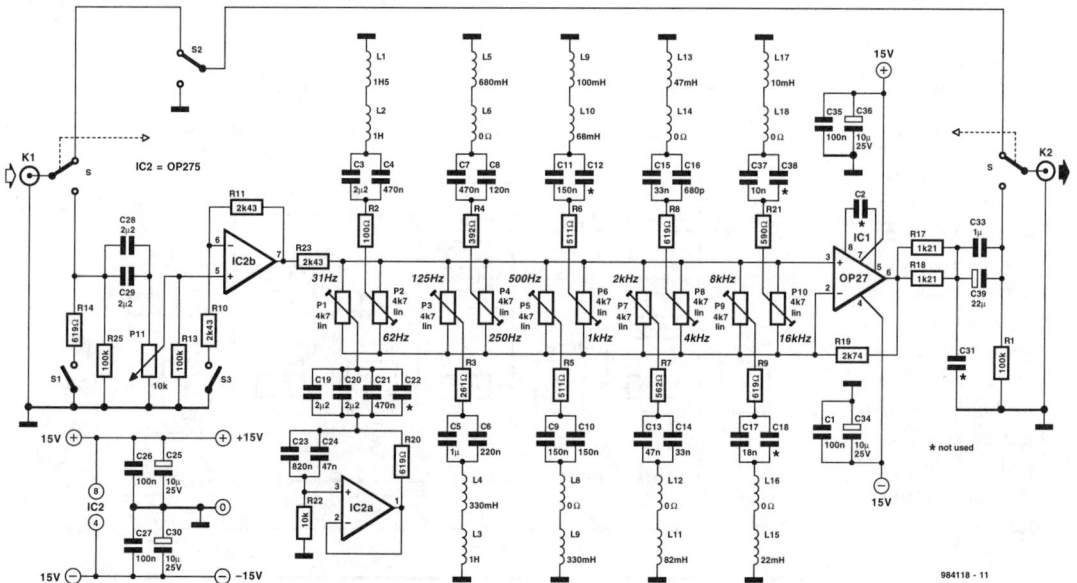

984118 - 11

Ten bands require ten filters, of which nine are passive and one active. The passive filters are identical in design and differ only in the value of the relevant inductors and capacitors. The requisite characteristics of the filters are obtained by series and parallel networks. The filter for the lowest frequency band is an

With component values as specified, the control range is about ±11 dB, which in most cases will be fine. A much larger range is not attainable without major redesign.

The input level can be adjusted with P_1, which may be necessary for adjusting the balance between the channels or when a loudness control is used in the output amplifiers.

Several types of op amp can be used: in the prototype, IC_1 is an LT1007, and IC_2, an OP275. Other suitable types for IC_1 are OP27 or NE5534; and for IC_2, AD712, LM833 and NE5532. If an NE5534 is used for IC_1, C_2 is needed; in all other cases, not.

The circuit needs to be powered by a regulated, symmetrical 15 V supply. It draws a current of not more than about 10 mA.

4-bit analogue-to-digital converter

The operation of the converter is based on the weighted adding and transferring of the analogue input levels and the digital output levels. It consists of comparators and resistors. In theory, the number of bits is unlimited, but each bit needs a comparator and several coupling resistors. The diagram shows a 4-bit version.

The value of the resistors must meet the following criteria:

$R_1:R_2 = 1:2;$
$R_3:R_4:R_5 = 1:2:4;$
$R_6:R_7:R_8:R_9 = 1:2:4:8.$

The linearity of the converter depends on the degree of precision of the value of the resistors with respect to the resolution of the converter, and on the accuracy of the threshold voltage of the comparators. This threshold level must be equal, or nearly so, to half the supply voltage. Moreover, the comparators must have as low an output resistance

In the present converter, complementary metal-oxide semiconductor (CMOS) inverters are used, which, in spite of their low gain, give a reasonably good performance.

If standard comparators are used, take into account the output voltage range and make sure that the potential at their non-inverting inputs is set to half the supply voltage.

If high accuracy is a must, comparators Type TLC3074 or similar should be used. This type has a totem-pole output. The non-inverting inputs should be interlinked and connected to the junction of two 10 kΩ resistors connected in series across the supply lines.

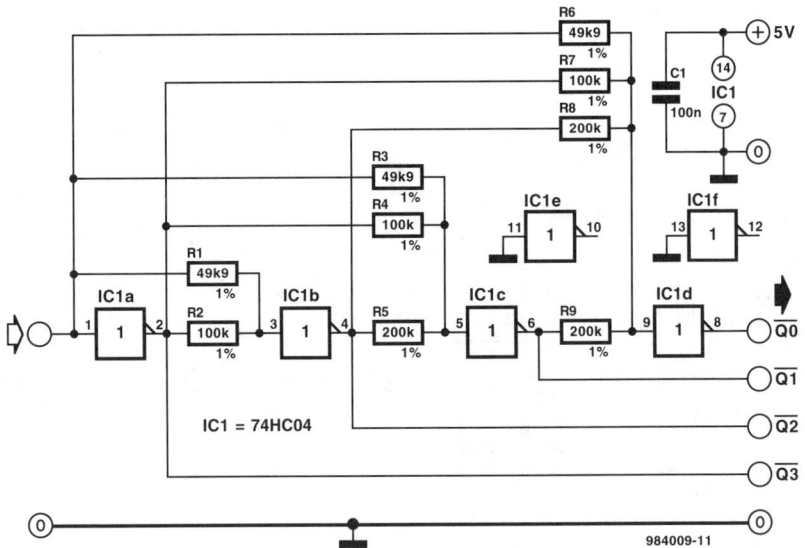

984009-11

as possible and as high an input resistance with respect to the load resistors as feasible. Any deviation from these requirements affects the linearity of the converter adversely. If the value of the resistors is not too low, the use of inverters with an FET (field-effect transistor) input leads to a near-ideal situation.

It is essential that the converter is driven by a low-resistance source. If necessary, this can be arranged via a suitable op amp input buffer.

The converter draws a current not exceeding 5 mA.

Accurate bass tone control

A difficult problem in the design of conventional stereo tone controls is obtaining synchronous travel of the potentiometers. Even a slight error in synchrony can cause phase and amplitude differences between the two channels. Moreover, linear potentiometers are often used in such controls, and these give rise to unequal performance by human hearing. Special potentiometers that counter these difficulties are normally hard to obtain in retail shops.

A good alternative is a control based on a rotary switch and a discrete potential divider. The problem with this is that for good tone control more than six steps are needed, and switches for this are not readily available. Fortunately, electronic circuits can overcome these difficulties.

The analogue selectors used may be driven by mechanical switches, standard logic circuit or a microcontroller. The selectors used in the present circuit are Type SSM2404 versions from Analogue Devices, which switch noiselessly. Each IC contains four selectors, so that a total of eight are used. The step size is 1.25 dB at 20 Hz with a maximum of 10 dB.

The circuit can be mirrored with S1 to enable a selection to be made of amplification or attenuation of bass frequencies. The control can be short-circuited with switch S2.

To prevent the output impedance of the circuit having too much effect on the operation of the circuit, it must be ≤ 10 Ω. Resistor R_{12} protects the circuit against too small a load.

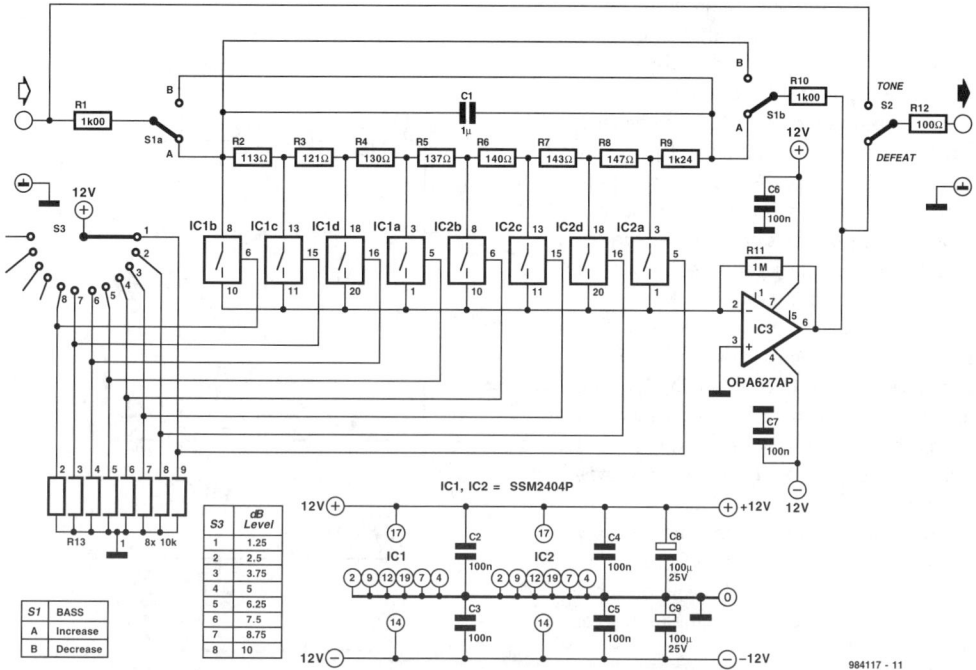

At maximum bass amplification at U_{in} = 1 V r.m.s., the THD+N <0.001% for a frequency range of 20 Hz to 20 kHz and a bandwidth of 80 kHz.

The circuit draws a current of about 10 mA.

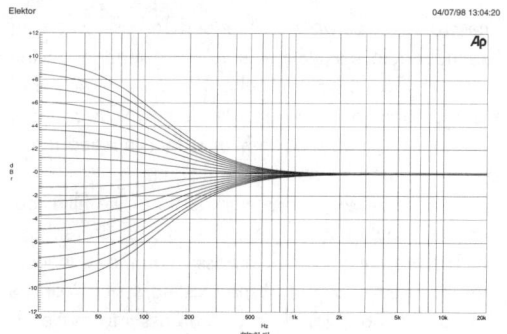

Balanced microphone preamplifier

The preamplifier is intended for use with moving coil (MC) microphones with an impedance of up to 200 Ω and balanced terminals. It is a fairly simple design, which may also be considered as a single stage instrument amplifier based on a Type NE5534 op amp.

To achieve maximum common-mode rejection (CMR) with a balanced signal, the division ratios of the dividers (R_1-R_4 and R_2-R_5 respectively) at the inputs of the op amp must be identical. Since this may be difficult to achieve in practice, a preset potentiometer, P_1, is connected in series with R_5. The preset enables the common-mode rejection to be set optimally.

Capacitor C_1 prevents any direct voltage at the input, while resistor R_7 ensures stability of the amplifier with capacitive loads. Resistor R_3 prevents the amplifier going into oscillation when the input is open circuit. If the microphone cable is of reasonable length, R_3 is not necessary, since the parasitic capacitance of the cable ensures stability of the amplifier. It should be noted, however, that R_3 improves the CMR from >70 dB to >80 dB.

Performance of the preamplifier is very good. The THD+N (total harmonic distortion plus noise) is smaller than 0.1% with an input signal of 1 mV and a source impedance of 50 Ω.

Under the same conditions, the signal-to-noise ratio is –62.5 dBA.

With component values as specified, the gain of the amplifier is 50 dB (x316).

After careful adjustment of P_1 at 1 kHz, the CMR, without R_3, is 120 dB.

The supply voltage is ±15 V. The amplifier draws a current at that voltage of about 5.5 mA. Note the decoupling of the supply lines with L_1, L_2, C_2-C_5.

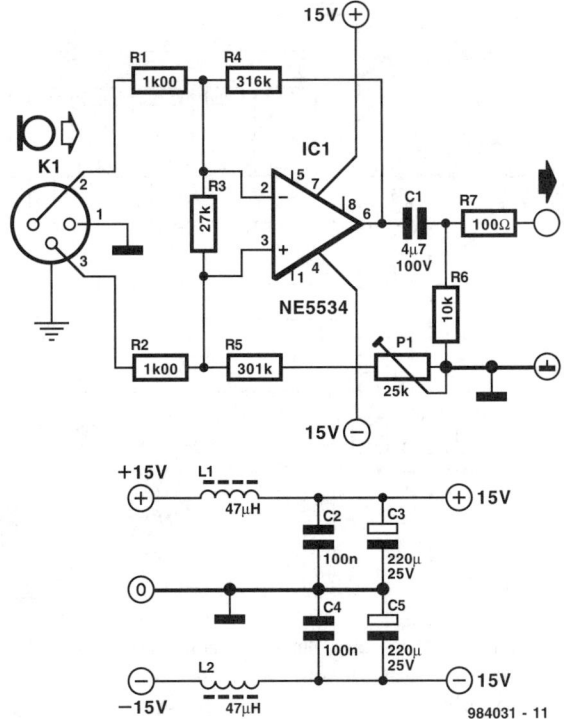

984031 - 11

Crossover for subwoofer

The crossover network is intended for use when an existing audio installation is to be extended by the addition of a subwoofer. If the frequency response of the additional loudspeaker extends down far enough, all is well and good, but a filter is then needed to cut off any frequencies above, say, 150 Hz.

Often, a subwoofer network is an active filter, but here this would necessitate an additional power supply. The present network is a passive one, designed so that the speaker signal of the existing system can be used as the input signal. Since the bass information is

present in both (stereo) loudspeakers, the signal for the subwoofer can simply be tapped from one of them.

The network is a 1st order low-pass filter with variable input (P_1) and presettable cut-off frequency (P_2).

The signal from the loudspeaker is applied to terminal 'LSP'. Voltage divider R_1-R_2-P_1 is designed for use with the output signal of an average output amplifier of around d 50 W.

The crossover frequency of the network may be varied between 50 Hz and 160 Hz with P_2. The values of R_3, P_2, and C_1, are calculated on the assumption that the subwoofer amplifier to be connected to K_1 has a standard input resistance of 47 kΩ. If this figure is lower, the value of C_1 will need to be increased slightly.

It is advisable to open the volume of the subwoofer amplifier fully and adjust the sound level with P_1. This ensures that the input of the subwoofer amplifier cannot be overloaded or damaged.

Make sure that the ground of the loudspeaker signal line is linked to the ground of the subwoofer amplifier. If phase reversal is required, this is best done by reversing the wires to the subwoofer.

If notwithstanding the above additional protection is desired at the input of the subwoofer amplifier, this is best effected by the 'overload protection' elsewhere in this chapter.

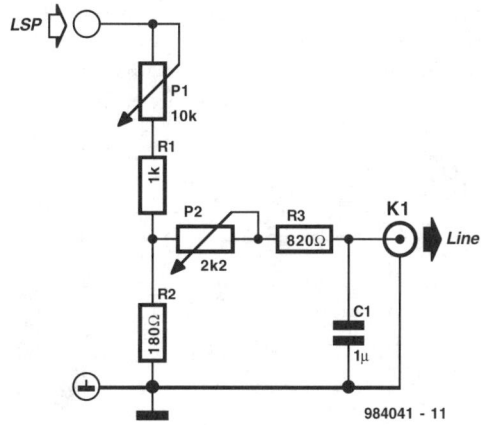

984041 - 11

Mains splitter for AF power amps

In many home-made AF power amplifiers, the primaries of the mains transformers are simply connected in parallel and protected by a single, large, fuse. There may be one or two transformers inside the case, or even three, where a smaller one is used to power an ancillary circuit like a protection circuit. The use of a single fuse to protect the lot is undesirable because this fuse has to be rated for the rush-in current of the large transformers. Moreover, when the fuse burns out it may be difficult to find out which monoblock, or indeed which other part of the amplifier, is the culprit.

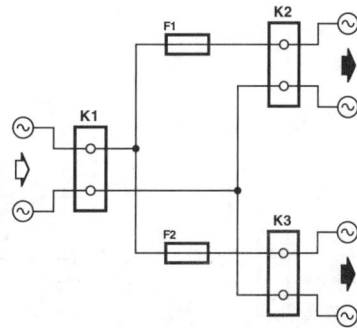

984026 - 11

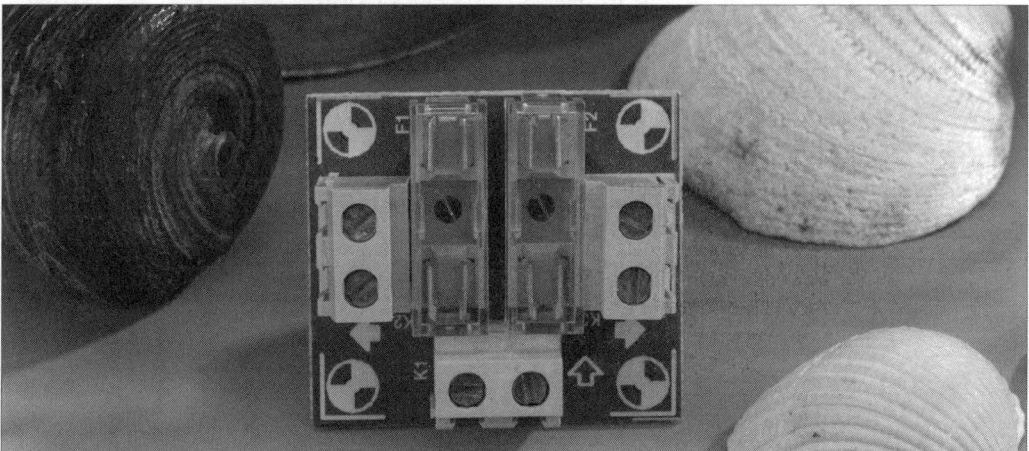

The small circuit board shown here (which may be made with the aid of the track layout shown here) allows the mains input voltage to be distributed in a safe manner to two loads, each with its own (properly rated) fuse. Because the 'circuit' does not include an earth line, it may not be used as an external unit, that is, outside an earthed enclosure—see the safety page (p. vii) at the front of this book.

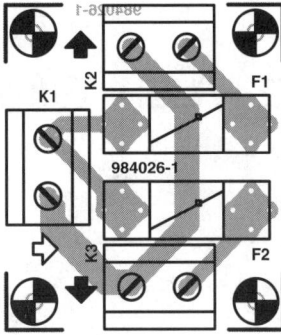

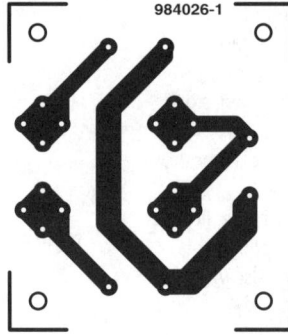

Parts list

K1, K2, K3 = 2-way terminal block, pitch 7.5 mm

F1, F2 = fuseholder for board mounting with cap and fuse rated as needed by application

Microphone valve preamplifier

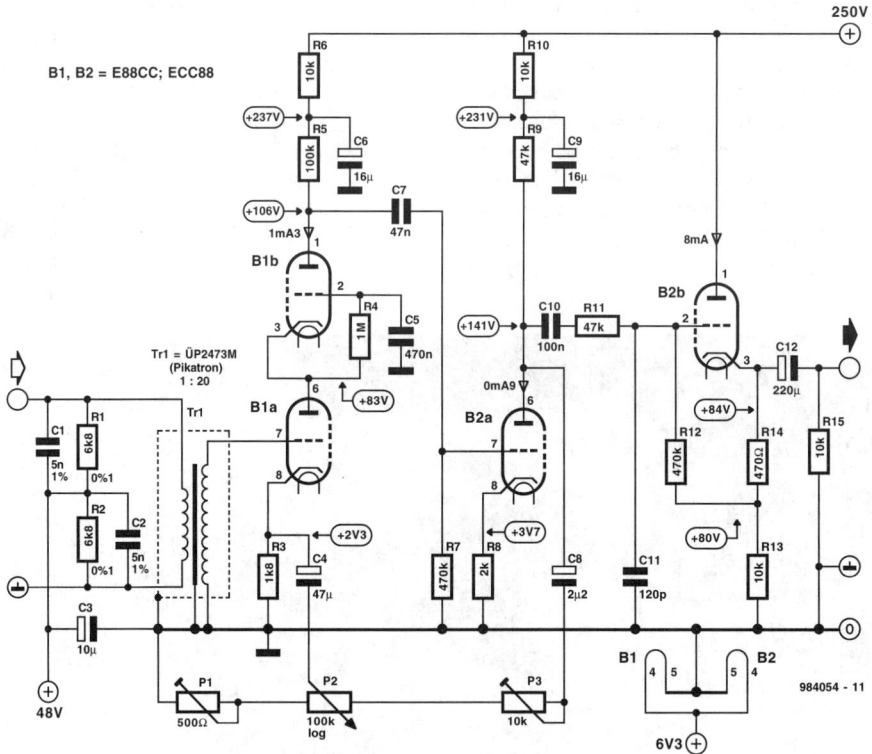

B1, B2 = E88CC; ECC88

Tr1 = ÜP2473M (Pikatron) 1 : 20

984054 - 11

To many hi-fi enthusiasts and musicians, the sound of a valve amplifier cannot be bettered by that of a solid-state amplifier. To satisfy that conviction, here is a microphone preamplifier based on valves.

The circuit in the diagram is intended for use with a studio microphone with or without phantom supply.

The microphone output signal is applied to the control grid of V_{1A} via transformer Tr_1, which has a transformation ratio of 1:20. The input double triode, a Type ECC88 or E88CC is configured as a cascode circuit. This type of circuit has the high amplification of a pentode and the low noise of a triode.

The grid bias for V_{1B} is the potential drop across R_4, so that the operating point is established automatically.

The output of V_{1B} is applied to the control grid of V_{2A} This half of another double triode Type E88CC or ECC88, is arranged as a straightforward voltage amplifier. Its output is applied to grounded-anode amplifier V_{2B}.

Power is supplied by a traditional valve circuit providing an anode voltage of 250 V and a heater (direct) voltage of 6.3 V. The heater voltage is stabilized by a solid-state variable regulator. The heater current is 600 mA. The anode current is about 15 mA.

Building the amplifier is not specially difficult, as long as the usual care is taken, such as correct balancing and low-capacitance link from the secondary of Tr_1 to the grid of V_{1A}. The ratings of the various resistors and capacitors can be derived readily from the voltage and current values shown in the diagram.

Preset P_1 is adjusted for an amplification $a = \times10^4$ (80 dB) and P_3 for $a = \times10^2$ (40 dB). The potential divider may also be constructed from fixed resistors, when $P_1 = 130\ \Omega$; $P_3 = 8640\ \Omega$; and $P_2 = 549\ \Omega + 9760\ \Omega + 68100\ \Omega + 24100\ \Omega$. Successive nodes in the chain represent 10 dB gain intervals so that gains of 40 dB and 80 dB respectively can be set without any difficulty.

E88CC
ECC88

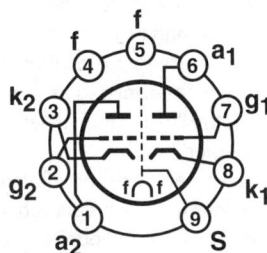

The microphone amplifier does not just meet the requirements of a good hi-fi unit, but satisfies those of professional audio equipment. In this context, measurements on the prototype were taken under the rules of professional equipment (source impedance = 200 Ω, load resistance = 5 kΩ). The results are shown in the table on the next page.

Main parameters

Frequency range	30–20,000 Hz ±3 dB
Distortion	
Maximum drive, a = 80 dB, 30–10,000 Hz	≤ 0.1%
Maximum drive, a = 40 dB, 30–80 Hz	≤ 0.2%
Maximum drive, a = 40 dB, 80–10,000 Hz	≤ 0.1%
Output voltage at maximum drive	8 V r.m.s.
Drive limit for k = 1%	
With a = 80 dB	+34 dBm (38 V r.m.s.)
With a = 40 dB	+25 dBm (13.5 V r.m.s.)
Noise output (input terminated by 200 Ω impedance)	
Weighted, a = 80 dB	–117 dB
Weighted, a = 40 dB	–116.6 dB
Unweighted, a = 80 dB	–128 dB
Unweighted, a = 40 dB	–127 dB
Common mode rejection	60 dB
Output impedance	
30–80 Hz	about 200 Ω
80–20,000 Hz	about 120 Ω

Overload protection

Although the protection circuit is fairly simple, it forms an effective guard against overload of the input of amplifiers and loudspeakers. Why these inputs may need protection now that line levels have been standardized is because there are signal sources on the market that generate outputs of several volts instead of the standardized 1 V r.m.s. Also, in some applications, the loudspeaker signal is applied to the line output of a separate amplifier via a voltage divider, in which case the levels may be well above 1 V r.m.s.

The diagram shows a circuit that resembles the familiar series resistor and zener diode. Here, however, the zener is constructed from a small rectifier and a transistor, since commercial zeners appear to start conducting way below their rated values, which gives rise to unwanted distortion.

The constructed zener makes a well-defined limitation possible and does not affect signals below the critical level. Configuring T_1 as a diode reduces the number of components needed to a minimum: not even a voltage divider or potentiometer is required.

Measurements on the prototype show that the input signal remains virtually undistorted at levels up to 700 mV r.m.s. At the threshold of 1 V r.m.s., the distortion is about 0.02%. Above this level, limiting is well-defined. The peak output voltage of the circuit is

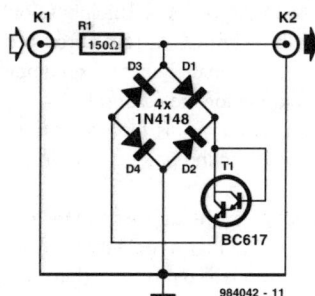

about 3 V with an input voltage of about 13 V r.m.s. If the limiting level is required to be slightly higher, consideration should be given to replacing T_1 by three or four cascaded diodes.

Pan pot

A pan pot enables a monophonic input signal to be positioned where desired between the stereo loudspeakers. When P_1 (see diagram) is in the centre position, there is no attenuation or amplification between the input and output. When the control is turned away from the centre position, the signal in one channel will be amplified 3 dB more than the other.

Circuit IC_1 at the input is a buffer stage. It is arranged as an inverter to ensure that the phase of the input signal is identical to that of the output signal. The input impedance is set with R_1 (10 kΩ).

The output of the buffer is applied to stereo amplifiers IC_2 and IC_3. A special arrangement here is the position of P_1, in conjunction with R_3, R_4, R_8, and R_9, in the feedback circuits of both amplifiers. This means that any adjustment of the potentiometer will have opposite effects in the amplifiers.

Series resistors R_7 and R_{12} serve to ensure that the outputs can handle capacitive loads.

Coupling capacitors C_3, C_6, and C_9, may be omitted if an offset voltage of 20–30 mV is of no consequence in the relevant application.

Capacitors C_2, C_5, and C_8, ensure that the op amps remain stable even at unity gain.

Capacitors C_1, C_4, and C_7, minimize any r.f. interference, resulting in a usable bandwidth of 2.5 Hz to 200 kHz.

The performance of the circuit is of sufficiently high quality to allow the pot being incorporated in good-quality control panels.

Total harmonic distortion plus noise (THD+N) at a frequency of 1 kHz and a bandwidth of 22 kHz is 0.0014%. Over the band 20 Hz to 20 kHz and a bandwidth of 80 dB, this figure is still only 0.0023%.

The circuit needs a power supply of ±18 V, from which it draws a current of about 16 mA.

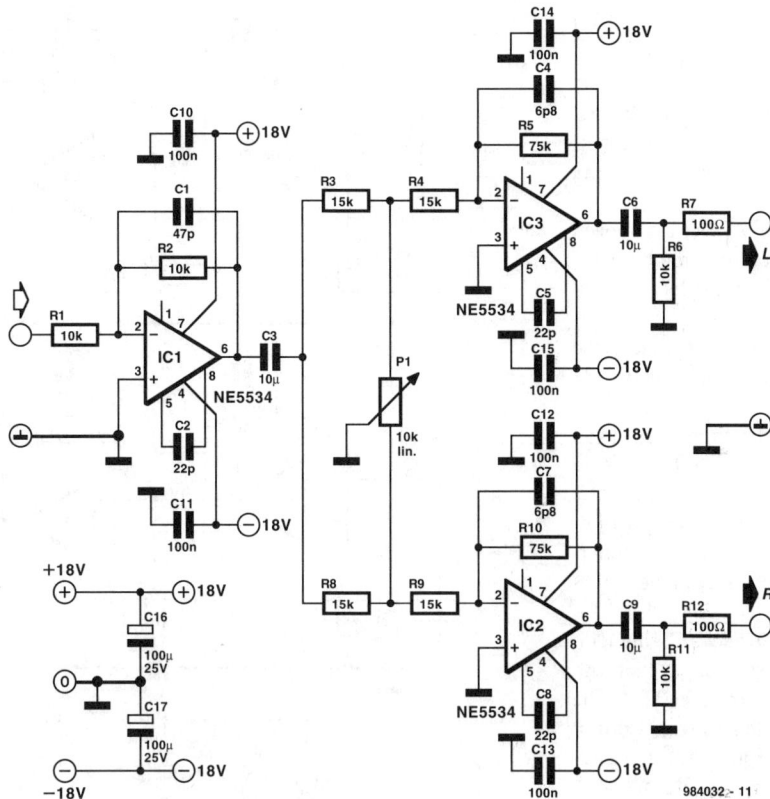

C14
100n
C4
6p8
R5
75k
C10
100n
C1
47p
R2
10k
R3
15k
R4
15k
IC3
NE5534
C6
10µ
R7
100Ω
R6
10k
L
C5
22p
C15
100n
R1
10k
IC1
NE5534
C3
10µ
P1
10k lin.
C2
22p
C11
100n
C12
100n
C7
6p8
R10
75k
+18V
C16
100µ 25V
C17
100µ 25V
−18V
R8
15k
R9
15k
IC2
NE5534
C9
10µ
R12
100Ω
R11
10k
R
C8
22p
C13
100n
984032 - 11

Playback amplifier for cassette deck

For some time now, there have been a number of tape cassette decks available at low prices from mail order businesses and electronics retailers. Such decks do not contain any electronics, of course. It is not easy to build a recording amplifier and the fairly complex magnetic biasing circuits, but a playback amplifier is not too difficult as the present one shows.

The stereo circuits in the diagram, in conjunction with a suitable deck, form a good-quality cassette player. The distortion and frequency range (up to 23 kHz) are up to good standards. Moreover, the circuit can be built on a small board for incorporation with the deck in a suitable enclosure.

Both terminals of coupling capacitor C_1 are at ground potential when the amplifier is switched on. Because of the symmetrical ±12 V supply lines, the capacitor will not be charged. If a single supply is used, the initial surge when the capacitor is being charged causes a loud click in the loudspeaker and, worse, magnetizes the tape.

The playback head provides an audio signal at a level of 200–500 mV. The two amplifiers raise this to line level, not linearly, but in accordance with the RIAA equalization characteristic for tape recorders. Broadly speaking, this characteristic divides the frequency range into three bands:

• Up to 50 Hz, corresponding to a time constant of 3.18 ms, the signal is highly and linearly amplified.
• Between 50 Hz and 1.326 kHz, corresponding to a time constant of 120 µs, for normal tape, or 2.274 kHz, corresponding to a time constant of 70 µs, for chromium dioxide tape, the signal is amplified at a steadily decreasing rate.
• Above 1.326 kHz or 2.274 kHz, as the case may be, the signal is slightly and linearly amplified.

29

This characteristic is determined entirely by A_1 (A_1'). To make the amplifier suitable for use with chromium dioxide tape, add a double-pole switch (for stereo) to connect a 2.2 kΩ resistor in parallel with R_3 (R_3').

The output of A_1 (A_1') is applied to a passive high-pass rumble filter, C_3-R_5 (C_3'-R_5') with a very low cut-off frequency of 7 Hz. The components of this filter have exactly the same value as the input filter, C_1-R_1 (C_1'-R_1').

The second stage, A_2 (A_2') amplifies the signal x100, that is, to line level (1 V r.m.s.).

Capacitor C_4 limits the upper frequency range to avoid r.f. interference and any tendency of the amplifier to oscillate.

The amplifier needs a symmetrical ±12 V power supply that can provide a current of up to 0.5 A. The greater part of this current is drawn by the motor of the deck; the electronic circuits draw only 15 mA.

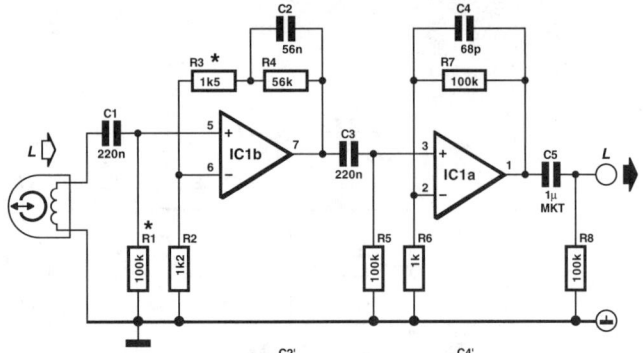

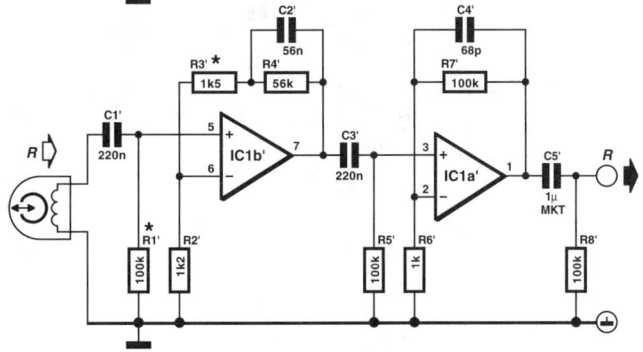

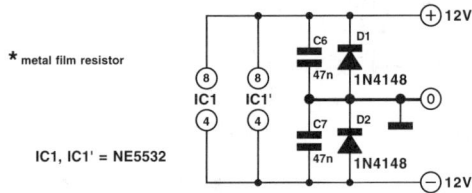

Presence filter

To make a certain musical instrument in a group stand out, a so-called presence filter is normally used. Unfortunately, the types usually found in amplifiers and mixers can only raise the level of the instrument output, but not attenuate it.

The filter in the diagram provides amplification (15 dB) as well as attenuation (15 dB) over the presence range (see Figure 1). When potentiometer P_1 is at its centre position, the signal is unaltered.

The input signal (see Figure 2) is applied to impedance converter A_1.

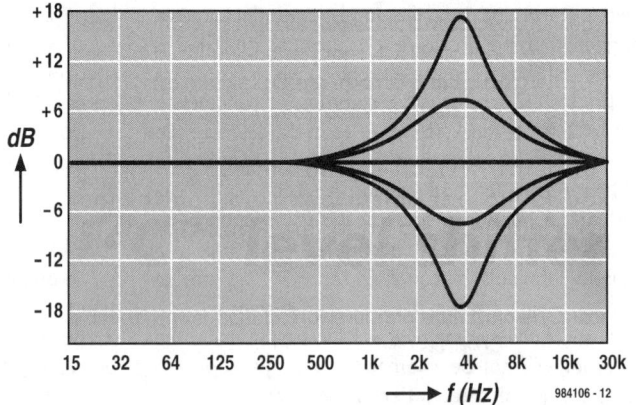

Capacitor C_1 blocks any d.c. on the signal. Resistor R_1 sets the input resistance of the circuit. Diodes D_1 and D_2 protect the input against high voltages. Resistor R_2 limits the current to the input of the impedance converter.

The actual filter process is carried out by op amps A_2 and A_3 and associated components. The filter behaves as a frequency-dependent resistance whose value is a minimum at about 3.5 kHz. At very high and very low frequencies, the resistance of the filter is high. Depending on the setting of P_1, the filter forms a potential divider with R_3 or part of the feedback loop with R_4. When P_1 is at its centre position, the filter attenuates the signal to the same degree as it is amplified by A_2.

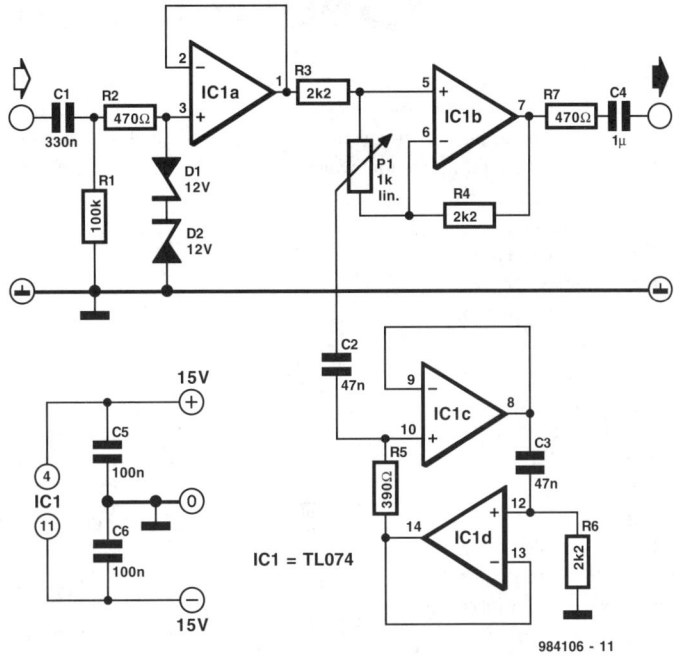

984106 - 11

Pulse/frequency modulator

The pulse width of the compact pulse/frequency modulator can be varied by altering the change-over point of comparator IC_1 with a control voltage via resistor R_1. The hysteresis of the IC is determined by resistors R_3 and R_4. The control voltage also causes the frequency of the present circuit to be altered. When the input voltage is 0 V, the frequency is a maximum: in the present design this is about 3.8 kHz. The level of the output voltage is ±12–13 V.

The more the control voltage shifts the change-over point, the longer it takes before the potential across capacitor C_1 has reached the level at which IC_1 is enabled. When the control voltage is larger than the zener voltage, the oscillator ceases to work. The maximum period is 25 ms, which may be adapted by altering the value of C_1. This will, of course, also alter the maximum frequency.

The duty cycle is inversely proportional to the control voltage. The minimum pulse width attainable at the lowest frequency is about 6 μs.

The modulator draws a current not exceeding 5 mA.

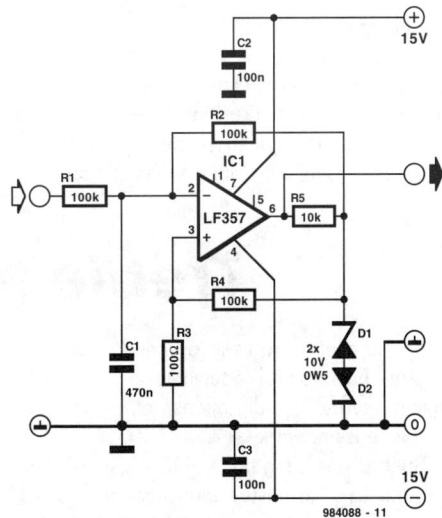

984088 - 11

Speech modifier

Nowadays, the speech quality on our telephone systems is generally very good, irrespective of distance. However, there are occasions, for instance, in an amateur stage production, or just for fun, when it is desired to reproduce the speech quality of yesteryear.

The modifier circuit accepts an acoustic (via an electret microphone) or electrical signal. The signals are applied to the circuit inputs via C_1 and C_2, which block any direct voltage. The input cables should be screened.

The signals are brought to (about) the same level by variable potential dividers P_1-R_1-R_4 and P_2-R_2-R_3, and then applied to the base of transistor T_1. The level of the combined signals is raised by this pre-amplifier.

The preamplifier is followed by an active low-pass filter consisting of T_2–T_4, C_3, C_4, R_6–R_8, and P_4.

Although, strictly speaking, P_3 serves merely to adjust the volume of the signal, its setting does affect the filter characteristic. Note, by the way, that the filter is a rarely encountered current-driven one in which C_3 and C_4 are the frequency-determining elements. It has a certain similarity with a Wien bridge.

Transistors T_3 and T_4, in conjunction with resistors R_8 and P_4, form a variable current sink. The position of P_4 determines the slope of the filter characteristic and the degree of overshoot at the cut-off frequency.

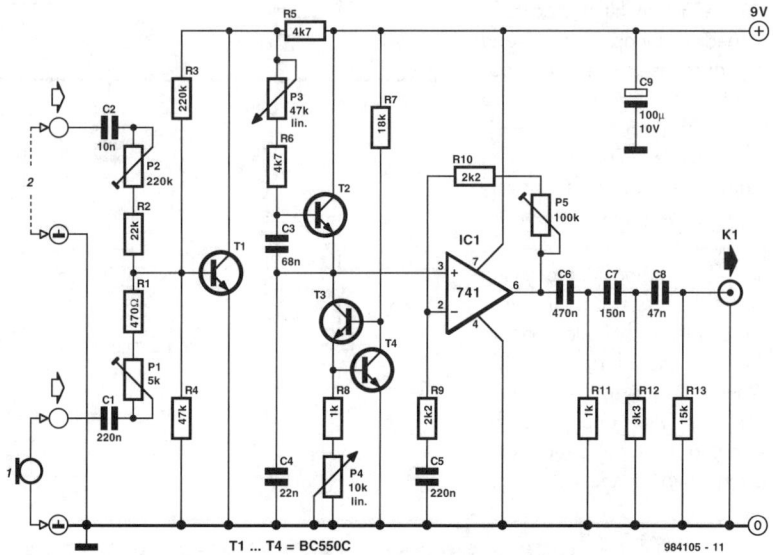

T1 ... T4 = BC550C   984105 - 11

The low-pass filter is followed by integrated amplifier IC_1 whose amplification is matched to the input of the electronic circuits connected to the modifier with P_5.

The final passive, third-order, high-pass filter is designed to remove frequencies above about 300 Hz.

The resulting output is of a typical nasal character, just as in telephones of the past.

Treble tone control

The treble control works in a similar manner as the 'Accurate bass control' elsewhere in this chapter, but contains several modifications, of course. One of these is series network C_1-C_2-R_1-R_{11}.

The d.c. operating point of IC_3 is set with resistors R_{12} and R_{13}. To ensure that these resistors do not (adversely) affect the control characteristics, they are coupled to the junction of R_9 and R_{10}. In this way they only affect the low-frequency noise and the load of the

op amp. Their value of 10 kΩ is a reasonable compromise.

The functions of switches S_1–S_3 are identical to those of their counterparts in the bass tone control; their influence is seen clearly in the characteristics.

The left-hand and right-hand channels are well balanced if 1% versions of R_1–R_{13} and C_1, C_2 are used.

The value of resistors R_2–R_{10} is purposely different from that of their counterparts in the bass tone control. In the present circuit, the control range starts above 20 kHz. To make sure that a control range of 10 dB is available at 20 kHz, the nominal amplification is x3.5 (11 dB).

The control circuit draws a current of about ±10 mA.

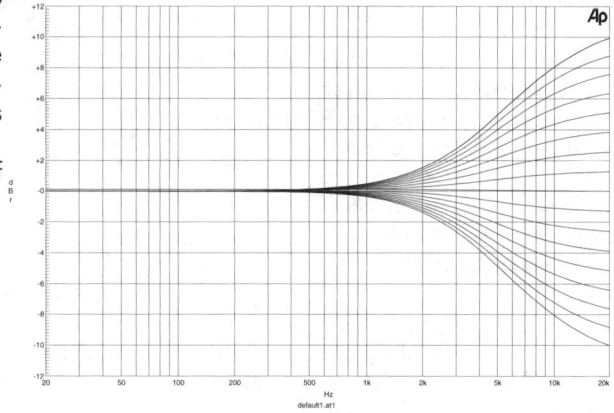

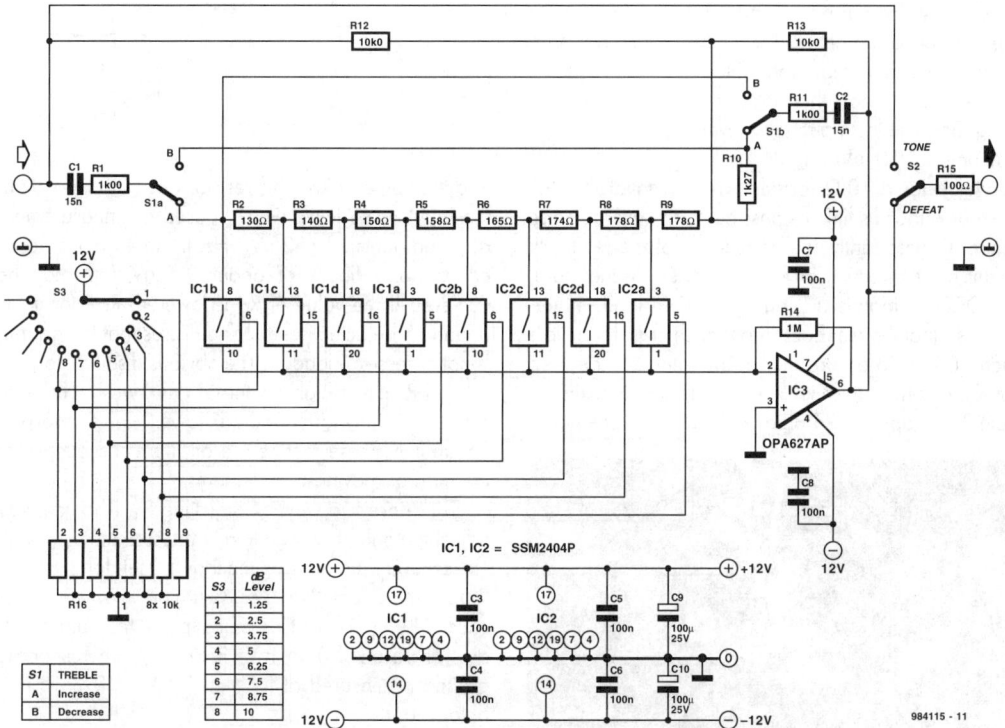

Ultra-low-noise preamplifier

The preamplifier is intended for use with low-impedance signal sources like moving-coil pick-up cartridges used in high-end record players. The input impedance of the preamplifier is 100 Ω. To keep the input noise as low as possible, three dual transistors type

SSM2220 or MAT03 transistors are connected in parallel to form a discrete differential amplifier. By connecting this amplifier ahead of IC_1, the input noise of the op amp becomes im-material. The base connections of the discrete amplifier then function as the

33

inputs of a super op amp with a very low input noise level. The p-n-p transistors used score over their n-p-n counterparts in their much lower low-frequency noise level. A fairly large bias current of about 5.5 mA is created at the input. This is the result of the 2 mA setting of each transistor in combination with the relatively low gain of the p-n-p devices.

Preset P_1 and resistors R_7-R_8 enable any tolerances of R_4 and R_5 in the differential amplifier output to be compensated. Transistor T_4 and light-emitting diode D_1 ensure a stable current setting for the differential amplifier. The LED should be a flat, red, type fitted securely to T_4 for thermal coupling. Because the input noise level amounts to 0.4 nV/√Hz (theoretical value for a 10 Ω resistor), it is essential that the feedback adds as little as possible to the overall noise figure. Consequently, the impedance of the feedback circuit must be much lower than 10 Ω. Furthermore, the OP27 demands a certain minimum load impedance, so that the feedback impedance may not be less than 600 Ω. To ensure that a low value can be used for R_9, a compromise is necessary between maximum gain (here, approx. 24 dB or x15.7) on the one hand,

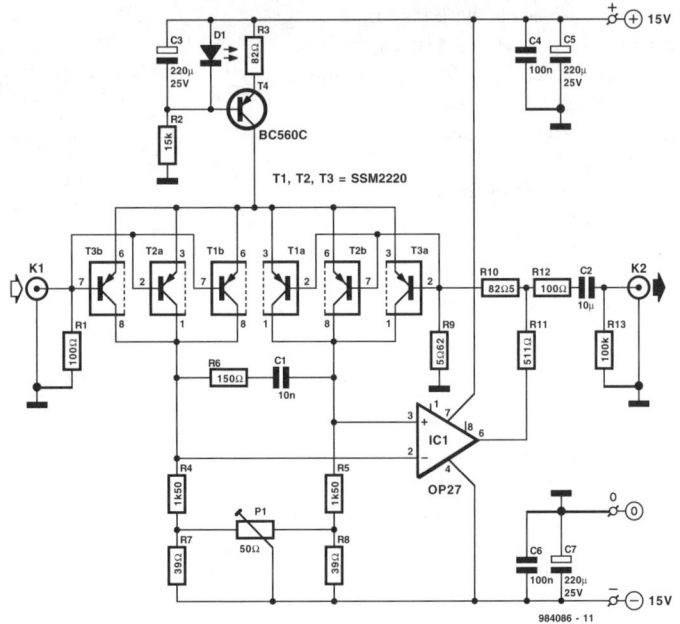

and the value of R_9. This resistor adds 0.3 nV/√Hz to the input noise level, which, based on measurement data, amounts to 0.52 nV/√Hz. If more gain is needed, a noise figure of about 0.4 nV/√Hz may be achieved if the value of R_9 is lowered. Resistor, R_{11}, ahead of the actual feedback, ensures that the opamp is not excessively loaded. The obvious disadvantage of adding R_{11} is a higher internal gain, which causes a smaller bandwidth and a lower drive margin. Fortunately, these factors are of little consequence in the case of moving-coil elements.

There are two ways of adjusting P_1. The first is to adjust the output voltage to nil (measure at IC_1 pin 6). The second is to measure the input offset, for example, 0.55 mV across 100 Ω. Assuming that the offset caused by T_1, T_2 and T_3 is negligible, the output voltage of 15.68 x 0.55 mV (8.62 mV with respect to ground), measured at junction R_{10}-R_{11}-R_{12}, is well balanced.

Readrs may like to try the effects of reducing the number of input transistors from three to just one to reduce the input bias current. The value of resistor R_3 must then be changed to 249 Ω. Bear in mind, however, that the input noise level then rises by 2.5 dB!

Capacitor C_2 prevents any offset voltage from being applied to the input of the following amplifier.

The preamplifier is powered by a symmetrical, regulated 15-V supply, and draws about 16 mA from each rail.

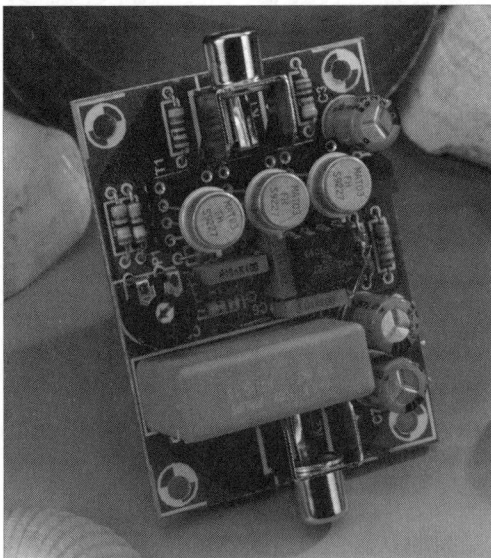

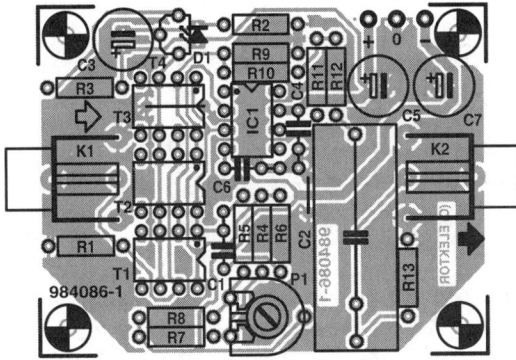

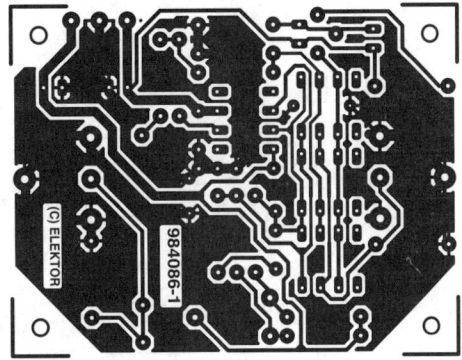

Parts list

Resistors:
$R_1, R_{12} = 100\ \Omega$
$R_2 = 15\ k\Omega$
$R_3 = 82\ \Omega$
$R_4, R_5 = 1.50\ k\Omega$
$R_6 = 150\ \Omega$
$R_7, R_8 = 39\ \Omega$
$R_9 = 5.62\ \Omega$
$R_{10} = 82.5\ \Omega$
$R_{11} = 511\ \Omega$
$R_{13} = 100\ k\Omega$
$P_1 = 50\ \Omega$, preset, horizontal

Capacitors:
$C_1 = 0.01\ \mu F$
$C_2 = 10\ \mu F$, metallized polyester (MKT), pitch 22.5 mm or 27.5 mm

$C_3, C_5, C_7 = 220\ \mu F$, 25 V, radial
$C_4, C_6 = 0.1\ \mu F$

Semiconductors:
D_1 = red LED, flat
T_1, T_2, T_3 = SSM2220 or MAT03 (Analog Devices)
T_4 = BC560C

Integrated circuits:
IC_1 = OP27GP (Analog Devices)

Miscellaneous:
K_1, K_2 = phono socket, gold-plated, for PCB mounting
PCB may be made with the aid of the track layout.

Up/down drive for tone control

The up/down drive is intended primarily for use with the tone controls described on pages 36 and 52.

The tone controls use electronic switches that are operated by a multi-position selector. The present circuit is intended as a replacement for this selector and has facilities for operating the tone controls via an up key and a down key. A third key enables the user to switch over rapidly to a preprogrammed position of the relevant tone control.

The electronic switches are driven by a BCD-to-decimal decoder Type 4028 (IC_3), which in turn is controlled by a 4-bit preset up/down counter (IC_2). The counter uses the three lowest bits only. The MSB of decoder IC_3 is permanently low. Only the eight lowest

outputs of the decoder are used and these are linked via K_1 to the control inputs of IC_1 and IC_2 in the tone controls.

The circuit is operated with S_1 and S_2. Switch S_3 is the earlier mentioned preset key. The data for the preset inputs are set with DIP switch S_4. Capacitor C_3 ensures that when the supply voltage is switched on, the preset data are automatically adopted by the counter.

Each of switches S_1 and S_2 drives an S/R bistable (US: flip-flop), which determines the level at the U/D input of counter IC_2.

Networks R_3-C_1 and R_4-C_2, in conjunction with Schmitt trigger IC_{1b} provide debouncing and delay

35

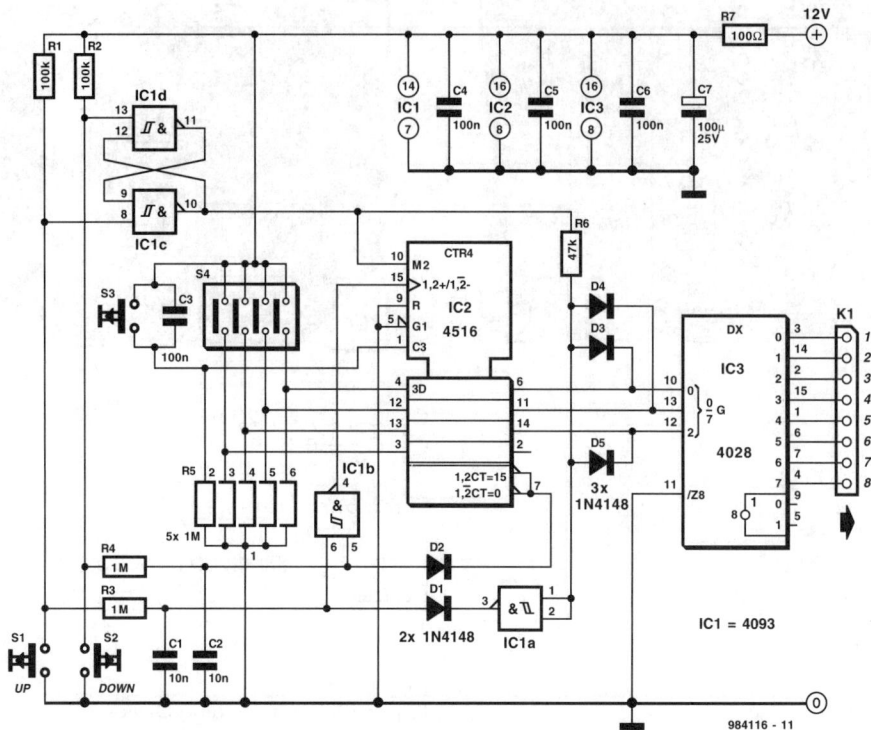

984116 - 11

the clock pulse slightly. This delay guarantees that the clock pulse (output of IC_{1b}) arrives after the state of the counter has been defined.

To prevent the counter jumping from minimum to maximum or vice versa, the clock pulse is disabled in the outermost positions. In the minimum state, this is achieved simply by the carry-out terminal (pin 7) of the counter. In the maximum state, an auxiliary network, consisting of R_6, D_3, D_4, D_5/IC_{1a}, and D_1, is used.

Diode D_2 ensures that when the minim state is reached, pin 5 of IC_1 remains low until S_1 is pressed. Diode D_1 does the same in regards of pin 6 of the IC when the maximum state is reached. Resistor R_6 serves to reset the clock, which is disabled during

down counting; when the down key is pressed, the output of IC_{1a} goes high again.

If an indication is desired of the actual state of the up/down drive, eight high-efficiency LEDs may be added at the output of IC_3 (anodes to the output, cathodes via a common $10 \text{ k}\Omega$ resistor to ground).

Whether the circuit amplifies or attenuates is indicated by an additional LED at the output of IC_{1c} or IC_{1d}.

During quiescent operation, the circuit draws a current of $20 \, \mu A$, which rises to about $140 \, \mu A$ when S_1 or S_2 is pressed. Network R_7-C_7 provides decoupling of the digital circuit from the analogue supply.

Musical touch-tones

Electronic Touch-Tones offers both the sight and sound of musical notation. Although this project is limited to a one-octave major scale, it indicates a method of transforming written notation directly into sound. Moreover, it provides a useful teaching aid for young-

sters learning the basics of scales, intervals and primary chords.

Eight oscillators, their tuning controls and two quad analogue switch ICs feed a transistor output amplifier—see block diagram. As indicated on the

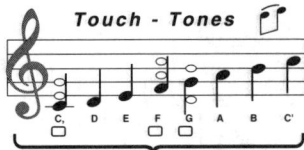

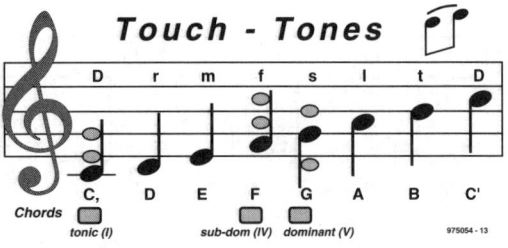

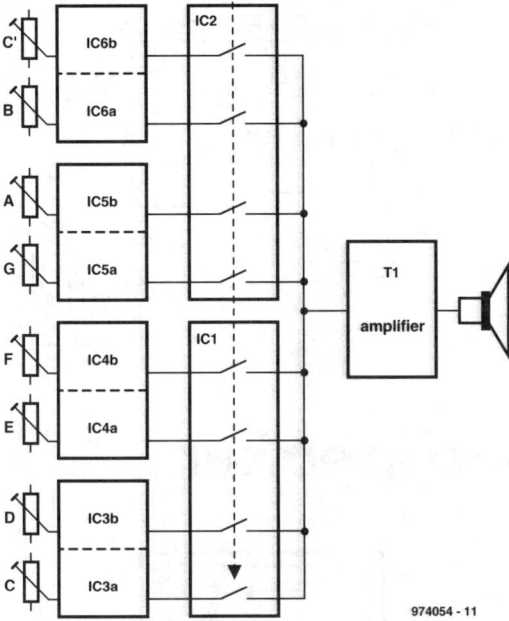

Touch - Tones

C, D E F G A B C'

C'	IC6b	IC2
B	IC6a	
A	IC5b	
G	IC5a	
F	IC4b	IC1
E	IC4a	
D	IC3b	
C	IC3a	

T1
amplifier

974054 - 11

Touch - Tones

D r m f s l t D

C, D E F G A B C'

Chords — tonic (I) — sub-dom (IV) — dominant (V)

975054 - 13

staff, three primary chord touch plates are available, providing a basic accompaniment for any of the notes in this major scale.

In the circuit diagram, things may look complex at first, until it is realized that there are eight identical oscillators, two for each IC. With reference to, for instance, IC_{3a}, the frequency of oscillation is controlled by network P_1-R_1-C_1. With the values indicated, several octaves are available by adjusting P_1. The oscillator output signal is capacitively coupled to pin1 of IC_1, i.e. one of four identical analogue switches. When the C, note on the front panel scale is touched by finger, the skin resistance enables the analogue switch. Each note has two adjacent wires, one to 0 V and the other to the enable pin on the analogue switch. This routes IC_1 pin1 to pin2, and the low-C oscillator output is applied via R_{16} to the base of output transistor T_1.

All eight oscillators operate in identical manner.

Note that diodes D_1-D_9 connect the three front panel chord touch-plates to the appropriate analogue touch controls that sound the three primary triads; tonic (I) to C,, E and G; sub-dominant (IV) to F, A and C'; and dominant (V) to G, B and D. The diodes serve

to connect and isolate the touch-plates; the cathodes of the diodes must connect with the touch-tone wires associated with the analogue switches.

The circuit is best built on the PCB shown, whose track layout is given in the Appendix. Use flexible ribbon cable for connecting the analogue switches to the front-panel touch wires, the battery and switch connections, and the speaker leads. In the prototype, the computer-printed scale is mounted between a paxolin panel and a sheet (260x110 mm) of thin perspex for protection. Each touch-tone consists of two short parallel gold-plated wires, the upper connected to the appropriate analogue switch enable contact, and the other to the +supply rail. The front panel is secured with wood screws to a shallow plywood box (40 mm deep) that houses the PCB and the battery. The miniature loudspeaker is glued behind the treble clef on the front panel with small holes drilled to emit the sound. Current drain is approximately 20 mA, so a 9 V PP3 battery is adequate.

The eight presets in the circuit should be tuned against a keyboard or other musical instrument that is available.

Finally, this simple eight-note version may easily be extended to the chromatic scale of twelve semitones and further.

IC1, IC2 = 4016
IC3 ... IC6 = 556

974054 - 12

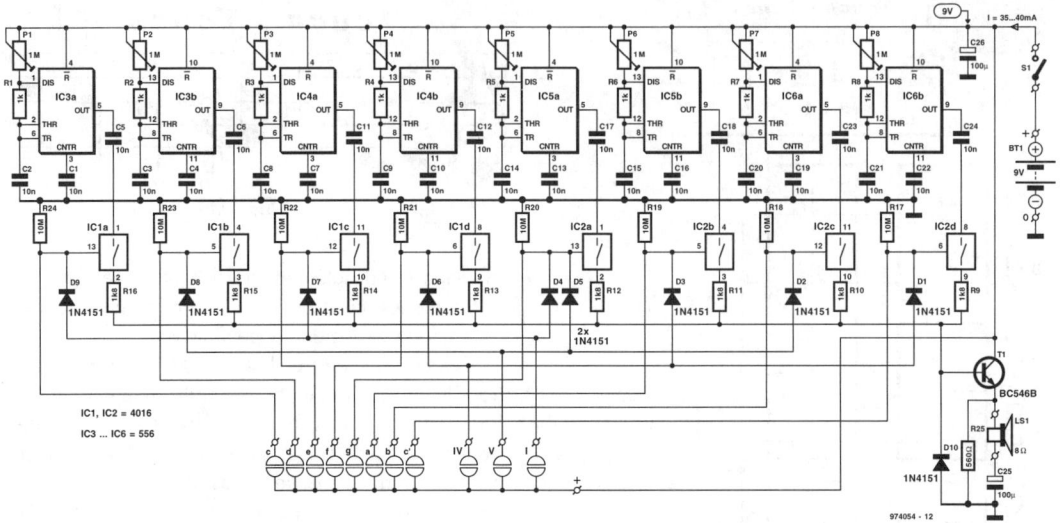

Two-way loudspeaker

In this design of a simple loudspeaker enclosure an attempt has been made to achieve reasonably good quality of reproduction with a minimum of material and inexpensive drivers.

The cross-over filter has a 6-dB roll-off, which means only one component per driver: L_1 for the woofer and C_2 for the tweeter. There is also an impedance-correction network, R_1-C_1, for the woofer, which 'flattens' the rising impedance of this driver.

There is an attenuation network, R_2-R_3, to match the volume level of the tweeter to that of the woofer.

Note that owing to the position of the drivers, the polarization of the tweeter must be the opposite of that of the woofer.

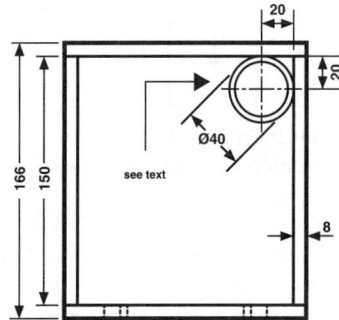

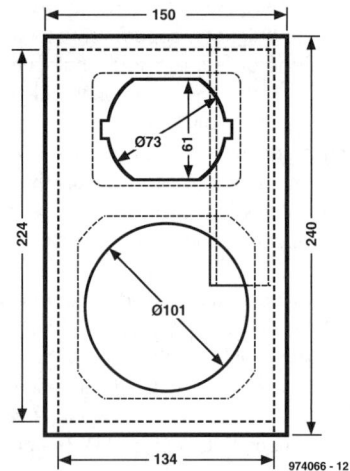

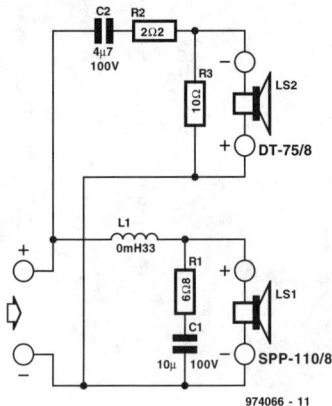

974066 - 11

974066 - 12

The unit may be used as a rear speaker in a surround-sound system or with a multimedia computer. In the latter case, it must be placed well away from the monitor since the magnets of most inexpensive drivers are not screened.

The bass-reflex enclosure, made from 8 mm thick chip/fibre-board, has a volume of 4.5 litres. The bass-reflex port is a standard 40 mm dia. PVC pipe, 175 mm long (if its walls are 2 mm thick; if they are 3 mm thick, the length must be 150 mm).

The nominal impedance of the system is 6 Ω. Maximum power input is 30 W. The cross-over frequency is 4 kHz. The frequency characteristic of the loudspeaker is shown below.

If the coil is not obtainable ready-made, it may be wound on a non-metallic former, 28 mm dia and 28 mm long. The winding consists of seven layers of 1.5 mm dia. enamelled copper wire.

Active two-way system

The amplifier in this system may be used to drive the two-way loudspeaker described in the previous article. The overall system consists of an active two-way crossover network with a cut-off frequency of 3 kHz, a small output amplifier, and a power supply.

The output amplifier is based on a dual integrated bridge amplifier, IC_2, which was developed for use in cars. With a supply voltage of 16 V, the maximum power output is 2x20 W into 4 Ω or 2x12 W into 8 Ω. In both cases a heat sink of 2 K/W will be sufficient, but bear in mind the requisite isolation.

Important advantages of the IC are that few external components are needed, and that the device is provided with all kinds of protection circuit. A drawback is that it provides poor ripple suppression, but this is countered in the present circuit by the rel-

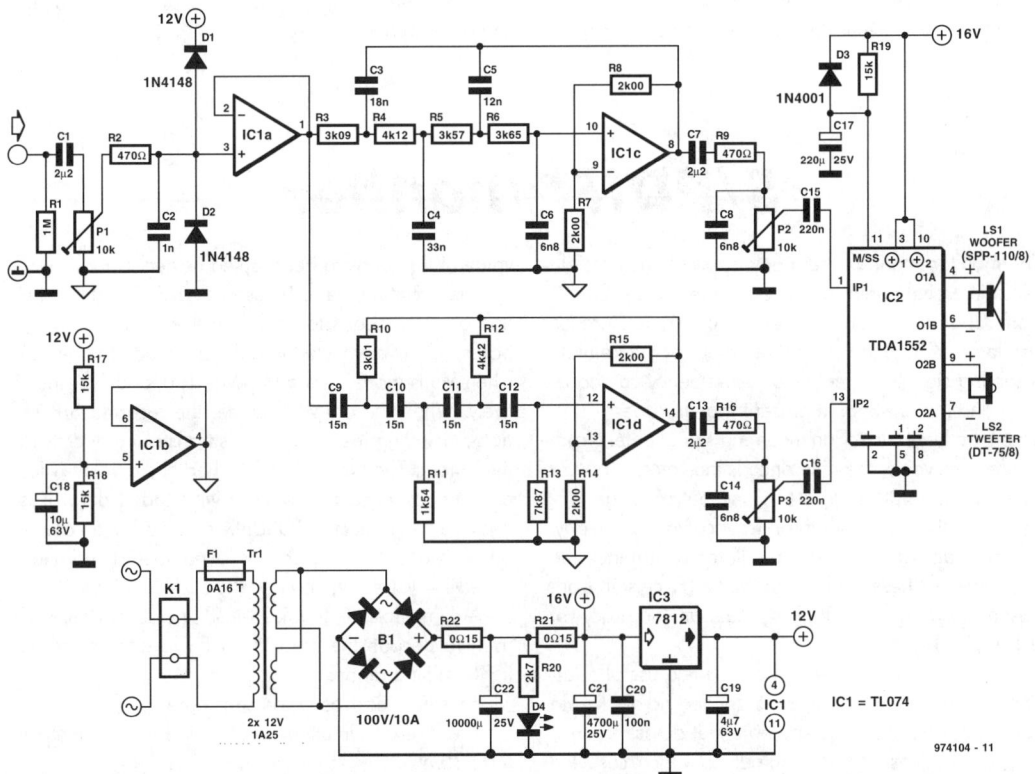

974104 - 11

atively large-value capacitors, C_{21} and C_{22}, in series with resistors R_{21} and R_{22} respectively, in the power supply.

Network R_{19}-C_{17} at pin 11 of IC_2 suppresses any switch-on clicks. Diode D_3 prevents C_{17} being discharged via IC_2 when the supply voltage fails.

The active crossover network is a 4th-order Linkwitz-Riley configuration and is known for its homogeneous radiation and constant amplitude. Its response with the conventional cut-off frequency of 3 kHz is illustrated in Figure 2.

The low-pass section of the filter is based on IC_{1c} and the high-pass section on IC_{1d}. The tolerance of the frequency-determining components has great influence on the response, so that it is recommended to use 1% components for R_3-R_8, R_{10}-R_{15}, C_3-C_6, and C_9-C_{12}.

Because the filter uses an asymmetric 12 V power supply, regulated by IC_3, a virtual earth, half the supply voltage above the real earth, has to be provided for the signal path. This is done here by IC_{1b} and associated components.

Operational amplifier IC_{1a} is used as input buffer, which is protected against overdrive by diodes D_1 and D_2. The input sensitivity is set with P_1.

Differences in efficiency of the two loudspeakers

are nullified with P_2 and P_3. This is done by setting the preset linked to the loudspeaker with the lower efficiency (usually the woofer) to maximumum and the other as necessary.

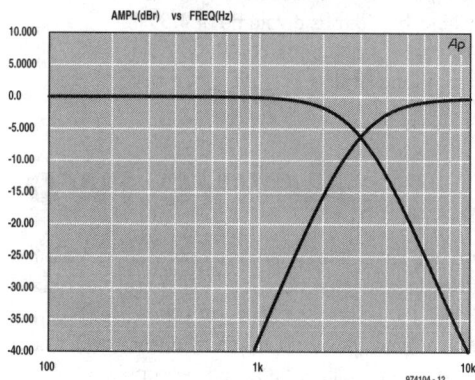

If the amplifier is used to drive the two-way loudspeaker described elsewhere in this section, the signal to the tweeter must be attenuated by about 4.5 dB.

Finally, it was noted in the prototypes that increasing the value of C_9-C_{12} to 0.022 μF gave a rather smoother frequency response.

S/PDIF* monitor

The monitor is one of the many applications possible with the digital audio interface receiver Type CS8412 from Crystal. The addition of an external reference oscillator, IC_4, enables the receiver to differentiate incoming signals by means of a frequency comparator – and this is what the present monitor does

When the frequency of an incoming signal differs from a reference value, the difference is indicated in one of three ways: <400 ppm; <4%; and out of range (differs more than 4% from the reference value). Clearly, the accuracy of the crystal oscillator determines the precision of these limits (the SG531P crystal from Epson used in the diagram has an accuracy of ±100 ppm).

The optical input provided by IC_2 is a useful addition. The output of this circuit is applied across R_1 via C_4, R_3 and jumper JP_1. The potential across R_1 may also be used as a digital output, in which case the

value of R_3 needs to be adapted as necessary.

The circuit may also be used as a kind of relay station or as a means for reducing jitter. For these purposes, IC_1 is connected in a special mode (mode 13) when M_3 is made 1, and M_0-M_2, 1, 0, and 1, respectively. When these levels are set the received S/PDIF data, including the preamble, is transferred directly to the output. The bit clock, SCK, then has a value twice as high as would be the case with coded data. It is possible to connect a TOSLINK module, or a coaxial output via a buffer (such as a number of parallel-linked 74HC04 inverters), to the SDATA output.

A demultiplexer, that is, 3-to-8 line decoder IC_3, is used to decode the data at F_0-F_2 to eight separate light-emitting diodes.

whether IC_1 receives no or poor data.

The overall circuit draws a current of not more than 35 mA.

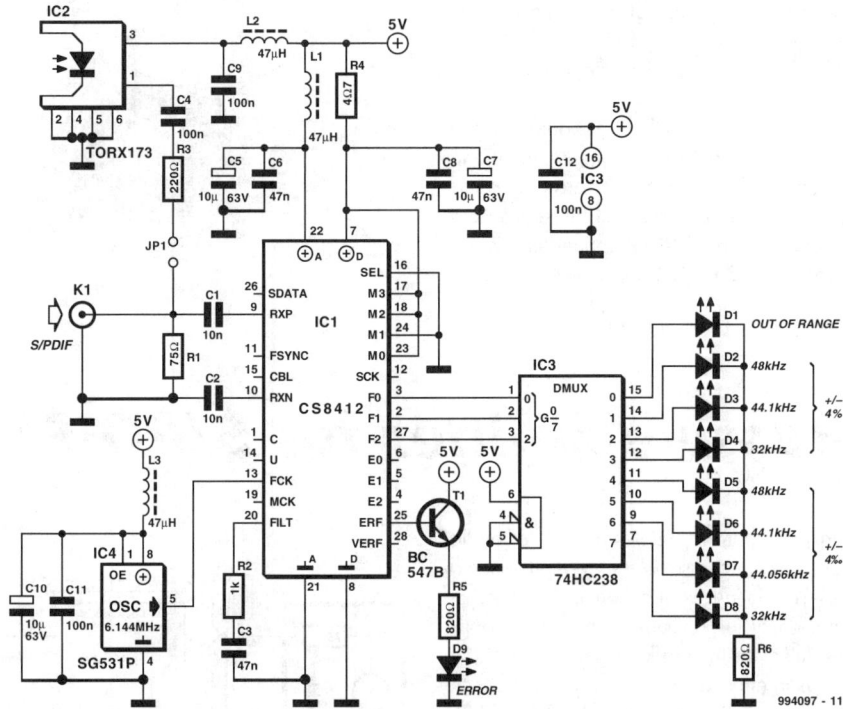

994097 - 11

Isolating transformer for S/PDIF*

WARNING: do not use this transformer for mains isolating, because its insulation is not capable of handling this. It is intended to prevent earth loops arising or undesired signals being applied to the input of an appliance. For instance, a tape recorder is to be linked to a sound card in a computer which has an S/PDIF (Sony/Philips Digital Interface Format); the computer is not connected to protective earth (which in the UK is next to impossible and certainly not advisable). Owing to the mains filter, half the mains voltage will be present on the enclosure and thus on the input earth. The linking of this potential to the tape recorder is prevented by the isolating transformer in the diagram.

To ensure a good bandwidth, the coupling factor of the transformer must be good (low stray self-inductance), so that a core with a high μ_t is needed. The

prototype uses a Philips Type TN13/7.5/5-3E25, which has a μ_t of 4500. The primary and secondary windings, each consisting of six turns of 0.5 mm dia.

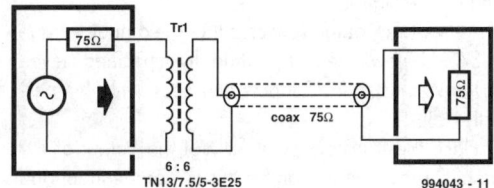

994043 - 11

enamelled copper wire, are laid on opposite sides of the toroid. The windings should be covered with insulating tape. If heavy-duty, insulated wire is used for the windings, they can be laid over one another, which

---

* The S/PDIF – Sony/Philips Digital Interface Format – is the consumer version of the AES/EBU professional standard. It was devised by the AES and EBU to define the signal format, electrical characteristics, and connectors, to be used for digital interfaces between professional audio products. AES is the American Audio Engineering Society and EBU is the European Broadcasting Union.

41

improves the coupling factor. But even with the first method, the bandwidth ranges from 50 kHz to 17 MHz, which is more than adequate for an S/PDIF link.

Place the transformer directly at the output of the signal source. The reason for this is that the input and output impedances of the transformer are not exactly 75 Ω. With the transformer directly at the source and provided the coaxial cable at the computer end is terminated correctly into 75 Ω, all will be well.

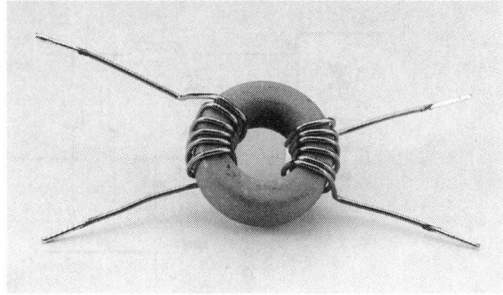

Passive splitter for S/PDIF*

The circuit in the diagram enables the digital audio output of, say, a compact-disc (CD) player to be linked to two different appliances simultaneously. It is, of course, considerably less expensive than the proprietary active splitters on the market.

The circuit is in effect a small transformer that can be wound easily on a Philips Type TN13/7.5/5-3E25 toroidal core. The wire should be 0.5 mm dia. enamelled copper wire. The primary winding is seven turns and there are two secondary windings, each of five turns. The bandwidth of the transformer is 40 kHz to 16 MHz. When both outputs are loaded, there is a voltage of 0.33 V_{p-p} at each output. When one of the outputs is open-circuited, the voltage at the other output rises to 0.43 V_{p-p}, which is caused by the slightly higher primary impedance and the slightly smaller load on the signal source.

A drawback of the splitter is that the output voltage is 34% below the internationally specified level. However, most S/PDIF inputs can cope with this perfectly well.

Place the transformer directly at the output of the signal source. The reason for this is that the input and output impedances of the transformer are not exactly 75 Ω. With the transformer directly at the source and provided the coaxial cable at the computer end is terminated correctly into 75 Ω, all will be well.

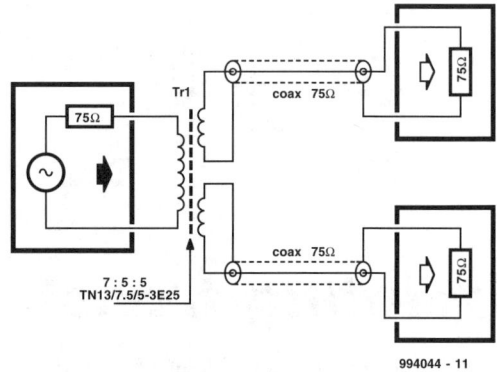

994044 - 11

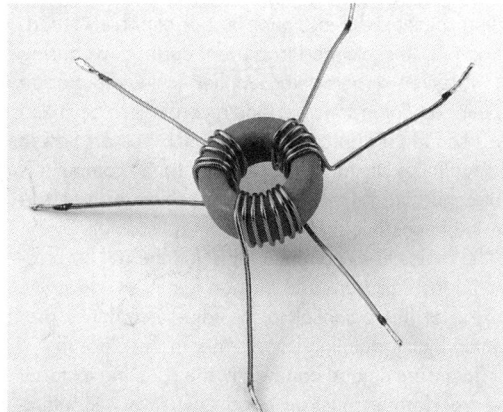

* The S/PDIF – Sony/Philips Digital Interface Format – is the consumer version of the AES/EBU professional standard. It was devised by the AES and EBU to define the signal format, electrical characteristics, and connectors, to be used for digital interfaces between professional audio products. AES is the American Audio Engineering Society and EBU is the European Broadcasting Union.

Fifth-order low-pass filter

The LTC1062 is an integrated fifth-order low-pass filter that stands out by the absence of any d.c. error. This is achieved by keeping the actual filter outside the d.c. range to eliminate matters like d.c. offset and low-frequency interference. It therefore makes the LTC1062 eminently suitable for filter applications where d.c. errors cannot be tolerated. A pivotal role in this is played by exter-nal resistor R_5. The output signal is fed back to the input of the filter via external capacitor C_4. Network R_5-C_4, in association with the internal switched-capacitor network, provides the fifth-order low-pass function.

The cut-off frequency of the filter is determined by an internal clock that can be controlled externally. This control is provided by oscillator IC_1, which is configured as an astable multivi-brator (AMV). The filter has an internal divider that can be set for a scaling factor of 1, 2 or 4. In the present circuit, this factor is set to 4, resulting in a cut-off frequency equal to 1/400 of the clock frequency. Since the clock here is 4 kHz, the cut-frequency is 10 Hz. If a different cut-off frequency is wanted, R_5, R_6, C_4 and C_5 must meet the following requirement to retain a smooth pass band.

$$1/2\pi R_5 C_4 = f_C/1.84,$$

provided that $C_4 = C_5$ and $R_6 = 12R_5$.

R_1, R_2, P_1 and C_2 must satisfy the following requirement

$$R_1 C_2 = 1.4/3400 f_C,$$

provided that $R_2 = R_1/2$ and $P_1 = R_1$.

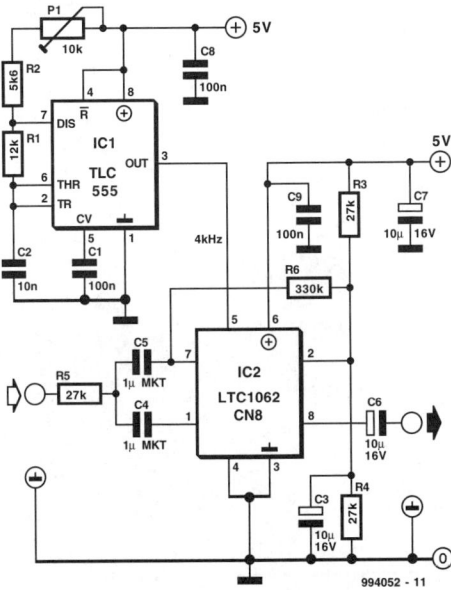

994052 - 11

43

2
Circuit ideas

0.5–6 GHz low-noise amplifier

The MGA-86563 from Hewlett-Packard is a three-stage, GaAs, MMIC (monolithic microwave integrated circuit) that offers low noise figure and excellent gain for applications from 0.5 GHz to 6 GHz. The device

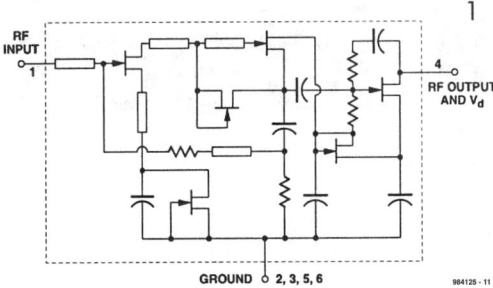

1

RF INPUT

RF OUTPUT AND V_d

GROUND ○ 2, 3, 5, 6

984125 - 11

uses internal feedback to provide wideband gain and impedance matching. It is housed in an ultra-miniature SOT-363 package, which requires half the board space of the SOT-143. See internal diagram in Figure 1.

The MGA-86563 may be used without impedance matching as a high performance 2 dB NF (noise fig-

50 Ω. Below 1.5 GHz, gain can be increased by using conjugate matching.

The input of the circuit in Figure 2 is fixed tuned for a conjugate power match (maximum power transfer or minimum VSWR – voltage standing wave ratio) at 2 GHz.

The 3.3 nH inductor, L_1, in series with the input of the amplifier matches the input to 50 Ω at 2 GHz.

Inductor L_2 prevents any tendency to resonance over the operating range (2 GHz). When operation takes place at lower frequencies, its value may have to be increased accordingly.

A circuit for operation up to 6 GHz is shown in Figure 3. A 50 Ω microstripline with a series d.c. blocking capacitor, C_1, is used to feed r.f. to the MMIC. The input of the device is already partially matched for noise figure and gain to 50 Ω. The use of a simple input matching circuit, such as a series inductor, will minimize the amplifier noise figure.Since the impedance match for NF_0 (minimum noise figure) is very close to a conjugate power match, a low noise figure can be realized simultaneously with a low input VSWR.

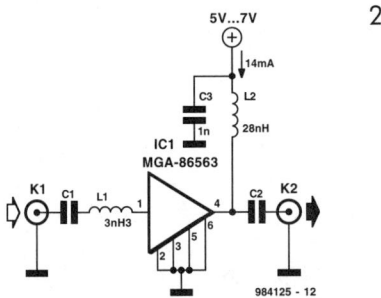

2

5V...7V

14mA

C3 1n

L2 28nH

IC1 MGA-86563

K1 C1 L1 3nH3

C2 K2

984125 - 12

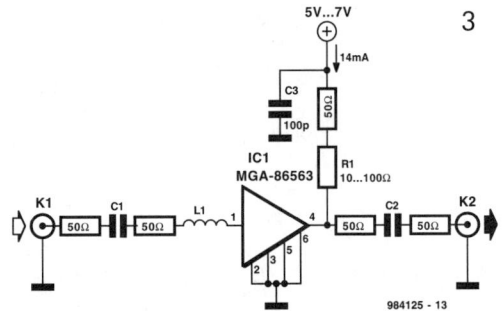

3

5V...7V

14mA

C3 100p

50Ω

IC1 MGA-86563

R1 10...100Ω

K1 C1 50Ω 50Ω L1

C2 50Ω 50Ω K2

984125 - 13

ure) amplifier. Alternatively, with the addition of a simple shunt-series inductor at the input, the noise figure can be reduced to 1.6 dB at 2.4 GHz. For 1.5 GHz applications and above, the output is well matched to

DC power is applied to the MMIC through the same pin tat is shared with the r.f. output. A 50 Ω microstripline is used to connect the circuit to the following stage.

100-watt single-IC amplifier

The LM3886 from National Semiconductor is a high-performance 150 W audio power amplifier with mute. Its performance, using the self peak instantaneous

temperature (K) (SPIKe) protection circuitry, is better than discrete and hybrid amplifiers since it provides an inherently, dynamically protected safe operating area

(SOA). The LM3886T comes in an 11 (staggered-) lead non-isolated TO220 package.

For test purposes, the prototype of the amplifier was powered by a stabilized ±35 V supply. A maximum undistorted output power of about 63 watts into 8 ohms was obtained at a drive level of 1 Vrms. Lowering the load impedance to 4 ohms raises the output power to 108 watts. Bear in mind that the amplifier will not normally be powered from a regulated supply!

Great attention should be paid to the cooling of the amplifier IC. The cooling capacity offered by a heat sink as specified in the parts list is really only sufficient for load impedances of 6 ohms or more. Even if a heat sink with a thermal resistance lower than 1 K W^{-1} is employed, the amplifier IC will cause a 'hot spot' on the heat sink surface where the actual thermal resistance is much higher locally than the specification! With this in mind, it is recommended to drop the supply voltage to about ±30 V if the amplifier is used to drive a 4 ohm load. Also, bear in mind that heatsink isolating materials like mica and even ceramics tend to raise the thermal resistance by 0.2 K W^{-1} to 0.4 K W^{-1}. The metal tab at the back of the IC is at the negative supply potential.

Boucherot network C_6-R_6 is not normally required in this application, and should be omitted unless the amplifier is found to be unstable as a result of an application which is widely different from the one shown here.

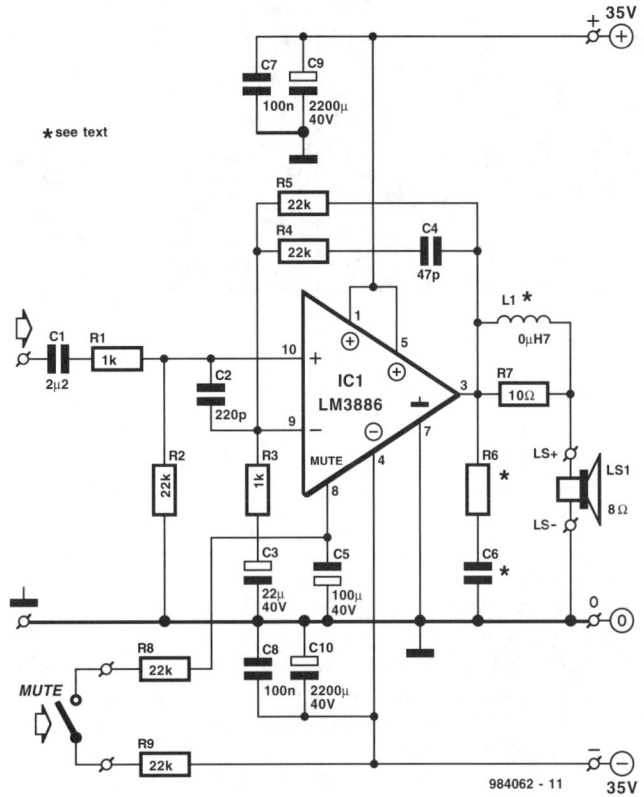

Populating the amplifier board is straightforward, and most of the time required to build the amplifier will be spent in drilling, cutting, mounting and isolating the heat sink. Since the electrolytic capacitors are rated at 40 volts, the supply voltage must not exceed that level. The PCB may be made with the aid of the track layout.

Brief parameters

Input sensitivity:	1 Vrms (63 W into 8 W)
Output power, 8 Ω:	63 W (THD <1%)
Output power, 4 Ω:	108 W (THD <1%)
Damping factor (8 Ω)	>450 at 1 kHz
	>170 at 20 kHz
Slew rate:	>10 V/ms (rise time = 5 ms)
Power bandwidth:	8 Hz to 90 kHz
Signal/noise ratio:	94 dBA (1 W into 8 W)
Quiescent current:	50 mA

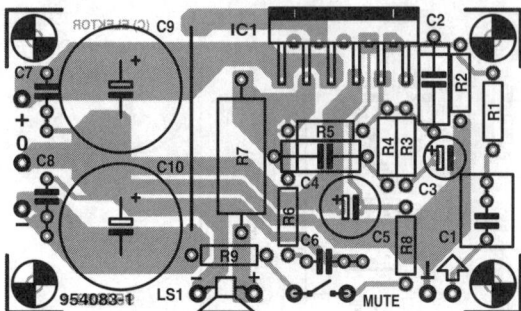

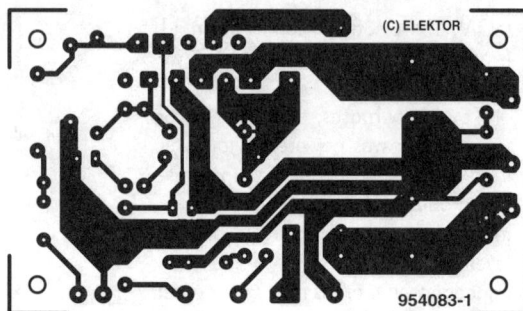

Parts list

Resistors:
R_1, R_3 = 1 kΩ
R_2, R_4, R_5, R_8, R_9 = 22 kΩ
R_6 = not fitted, see text
R_7 = 10 Ω, 5 W

Capacitors:
C_1 = 2.2 µF, metallized polyester (MKT), pitch 5 mm or 7.5 mm
C_2 = 220 pF, 160 V, axial, polystyrene
C_3 = 22 µF, 40 V, radial
C_4 = 47 pF, 160 V, axial, polystyrene
C_5 = 100 µF, 40 V, radial
C_6 = not fitted, see text
C_7, C_8 = 0.1 µF
C_9, C_{10} = 2200 µF, 40 V, radial, max. diameter 16 mm

Inductors:
L_1 = 0.7 mH, 13 turns of 1.2-mm diameter (18 SWG) enamelled copper wire, 10 mm internal diameter, wound around R_7.

Integrated circuits:
IC_1 = LM3886T (National Semiconductor)

Miscellaneous:
Heatsink for IC_1: Rth < 1 K W^{-1}

78xx replacement

The circuit in the diagram may be useful when in a certain application it is found necessary to replace a 78xx voltage regulator by a better quality type for which there is not enough space. It may also prove useful when a slightly different voltage is required.

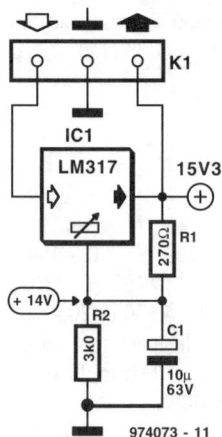

The replacement circuit is based on a Type LM317 regulator from National Semiconductor. The IC and the three requisite external components are fitted on a tiny PCB whose terminals coincide with those of the terminals of a 78xx device. In other words, the boards fit exactly where the 78xx used to be – it is, however, slightly higher.

The LM317 offers three advantages over a 78xx: (a) the ripple suppression is better; (b) the input voltage range is larger; (c) the output voltage can be arranged at any desired value with the aid of two standard resistors.

The resistors are calculated from

$$U_0 = U_{REF}(1 + R_2/R_1).$$

In case of the LM317, U_{REF} is 1.25 V. The values of the resistors must ensure that the output current does not drop below 3.5 mA. With values as specified in the diagram, the output voltage is 15.3 V and the quiescent current is 4.6 mA.

The LM317 can provide output currents of up to 1.5 A. If a larger current is required, the pin-compatible Type LM350 may be used, which can provide currents of up to 3 A. Bear in mind, however, that the board is not designed for continuous currents >3 A.

In either case, it may be necessary, depending on the dissipation, to mount the IC on a small heat sink.

Although not shown in the diagram, the IC needs decoupling capacitors of $\geq 0.1\,\mu F$ at the input and $1\,\mu F$ at the output. Since these are also required for the 78xx, it is assumed that these capacitors are already present. The PCB may be made with the aid of the track layout.

Parts list:
Resistors ($U_o = 15.3$ V):
$R_1 = 270\,\Omega$
$R_2 = 3.0\,k\Omega$

Capacitors:
$C_1 = 10\,\mu F$, 63 V

Integrated circuits:
$IC_1 = LM317$ (or LM350: see text)

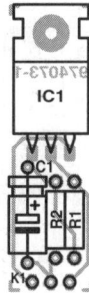

974073-1

79xx replacement

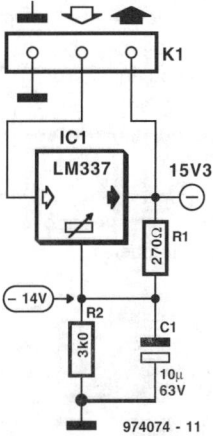

The circuit in the diagram may be useful when in a certain application it is found necessary to replace a 79xx voltage regulator by a better quality type for which there is not enough space. It may also prove useful when a slightly different voltage is required.

The replacement circuit is based on a Type LM337 regulator from National Semiconductor. The IC and the three requisite external components are fitted on a tiny PCB whose terminals coincide with those of the terminals of a 79xx device. In other words, the board fits exactly where the 79xx used to be – it is, however, slightly higher.

The LM337 offers three advantages over a 79xx: (a) the ripple suppression is better; (b) the input voltage range is larger; (c) the output voltage can be arranged at any desired value with the aid of two standard resistors.

The resistors are calculated from

$$U_o = U_{REF}(1 + R_2/R_1).$$

In case of the LM337, U_{REF} is 1.25 V. The values of RF_1 and R_2 must ensure that the output current does not drop below 3.5 mA. With values as specified in the diagram, the output voltage is -15.3 V and the quiescent current

49

is 4.6 mA.

The LM337 can provide output currents of up to 1.5 A. It may be necessary, depending on the dissipation, to mount the IC on a small heat sink.

Although not shown in the diagram, the IC needs decoupling capacitors of $\geq$ 100 nF at the input and 1 μF at the output. Since these are also required for the 78xx, it is assumed that these capacitors are already present. The PCB may be made with the aid of the track layout.

Parts list:

Resistors
(U_o = −15.3 V):
R_1 = 270 Ω
R_2 = 3.0 kΩ

Capacitors
C_1 = 10 μF, 63 V

Integrated circuits:
IC_1 = LM337

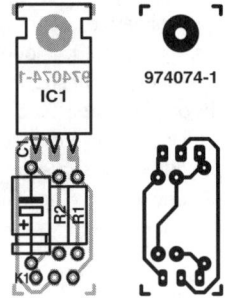

974074-1

Active Bessel filter

A Bessel filter is typified by the complete absence of any ringing. However, its frequency characteristic is less steep around the cut-off point than that of a Butterworth section.

The table in the diagram gives six different values for R_4 and R_5 resulting in varying amplification factors. Since the amplification has a direct influence on the filter response, the values of several frequency-determining components must be carefully calculated. As an aid to this, tables 1 and 2 show the values of R_1–R_3 (in Table 1: R_1 = R_2 = R_3 = 10.0 kΩ) and C_1–C_3 for a cut-off point of 1 kHz. Table 1 is based on standard values for the resistors and Table 2 on those for the capacitors. In practice, the latter is more convenient since the resistor values shown are close to the standard E-96 values.

The prototype used a Type TL081 op amp, but if high amplification factors or high cut-off frequencies are wanted, it is advisable to use an AD847.

The circuit draws a current of only a few milliamperes.

Table 1

Amplification	Gain	C_1 (nF)	C_2 (nF)	C_3 (nF)
×1	(0 dB)	15.7780	22.734	4.0546
×1.7783	(5 dB)	19.1130	9.6020	7.9252
×2	(6 dB)	19.7380	8.6605	8.5084
×3.1632	(10 dB)	22.3110	6.1051	10.6780
×5	(14 dB)	25.1900	4.4843	12.8760
×10	(20 dB)	30.2550	2.8955	16.6020

$f_{-3\ dB}$ = 1 kHz.

Table 2

Amplification	Gain	C_1 (nF)	R_1 (kΩ)	C_2 (nF)	R_2 (kΩ)	C_3 (nF)	R_3 (kΩ)
×1	0 dB	15	10.5030	22	10.4810	3.9	10.2660
×1.7783	5 dB	18	10.8380	10	9.8479	8.2	9.2323
×2	6 dB	18	10.8860	8.2	10.8440	8.2	10.1800
×3.1632	10 dB	22	9.7017	5.6	11.0810	10	10.9810
×5	14 dB	27	10.3280	4.7	8.5890	12	10.7670
×10	20 dB	33	8.4821	2.7	10.7750	15	11.9070

$f_{-3\ dB}$ = 1 kHz

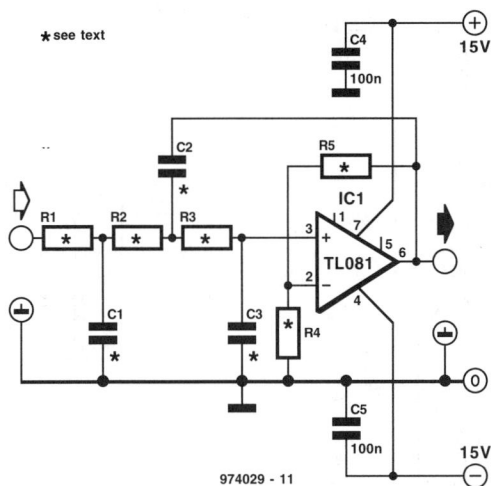

* see text

A	dB	R4	R5
1	0	----	0
1.7783	5	1k00	0.7783 x R4
2	6	1k00	1k00
3.1623	10	1k00	2.1623 x R4
5	14	1k00	4 x R4
10	20	1k00	9 x R4

974029 - 11

Active Butterworth filter

Active filters are invariably designed with a unity-gain buffer. In fact, this has become such a habit that one is inclined to think that it is obligatory, which of course it is not. An active filter can easily be designed with an

amplifier without making it less accurate. This has real benefits for in many cases it means that an entire stage may be omitted from the relevant amplifier.

It is a fact, however, that the degree of amplification has a direct effect on the filter characteristic. This means that the values of the filter components must be in accord with the amplification factor.

A further, slight, drawback is that as the amplification increases, the properties of the op amp used will have a greater effect on the signal transfer. Designers therefore use a high-speed op amp if the amplification is greater than, say, ×3. Since, however,

Table 1

Amplification	Gain	C_1 (nF)	C_2 (nF)	C_3 (nF)
×1	(0 dB)	22.1630	36.4490	3.2210
×1.7783	(5 dB)	26.4230	15.4490	9.8430
×2	(6 dB)	27.1490	13.8000	10.7600
×3.1632	(10 dB)	30.1370	9.4930	14.1720
×5	(14 dB)	33.3670	6.2040	17.6090
×10	(20 dB)	39.8830	4.3340	23.3210

$f_{-3\,dB} = 1$ kHz

Table 2

Amplification	Gain	C_1 (nF)	R_1 (kΩ)	C_2 (nF)	R_2 (kΩ)	C_3 (nF)	R_3 (kΩ)
×1	0 dB	22	9.9800	56	10.2670	3.3	9.6776
×1.7783	5 dB	27	9.3281	13	10.6770	10	9.9944
×2	6 dB	27	10.1810	13	9.5328	12	8.5469
×3.1632	10 dB	33	8.9814	10	9.4473	15	9.5983
×5	14 dB	33	9.9922	6.8	10.2730	18	9.7229
×10	20 dB	39	8.8965	3.9	12.0930	22	11.1970

$f_{-3\,dB} = 1$ kHz

★ see text

C4 100n

(+) 15V

C2

R5 ★

IC1

R1 ★ R2 ★ R3 ★

3 + 1 7
5 6
TL081
2 − 4

C1 ★ C3 ★ R4 ★

C5 100n

15V (−)

(0)

974033 - 11

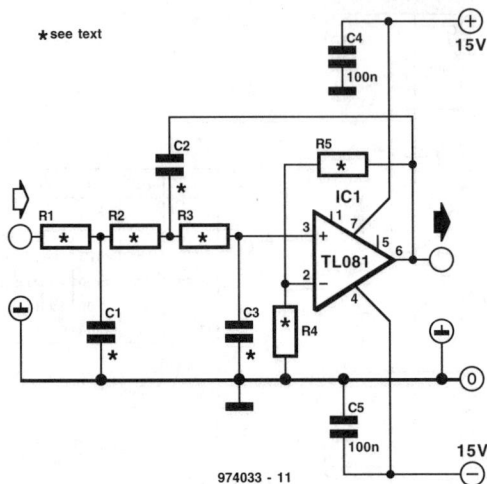

A	dB	R4	R5
1	0	----	0
1.7783	5	1k00	0.7783 x R4
2	6	1k00	1k00
3.1623	10	1k00	2.1623 x R4
5	14	1k00	4 x R4
10	20	1k00	9 x R4

the effect of the op amp at frequencies below 1 kHz is not great anyway, in most cases the op amp specified (TL081) will give excellent service.

The amplification of the op amp in the present circuit is $A = 1 + R_5/R_4$. The table in the diagram gives values for R_4 and R_5 for a number of amplification factors.

To make the arithmetic a little easier, the tables in the text give the values of the frequency-determining components for a 3rd order Butterworth filter with a cut-off frequency of 1 kHz for the same amplification factors as in the table in the diagram.

Table 1 is based on the assumption that $R_1 = R_2 = R_3 = 10.0$ kΩ, which results in awkward values of C_1–C_3 that will have to be resolved by series and parallel connecting of 1% capacitors.

Table 2 is based on standard values for C_1–C_3, which results in non-standard values for R_1–R_3, which are, however, fairly close to E96 values.

The filter in the diagram is a low-pass section, which may be converted to a high-pass section by interchanging C_1–C_3 and R_1–R_3 (which means, of course, that the values of the components will have to be recalculated). The ratios of the component values indicated in the tables will remain the same.

The filter draws a current of only a few milliamperes.

AHC(T) CMOS circuit

Designers have been using 74HCTxx, that is, high-speed CMOS, ICs instead of standard circuits since the late 1980s. These ICs have made the 74LSxx series obsolescent and the 74xx series obsolete. The only LS (Low-speed CMOS) types still used in rare cases are the 74LS05, 74LS06, and 74LS07, since these can provide higher currents and voltages than the HC equivalents.

The HC series has been edged out by the faster AC (Advanced CMOS) series, and now this is being superseded by the even faster, pin-compatible AHCT (Advanced High-speed CMOS) series.

Circuits in the AHCT series are about three times as fast as their HCT counterparts and draw only half the current. Moreover, they are suitable for use on 5 V as well as 3.3 V supply lines. This is an important advantage, since 3.3 V lines are becoming very popular. Clearly, the ICs are not as fast on 3.3 V lines as on 5 V lines, but still 50 per cent faster than HCT circuits on 5 V.

Great attention has been paid in the new devices to minimize the ground bounce, which is not very good in the AC series. In spite of the much higher speed, the ground bounce in the new devices is lower than that in HCT circuits.

An interesting aspect is that the new ICs may be used as an interface between 5 V and 3.3 V logic.

Finally, HC circuits have switching thresholds that

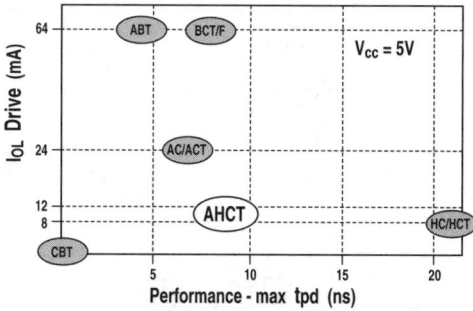

ABT	Advanced BiCMOS Technology
BCT	BiCMOS Technology
F	74F Bipolar Technology
AC/ACT	Advanced CMOS
HC/HCT	High Speed CMOS
CBT	Cross Bar Technology

974107 - 12

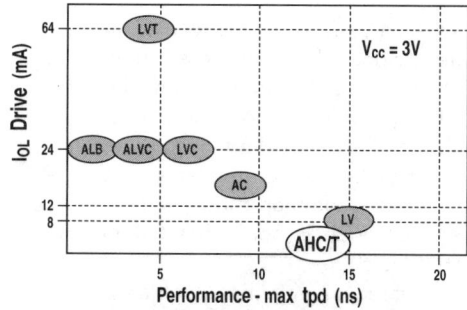

LV	Low Voltage (CMOS)
LVC	Low Voltage CMOS
LVT	Low Voltage Technology
AC	Advanced CMOS
ALVC	Advanced Low Voltage CMOS
ALB	Advanced Low Voltage BiCMOS

974107 - 13

are symmetrical w.r.t. the earth and supply lines, while HCT circuits are TTL-compatible. This latter means that voltages lower than 0.8 V are treated as 0 (logic low) and those above 2 V as 1 (logic high).

Datasheets can be obtained from Texas Instruments at web site http://www.ti.com/sc/asl/lit/lit.htm General information is available at http://www.ti.com

AVM with auxiliary start

Most constructors will have no questions when they have a first look at the diagram. And, indeed, the section based on IC_{1c} and IC_{1d} is a fairly standard astable multivibrator (AVM).

However, the special aspect of the design is the addition of feedback network IC_{1a}-IC_{1b}-D_1. This unusual circuit (see diagram on next page) ensures that the AVM always starts from the same position when the power is switched on. This is, of course, very useful in a number of timer and counter applications.

The feedback network also functions as the start-up circuit. If it were removed, nothing would happen on power-up and both outputs would remain high.

With component values as specified in the diagram on the next page, the AVM is arranged for a frequency of 9 kHz.

The circuit draws a current not exceeding 0.15 mA.

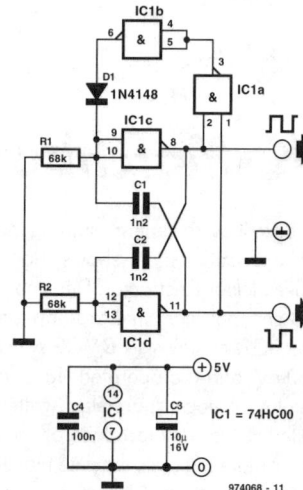

974068 - 11

Balanced amplifier for photo diode

A photo-diode is a p-n diode whose reverse current depends on the amount of light falling on its junction. The reverse current is greatly dependent on the temperature since heat can liberate more covalent bonds. As light can also do this, the diode can be housed in a transparent case.

53

When a photo-diode is located at some distance from the associated electronic circuits, noise may be picked up in the connecting cable, even when this is screened. Such noise can, fortunately, be suppressed easily, provided it is common mode, that is, when the diode is not connected to earth ('floats').

A differential amplifier enables a feedback signal to be amplified, but does not respond to common-mode signals. In the diagram, the differential amplifier consists of two op amps, IC_{1b} and IC_{1c}, which convert the diode current into a voltage. The current-to-voltage conversion depends on R_1 and R_2, so that gain setting in amplifier IC_{1d} is not necessary.

The output voltage, U_o, of the differential amplifier is

$$U_o = (U_{in1} - U_{in2}) \cdot R_4/R_3.$$

When $R_3 = R_5 = R_4 = R_6 + P_1$, the amplification is unity. In that case,

$$U_o = (R_1 + R_2) \cdot I_D,$$

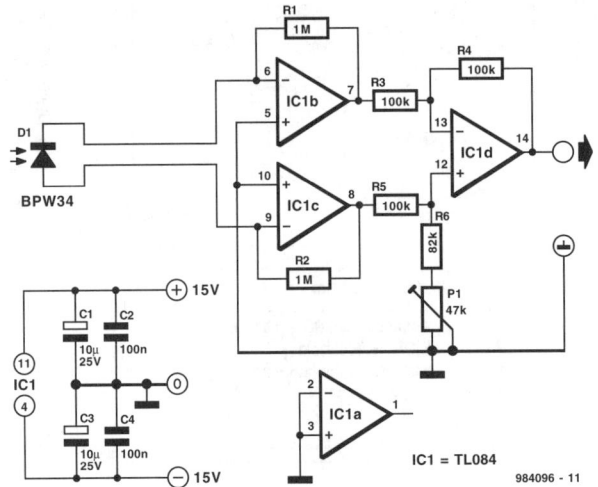

where I_D is the diode current.

The Common Mode Rejection, CMR, depends on the equality of the resistors as stipulated earlier. Their tolerances, and those of R_1 and R_2, can be nullified with P_1 so as to achieve optimum CMR. A Common Mode Rejection Ratio, CMRR, of >60 dB is obtained when the specified op amps are linked to the photo-diode by a twisted pair.

The circuit draws a current of about 10 mA.

BASIC-Matchbox-driven inductive loads

This article describes a general starting point for a number of applications: it shows how a Basic Matchbox (See *Elektor Electronics*, December 1995) is linked to a 6-channel serial interface low-side driver.

The driver, IC_1, contains six DMOS switches which can sink 350 mA and are operated via a serial interface. The status of each channel is written into the FAULT register at the trailing edges of the incoming pulses. One or more high bits indicate that the outputs are short-circuited to the supply line or ground, or are open. The output voltages are monitored: if a level of 35 V is exceeded, an error condition is signalled. There is also an additional OT bit to indicate whether the chip temperature is high.

When an output is short-circuited, the IC automatically shifts to the PWM mode; the output is re-engaged when the error condition has been removed. In this manner, it is possible to drive

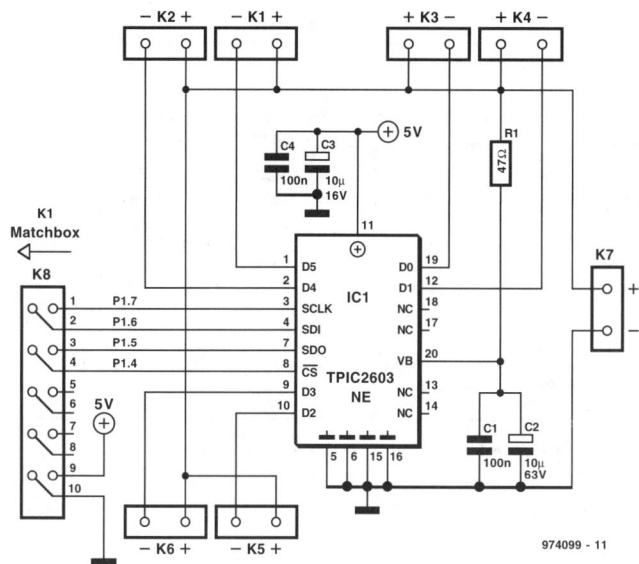

974099 - 11

Parts list

Resistors:
$R_1 = 47\,\Omega$

Capacitors:
$C_1, C_4 = 0.001\,\mu F$
$C_2, C_3 = 10\,\mu F$, 63 V, radial

Integrated circuits:
IC_1 = TPIC2603NE
   (Texas Instruments)

Miscellaneous:
K_1–K_7 = 2-way terminal block for board mounting, pitch 5 mm
K_8 = 10-pin box header
PCB may be made with the aid of the track layout.

inductive loads, such as motors, relays, coils and lamps, since these form a short-circuit at power-on.

During the making of the printed-circuit board, a large ground plane should be provided on the component side if large loads are to be controlled. The IC is then best soldered directly to the board without a socket.

The board is linked to the Matchbox via a 10-core flatcable, terminated at one end into a 10-pin socket and at the other end into a 20-pin socket of which only pins 1–10 are used. The 20-pin socket mates with K_1 on the Matchbox.

An external supply of up to 35 V should be connected to K_7. This supply should, of course, be able to handle the maximum loads that may be connected to the six channels.

In the sample program note that the status of the six channels is read only when the state of an output changes. In practice, this is, of course, insufficient and use should be made of a timer-controlled interrupt routine to regularly check whether a short-circuit has occurred.

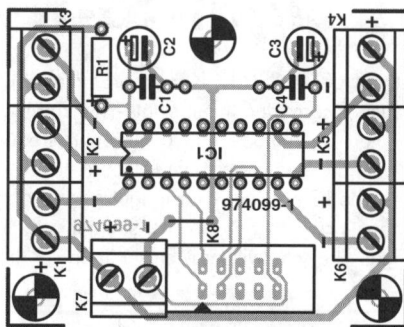

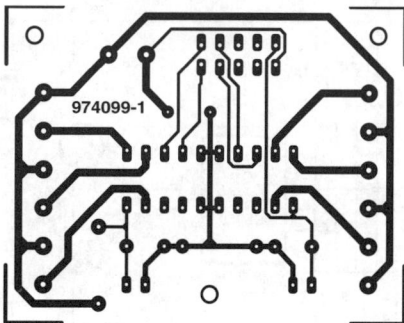

974099-1

Illuminance* monitor

The monitor actuates a buzzer when the illumination* drops below a certain level. It is based on light-dependent resistor (LDR) R_2 which, in conjunction with R_1 and P_1, forms a potential divider.

The threshold of illuminance* is set with P_1. When the illumination drops below this limit, the resistance of R_2 rises to a level that causes the output of IC_{1a} to go low. This results in a short negative pulse being applied to the set input of SR bistable IC_{1b}-IC_{1c} via network R_3-C_1 and diode D_1. This causes the output of IC_{1c} to go high, which results in oscillator IC_{1d} being enabled, whereupon the buzzer sounds.

When press-button switch S_1 is pressed, the bistable is reset, which disables IC_{1d} and, consequently, the buzzer.

The circuit draws a current of only a few milliamperes, which rises to 15 mA, however, when the buzzer is actuated.

* Illuminance=illumination=the quantity of light falling on unit area of a surface.

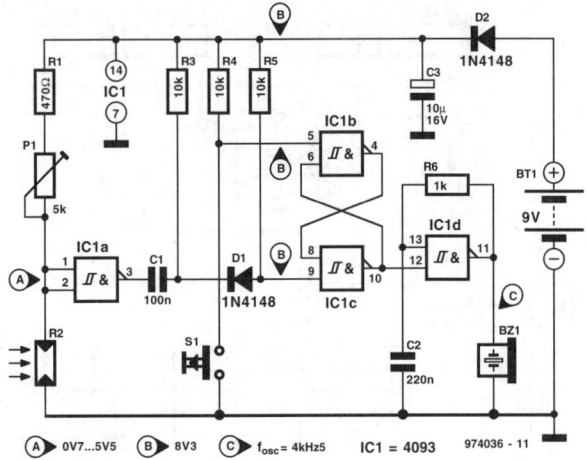

Light-to-frequency converter

The Type TSL230 IC is a programmable light-to-frequency converter. It is a single-chip combination of a silicon photodiode and a current-to-frequency converter, housed in an 8-pin DIL case.

The IC provides a rectangular-wave signal whose frequency depends on the strength of the incident light. The sensitivity may be set in one of three ranges

measuring instruments, in the present circuit it is used as a light-dependent squeaker. All this needs is the addition of a small push-pull amplifier and a tiny loudspeaker.

The amount of incident light determines the frequency emanated by the loudspeaker. So, if the incident light is made to vary, the reproduced sound varies. It is not known whether a melody can be generated!

The sensitivity and divisor of the output frequency are set with quad DIP switch S_2. If the four switches are called S_{2a}–S_{2d} from top to bottom, the following functions are obtained.

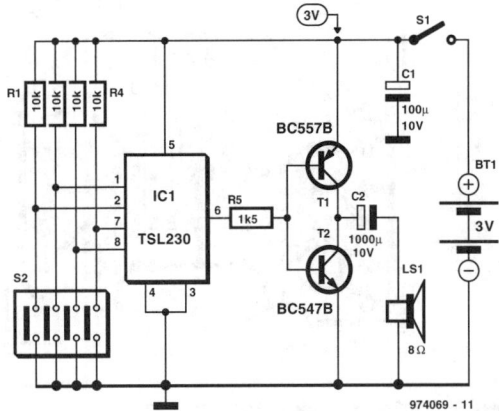

S_{2a}	S_{2b}	Sensitivity
0	0	power down
0	1	×1
1	0	×10
1	1	×100

S_{2c}	S_{2d}	Divisor
0	0	1
0	1	2
1	0	10
1	1	100

via pins 1 and 2. The divisor of the output frequency may be set in one of four ranges via pins 7 and 8.

Although the IC is primarily intended for use in

As usual, a 1 represents an open switch and a 0 a closed one.

Depending on the ambient light, there is a need for some (or much) experimenting with the switches before the output frequency falls within the audible range.

Power-down is a kind of standby position in which the IC draws a current of only 10 μA. In normal operation, the current does not exceed 10 mA. Power may therefore be obtained from two AA alkaline batteries. The IC needs a simple power supply of only 2.7 V.

Low-power crystal oscillator

The Type HA7210 IC from Harris Semiconductor is a complete integrated low-power crystal oscillator that can be programmed externally to generate output frequencies between 10 kHz and 10 MHz.

The oscillator is a Pierce type that is arranged to draw as small a current as possible. Only a decoupling capacitor, a crystal and frequency-determining components are required externally.

The circuit is highly stable over a wide range of supply voltages and a wide temperature range.

The application shown in the diagram is a basic circuit suitable for frequencies between 10 kHz and 10 MHz. The position of jumpers JP_1 and JP_2 depends on the chosen crystal frequency. In the table a '1' indicates that the jumper is left open and a '0' that the jumper must be placed.

The crystal must be cut for parallel resonance. In the present application, the load capacitance is 7.5 pF for the bottom range and 2.5 pF for the other ranges. If this does not suffice for the relevant crystal, ceramic capacitors of twice the value of the specified load capacitance must be placed between pin 2 and earth and between pin 3 and earth.

The oscillator draws a current of 0.5 mA in the bottom range and 7 mA at 10 MHz.

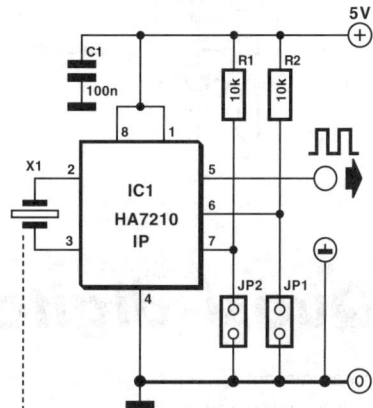

X1	JP2	JP1
10kHz – 100kHz	1	1
100kHz – 1MHz	0	1
1MHz – 5MHz	1	0
5MHz – 10MHz	0	0

974045 - 11

MMV from 4093

Most electronics constructors will at some time have made an MMV (monostable multivibrator) from a Schmitt trigger NAND gate, a capacitor and a resistor. Strictly, in this setup, the NAND gate is used as an inverter, since one input is permanently linked to the positive supply line.

It is possible to construct the same logic function by interlinking the two inputs (note, however, that in some data books this is not recommended if the current drain is to be kept low. For instance, the 4093 from SGS Thomson does not work well with its inputs interlinked). Research into this showed that the switching levels of the inputs are raised when the inputs are interlinked. With a supply voltage of 9 V, the lower and upper levels were 3.5 V and 5.5 V and 4.5 V and 6.5 V respectively. This is a difference in hysteresis of 1 V.

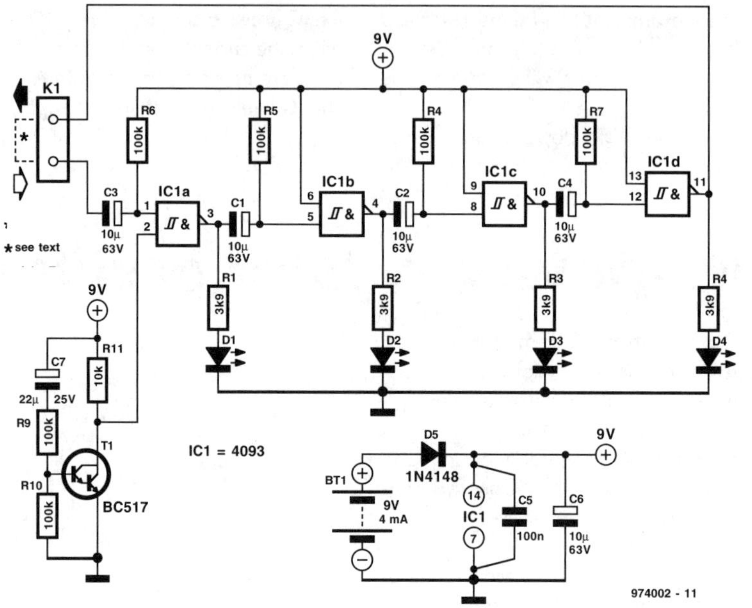

974002 - 11

Quasi-digital bandpass filter

The filter ensures that TTL signals are transferred within a certain frequency range, albeit at half their original frequency. Below and above this range, the filter output is a stable logic level.

The square wave signal at the input is delayed by networks R_2-C_1 and R_3-C_2. So, a logic 1 at the input appears at the output of both gates in the 4093 after some delay, whereas, owing to the use of diodes, the transfer of a 0 is not, or hardly, delayed. Provided the circuit is set up correctly, its operation is as follows.

Well below the cut-off point, the clock as well as JK signal are delayed. The delay of the JK signal is somewhat longer, however, owing to network R_2-C_1.

The level at the JK inputs written by the bistable at the leading edge of the clock signal is low. The outputs then remain stable.

Just below the cut-off point, network R_2-C_1 can no longer follow the signal. The level at the input of IC_{1a} is then low and that at the JK inputs of IC_2 is high.

Since the clock signals continue to be transferred, the bistable receives clock pulses and becomes a binary scaler.

If the frequency rises further, the clock will no longer reach the bistable. The last attained position is then stored.

With values as specified in the diagram, the signals that are transferred lie in the frequency range

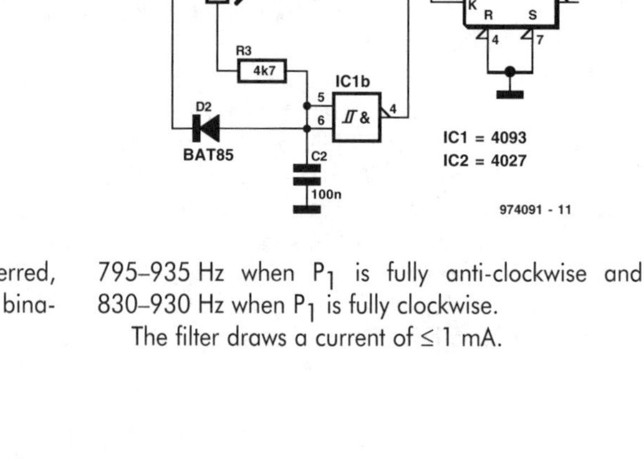

974091 - 11

795–935 Hz when P_1 is fully anti-clockwise and 830–930 Hz when P_1 is fully clockwise.

The filter draws a current of ≤ 1 mA.

Random clock generator

For many test and experimental purposes a random instead of a periodic clock frequency is required.

In the present circuit, the random operation is provided by transistor T_1, whose base-emitter junction is reverse biased so that it operates as a noise generator. In this mode, the junction behaves as a zener diode with an operating voltage of about 8 V. The resulting noise manifests itself as an alternating noise voltage at the collector of the transistor with a peak-to-peak value of 150 mV. The bandwidth of the noise signal is limited by parallel network R_1-C_1.

The noise signal is applied to the non-inverting input of comparator IC_{1a} directly and to the inverting input via low-pass filter R_2-C_2. The output of the filter is the average value of the signal, so that IC_{1a} is fed with the instantaneous as well as the average value of the signal.

The output of IC_{1a} is applied to another comparator, IC_{1b}, which, because of its hysteresis, outputs a

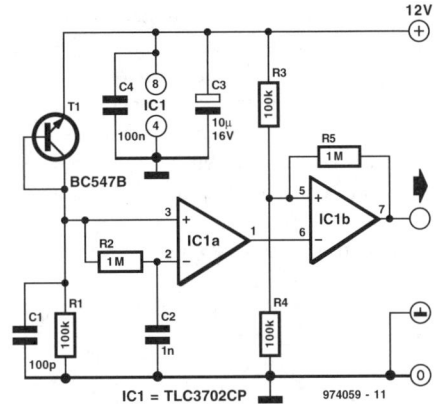

random clock signal with very steep edges. The period of the output signal varies statistically (Gaussian distribution) between 10 μs and 50 μs.

Single-chip ac-dc inverter

The Type α10777APA IC is an integrated ac-dc inverter that can handle inputs from 18 V r.m.s. to 276 V r.m.s. It contains a switch-mode amplifier and a rectifier bridge. It is able to provide a very compact, light and inexpensive power supply with a minimum of external components. The peak output current is 50 mA. The output voltage may be set at voltages up to +70 V with the aid of zener diode D_1.

The inversion process depends on charging and discharging a capacitor during each ac input cycle. At the start of the cycle capacitor C_3 is linked to the internally rectified direct voltage via a switch and is then charged to +70 V (equivalent to the internal zener voltage) or to the zener voltage of D_1. After the positive half-cycle the switch opens, whereupon C_3 is discharged via the load during the subsequent negative half-cycle. The process then repeats itself, so that C_3 is charged during each positive half-cycle. Charging does not start, however, until the input voltage is about 1 V higher than the potential across C_3.

The IC can handle input frequencies between 48 Hz and 200 Hz. The switching rate and thus the charge-discharge frequency is always twice the input frequency.

Note that the circuit is electrically connected with the mains supply, so that the supply as well as the circuit being supplied must be housed in a non-metallic enclosure.

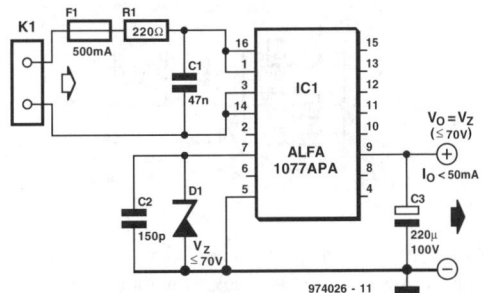

Single-transistor astable

An astable multivibrator (AMV) consists by definition of two transistors but, since these operate purely as switches, it is possible to replace one of them by a mechanical switch or relay as is done in the present circuit. The switch-on time is determined by the values of R_1 and C_1, while the switch-off time depends on the inertia of the relay. With values and relay as specified, the switch-on time is 2.8 s and the switch-off time 1.3 ms.

The second relay contact may be used to operate an LED, buzzer, or similar.

The circuit draws a current of about 45 mA, which is largely the relay current.

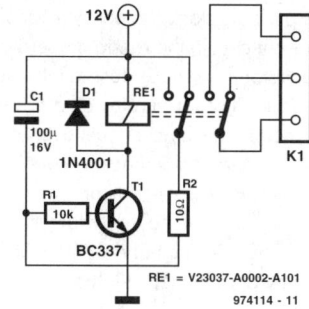

Supply board for output amplifiers

Apart from their electronic configuration, all output amplifiers comprise the same elements: an amplifier board, a mains transformer, a bridge rectifier, and the

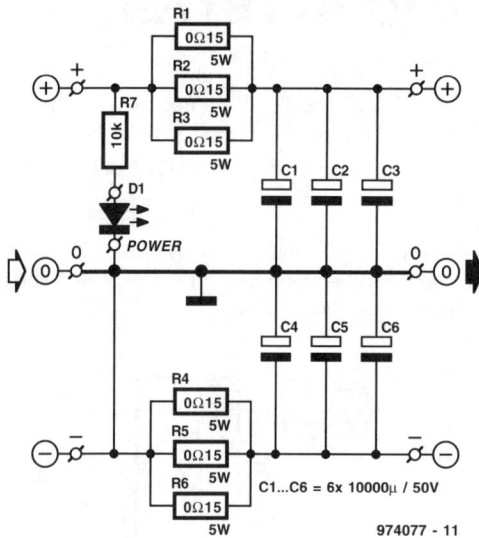

electrolytic smoothing capacitors. The board is normally screwed to the heat sink, while the transformer and bridge rectifier are fixed to the bottom of the

enclosure. Often, there is no such defined location for the electrolytic capacitors. They are mounted on a piece of prototyping board, or to the bottom of the enclosure with suitable brackets, or ...

Since this is a recurring difficulty, many constructors will be pleased with the board design shown here. Its layout is such that it is suitable for use with almost any type of output amplifier operating from a symmetrical power supply.

The board can accommodate six electrolytic capacitors with a value of up to 10,000 μF and a rating of 50 V. These are assumed to have a pitch of 10 mm and a maximum diameter of 30 mm.

The board also has space for 'soft switch-on' resistors with a value of 0.15 Ω and rated at 5 W. These resistors damp the peaks in charging current and also aid in smoothing spurious current peaks on the supply voltage.

Finally, the board has an on/off indicator in the shape of a high-efficiency LED and requisite series resistor.

Connections to the board are via car-type flat-connectors, which guarantee good contacts and can handle large currents. The PCB may be made with the aid of the track layout.

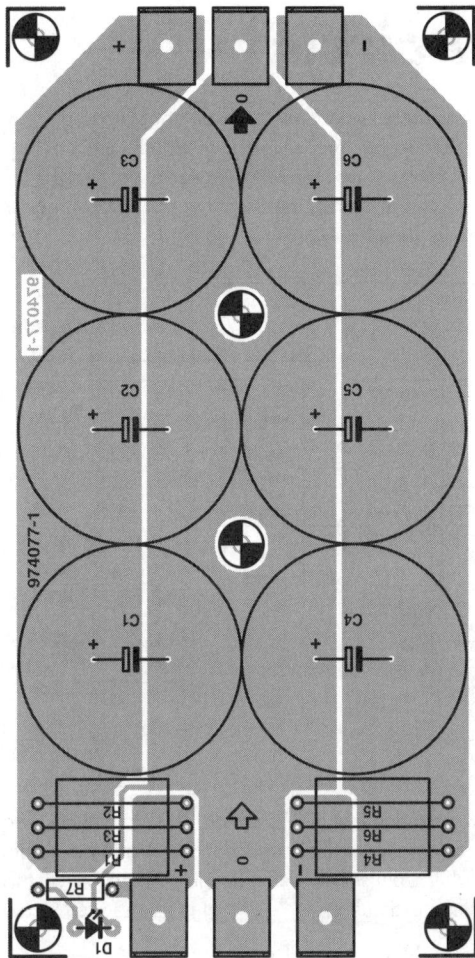

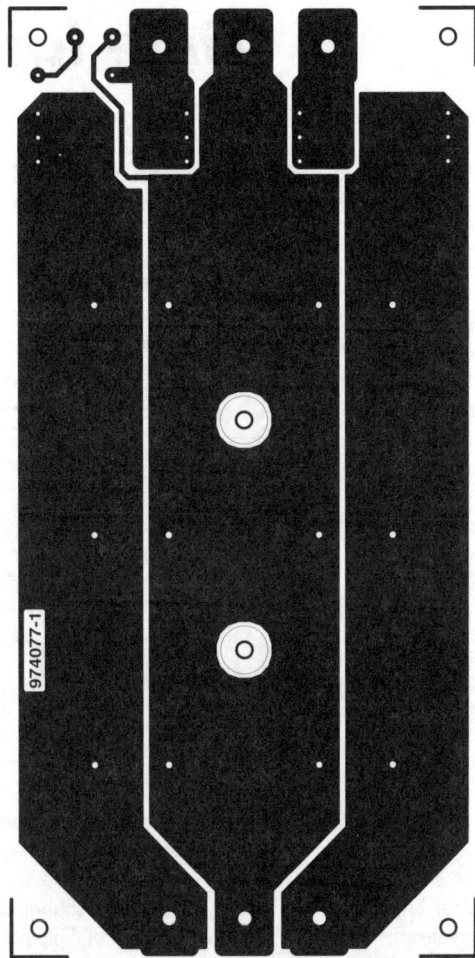

Parts list

Resistors:
R_1–R_6 = 0.15 Ω, 5 W
R_7 = 10 kΩ

Capacitors:
C_1–C_6 = 10,000 μF, 50 V, pitch 10 mm,
max. dia. 30 mm

Semiconductors:
D_1 = LED, high efficiency

Miscellaneous:
6 off car-type flat-connectors with screw
mountings

Time lapse circuit

Time intervals play an important role in electronics and more particularly in test and measurement technology. Normally, a digital circuit is used in which an oscillator provides good resolution and reproducibility. The present circuit is no exception: it consists of a retriggerable monostable multivibrator (MMV), a Clear circuit, a decimal counter and a BCD-to-decimal counter (BCD=binary coded digit). The gate at its output enables it to be operated from an external source.

The input of the MMV is linked to the TTL output of a clock generator, which may be an oscillator or a switch (for inputting single pulses).

The MMV is triggered at each leading edge of the clock and generates a pulse of about 7 ms, which is long enough to allow any switch bounce to be eliminated. The pulse width may be altered by changing the values of R_1 and C_1. The trailing edge of the pulse increases the counter state by one.

When the counter state is zero, pin 1 of IC_3 is low and all other outputs are high. Diode D_1 then lights. Output gate IC_4 is disabled, irrespective of the levels at its other inputs.

Incoming pulses raise the decoder state and make the outputs of IC_3 high sequentially. Diode D_1 goes out and the counter is enabled.

The desired time interval is set with one (and only one) of the DIP switches in S_2. There is then a low level across the relevant switch, which is applied to pin 2 of the output gate, diode D_2 and reset input pin 3 of IC_1. The LED lights and the MMV is disabled, so that all further input pulses are ignored. The output of IC_4 goes

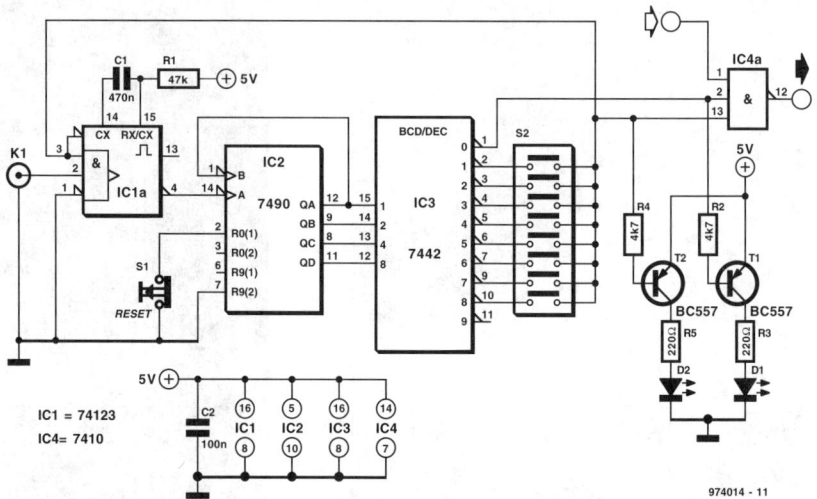

974014 - 11

low, provided that pin 13 is high. This pin may be used for external control or as window input. If these are not used, the pin should be held high permanently. In this state, the circuit is stable and may be restarted with a reset.

The current drawn by the circuit depends largely on the type of ICs used. With TTL circuits, it is about 50 mA, whereas if LS chips are used, it is only about 10 mA. If CMOS ICs (either HC or HCT) are used, the current is determined almost exclusively by that through the LEDs. Note that CMOS devices need additional pull-up resistors: 4.7 kΩ to the +ve supply line from pins 2, 3 and 6 of IC_2 and pin 2 of IC_1. These resistors prevent undefined levels and consequent errors.

Bidirectional I²C™ level shifter

The advantages and ease of use of an I²C link are well known. Since its introduction almost 20 years ago, it has been applied in more than 1000 types of IC. All these ICs are designed for operation from a 5 V power supply. However, more and more ICs coming on to the market use a 3.3 V supply. The bi-directional shifter described in this application note is intended to interconnect two sections of an I²C-bus system, each section with a different supply voltage and different logic levels. In the bus system in the diagram, the left section has pull-up resistors and devices connected to a 3.3 V supply voltage, whereas the right section has pull-up resistors and devices connected to a 5 V supply voltage. The devices of each section have I/Os with supply voltage related logic input levels and an open drain output configuration.

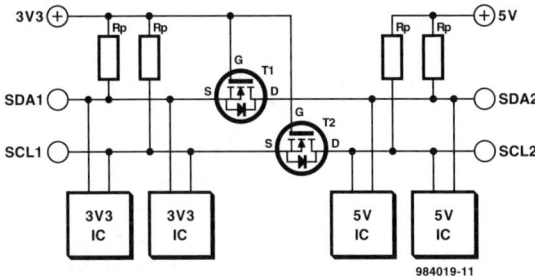

984019-11

The level shifters for each bus line are identical and consist of one discrete n-channel enhancement MOSFET, T_1, for the serial data line, SDA, and T_2 for the serial clock, SCL. The gates (g) have to be connected to the lowest supply voltage, V_{DD1}, the sources (s) to the bus lines of the lower voltage section, and the drains (d) to the bus lines of the higher voltage section. Many MOSFETs have the substrate internally already connected with the source; if not, it should be done externally. The diode between the drain and substrate is inside the MOSFET and represents the n-p junction of drain and substrate.

Three situations may arise.
• The bus is not used and is, therefore, not made low by one of the ICs. The gate and source are both at 3.3 V so that the transistor is cut off. The I^2C bus at the right section is not affected and this line is also high, but here the high level is 5 V (in the left section, it is 3.3 V).
• An IC in the left section makes the bus low. The level at the source of the transistor is 0 V and that at the gate, 3.3 V. The transistor conducts, so that the 5 V section is pulled low by the transistor and the relevant IC. This means that the low level at the left section is transferred to the right section.
•An IC in the right section makes the bus low. The left section is pulled low via the diode in the transistor, not necessarily to zero but to a level a diode drop above zero. This level is, however, low enough to switch on the transistor, since the potential at the source is a few volts below that at the gate. Since FETs can conduct in two directions, the left section is made low via the transistor and the relevant IC in the right section. So, again the low level is transferred.

It is obvious that the FETs used must have some specific properties. One of the most important of these is that the transistor must conduct when the gate-source potential is less than 2 V. Also, its channel resistance must be lower than 100 Ω, and it must be able to carry a current of at least 10 mA. Its input capacitance should not exceed 100 pF and it should be capable of switching within 5 ms. Suitable Philips types are the BSN10, BSN20, BSS83, and BSS88.

™ Philips Trade Mark

Celsius thermometer

The circuit of the Celsius thermometer in the diagram below is based on the well-known Type LM334 from National Semiconductor. This IC is a sensor that provides a current which is directly proportional to the temperature in kelvin (K). Unfortunately, this is a quantity that is not suitable for use in most practical applications. In the circuit, therefore, the sensor is set to $1 \mu A K^{-1}$ with P_1 and the offset of 273 K removed with P_2. This renders the output voltage of the sensor

directly proportional to the temperature in degrees Celsius (°C) and this makes the circuit suitable for a great many applications (since 1 K = 1 °C).

Circuit IC_2 is arranged as a 2.5 V reference voltage source. The current setting of the sensor is determined by the resistance between the adj(ust) pin and earth. If the earth is made virtual by raising the potential at the adj pin, the zero point can be shifted as desired.

Calibration is best done by using a good domestic thermometer as reference. Start by short-circuiting IC2 and adjusting P1 until the reading of meter M1 shows a current value numerically equal to the ambient temperature plus 273. If, say, the room temperature is 25 °C, adjust P1 until the meter reads 298 μA. Then, remove the short-circuit from IC2 and adjust P2 until the meter reads a current whose numerical value is equal to the room temperature, that is, 25 μA.

The circuit draws a current not exceeding 1 mA, so using two AA size (AM3, MN1500, LR6, SP/HP7) batteries as power source will give a life of a couple of years.

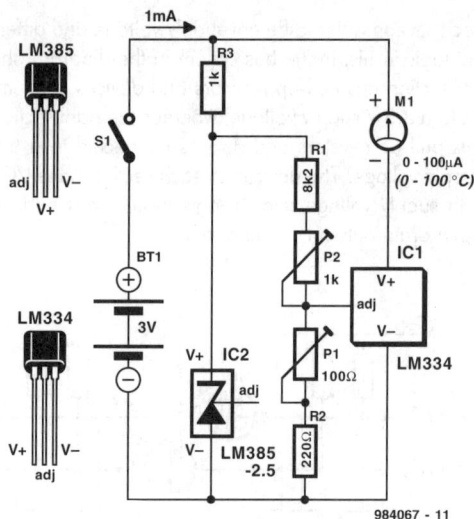

984067 - 11

Some designs based on the NE612

The NE612 is an active mixer/ oscillator that is used in numerous r.f. circuits. Three unusual applications of this versatile building block are described in this article.

The IC can be arranged as a frequency doubler (Figure 1). In this application, pin 6, which is normally linked to the tuned oscillator circuit, is connected to the input. The internal oscillator transistor (base at pin 6; emitter at pin 7) then functions as a linear amplifier. The frequency of the output signal is twice that of the input signal.

The fundamental frequency, f, and harmonics $3f$, $4f$, and so on, are only 10 dB away from the output frequency, $2f$, if the output is taken from C7. It is, therefore, advisable to use a bandpass filter at the output if the circuit works permanently with a fixed input frequency.

The optional bandpass filter consists of two inductively coupled tuned circuits, L_1-C_5 and L_2-C_6. If losses at higher harmonics are acceptable, these circuits may be tuned for use with such harmonics.

The NE612 can be configured as an overtone oscillator (Figure 2). The internal oscillator is normally not accessible and mixes or multiplies the input signal with the oscillator signal, so yielding an output $f_{in}+f_{lo}$. If, however, the r.f. input of the multiplier, pin 1, is linked to pin 8 via resistor R_1, the mixer produces a high-level output at the oscillator frequency.

The maximum output level is a function of the value of R_1 and the supply voltage. It has been found by trial and error that a resistor value of 560 Ω is optimal. The desired output level is optimal at a supply voltage of +5 V.

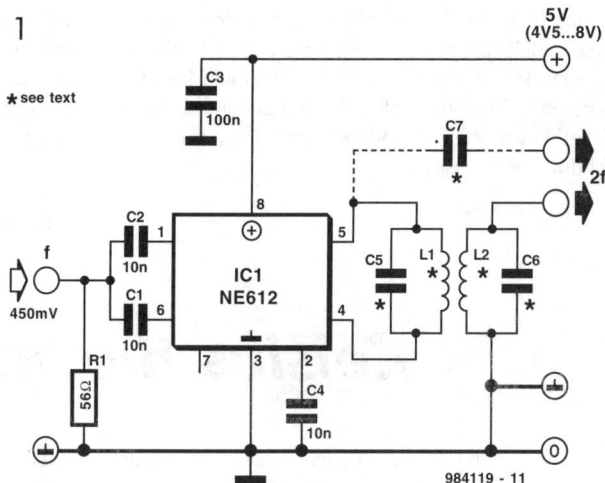

1

★ see text

984119 - 11

Although the first harmonic (72 MHz) is only 10 dB away from the fundamental (36 MHz), the higher harmonics are more than 25 dB lower. Greater distances may be obtained by the use of a bandpass filter at the output.

If a fundamental-frequency crystal is used, circuit L_1-C_3, which is tuned to the first harmonics, as well as

2

5V (4V5...8V)

R1 560Ω

★ see text

C5 100n

C8

IC1 NE612

C7 10n   36MHz 800mV$_{tt}$

C2 4p7

C4 ★ 10n

X1 ★ 36MHz (3rd overtone)

C1 22p   C3 ★ 22p   L1 ★ 1μH   C6 10n

984119 - 12

3

5V (4V5...8V)

R1 10k

★ see text

C5 100n

C8

IC1 NE612

C7 10n   72MHz 50mV$_{tt}$

C2 4p7

C4 ★ 10n

X1 ★ 36MHz (3rd overtone)

C1 22p   C3 ★ 22p   L1 ★ 1μH   C6 10n

984119 - 13

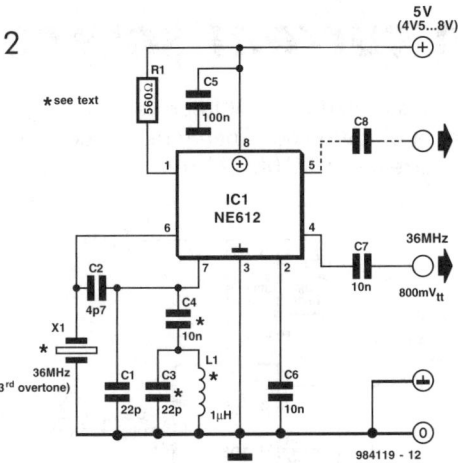

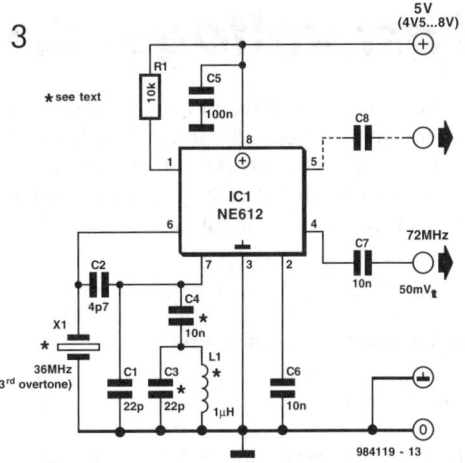

coupling capacitor C_4, may be omitted.

The two applications just discussed may be combined into a third: an overtone oscillator with frequency doubler. In this, the mixer input (pin 1) is linked to the emitter of the oscillator transistor (pin 7) via resistor R_1. In this application, a value of 10 kΩ for this resistor proved optimal.

It should be noted that the output voltage possible with an optimal output range is lower than in the previous application: about 50 mV peak-to-peak. On the other hand, the harmonics in the output range are 20 dB away from the output frequency of 72 MHz. This means that in most cases there is no need for a bandpass filter at the output.

Electrical isolation for I²C™ bus

When the SDA (Serial DAta) lines on both the left and right lines are 1, the circuit is quiescent and optoisolators IC_1 and IC_2 are not actuated. When the SDA line at the left becomes 0, current flows through the LED in IC_1 via R_2. The SDA line at the right is then pulled low via D_2 and IC_1. Optoisolator IC_2 does not transfer this 0 to the left, because the polarity of the LED in IC_2 is the wrong way around for this level. This arrangement prevents the circuit holding itself in the 0 state for ever.

As is seen, the circuit is symmetrical. So, when the SDA line at the right is 0, this is transferred to the left.

The lower part of the diagram, intended for the SCL (Serial CLock) line, is identical to the upper part.

Resistors R_1, R_4, R_5, and R_8, are the usual 3.3 kΩ pull-up resistors that are obligatory in each I²C line. If these resistors are already present elsewhere in the system, they may be omitted here.

The current drawn by the circuit is slightly larger than usual since the pull-up resistors are shunted by the LEDs in the optoisolators and their series resistors. Nevertheless, it remains within the norms laid down in the I²C specification.

™ Philips Trade Mark

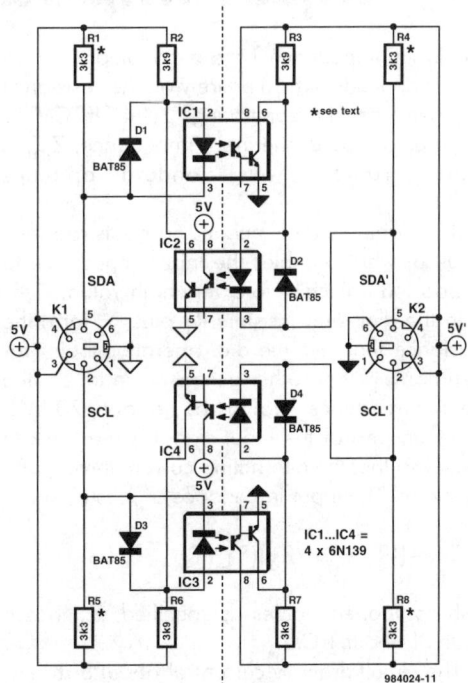

R1 3k3 ★   R2 3k9   R3 3k9   R4 3k3 ★

IC1   ★ see text

D1 BAT85

IC2   D2 BAT85

SDA   SDA'

K1   5V ⊕   K2   5V' ⊕

SCL   D4 BAT85   SCL'

IC4   ⊕ 5V

D3 BAT85   IC1...IC4 = 4 x 6N139

IC3

R5 3k3 ★   R6 3k9   R7 3k9   R8 3k3 ★

984024-11

Fast voltage-driven current source

The current source in the diagram, which reacts very fast to changes in the input signal, may be used, for instance, in certain measurements.

Differential amplifier IC_1 ensures that the potential across R_2 is equal to the input voltage:

$$I_{out} = U_{in}/R_2.$$

The bandwidth, B, is given by

$$B = R_2 \, f \, /R_L,$$

where $f = 80$ MHz, and the load impedance $R_L \geq R_2$ (both in ohms).

The input is terminated into R_1 to give the usual 50 Ω impedance required by measuring instruments. At the same time, this resistor sets the d.c. operating point. If the link to the driving signal source is short and d.c. coupled, R_1 may be omitted.

The peak voltage between pins 1 and 2 of the IC is limited to 2.1 V to prevent too large a current at the output. Therefore, the peak output current is 2.1/100=21 mA.

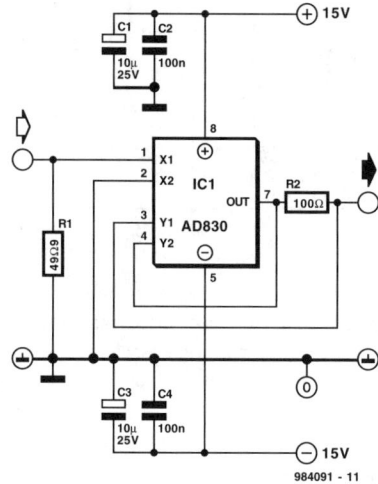

984091 - 11

Input impedance booster

The input impedance of a.c.-coupled op amp circuits depends almost entirely on the resistance that sets the d.c. operating point. If CMOS op amps are used, the input impedance, Z_{in}, is high (≥ 10 MΩ), with standard op amps ≤ 10 MΩ.

If a higher value is needed, a bootstrap may be used, which enables the input impedance to be boosted artificially to a very high value.

In the diagram, resistors R_1 plus R_2 form the resistance that sets the d.c. operating point for op amp IC_1. If no other actions were taken, the input impedance would be about 20 MΩ. However, part of the input signal is fed back in phase, so that the alternating current through R_1 is smaller. The input impedance, Z_{in}, is then:

$$Z_{in} = [(R_2 + R_3)/R_3](R_1 + R_2).$$

With component values as specified, Z_{in} has a value of about 1 GΩ.

The circuit draws a current of about 3 mA.

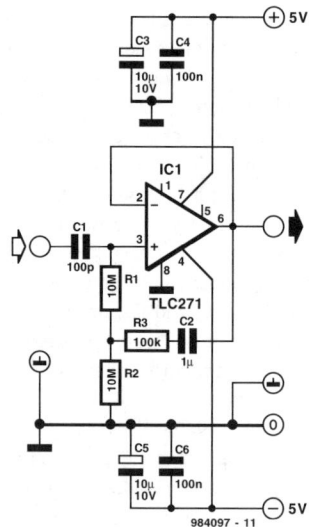

984097 - 11

Latch based on 555 in memory mode

The familiar Type 555 can be used to switch currents up to 200 mA. Less well-known is its use as a latch

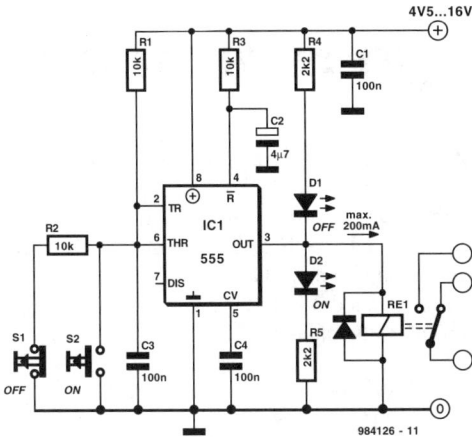

984126 - 11

with control input. When the input pins 2 (trigger) and 6 (threshold) are linked and connected to half the supply voltage, the output can be switched as follows. When the potential at pins 2 and 6 is raised to full supply voltage level, the output is switched to ground. When pins 2 and 6 are linked to ground potential, the output assumes supply voltage level.

The circuit in the diagram uses this mode of operation of the 555 to obtain a two-wire on/off switch. The combination S_1 (closed), R_2 and R_1 provides half the supply voltage to the input (pins 2, 6) of IC_1. When S_2 is closed, the output, pin 3, goes high so that D_2 (on) lights. When S_1 is opened, the input at pins 2, 6 of IC_1 rises to above 2/3 of the supply voltage, whereupon IC_1 is disabled and the output goes low. Diode D_1 (off) then lights.

Network R_3-C_1 at the reset input, pin 4, forces the latch to come up in the off state when power is first applied.

Low-drop 5 V regulator

A 4-cell pack is a convenient, popular battery size. Alkaline manganese batteries are sold in retail stores in packs of four, which usually provide sufficient energy to keep battery replacement frequency at a reasonable level.

Generating 5 V from four batteries is, however, a bit tricky. A fresh set of four batteries has a terminal voltage of 6.4 V, but at the end of their life, this voltage is down to 3.2 V. Therefore, the voltage needs to be stepped up (boosted) or down (bucked), depending on the state of the batteries.

A flyback topology with a costly, custom designed transformer could be used, but the circuit in the diagram gets around the problem by using a flying capacitor together with a second inductor.

The circuit also isolates the input from the output, allowing the output to drop to 0 V during shutdown.

The circuit can be divided conceptually into boost and buck sections. Inductor L_1 and switch IC_1 comprise the boost or step-up section, and inductor L_2, diode D_1 and capacitor C_3 form the buck or step-down section.

Capacitor C_2 is charged to the input voltage, V_{in}, and acts as a level shift between the two sections. The

switch toggles between ground and $V_{in}+V_{out}$, while the junction of L_2, C_2 and D_1 toggles between $-V_{in}$ and $V_{out}+V_{d1}$.

Efficiency is directly related to the quality of the capacitors and inductors used. Better quality capaci-

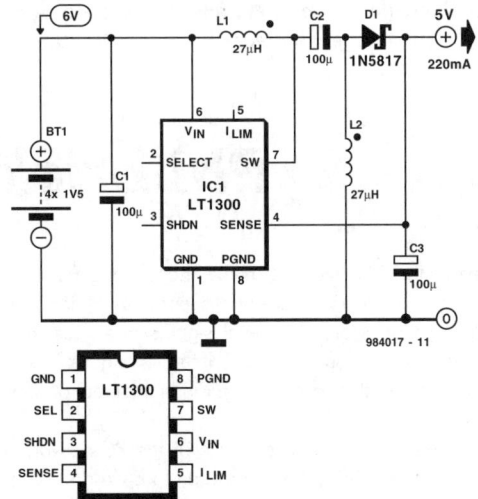

984017 - 11

tors are more expensive. Better quality inductors need not cost more, but normally take up more space. The Sanyo capacitors used in the prototype (C_1–C_3) specify a maximum ESR (effective series resistance) of 0.045 Ω and a maximum ripple current rating of 2.1 A. The inductors used specify a maximum DCR (direct current resistance) of 0.058 Ω.

Worst-case r.m.s. current through capacitor C_2 occurs at minimum input voltage, that is, 400 mA at full load with an input voltage of 3 V.

Low-power voltage reference

The present reference is a special application of current source IC Type LM334. It has a tiny temperature coefficient and draws only a minute current: at room temperature, only 10 μA, which increases with large rises in temperature by only a few μA.

The circuit is basically a bandgap reference, because the positive temperature coefficient of the LM334 is combined with the negative temperature coefficient of the base-emitter junction of a transistor (which ensures good thermal coupling).

To obtain a temperature coefficient of zero, or very nearly so, the output voltage of the circuit is adjusted to exactly 1.253 V with P_1. It is, therefore, advisable to measure the set value of P_1 accurately after it has been adjusted and to replace the combination of R_1+P_1 by a fixed resistor of the precise value. Use a 1% metal-oxide film type from the E96 series.

Since current source IC_1 is tapped at the control input, a reference source with a negative output resistance of about 3.8 kΩ ensues. Resistor R_3 ensures that the ultimate output resistance is about 400 Ω. The load current is then limited to not more than 5 μA.

The performance of the reference is good: when the input voltage is increased from 5 V to 30 V, the output voltage varies by only 0.6 μV (from 1,2530 V to 1.2536 V).

The temperature coefficient stays below 50 ppm $^\circ C^{-1}$, and, with careful adjustment, may even come down to 5 ppm $^\circ C^{-1}$.

The current drawn by the prototype is 9.8 μA at an ambient temperature of 22 $^\circ$C.

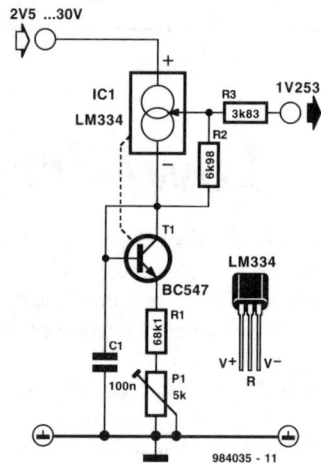

984035 - 11

Philbrick oscillator

The Philbrick oscillator is a little known design, patented by the American scientist George A Philbrick in 1956. It generates low-level signals and uses fairly standard components. The circuit in Figure 1, was originally used for d.c. decoupling at the input of oscilloscopes.

Since the transient response of the RC network is greater than 1000, it may be used to build an oscillator by feeding back the output signal to the input via a

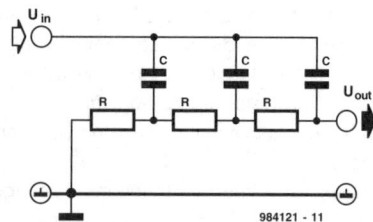

984121 - 11

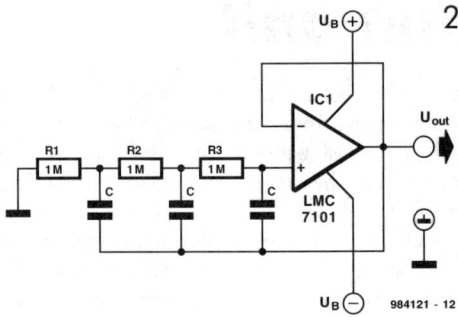

984121 - 12

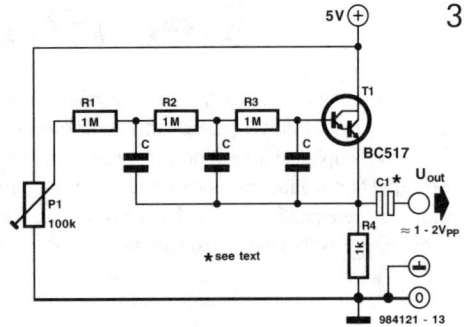

984121 - 13

high-resistance voltage follower. The resulting oscillator can generate even very low audio frequencies.

The diagrams in Figures 2 and 3 show two versions of the oscillator. The one with the op amp has the disadvantage that it needs a symmetrical power supply of ±1.5 V to ±7.5 V. If that is a problem, the circuit based on a transistor can be used. This operates from a single power supply of +5 V.

The operating point of the emitter follower circuit in Figure 3 is set with P_1 so that oscillations and maximum output voltage are guaranteed.

When the output of the transistor version contains very low near-sinusoidal frequencies, it should be

applied to the following stage via an electrolytic capacitor. This capacitor may have to be polarized, depending on whether there is any direct voltage at the input of the following stage.

In the transistor oscillator with resistor values as specified, the following frequencies were measured with the stated capacitor values.

C (μF)	f (Hz)
0.15	5
0.01	50
0.001	500

Polarity reverser

There are systems in which it is imperative that the supply voltage of, say, a motor, always has the correct polarity. It is, of course, possible to use a bridge rectifier for this, but if large currents are involved, this is not always possible. This may be because large voltage drops across diodes result in appreciable heat dissipation, or that the peak current exceeds the current rating of a diode. Fortunately, a good, inexpensive mechanical rectifier may be constructed with the aid of a relay.

In the diagram, the supply voltage is applied to K_1, while the motor that needs a supply with correct polarity is linked to K_2. Provided fuse F_1 is intact, a positive potential at terminal a of K_1 will be applied to the positive terminal of K_2. Diode D_2 prevents the relay being energized. When the polarity at K_1 is reversed, the relay will be energized via D_2. The relay contacts then interchange the connections to the terminals of K_2 to ensure that the previous polarity of the supply to

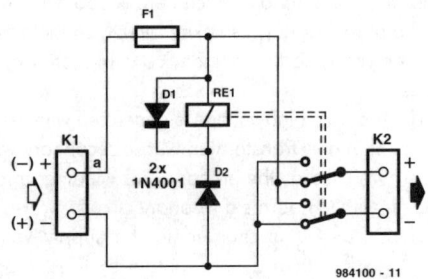

984100 - 11

the load is retained.

Diode D_1 is a freewheeling diode for the relay coil.

The type of relay to be used depends on the requisite operating voltage and the current through its contacts. Other parts of the circuit are not critical.

It stands to reason that the circuit is not suitable for use with a small battery, since the relay coil draws a fairly large current.

Electrification unit

The circuit is intended for carrying out harmless experiments with high-voltage pulses and functions in a similar way as an electrified-fence generator. The p.r.f. (pulse repetition frequency) is determined by the time constant of network R_1-C_3 in the feedback loop of op amp IC_{1a}: with values as specified, it is about 0.5 Hz.

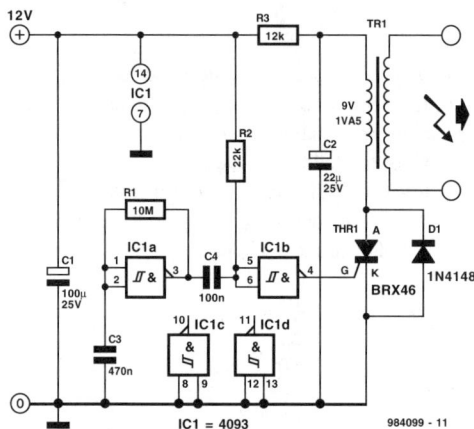

The stage following the op amp, IC_{1b}, converts the rectangular signal into narrow pulses. Differentiating network R_2-C_4, in conjunction with the switching threshold of the Schmitt trigger inputs of IC_{1b}, determines the pulse period, which here is about 1.5 ms.

The output of IC_{1b} is linked directly to the gate of thyristor THR_1, so that this device is triggered by the pulses.

The requisite high voltage is generated with the aid of a small mains transformer, whose secondary winding is here used as the primary. This winding, in conjunction with C_2, forms a resonant circuit.

Capacitor C_3 is charged to the supply voltage (12 V) via R_3. When a pulse output by IC_{1b} triggers

the thyristor, the capacitor is discharged via the secondary winding. The energy stored in the capacitor is, however, not lost, but is stored in the magnetic field produced by the transformer when current flows through it.

When the capacitor is discharged, the current ceases, whereupon the magnetic field collapses. This induces a counter e.m.f. in the transformer winding which opposes the voltage earlier applied to the transformer. This means that the direction of the current remains the same. However, capacitor C_2 is now charged in the opposite sense, so that the potential across it is negative.

When the magnetic field of the transformer has returned the stored energy to the capacitor, the direction of the current reverses, and the negatively charged capacitor is discharged via D_1 and the secondary winding of the transformer. As soon as the capacitor begins to be discharged, there is no current through the thyristor, which therefore switches off. When C_2 is discharged further, diode D_1 is reverse-biased, so that the current loop to the transformer is broken, whereupon the capacitor is charged to 12 V again via R_3.

At the next pulse from IC_{1b}, this process repeats itself.

Since the transformer after each discharge of the capacitor at its primary induces not only a primary, but also a secondary voltage, each triggering of the thyristor causes two closely spaced voltage pulses of opposite polarity. These induced voltages at the secondary, that is, the 230 V, winding, of the transformer are, owing to the higher turns ratio, much higher than those at the primary side and may reach several hundred volts. However, since the energy stored in capacitor C_2 is relatively small (the current drain is only about 2 mA), the output voltage cannot harm man or animal. It is sufficient, however, to cause a clearly discernible muscle convulsion.

Sine wave to TTL converter

As the title implies, the present circuit is intended to convert sinusoidal input signals to TTL output signals. It can handle inputs of more than 100 mV and is suit-

able for use at frequencies up to about 80 MHz.

Transistor T_1, configured in a common-emitter circuit, is biased by voltage divider R_3–R_5 such that the potential across output resistor R_1 is about half the supply voltage. When the circuit is driven by a signal whose amplitude is between 100 mV and TTL level (about 2 V r.m.s.), the circuit generates rectangular signals. The lowest frequencies that could be processed by the prototype were around 100 kHz at an input level of 100 mV, and about 10 kHz when the input signals were TTL level.

Resistor R_6 holds the input resistance at about 50 Ω, which is the normal value in measurement techniques. It ensures that the effects of long coaxial cables on the signal are negligible.

If the converter is used in a circuit with ample limits, R_6 may be omitted, whereupon the input resistance rises to 300 Ω.

984120 - 11

Single-supply operation of the AD736

In dual-supply operation, the output (pin 6) of the AD736 is at 0 V, that is, halfway between the supply lines. But in single-supply operation, the output is at

984090 - 11

1/2 V_{CC}. By adding a single-supply op amp as a differential amplifier, however, a true single supply circuit with a ground-referenced output is obtained as shown in the

diagram. For this circuit, $V_{RMS} = 0$ V when $V_{IN} = 0$ V, and $V_{RMS} = 200$ mV d.c. when $V_{IN} = 200$ mV r.m.s.

In the circuit, a single 9 V positive supply powers the AD736. Resistors $R_7 = R_8 = 100$ kΩ form a potential divider across the 9 V battery that establishes a local 'ground' rail at 1/2 V_{CC}, or 4.5 V. The AD736's 'common' pin, its 22 MΩ input bias resistor, and the inverting input of U_2 (via R_4 and R_5) are all connected to this rail. The quiescent output voltage of the AD736, which is referenced to its 'common' pin, is 4.5 V.

A single-supply op amp, IC_2, is arranged as a unity-gain differential amplifier. Large value feedback resistors, R_2–R_5, are used to minimize loading of the 4.5 V rail. The op amp amplifies the difference between local ground at 4.5 V and the output of the AD736, which is also at 4.5 V for 0 V r.m.s. input. As the r.m.s. input to the AD736 increases from 0 mV to 200 mV, the AD736's output increases from 4.5 V to 4.7 V. The output of op amp IC_2 is the difference between the AD736's output and 4.5 V, or 0 mV to 200 mV d.c.

The remainder of the circuit works as follows. The AD736's output is a.c. coupled; R_1 provides a path for the BiFET op amp's input bias current (typically 1 pA) to flow. The offset voltage caused by the bias current flowing through R_1 is negligible.

Capacitor C_3 between pins 1 and 8 of IC_1 provides a low frequency cutoff of 2 Hz. Other cutoff frequencies, f_C, can be calculated from

$$f_C = 1/2\pi RC$$

where f_C is in Hz, R is in ohms, and C is in farads.

Optional capacitor C_F, in parallel with an 8 kΩ feedback resistor, fixed internally by IC_1, forms a single-pole low-pass filter with a 2 Hz cutoff frequency. The value of C_F in farads is given by

$$C_F = 1/2\pi Rf_C$$

where f_C is in Hz and R=8 kΩ.

Soft start for switching power supply

A switching power supply whose output voltage is appreciably lower than its input voltage has an interesting property: the current drawn by it is smaller than its output current. However, the input

974074 - 11

power (UI) is, of course, greater than the output power. There is another aspect that needs to be watched: when the input voltage at switch-on is too low, the regulator will tend to draw the full current. When the supply cannot cope with this, it fails or the fuse blows. It is, therefore, advisable to disable the regulator at switch-on (via the on/off input). until the relevant capacitor has been charged. When the regulator then starts to draw current, the charging current has already dropped to a level which does not overload the voltage source.

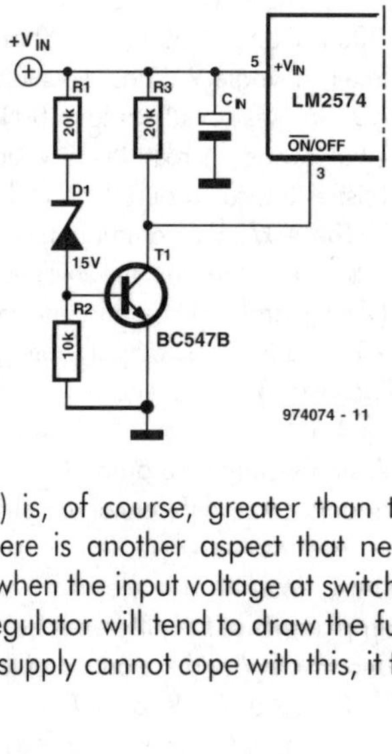

The circuit in the diagram provides an output voltage of 5 V and is supplied by a 24 V source. The regulator need not be disabled until the capacitor is fully charged: when the potential across the capacitor has reached a level of half or more of the input voltage, all is well. This is why the zener diode in the diagram is rated at 15 V.

Many regulators produced by National Semiconductor have an integral on/off switch, and this is used in the present circuit. The input is intended for TTL signals, and usually consists of a transistor whose base is accessible externally. This means that a higher switching voltage may be applied via a series resistor: the value of this in the present circuit is 22 kΩ. When the voltage across the capacitor reaches a level of about 17 V, transistor T_1 comes on, whereupon the regulator is enabled.

Switch-mode power supply

The switch-mode power supply is based on the LM2671 or LM2674 from National Semiconductor. The components for it are available for outputs of 3.3 V, 5 V and 12 V. There is also a version providing a presettable output voltage.

Within the specified application, the supplies can deliver currents of up to 500 mA. Noteworthy is the high switching frequency of 260 kHz. This has the advantage that only low-value inductor and capacitors are needed, and this results in excellent efficiency and small dimensions. In normal circumstances, the efficiency is 90% and may even go up to 96%.

Both ICs provide protection against current and temperature overloads. The LM2671 has a number of additional facilities such as soft start and the option to work with an external clock. The latter enables several supplies to be synchronized so as to give better control of any EMC (Electro Magnetic Compatibility).

The application shown in the diagram provides an output voltage of 5 V and an output current of up to 500 mA. Diode D_1 is a Schottky type ($U_{co} \geq 45$ V and $I_{max} \geq 3$ A).

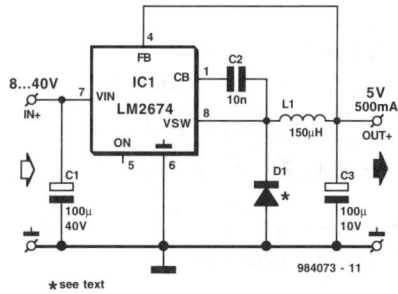

Thrifty light-controlled switch

The circuit is intended mainly for use in low-power battery-powered equipment and is capable of switching up to 25 milliamps. The SFH309-4, T_1, is a phototransistor from Siemens; its pinout is included in the circuit diagram. In this application, T_1 draws a current of 20–25 mA. At a certain ambient light intensity level, the voltage at the input of gate IC_{1f} drops below the

The ambient light intensity at which the output changes state is adjusted to individual requirements with preset P_1. The supply voltage should be reasonably clean and not exceed 16 volts d.c. The circuit is best built on the miniature printed circuit board shown here. When fitting the phototransistor, make sure it is connected the right way around — the shorter pin is

switching threshold of the Schmitt trigger, and the output consequently toggles to logic high. This level is again inverted by the five remaining gates in IC_1 which are connected in parallel to boost their output drive capacity. The effect of stray light picked up from remote controls and other infra-red transmitters is suppressed to some extent by R_1-C_1. If interference is still a problem, the value of C_1 may be increased a little.

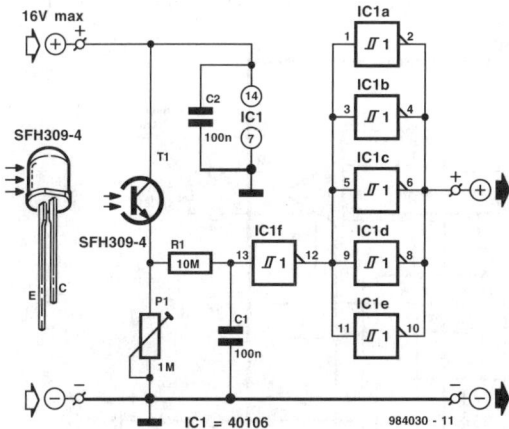

the collector.

The circuit draws a current of 1–2mA in the dark, and about 20 mA when light is detected (with a 9 V supply and with P_1 set to mid-travel). Finally, the switching function of the circuit may be reversed by interchanging P_1 and T1, and connecting R_1 to the collector.

Parts list

Resistors:
$R_1 = 10\ M\Omega$
$P_1 = 1\ M\Omega$ preset, horizontal

Capacitors:
$C_1, C_2 = 0.1\ \mu F$

Semiconductors:
$T_1 = SFH309\text{-}4$ (Siemens)

Integrated circuits:
$IC_1 = 40106$

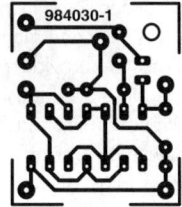

Two-way AF amplifier LM4830

The LM4830, whose internal circuit is shown in Figure 1, is an integrated solution for two-way audio amplification. It contains a bridge-connected audio power amplifier capable of delivering 1 W of continuous average power to an 8 Ω load with less than 1% total harmonic distortion (THD) operating from a 5 V power supply.

The LM4830 can also deliver 100 mW into a single-ended 32 Ω im-pedance (headset operation). There is a 30 dB attenuator in front of a bridged power amplifier with 6 dB of gain. The attenuation is controlled through a 4-bit parallel digital control; 15 steps of 2 dB each.

The device also contains a microphone preamplifier with two selectable inputs. Mic_1 is selected when HS is high and A_1 is in single-ended mode. Mic_2 is selected when HS is low and A_1 is in bridged mode. This configuration is ideal for switching between the internal

speaker and an external headset with microphone.

The LM4830 also incorporates a buffer for

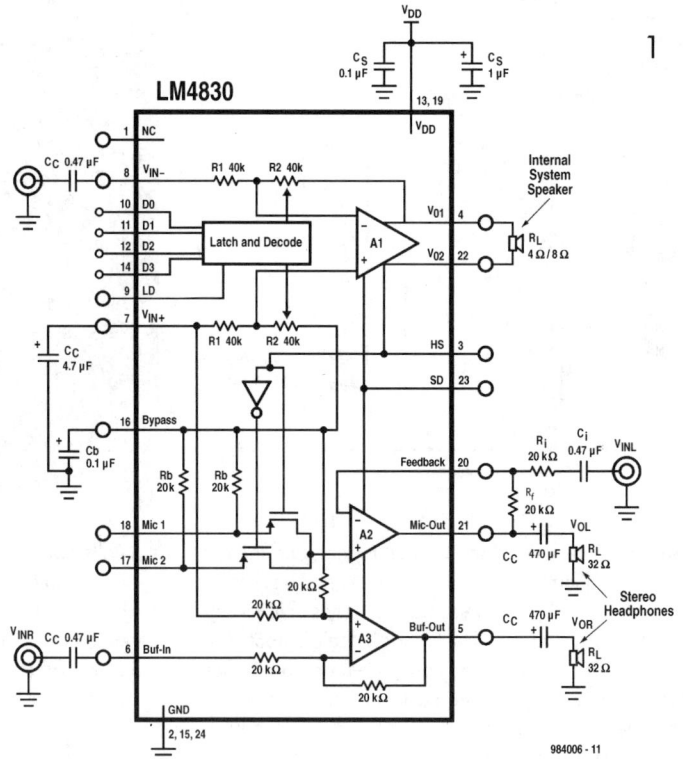

984006 - 11

driving capacitive loads.

The device has a low-current shut-down mode making it suitable for low-power portable systems. In addition, it has internal thermal shut-down protection.

As shown in Figure 2, amplifier A_1 can be used in one of two modes, bridged output or single-ended output. This also allows head-

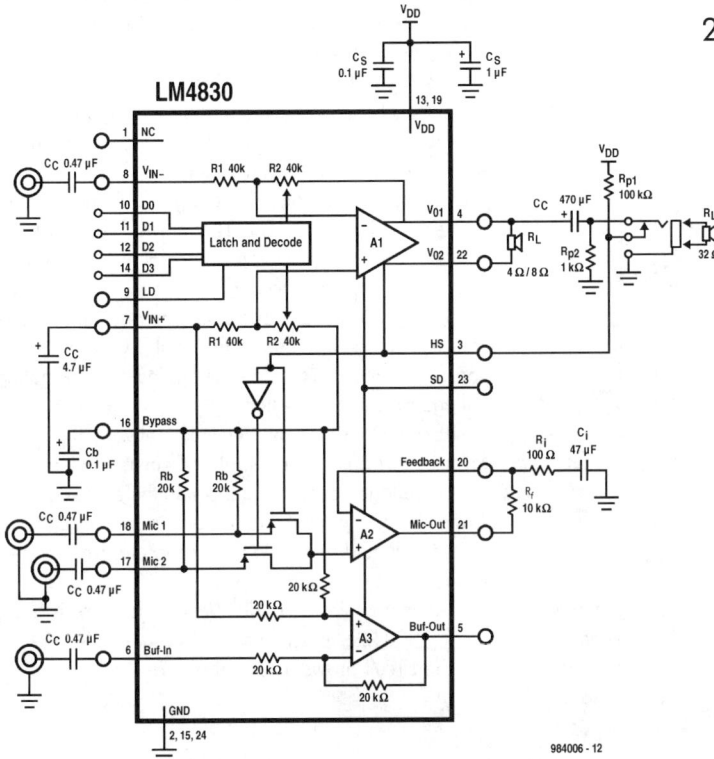

984006 - 12

phones to be driven single-endedly. The output can be switched automatically from bridged speaker drive to single-ended headphone drive using a control pin in the headphone jack that is tied to the HS (headset) pin 3. When the voltage at the HS pin input changes from 0 V to 5 V, V_{O2} of the bridged amplifier output becomes high impedance. This allows the permanently connected internal speaker to be disabled when a headphone is plugged into the headphone socket. Output V_{O1} then drives the headphone single-endedly via output coupling capacitor C_C. This capacitor should be chosen to allow the full audio bandwidth to be amplified. Since C_C and R_L form a high-pass filter, the value of C_C must be high enough to allow

frequencies down to 20 Hz to be amplified. The value is calculated from

$$C_C = 1/[2\pi 20 R_L] \qquad [1]$$

where $16\ \Omega \leq R_L \leq 600\ \Omega$.

The LD (load) pin 9 has two modes of operation. When this input is high, the power amplifier's attenuation control is in transparent mode, where the voltage on bits D_0–D_3 will cause the appropriate attenuation level to be latched and decoded within the IC. For normal attenuation, pin 9 should be at 5 V. When the pin is low, the power amplifier's attenuation control is locked out, so that any change in the input bits will not cause a subsequent change in the amplifier's attenuation level.

The attenuation level is preset to –16 dB when the IC is first powered up, ensuring that pin 9 is low until the IC is fully biased.

The preamplifier in this IC is intended for use as a microphone amplifier. Depending on the frequency response of the microphone, the preamplifier's response can be configured to suit the microphone. Capacitors may simply be used to limit the bandwidth of the frequency response of the preamplifier and improve the system's performance. Once the gain of the amplifier is chosen, the values of the resistors and capacitors may be determined with the following equations.

$$A_{VCL} = 1 + R_f/R_i \qquad [2]$$

$$f_{LP} = 1/2\pi R_f C_i \qquad [3]$$

$$f_{HP} = 1/2\pi R_f C_i \qquad [4]$$

Two-wire temperature sensor I

Remote temperature measurements have to be linked by some sort of cable to an appropriate display or monitort. Normally, this is a three-core cable: one core for the signal and the other two for the supply

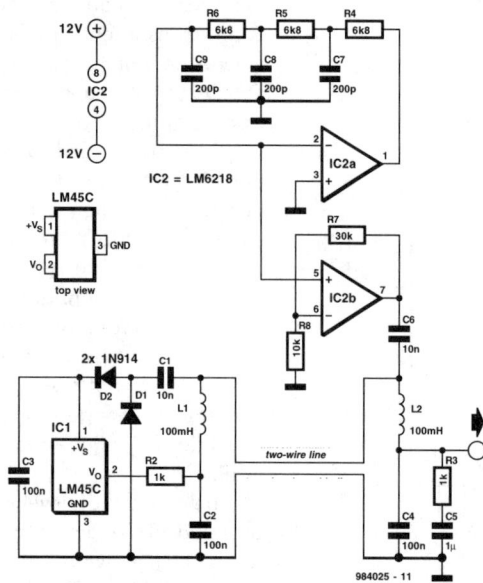

lines. If the link is required to be a two-core cable, one of the supply lines and the signal line have to be combined. This is possible with, for instance, temperature sensors LM334 and LM335 from National

Semiconductor. However, these devices provide an output that is directly proportional to absolute temperature and this is not always a practical proposition.

If an output signal that is directly proportional to the Celsius temperature scale is desired, the present circuit, which uses a Type LM45 sensor, offers a good solution. The LM45 sensor is powered by an alternating voltage, while its output is a direct voltage.

The supply to the sensor is provided by a sine-wave generator, based on A_1 and A_2 (see diagram). The alternating voltage is applied to the signal line in the two-core cable via coupling capacitor C_6.

The sensor contains a voltage-doubling rectifier formed by D_1-D_2-C_1-C_2. This network converts the applied alternating voltage into a direct voltage. Resistor R_2 isolates the output from the load capacitance, while choke L_1 couples the output signal of the sensor to the signal line in the cable. Choke L_1 and capacitor C_2 protect the output against the alternating voltage present on the line.

At the other end of the link, network R_3-L_2-C_4 forms a low-pass section that prevents the alternating supply voltage from combining with the sensor output. Capacitor C_5 prevents a direct current through R_3, since this would attenuate the temperature-dependent voltage.

The output load should have a high resistance, some 100 kΩ or even higher.

The circuit draws a current of a few mA.

Four-state bistable

Bistables (US: flip flops) are well-known and widely used building blocks performing control, register, memory and toggle functions in logic circuits. Some popular ones are the CMOS 4013 (dual D type), the 4027 (dual J-K type), and the TTL 7474 (dual-D type). The latter also comes in LS-TTL, HC and HCT versions. Although D, J-K and S-R bistables have slightly different truth tables, they all share a common characteristic: two stable states.

The circuit shown here is intended for applications where four instead of two logic states are required. The 4028 CMOS BCD-to-decimal decoder at the heart of the circuit has four binary inputs and ten dec-

imal outputs. (BCD=binary coded decimal). Any allowed BCD input combination (0000 through 1001) will set the corresponding output to logic '1'. Each of the remaining six input combinations (1010 through 1111) resets all of the decoder outputs to '0'. In this application, only combinations are used which contain one logic '1' and three logic '0's, that is, 0001, 0010, 0100 and 1000. These actuate decoder outputs 1, 2, 4 and 8 respectively, to which a kind of feedback is applied via diodes D_2–D_5. When, for instance, decoder output1 is set to logic '1', this state is transferred to input 2^0 (1) via diode D_2, while the other three inputs remain at '0' because of resistors R_2, R_3

and R_4. This state remains stable until one of the push-buttons is pressed. Switch S_2, for example, then sets a '1' at the 2^1 (2) input of the decoder, which responds by resetting output1 (as well as input 2^0),

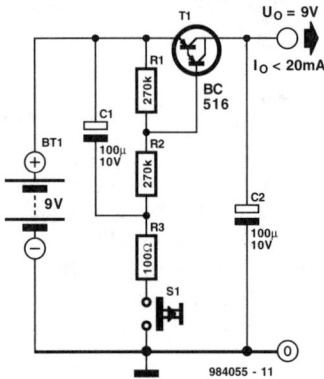

984055 - 11

and actuating output2. This state is 'latched' by diode D_3, which starts to conduct as soon as S_2 is released.

Components C_1, R_5 and D_1 produce a short positive pulse at power-on, thus defining decoder state1 as the initial state of the circuit.

The inverters in IC_2 act as current boosters to enable the decoder to control four LEDs which flag the state of the 'fourstable'. Based on standard CMOS circuits, the circuit may, in principle, be run off any supply voltage between 3V and 18V. However, the indicated value of R_6 is for a 5 volt supply. With different supply voltages, the value should be recalculated from

$$R_6 = (U_b - 2)/I$$

where I is the LED current in mA and the value of R_6 is in kilo-ohms.

Although it is tempting to simplify the circuit by omitting the LED driver and substituting LEDs for D_2, D_3, D_4 and D_5, note that such a modification may not work because the 4028 used may not be able to source the required current. The widely used HCF-type CMOS ICs from Thomson, for example, can not supply more than about 2.6mA (typ.) without degradation of the output voltage level, if the supply voltage is higher than 10V. The solution is to use high-efficiency (low-current) LEDs, and lower the value of resistors R_1–R_4 to 4.7 kΩ for a 10 V supply, or about 10 kΩ for an 18 V supply. Also, the value of C_1 should be increased to 1 µF. Note that these changes will only work if the supply voltage is higher than 10 V.

Temperature sensor

The AD22100 from Analog Devices is a monolithic temperature sensor with on-chip signal conditioning, which can be used over the temperature range–50 °C to +150 °C. The signal conditioning eliminates the

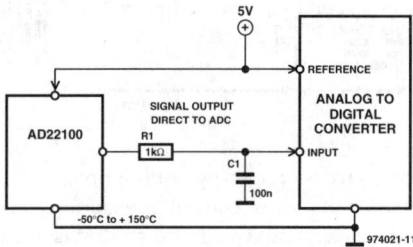

974021-11

need for any trimming, buffering or linearization circuitry, greatly simplifying the system design and reducing overall system cost.

The output voltage is proportional to the temperature times the supply voltage (ratiometric). The output swings from 0.25 V at–50 °C to +4.75 V at +150 °C

when a single +5.0V supply is used.

The ratiometric characteristic of the AD22100 allows it to be easily interfaced to an A-D converter: a precision reference is not required.

The heart of the sensor is a proprietary temperature-dependent resistor, R_T, built into the IC. This resistor undergoes a change in resistance that is nearly proportional to temperature. It is excited with a current source that is proportional to the supply voltage. The resulting voltage across R_T varies linearly with

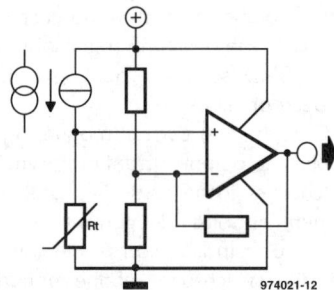

974021-12

temperature in proportion to the supply voltage. The remainder of the AD22100 consists of a signal conditioning block that amplifies the voltage across R_T and applies the proper offset according to:

$$V_{out} = (V_+/5)(1.375+22.5T_A),$$

where T_A is the absolute temperature (K).

The AD22100 output pin is capable of withstanding an indefinite short circuit to either ground or the power supply. These characteristics are provided by the output stage structure shown in Figure3. The active portion is a p-n-p transistor with its emitter connected to the V_+ supply and its collector tied to the output node. The transistor sources the required amount of output current. A limited pull-down capability is provided by a fixed current sink of about–80mA.

Owing to its limited current sinking ability, the

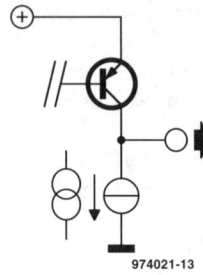

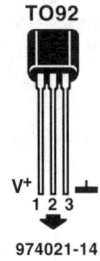

974021-13

TO92

974021-14

AD22100 is incapable of driving loads to the V_+ power supply and is instead intended to drive grounded loads. A typical value for short circuit limit is 7mA, so devices can reliably source 1mA or 2mA. However, for best output voltage accuracy and minimal internal self-heating, the output current should be kept below 1mA.

Voltage-independent 2 A battery charger

Linear Technology's LT1513 is an IC that is intended for use as a 500 mA current-switch-mode controller in a battery charger to provide constant voltage or constant current. A special property of the IC is that, depending on the input potential, it switches automatically from boost to buck mode or vice versa.

The IC may be used in chargers for Li-ion, NiMH or NiCd batteries as long as the nominal battery voltage is ≤ 20 V. The output voltage is accurate to within 1%, which is, of course, a must in the case of Li-ion batteries. Since the switching rate is 500 kHz and the IC is an SMD (surface-mount device), the circuit can be kept very small.

The IC contains a current-monitoring section that enables the output current of, for instance, a flyback charger to be controlled precisely. This arrangement enables the current to be monitored w.r.t. earth in isolation of the battery. This simplifies the switch-over of the batteries and

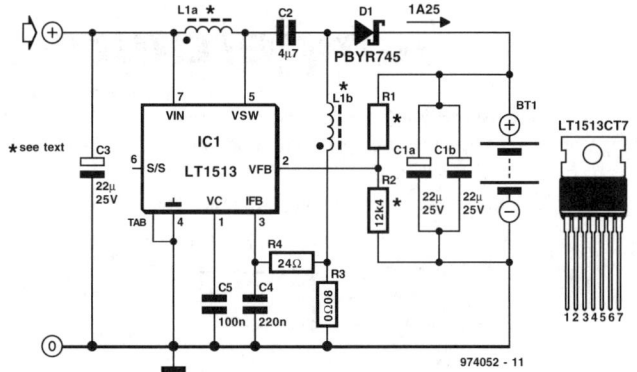

974052 - 11

prevents errors caused by earth loops.

Resistor R_3 is a current measuring device: with its value as specified, the current is 1.25 A.

The peak switching current of the IC is 3 A, which enables a charging current of up to 2 A for a single Li-ion cell. Curves relating the peak charging current and the input voltage of the IC are shown in the second diagram.

Diode D_1 should be a Schottky type.

The values of resistors R_1 and R_2 must cause

a potential of 1.245 V at pin V_{FB} when the charging voltage is maximum. The current through the potential divider should be about 100 μA.

Inductors L_1 and L_2 are wound on a common core; each must have a self-inductance of about 10 μH.

Finally, capacitor C_2 must in no circumstances be an electrolytic type.

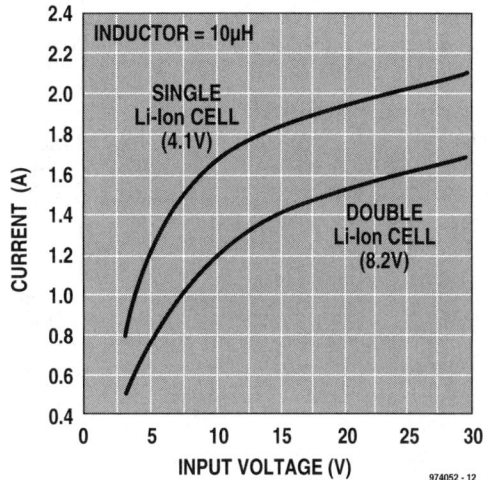

974052 - 12

PC control for two stepper motors

This circuit allows a PC or a suitably programmed microprocessor system to control a stepper motor. Rather than relying on the latest in dedicated motor controller ICs, this interface is based on ordinary CMOS logic, discrete transistors and one ULN power driver IC. The interface is designed with

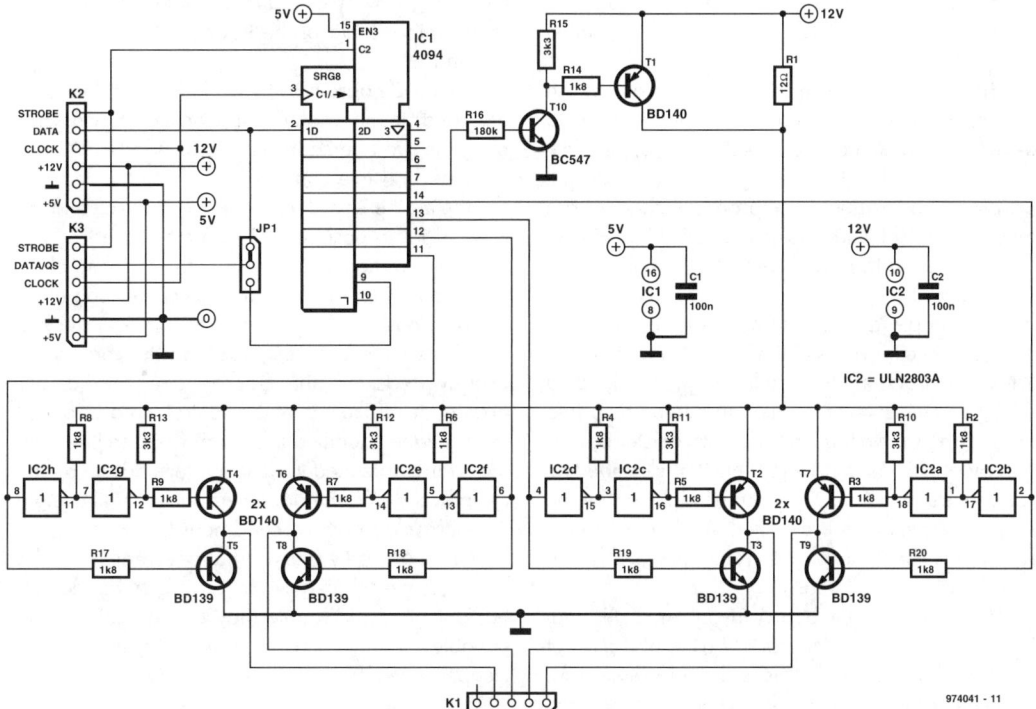

974041 - 11

extension in mind, because a second board may be added to control a second stepper motor. The Pascal program is a good starting point for further experiments. The program allows the interface to be driven by the PC's parallel port. Alternatively, a microcontroller's serial port may be used to control the interface.

Printer port data lines D1, D2 and D3 are connected to the strobe, data and clock inputs, respectively, of pinheader K_2, which also receives a 5 V supply voltage for the logic circuits, and 12 V for the motor drivers.

Circuit IC1 is a CMOS 8-bit latching shift register with three-state outputs. Only five of the eight shift register outputs are used here, the remaining ones are available for extension experiments. The motor driver transistor pairs, T_2–T_9, are controlled by shift register outputs Q5-Q8 via buffers/inverters which help to convert from 5 V to 12 V logic.

Resistor R_1 acts as a current limiter when the stepper motor is not active. When control signals are generated, R_1 is virtually short-circuited by T_1.

The program as printed here is based on the assumption that two interface boards are used to drive two motors. If only one motor is used, the first eight of the 16 bits transmitted to the interface will be lost (see below) unless appropriate changes are made to the software (basically, all references to mot_2 must be removed, and 'motors=2' altered to 'motors=1').

The program works basically as follows. Assume that a 1 arrives at the D (data) input of IC1. When the clock input is pulled high, the leading edge causes the 1 at the data input to be read into the shift register. In this way, the shift register is filled with eight bits (1s and/or 0s). Next, the strobe line (Centronics data line D1) is pulled high so that the shift register contents are copied and latched by the IC outputs, and the power transistors are driven accordingly.

The printed-circuit board layout indicates that K_1 is a 6-way pinheader block (2 rows of 3 pins) of which pin1 is removed. This is done to make sure the stepper motor connector can not be connected

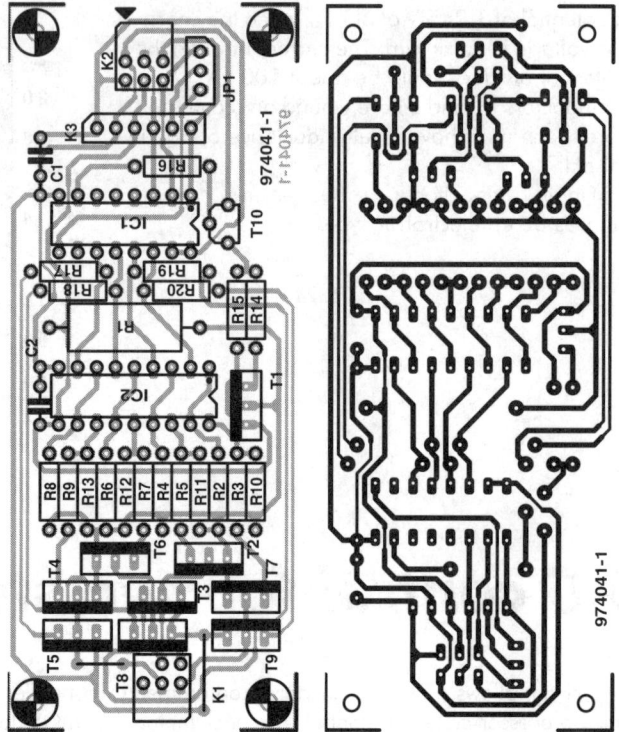

the wrong way around. Note that the PCB may be made with the aid of the track layout given in the Appendix.

A second board may be added for the control of a second stepper motor. This is achieved by assembling the two boards in a stacked ('sandwich') construction and interconnecting their pinheaders K_3. On the motherboard (i.e., the one connected to the PC), fit JP_1 in position 2-3. On the second board, fit JP_1 in position 1-2. This links the data pin of K_3 to the data input of IC1 on the second board. By the way, position 1 of the jumper block is marked by the bevelled edge as shown on the component overlay. Stacking the boards and the jumper arrangement allow a 16-bit control word (supplied by a suitably modified program) to be split up into two words intended for the motors driven by the main board and the doughterboard. The first eight bits always go to the second board.

Although only very small currents will be required, the 5 V supply for IC1 has to be regulated, but the 12-V supply may be unregulated but capable of furnishing all current required by the stepper motor(s).

```pascal
program steppermotor;

uses crt;

const portAddr=$3BC;
    motors=2;

var counter, a: integer;

procedure Low;
{Load one LOW bit in shift register}
begin
    port[portAddr]:=$4; { [0100]b }
    port[portAddr]:=$0; { [0000]b }
end;

procedure High;
{Load one HIGH bit in shift register}
begin
    port[portAddr]:=$2; { [0010]b }
    port[portAddr]:=$6; { [0110]b }
    port[portAddr]:=$0; { [0000]b }
end;

procedure Strobe;
{Create STROBE signal for shift registers
to latch contents of shift reg. to output}
begin
    port[portAddr]:=$1; { [0001]b }
    port[portAddr]:=$0; { [0000]b }
end;
procedure Init;
{Makes all outputs of shift register(s) LOW}
begin
    port[portAddr]:=$0; { [0000]b }
    for counter := 1 to (8*motors) do Low;
    Strobe;
end;

procedure Step1;
{Load pattern for Step1 [1000 1000]b }
begin
    High; Low; Low; Low; High; Low; Low; Low;
end;

procedure Step2;
{Load pattern for Step2 [0010 1000]b }
begin
    Low; Low; High; Low; High; Low; Low; Low;
end;

procedure Step3;
{Load pattern for Step3 [0100 1000]b }
begin
    Low; High; Low; Low; High; Low; Low; Low;
end;

procedure Step4;
{Load pattern for Step4 [0001 1000]b }
begin
    Low; Low; Low; High; High; Low; Low; Low;
end;

procedure Step2Res;
{Load pattern for Step2 with R3 in series
[0010 0000]b }
begin
    Low; Low; High; Low; Low; Low; Low; Low;
end;

procedure Step4Res;
{Load pattern for Step4 with R3 in series
[0001 0000]b }
begin
    Low; Low; Low; High; Low; Low; Low; Low;
end;

begin
    {User defined}
    ClrScr;
    Init;
    for a:= 1 to 50 do
    begin
        {Example causes one (slow) turn of both
        motors in opposite direction.
        mot_2; mot_1; strobe1+2; Delay
        --+------+-------+------+-----
          |      |       |      |
          V      V       V      V      }
        Step1; Step4; Strobe; delay(10);
        Step2; Step3; Strobe; delay(10);
        Step3; Step2; Strobe; delay(10);
        Step4; Step1; Strobe; delay(10);
    end;
    delay(1000);
    for a:= 1 to 50 do
    begin
        {Example causes one (fast) turn
        of both motors in opposite direction.
        mot_2; mot_1; strobe1+2; Delay
        --+------+-------+------+-----
          |      |       |      |
          V      V       V      V      }
        Step4; Step1; Strobe; delay(5);
        Step3; Step2; Strobe; delay(5);
        Step2; Step3; Strobe; delay(5);
        Step1; Step4; Strobe; delay(5);
    end;

    Step2Res;Step2Res;
    Step4Res;Step4Res;
    Strobe;
end.
```

Parts list
Resistors:
R_1 = 12 Ω, 5 watt
R_2–R_9, R_{14}, R_{17}–R_{20} = 1.8 kΩ
R_{10}–R_{13}, R_{15} = 3.3 kΩ
R_{16} = 180 kΩ

Capacitors:
C_1, C_2 = 0.1 μF

Semiconductors:
T_1, T_2, T_4, T_6, T_7 = BD140

T_3, T_5, T_8, T_9 = BD139
T_{10} = BC547

Integrated circuits:
IC_1 = 4094
IC_2 = ULN2803A (Sprague)

Miscellaneous:
K_1 = 6-way pinheader (2x3 pins) angled (see text)
K_2 = 6-way pinheader (2x3 pins) angled
K_3 = 6-way pinheader (1x6 pins)
JP_1 = 3-way pinheader with jumper

Single-supply instrumentation amplifier

The OP284 is a low noise dual op amp with a bandwidth of 4 MHz and rail-to-rail input/output operation. These properties make it ideal for low supply voltage applications such as in a two op amp instrumentation amplifier as shown in the diagram.

The circuit uses the classic two op amp instrumentation topology with four resistors to set the gain. The transfer equation of the circuit is identical to that of a non-inverting amplifier.

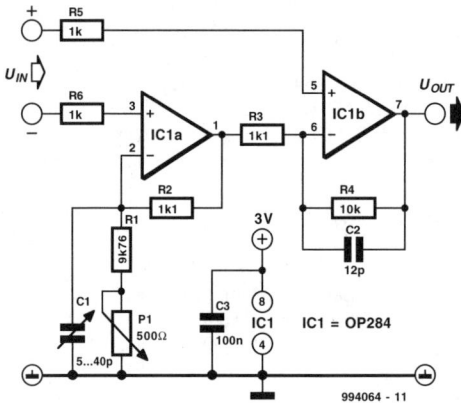

994064 - 11

Resistors R_2 and R_3 should be closely matched to each other as well as to resistors (R_1+P_1) and R_4 to ensure good common-mode rejection (CMR) performance. It is advisable to use resistor networks for R_2 an and R_3, because these exhibit the necessary relative tolerance matching for good performance. Potentiometer P_1 is used for optimum d.c. CMR adjustment, and capacitor C_1 is used to optimize a.c. CMR.

With circuit values as shown, circuit CMR is better than 80 dB over the frequency range of 20 Hz to 20 kHz. Circuit referred-to-input (RTI) noise in the 0.1 Hz to 10 Hz band is exemplary at 0.45 μV_{pp}.

Resistors R_5 and R_6 protect the inputs of the op amps against overvoltages. Capacitor C_2 may be included to limit the bandwidth. Its value should be adjusted depending on the required closed-loop bandwidth of the circuit. The R_4-C_2 time constant creates a pole at a frequency, f_{3dB}, equal to

$$f_{3dB}=1/2\pi R_4 C_2.$$

With a value of C_2 of 12 pF, the bandwidth is about 500 kHz.

The amplifier draws a current of about 2 mA.

Position sensor

The sensor converts the wiper position of a potentiometer (slide or rotary) into one of 11 binary values (0 through 10). The prototype of this 'digital potentiometer' was used to provide an interface between a microcontroller and the arm assembly of a robot.

The familiar LM3914 is used here as an analogue-to-digitall converter (ADC) which translates the (analogue) wiper voltage into a corresponding digital value. The LM3914 is used in bar mode in the circuit shown on the next page for reasons outlined below.

The ten open-collector outputs of the LM3914, L_1 through L_{10}, are connected to the inputs of a 10-to-4 priority encoder, IC_2, a 74147. Only the input bit with the highest significance appears in 4-bit binary code at the outputs of the encoder. The ten encoder inputs allow the codes 0000 through 1001 to be made (that's decimal 0 through 9). XOR gates IC_{3c} and IC_{3d} have been added to enable the value `10' or 1010

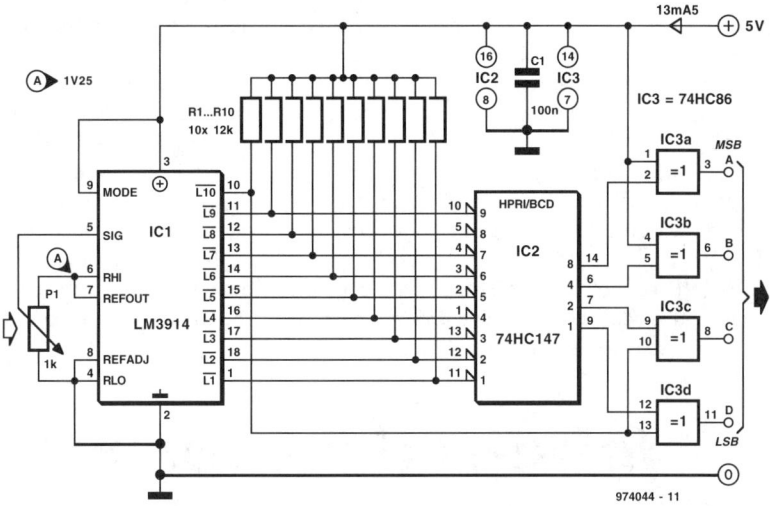

974044 - 11

to be produced also. Their function is as follows: if the '147 encoder is at the value 1001 (9), and L_{10} of the LM3914 goes low (active), then the XOR gates will invert the two least significant bits, creating the binary

word 1010. However, that only works if L_9 remains active when L_{10} is actuated, hence the use of the 'bar' mode rather than the 'dot' mode on the LM3914.

Temperature-compensated crystal oscillator

The clock in computers and many other electronic systems is normally provided by a simple crystal oscillator. Unfortunately, the electromechanical properties of such a clock are normally such that it may vary as much as 100 minutes per year from real time. This is,

even the cheap ones, are designed to work at body temperature, which is fairly constant. Clocks in computers and other electronic appliances, however, are designed to work at room temperature, say, 20–25 ∞C. When the ambient temperature is higher

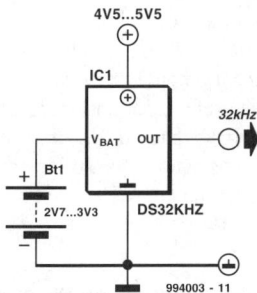

994003 - 11

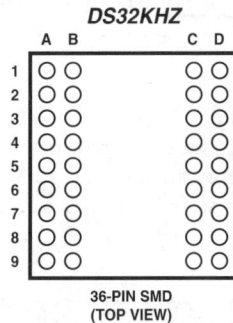

	VCC:	C2, C3, D2, D3
	VBAT:	A4, A5, B4, B5
	32KHZ:	C4, C5, D4, D5
	GND:	All Remaining Balls

36-PIN SMD
(TOP VIEW)

994003 - 12

of course, a highly unsatisfactory, and, to many people, inexplicable situation. After all, if a cheap watch can keep (reasonable) time, why can an expensive computer not?

Yet, the reason for this is fairly simple. Watches,

or lower, as happens in rooms and offices which are not constantly heated or cooled as the case may be, the clock will drift.

Dallas Semiconductor' catalogue includes a temperature-compensated oscillator (TXCO), which is eminently suitable for use in computers and other appliances where correct time-keeping is important. The very small integrated oscillator enables a clock to be constructed that does not vary by more than ±1 minute per year (±2 ppm) over a temperature range of 0–40 °C.

The IC type-coded DS32kHz is an accurate and affordable replacement for standard 32,768 kHz crystals and oscillators. Its output can drive virtually any RTC chip.

The SMD case of the DS32kHz contains a quartz crystak and a temperature-compensating circuit. In this circuit, use is made of a thermal sensing technol-ogy specially developed by Dallas. External components are not needed: the IC is calibrated at the factory.

The circuit diagram shows that connecting the TXCO is straightforward. The backup battery connected to V_{BAT} ensures that when the mains voltage fails, the clock remains on. If a backup battery is not used, V_{CC} should be connected to GND and a supply voltage of 2.7–5.5 V to V_{BAT}.

The device is available only in SMD format. Its pins are arranged as a 36-pin ball grid array. The pinout is shown alongside the circuit diagram.

Infra-red sensor/monitor

The sensor/monitor shown in the diagram 'wakes up' the host system on detection of infra-red (IR) signals. It draws so little supply current that it can remain on continuously in a notebook computer or PDA device. Its ultra-low current drain (4 μA maximum, 2.5 μA typical) is primarily that of the comparator/reference device, IC_1.

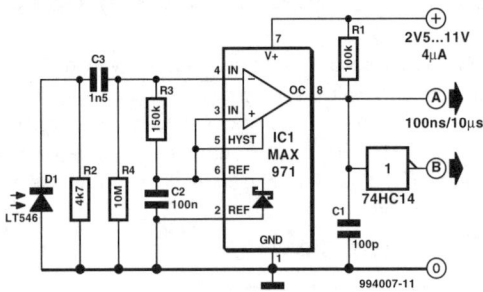

994007-11

The circuit is intended for the non-carrier systems common in infra-red Data Association (IrDA) applications. It also operates with carrier protocols such as those of TV remote controllers and Newton/Sharp ASK (an amplitude shift keying protocol developed by Sharp and used in the Apple Newton). The range for 115,000-baud IrDA is limited to about 6 in (15 cm), but for 2400-baud IrDA, it improves to more than 12 in (30 cm).

Immunity to ambient light is very good, although bright flashes usually cause false triggers. To handle such triggers, the system simply looks for IR activity after waking and then returns to sleep mode if none is present.

The sensor shown, D_1, a relatively large-area photodiode packaged in an IR-filter material, produces about 60 μA when exposed to heavy illumination, and 400 mV when open-circuited. Most photodiodes may be used.

Operation is in the photovoltaic mode without applied bias. This mode is slow and not generally used in photodiode circuits, but speed is not essential her. The photovoltaic mode simplifies the circuit and saves a significant amount of power. In a more conventional configuration, for instance, photoconductive, photo currents caused by ambient light and sources by the bias network would increase the quiescent current about ten times.

Three-phase sine wave generator

The diagram shows how a three-phase sine wave oscillator can be built with a single Type UAF42 state-variable filter and some resistors and diodes. Three output nodes are available: high-pass out, band-pass out, and low-pass out. The signal at the band-pass and low-pass out nodes is 90° and 180° out of phase, respectively, with that at the high-pass node. An on-chip auxiliary op amp is available for use as a buffer or amplifier stage.

The frequency of oscillation is set with resistors R_{F1} and R_{F2} according to

$$f_{OSC}=1/2\pi RC,$$
where
$$R=R_{F1}=R_{F2}$$
and
$$C=C_1=C_2=1000\ pF$$

The maximum frequency of oscillation obtainable with the UAF42 state-variable filter is 100 kHz. Distortion becomes a factor, though, for frequencies above 10 kHz. For frequencies of oscillation below 100 Hz, the use of external capacitors is recommended. These should be placed in parallel with the internal capacitors C_1 and C_2. This will reduce the requisite values of R_{F1} and R_{F2}. The external capacitors should preferably be NP0 ceramic or mica types.

To obtain the requisite output levels, resistors R_1–R_4 should meet the following requirement:

$$R_1/R_2=R_3/R_4=(V_O+V_S)/(V_O-0.15)-1$$

The values indicated in the diagram apply to a frequency of 1 kHz. At this frequency, the external capacitors may be omitted, since the internal ones are sufficient.

The actual output level may differ slightly from the calculated one owing to non-ideal operation of diodes and op amps. It may therefore be necessary to adapt the ratios R_1/R_2 and R_3/R_4 to some extent.

Positive feedback necessary for the onset of oscillation is provided by coupling the output of the bandpass section to the input of the summing amplifier via resistor R_{FB}. Suitable values for this resistor are 10 MΩ for $f>1$ kHz, 5 MΩ for $f=10$–1000 Hz, and 750 kΩ for $f<10$ Hz. Smaller values result in an increase in the output level and, consequently, distortion.

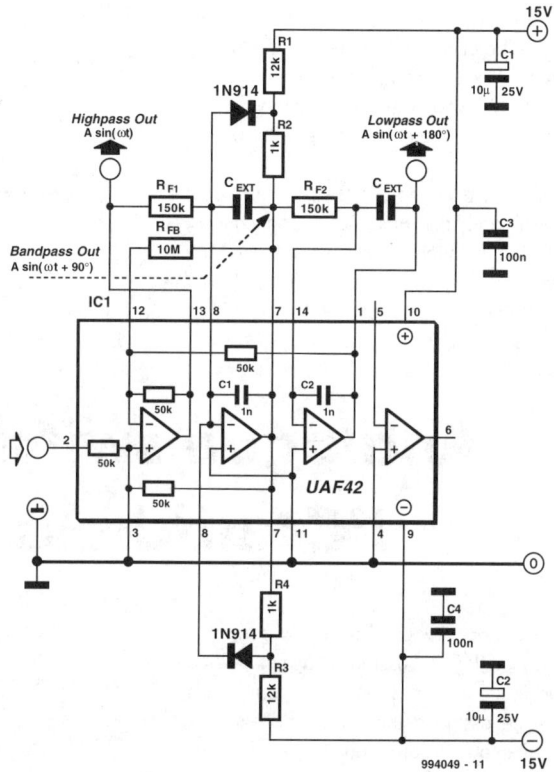

994049 - 11

Adjustable precision voltage source

Many applications require a precision voltage source which can be adjusted through zero to both positive and negative output voltage. An example is a bipolar power supply. Perhaps the most obvious implementation of a bipolar voltage source would be to use a bipolar voltage reference. However, a simpler solution is to use a single voltage reference and a precision unity-gain inverting amplifier. If a precision difference

amplifier is used for the unity-gain inverting amplifier, the circuit requires just two chips and a potentiometer.

In the present circuit a Type INA105 differential amplifier is used as the unity-gain amplifier. A potentiometer is connected between the input and ground. The wiper of the potentiometer is connected to the non-inverting input of the unity-gain amplifier. (The non-inverting input of a unity-gain amplifier is normally connected to ground.) With the wiper at the bottom of the potentiometer, the circuit is a normal precision unity-gain inverting amplifier with a gain of −1.0 V/V ±0.01% max. With the slider at the top of the potentiometer, the circuit is a normal precision voltage follower with a gain of +1.0 V/V ±0.001% max. With the wiper at the centre of its travel, there is equal positive and negative gain for a net gain of 0 V/V. The accuracy between −1.0 V/V and +1.0 V/V is normally limited by the accuracy of the potentiometer. Precision 10-turn potentiometers are available with 0.01% linearity.

The −1.0 V/V to +1.0 V/V linear gain control amplifier has many applications. With the addition of a precision +10.0 V reference, it becomes a −10 V to +10 V adjustable precision voltage source.

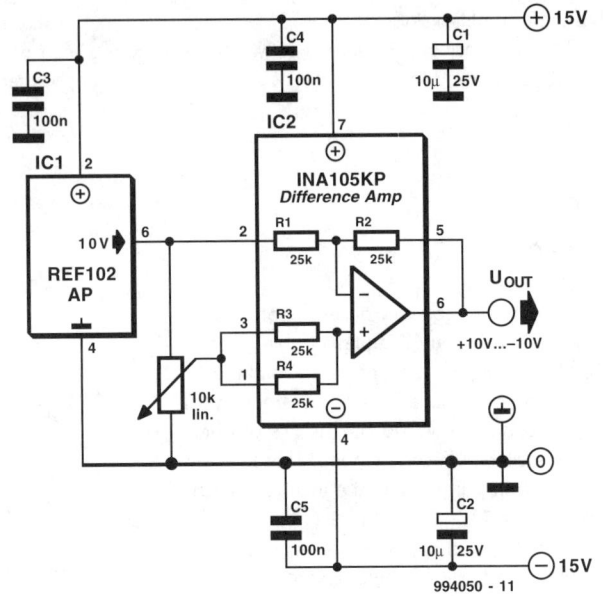

994050 - 11

Analogue switch alleviates I²C* address conflicts

To avoid address conflicts, every peripheral on an I²C™ bus must have a unique address. Sometimes, however, peripherals may be assigned the same address. The circuit shown resolves address conflicts by enabling the I²C bus to select between two peripherals with the same address.

The popular I²C bus is an open-collector, 2-wire interface that includes a clock line and a bidirectional data line. It allows a controller (the master) to select a particular device (the slave) by first issuing a serial address on the data line, then issuing appropriate commands or data. Master and slave can send data in both directions by pulling the data line low, and slaves can generate wait states by pulling the clock line low. Bus switching is complicated, however, by the open-collector architecture—it cannot be accomplished with the CMOS outputs of AND gates or

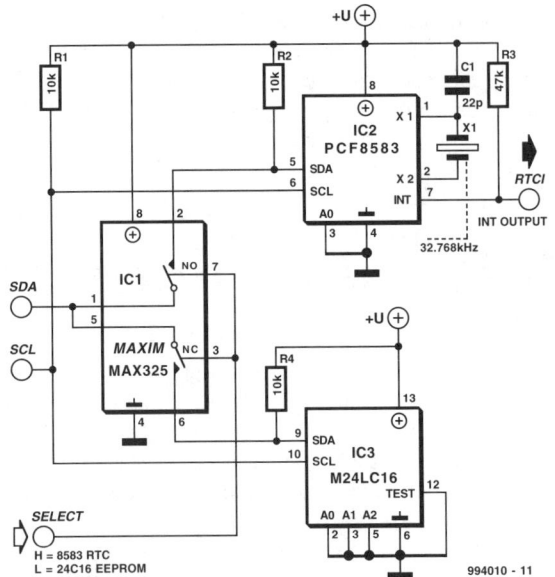

994010 - 11

74HC157 data selectors.

The peripherals shown in the diagram are a Philips I^2C real-time clock (PCF-8583) and a large I^2C EEPROM (Microchip M-24LC16). Both have an internal, hexadecimal slave address of A0. (The EEPROM takes up the entire address range, making it impossible to avoid.) The analogue switch connects either one device or the other. Selection involves the data line (SDA) only, because an I^2C start condition requires that the SDA signal goes low before the clock goes low. To select between the devices, the master device sets a port pin to control the state of the dual SPST analogue switch.

IC_1 is a CMOS chip well suited to this function. Its normally open switch and normally closed switch perform the 2:1 selector operation with no additional inverters or port line. It features low on-resistance (33 Ω) and low supply current (1 μA), and is specified for operation below 3 V. Also, its tiny 8-pin SOT package (μMAX) is only half the size of an SO-8 package.

* I^2C is a trademark of the Philips Corporation.

Buck-boost converter without magnetics

One of the problems that designers of portable equipment face is generating a regulated voltage whose level lies between those of a fully charged and a discharged battery. As an example, when a 3.3 V output is generated from a 3-cell battery, the regulator input

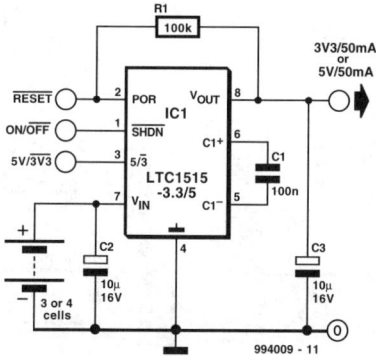

994009 - 11

voltage changes from about 4.5 V at full charge to about 2.7 V when the battery is discharged. At full charge, the regulator must step down the input voltage, and when the battery voltage drops below 3.3 V,

the regulator must step up the voltage. The same problem occurs when a 5 V output is required from a 4-cell input voltage that varies from about 3.6 V to 6 V. Normally, a flyback or SEPIC configuration is required to solve this problem.

The LTC1515 switched-capacitor DC/DC converter can provide this buck-boost function for load currents up to 50 mA with only three external capacitors. The circuit shown will provide a 3.3 V output from a 3-cell battery or a 5 V output from a 4-cell input. Connecting the 5/3 pin to V_{IN} will program the output to 5 V, whereas grounding the 5/3 pin programs the output to 3.3 V.

The absence of bulky magnetics provides another benefit: this circuit requires only 0.07 in^2 (0.45 cm^2) of board space in those applications where components can be mounted on both sides of the board.

The addition of resistor R_1 provides a power-on reset flag that goes high 200 ms after the output reaches 93.5% of its programmed value. The SHDN pin allows the output to be turned on or off with a 3 V logic signal.

Configurable clock generator

MicroClock have available a 28-pin generator IC Type ICS525, also called OSCaR (acronym for oscillator replacement), which may be used to replace virtually any kind of clock oscillator. The range of operation is wide although extensive hardware and/or software is not needed. The multipliers and divisors with which the crystal frequency may be manipulated are set with

pins, something that in most other similar ICs has to be done via a serial protocol and the aid of an additional processor.

In the diagram, the manufacturers' numbering of the jumpers given in the datasheets has been retained. The position of the jumpers for various combinations of input and output frequencies may be obtained via the Internet or by fax request. The table shows an example of how a 12 MHz clock may be converted to a 25.576 MHz one with 0 ppm error. Of course, not all combinations of input and output frequencies are error-free. If, for instance, the input for the 25.576 MHz clock is changed to 10 MHz, the error is not less than 34 ppm. Clearly, it pays to try out a number of different crystal frequencies.

The generator is best built on the printed-circuit board shown, which may be made with the aid of the track layout given in the Appendix. The IC and surface-mount technology (SMT) capacitors C_1 and C_2 should be soldered at the track side of the board. Since capacitor C_3 is at the component side of the board exactly opposite the IC, it is best to mount the capacitor first and then the IC.

The generator draws a current of about 15 mA from its +5 V supply.

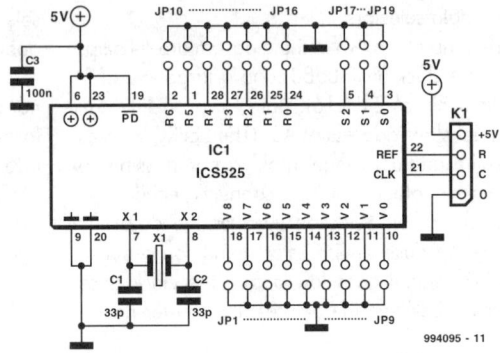

994095 - 11

INPUT FREQUENCY = 12 MHz									OUTPUT FREQUENCY = 24.576 MHz									
S2	S1	S0	R6	R5	R4	R3	R2	R1	R0	V8	V7	V6	V5	V4	V3	V2	V1	V0
PIN 5	4	3	2	1	28	27	26	25	24	18	17	16	15	14	13	12	11	10
1	0	0	0	0	1	0	1	1	1	0	0	1	1	1	1	0	0	0

ERROR ppm = 0.0

994095 - 12

Parts list

Capacitors:
C_1, C_2 = 33 pF SMT ceramic
C_3 = 0.1 μF ceramic
Integrated circuits:
IC_1 = ICS525-01R (MicroClock)

Miscellaneous:
X_1 = crystal (see text)
JP_1–JP_{19} = 2-way jumper
K_1 = 4-way header

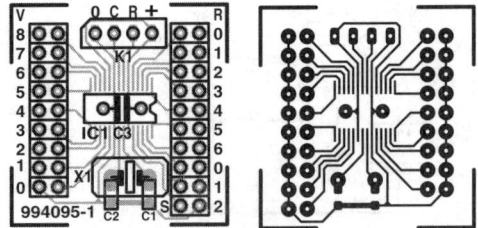

994095-1

Crystal frequency multiplier

Crystals for operation above 20 MHz are invariably cut to an overtone (harmonic), which may be the third, fifth, or seventh. In other words, the fundamental frequency of such a crystal is a third, fifth or seventh of the operating frequency. If, however, an overtone crystal is used in an oscillator designed for operation on the fundamental, it is doubtful whether the wanted frequency will be generated. In most such cases, there are difficulties that make it necessary for the oscillator circuit to be modified.

One practical solution for such problems is offered by the circuit in the diagram, which enables frequencies up to 160 MHz to be generated with the use of a fundamental-frequency crystal. The output signal is virtually free of any jitter.

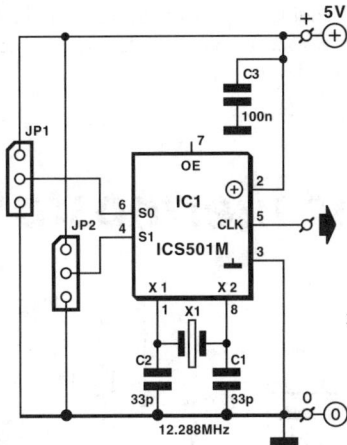

JP1	JP2		X1 min
0	0	4 x	
0	x	5.3125 x	20 MHz
0	1	5 x	
x	0	6.25 x	4 MHz
x	x	2 x	
x	1	3.125 x	8 MHz
1	0	6 x	
1	x	3 x	
1	1	8 x	

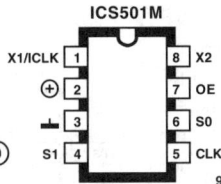

ICS501M

X1/ICLK 1		8 X2
(+) 2		7 OE
3		6 S0
S1 4		5 CLK

994067 - 11

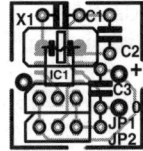

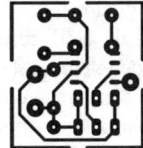

The circuit makes use of an IC that contains not only an oscillator but also a phase-locked loop (PLL) controlled frequency multiplier. The associated ROM stores nine different multipliers, which may be selected by appropriate placing of the jumpers in JP_1 and JP_2 as shown in the table. A 0 in the table indicates that the jumper is linked to earth; a 1, that is linked to the +5 V line; and a × that it should be left open.

The circuit may conveniently be built on the printed-circuit board shown, which is not available ready made.

The circuit draws a current of about 20 mA.

Parts list
Capacitors:
C_1, C_2 = 33 pF, ceramic
C_3 = 0.1 μF, ceramic

Integrated circuits:
IC_1 = ICS501M

Miscellaneous:
X_1 = crystal 5–20 MHz
JP_1, JP_2 = 3-way terminal strip with jumper

Current limiting at switch-on

In certain direct-current operated circuits, such as DC-to-DC converters, the switch-on current may be so high that the output voltage of the power supply cannot reach its nominal level. This difficulty may be prevented with a limiting circuit as shown in the diagram.

When the input voltage, V_{in}, is switched on, transistor T_1 is off since capacitor C_1 is not charged. The level of current I at the moment of switch-on is

$$I_{(t=0)} = V_{in}/R_2.$$

Capacitor C_1 is charged slowly via resistor R_1 until the gate–source threshold voltage, $U_{GS(th)}$, is exceeded, whereupon the transistor begins to conduct. The time interval between switch-on and the transistor coming on is determined by time constant R_1-C_1 and the ratio V_{in}:$U_{GS(th)}$. The gate-source voltage, U_{GS}, of

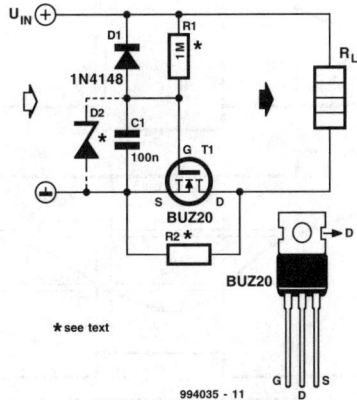

★ see text

994035 - 11

the BUZ20 is in the range ±20 V. If V_{in} is larger than these values, or a different transistor is used, U_{GS} must be limited to the range mentioned by zener diode D_2, for which a ZPD18 was used in the prototype.

Diode D_1 enables C_1 to be discharged via the load when V_{in} is switched off. The circuit is then ready for the next period of operation.

Dual-output, low-power thermostat

The LM56 from National Semiconductor is an accurate, dual-output, low-power thermostat contained in an 8-pin SMD case. It contains a 1.25 V bandgap voltage reference, two comparators with 5 °C hysteresis and a temperature sensor. The output voltage of the sensor is internally linked to the comparators. The supply voltage may lie in the range 2.7–10 V.

There are several variants: the LM56CIM has an accuracy of ±4 °C and the LM56BIM ±3 °C, both over the temperature range –40 °C to +125 °C. Typical applications are temperature monitoring in a variety of system, protection against low or high temperatures, and control of alarm systems or fans.

Outputs OUT1 and OUT2 are open-collector outputs, which can switch up to 5 mA (400 mV drop at 50 µA). If it is required to switch a higher current, a buffer stage is needed at the output. The outputs are enabled when the values set with R_1, R_2, and R_3 are exceeded—see the timing diagram. Resistors R_4 and R_5 are pull-up components.

The output voltage of the temperature sensor is 6.2 mV °C^{-1} plus a fixed offset voltage of 395 mV.

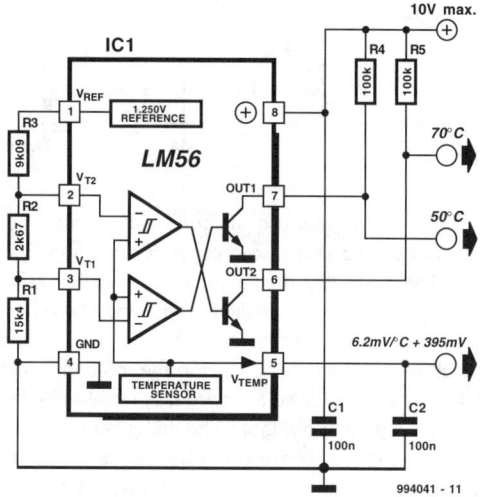

When the component values of potential divider R_1-R_2-R_3 are calculated, assume a load current of 50 µA for V_{REF}. This means that the total value of the three resistors is about 27 kΩ. If great accuracy is wanted, the bias current of the comparators should be taken into account, although the typical value of these is only 150 nA, so that they do not make much difference in practice. With values as specified in the circuit diagram, switching takes place at 50 °C and 70 °C.

Capacitor C_2 serves to decouple any interference. If such interference persists, the value of the capacitor may be increased to 1 µF without any adverse effect on the speed of reaction.

The LM56 draws a current of not more than 230 µA, which means that the total current drain, including the current through the potential divider, will not exceed 400 µA. This makes the IC particularly useful for systems operating from 3 V or 5 V battery packs.

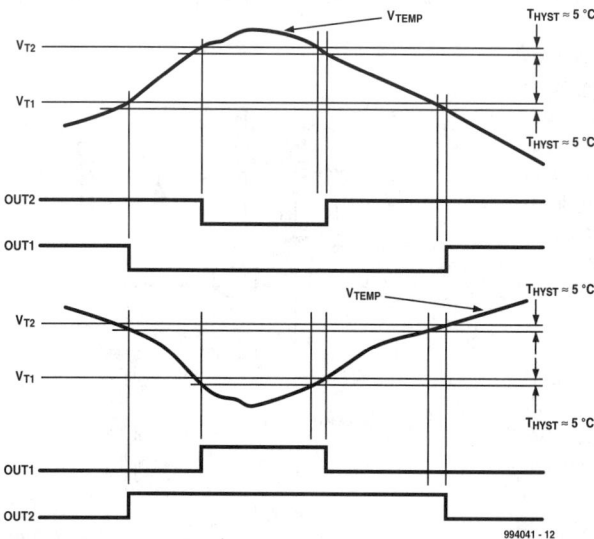

EEPROM protection in AVR controllers

AVR controllers have the unfortunate property of their data EEPROM being affected when the supply voltage drops below a certain level, which can, of course, be prevented by making the reset low in good time to disable the processor. Unfortunately, this requires a circuit for monitoring the supply voltage and for taking the requisite action automatically when needed.

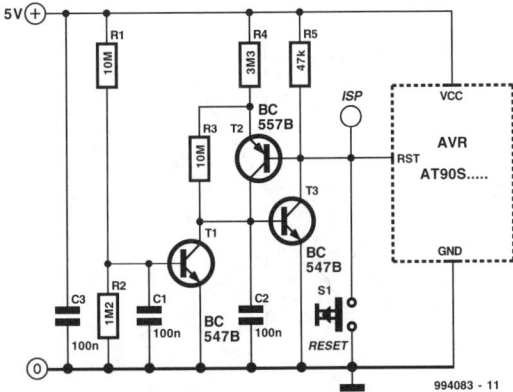

994083 - 11

This requirement is met by the circuit in the diagram, which draws a low enough current to enable it being powered by a battery. The circuit may be split into a detector, T_1, and an amplifier, T_2–T_3.

The trip voltage of the detector is determined by the values of R_1 and R_2. Normally, the transistor conducts, but as soon as the supply voltage drops below the trip level, it is cut off. The output of T_1 is applied to a low-power amplifier.

During normal operation, transistor T_3 is off, so that R_5 functions as pull-up resistor to retain the RST input of the AVR processor high. When the detector goes off, T_3 is switched on and the RST input goes low, a process that is enhanced by transistor T_2 being switched on, whereupon R_3 is shorted out. The resulting hysteresis required the supply voltage to exceed the trip voltage before this situation can change. A manual reset is possible at all times with witch S_1.

As stated earlier, the trip voltage is determined by the values of R_1 and R_2 (and, to some degree, the base-emitter potential of T_1 – about 540 mV), and the tolerance of these resistors should therefore be 1%. If the trip voltage needs to be altered, it is best to retain the value of R_1 at 10 MΩ and change the value of R_2 according to

$$R_2=0.54R_1/(U_b-0.54),$$

where R_2 is in ohms and U_b is the supply voltage.

The hysteresis is determined by the value of R_4: the smaller this is, the larger the hysteresis. The specified value of 3.3 MΩ is fine for most cases, but some experimentation does no harm.

Feedback circuit clamps precisely

A linear circuit consisting of an input buffer, IC_{1a}, and output-scaling amplifier, IC_{1b}, two zener diodes, D_1 and D_2, and several other components can supply sharp, precise, bipolar clamp levels with continuous variable control, from 0 V to ±11 V. A feedback loop enclosing the amplifiers and zener diodes generates the high clamping accuracy—see diagram.

Within the limit range of the clamp, $±V_L$, the zener diodes are off, and IC_{1b} feeds back its output to the inverting input of IC_{1a} via R_4. At the same time, IC_{1a} drives IC_{1b} via voltage divider R_V. The feedback forces the inverting input of op amp IC_{1a} to equal U_i at the non-inverting input terminal.

The circuit forces the inverting input of IC_{1b} also to follow U_i. There is no signal voltage drop across R_4,

because no current can flow from it into the inverting input of IC_{1a}. Consequently, the non-inverting input of IC_{1b}, which defines the potentiometer output at feedback equilibrium, must also track U_i. In fixed-level limiting applications, a resistive divider may replace potentiometer R_V. Amplifier IC_{1a} then delivers an output

$$U_O=(1+R_3/R_2)/U_i$$
when
$$-V_L<U_O<V_L$$
and
$$V_L=x[(1+R_3/R_2)](V_Z+V_F)$$

where x is the setting fraction of R_V and V_Z and V_F are

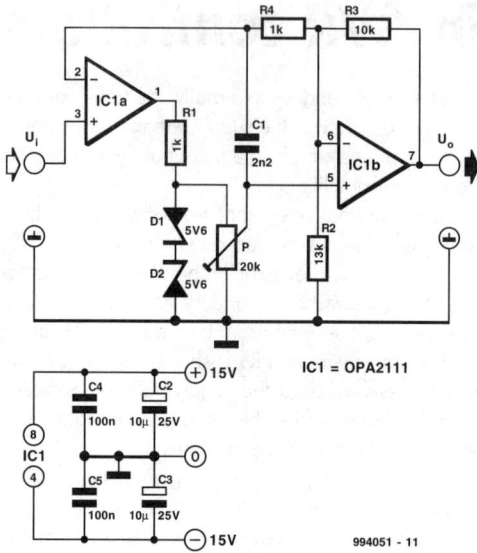

IC1 = OPA2111

994051 - 11

ideal response because the gain of IC_{1b}, $(1+R_3/R_2)$, magnifies any offset voltage and noise from IC_{1a}. Similarly, the loop gain mitigates the clamping error by sharpening its clamping response. The zener drive increases during the transition to the clamping state. The maximum clamping level depends on the zener voltage and the closed-loop gain of IC_{1b}. Zener diodes rated at 5.6 V provide a wide control range and good temperature stability. At higher zener voltages, the even wider control range is offset by increasing drift of the clamping level with temperature.

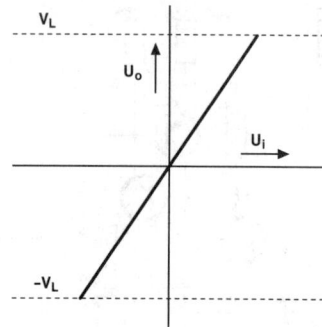

994051 - 12

the zener and forward voltages respectively. The overall circuit response, then, is simply that of a voltage amplifier when the output signal is within the limit boundaries.

Amplifier IC_{1a} generates small deviations from an

High-current, high-speed buffer

The circuit shown in the diagram is intended for applications in which relatively large pulse-shaped signals are to be applied to a standard impedance of, say, 50 Ω. The parallel circuit of two high-speed buffers prevents overheating during periods of high loads and high drive.

The operational amplifiers used, Analog Devices' Type BUF104, have a bandwidth of 110 MHz and a very high slew rate: 3000 V μs^{-1}. The peak output current is 65 mA, which is not sufficient to provide 5 V into 50 Ω, but the parallel combination is able to. Connecting 100 Ω resistors, R_3 and R_4, in series with the outputs of the op amps gives an overall output impedance of 50 Ω.

Owing to the large bandwidth, effective decoupling is imperative, which is why its is advisable to use tantalum capacitors for C_3, C_4, C_7, and C_8 and ceramic ones for C_5, C_6, C_9, and C_{10}.

With output currents exceeding 50 mA, the pulse response may be enhanced by damping the self-inductance of the tantalum capacitors with the aid of

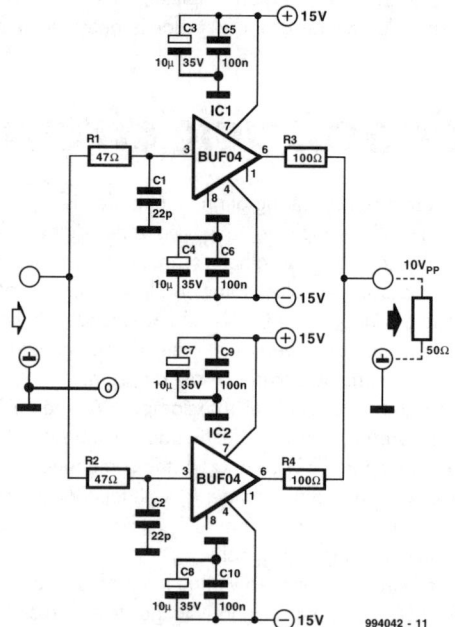

994042 - 11

series resistors of 1–4.7 Ω.

If the specified bandwidth of 110 MHz is to be attained, a central earth plane and SMT components must be used, since the parasitic self-inductance of standard components is too high. One of the prototypes constructed with standard components had a bandwidth of only 25 MHz. Also, with such a construction, the screening is less effective, so that at high frequencies positive feedback, and thus oscillations,

may occur. This is the reason that there is an RC filter at the input of the op amps: R_1-C_1, and R_2-C_2 respectively, which limits the bandwidth to 80–90 MHz.

The circuit draws a quiescent current of ±15 mA. When the output signal is rectangular at a level of 10 V_{p-p}, into 50 Ω, the current drain rises to 60–65 mA.

Modified mains switch

Sometimes it is necessary for the contacts of a mains switch to be isolated or for the switch to be able to handle a larger load than it was designed for. Normally, a relay is used for these situations, but this

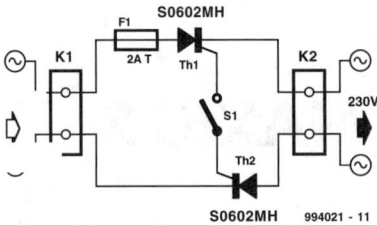

has the disadvantage of needing an auxiliary voltage for operating the relay.

The diagram shows that the wanted aim can also be achieved by two thyristors instead of a relay, which has the advantage of not requiring an auxiliary supply. The arrangement depends on the leakage current of one thyristor firing the other. It is important that the thyristors are sensitive types, otherwise there is a risk that the setup does not work. The thyristors used in the prototype are Type S0602MH from SGS-Thomson. These fire at an I_{GT} as low as 200 μA and can switch currents of up to 3.8 A. For safety's sake, the fuse is rated at 2 A (slow). It hardly needs mentioning that the switch must be a Class II approved type.

Overvoltage protection

Electronic circuits must never be operated with an excessive supply voltage. Such a situation may be prevented with the protection circuit shown in the diagram. If the current through the IC becomes excessive, or the IC overheats, an external silicon-controlled rectifier (thyristor), Th_1, is triggered, whereupon the supply voltage is short-circuited. This causes the current limiting in the power supply to be enabled or the relevant fuse in the supply to blow. Whatever, the circuit being supplied is protected.

In the diagram, the overvoltage protection comes into action when the supply voltage exceeds 5 V, but this may be set anywhere between 3.3 V and 9 V. Potential divider R_1-R_2 reduces the supply voltage to 1.19 V (nominal) at the ADJ(ust) pin of IC_1. As long as the level at this pin is ≤ 1.14 V, IC_1 remains in the standby mode and draws a current of about 70 μA.

When the potential at pin 5 rises above 1.19 V

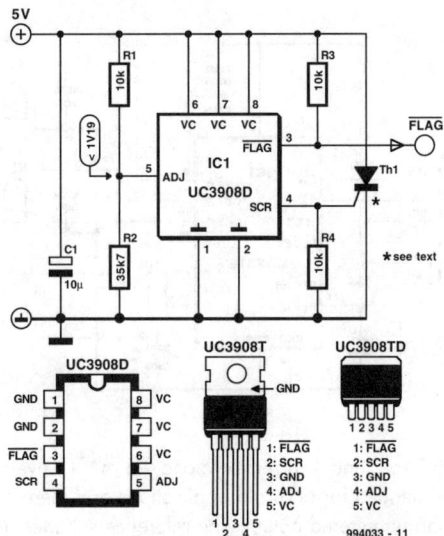

(maximum 1.24 V), IC_1 draws a current of up to 17 A so as to pull down the supply voltage – the flag signal is then actuated. If in this situation the peak current of 17 A is exceeded, or the body temperature of the IC rises above 165 ∞C, or when the internal shunt transistor goes into saturation, the external thyristor is triggered via pin 4 (SCR). This protects the IC itself and ensures that the overvoltage is negated. The rating of the thyristor must, of course, be in accordance with that of the power supply. In this situation, IC_1 shorts out its internal shunt transistor to minimize the internal dissipation.

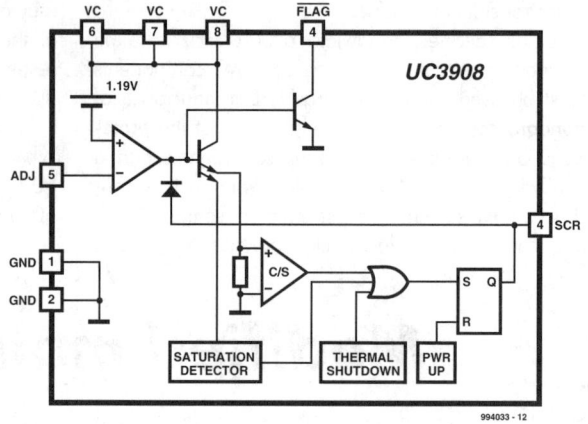

994033 - 12

The UCC3908 is available in three different enclosures. For situations in which large supply currents flow for long periods, the TO-220 version is recommended (if necessary with heat sink). When the load current is not large, the SO-8 version may be used. In that situation, it may even be possible to omit the thyristor, but the anticipated maximum temperature must then be calculated very carefully.

Further information from:
http://www.unitrode.com

Pascal for the MAX512

The MAX512 is a simple triple digital-to-analogue converter with a serial interface and a resolution of eight bits. Two of the three converters (DAC A and DAC B) provide a unipolar or bipolar buffered output voltage. Converter A can provide or sink currents of

applied separately (in contrast to the diagram) to converters A/B and C. Apart from the converter outputs, the MAX512 also has a digital output (1.6 mA) which, for instance, may be used for directly driving a high-efficiency LED.

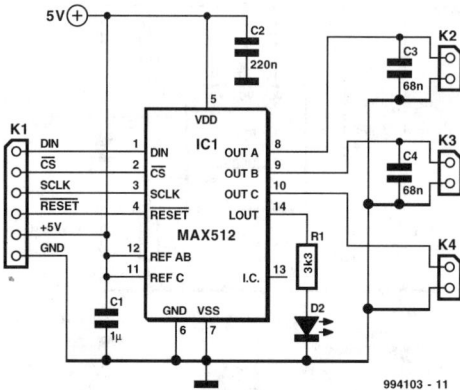

994103 - 11

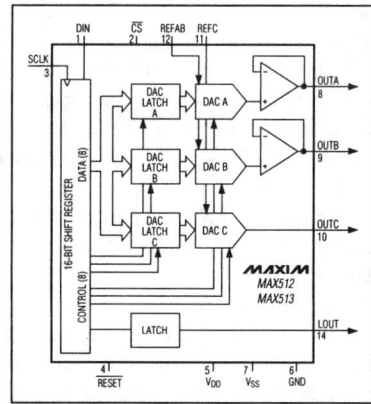

994103 - 12

up to 5 mA, and B currents of up to 0.5 mA. Converter C is intended for accurate applications and therefore has an unbuffered output. The reference voltages are

The data is applied to the converter via a 3-wire interface. The interface operates with frequencies up to 5 MHz and is compatible with standards such as

SPI, QSPI, and Microwire.

The serial shift register at the input is 16 bits wide: eight data bits and eight control bits. The latter enable a converter to be selected or switched off. In the shutdown mode, the R2R network of the relevant converter is isolated from the reference source. The DAC registers may be charged at the leading edge of CS either independently of one another or simultaneously.

The MAX512 may be operated from a single +5 V supply or a symmetrical ±2.5 V supply. It draws a current of about 1 mA and <1 µA in the shutdown mode.

The compact and well-documented Pascal program shows clearly how one can work with the MAX512. Serial lines D_{IN}, Chip-Select, CS, clock signal SCLK, and RESET are then to be connected to ports P4.0–P4.3 of an 87537 processor, but other devices may also be used for controlling the process. It may then be necessary to adapt the port addresses at the start of the program.

After the program has been started, all three converters generate a five-step staircase voltage at a frequency of about 5 Hz. By introducing variations in the FOR loop that determines the length of the steps, the voltage steps may be altered. The software is available from the Publishers (Ref. no. 996022) but it should be noted that the commentary is in German.

Polarity protection

In many cases, the battery or batteries in electronic equipment may be inserted with incorrect polarity. It is, therefore, advisable to use polarity protection such as shown in the diagrams. It should be noted that although a Schottky diode may be used, this causes a voltage drop of a few hundred millivolts, which in the case of a 3 V or 1.5 V battery supply is too much. The protection in the diagrams does not cause any reduc-

that the drain–source breakdown voltage, V(BR)DSS, must be larger than the battery voltage to ensure that the transistor survives an incorrectly connected battery. At the same time, the gate threshold voltage, VGS(th) must be small compared with the battery voltage to ensure that, provided the polarity is correct, the transistor can transfer the battery voltage to the load. Suitable types are certain HEXFETs from International

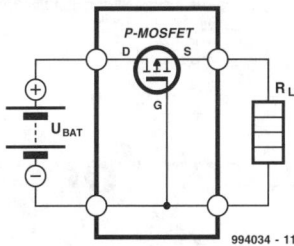

994034 - 11

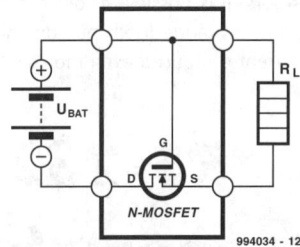

994034 - 12

tion in the supply voltage.

The use of a MOSFET, p-channel or n-channel, as the case may be, ensures that when the polarity is correct, the battery voltage is applied to the load without any loss. For good efficiency, it is best to use an n-channel MOSFET, although this has the disadvantage of having to be inserted in the negative supply line. In cases where this is impossible or impractical, a p-channel device must be used.

In the choice of MOSFET, it must be borne in mind

Rectifier. In the case of n-channel types, the IRF7401 in an SO-8 case, the IRF7601 in a Micro-8 case, and the IRLML in a Micro-3 case are suitable. Types suitable for p-channel operation are the IRF7404 in an SO-8 case, the IRF7604 in a Micro-8 case, and the IRML6302 in a Micro-3 case.

Data sheets are available at:
http://www.irf.com

Pull-up accelerator

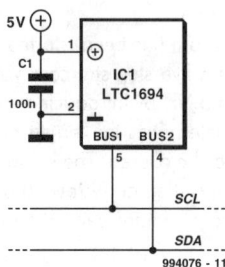

994076 - 11

The IC is intended for an I^2C bus frequency of 100 kHz. The 400 kHz and recently introduced 3.4 MHz versions are not supported. The IC is housed in a SOT-23 case.

The characteristic shows the difference of the leading edge obtained with a standard resistor and that resulting when the IC is used.

Systems like the SMBus or I^2C™ use a standard resistor to pull up the signal levels to the positive supply rail (normally 5 V). The bus goes low because an appliance connected to it pulls the signal to zero via its open-collector output. The well-known problem is this output can draw a much higher current than the pull-up resistor can compensate. This results in a steep trailing edge, but a much more gradually rising leading edge, whose transition in addition is not linear but exponential. This adversely affects the duty factor of the signals and also reduces the speed of the bus.

Linear Technology have available an IC (Type LTC1694) to replace the traditional pull-up resistor which can produce a current that is dependent on the changes in level taking place on the bus. When that level rises, the IC gives 2.2 mA, but when it falls, the current is only 275 μA.

Since the IC contains two circuits for replacing both pull-up resistors, it is possible to detect when the bus is in the quiescent mode (both pull-ups high). In this case, the current is reduced even more: to 100 μA.

LTC1694CS5

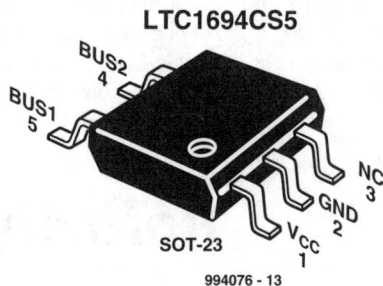

994076 - 13

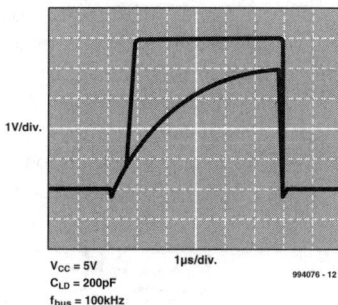

V_{CC} = 5V
C_{LD} = 200pF
f_{bus} = 100kHz

994076 - 12

RS-232 transceiver for portable applications

The ADM101E is a single-channel RS-232 driver and receiver in the Analog Devices Craft Port™ series, designed to operate from a single +5 V supply. A highly efficient charge-pump voltage inverter generates an on-chip –5 V supply, which eliminates the need for a negative power supply for the driver, and permits RS-232 compatible output levels to be developed using charge-pump capacitors as small as 0.1 μF.

A shutdown input disables the charge pump and puts the device into a low-power shutdown mode, in which the current drain is typically less than 5 μA.

An epitaxial BiCMOS construction minimizes power consumption to 3 mW and also guards against latch-up. Overvoltage protection is provided allowing the receiver inputs to withstand continuous voltage in excess of ±30 V. In addition, all pins have ESD protection to levels greater than 2 kV.

The transmitter converts 5 V logic input signals into RS-232-compatible output levels, whose average value is ±4.2 V. The receiver translates EIA-232 signals into 5 V logic levels.

The inputs are provided with 5 kΩ pull-down resistors. The guaranteed switching thresholds are 0.4 V minimum and 2.4 V maximum. The Schmitt trigger inputs have an hysteresis of 0.5 V, which ensures faultless reception in all conditions.

The ADM101E is available in a 10-pin micro-SO package, which makes it ideal for serial communications in small, portable applications, such as palmtop computers and mobile telephones, where a full RS-232 serial interface is not required, but compact size and low power drain are paramount.

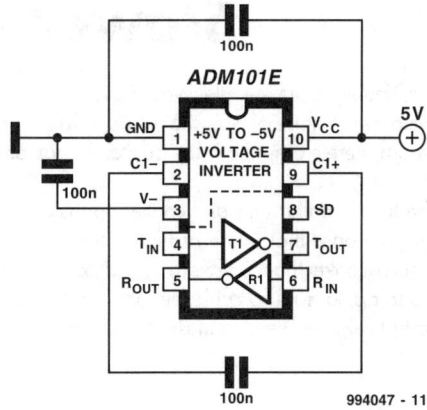

Single-supply precision rectifier

The precision full-wave rectifier circuit shown in the diagram accepts a.c. inputs of up to ±3 V, yet operates from a single +5 V supply voltage. The quiescent current is only 320 µA. Rectifier gain is unity, with the gain accuracy almost entirely dependent on the match between the two resistors $2R_1$. The frequency range is about d.c. to 2 kHz. The single-supply operation at very low quiescent current drain makes the circuit particularly useful for battery-powered equipment.

When the input voltage, V_{IN} is positive, A_1 drives T_1 and D_2 to make output voltage V_O equal to the input voltage. The output swing at V_O is about three diode drops below the supply voltage, so that the peak output voltage is around +3 V. The output of amplifier A_2 goes to negative saturation, which is about +0.8 V; T_2 is then reverse-biased and off.

When the input voltage is negative, the output of A_1 goes into negative saturation so that T_1 is switched off. Amplifier A_2 then serves as a unity-gain inverter. Since V_O is equal to V_{IN} in magnitude but opposite in polarity, V_O will be equal to the absolute value of V_{IN}.

The quiescent current is determined by the the the set current, I_{SET}. With a 5 V supply, the set current is $3.7/R_{SET}$. Slew rate and bandwidth vary directly with the set current.

Amplifier A_1 essentially operates with unity-gain feedback , while A_2 operates with a feedback gain of 0.5. The closed-loop gain-bandwidth is therefore made equal, and the frequency response symmetrical, by making the set current of A_2 twice that of A_1. Amplifier A_2 has a set current of $3.7/2 \times 10^5$, that is,

18.5 µA, and amplifier A_1 has a set current of $3.7/39 \times 10^4$, which is 9.5 µA. These set currents result in quiescent currents of 100 µA for amplifier A_1 and 220 µA for amplifier A_2.

The input stage of the amplifiers is a p-n-p Darlington, so that a negative input voltage can for-

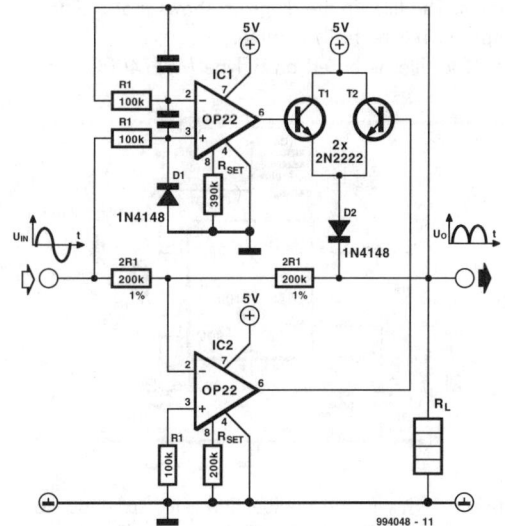

ward-bias the collector-base junction of the input transistor. This potential problem is prevented by adding resistor R_1 and diode D_1 at the A_1 input to limit the negative input voltage.

Slowed-down fan

If, like many other people, you have ever been annoyed by the noise of, say, the extractor fan in your bathroom, here's a tip that may quieten things down a bit.

The fans in bathrooms and cooker hoods are normally small ones that rotate at high speed (but note that many cooker hoods have a speed control). The idea is to displace many cubic feet of air at little cost.

Fortunately, the speed of these fans can be lowered fairly simply by placing a resistor in series with the motor. The impedance, Z, of the fan is calculated from

$$Z = U^2/P,$$

where P is the rating of the fan, and U is the working voltage, normally the mains voltage. If, for instance, the rating of the fan is 33 W, and the mains voltage is 230 V, its impedance is 1600 Ω, give or take an ohm.

The series resistor should have a value of about 1/3 of this value, that is, 470 Ω or 560 Ω. Since it will have to dissipate about 10 W, it is advisable to use two 1 kΩ resistors, rated at 10 W, in parallel. In view of the heat produced in them, it is advisable not to solder them to the motor connections, but to make the connections via a three-way terminal block.

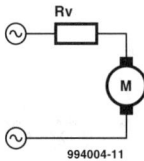

994004-11

Steep-skirted low-pass filter

When considering the design of a filter, one is inclined to think of a combination of inductors, resistors, and capacitors, perhaps in association with an active element. The filter in the diagram shows that a different approach is perfectly feasible.

The filter is based on a Type MAX7400CPA from paramount importance. According to the manufacturers' data sheet this is 82 dB: the frequency response diagram confirms this specification.

The IC may be provided with a clock signal in two different ways. The circuit as shown uses the internal oscillator, whose operating frequency may be adjust-

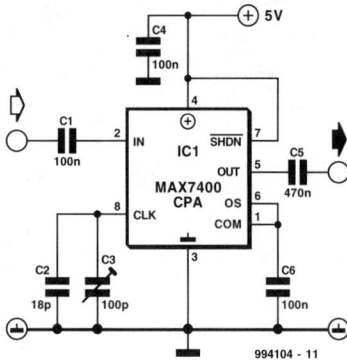

994104 - 11

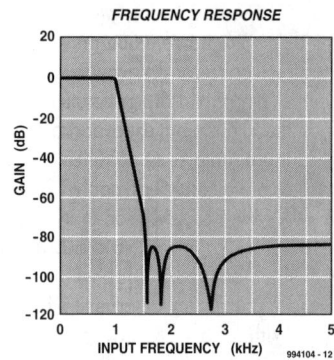

994104 - 12

Maxim. This IC, in conjunction with six capacitors, forms an elliptic, eighth order low-pass filter. It operates from a single +5 V supply from which it draws a current of only 2 mA.

The cut-off frequency may be set between 1 kHz and 10 kHz and depends solely on the clock frequency used. The obtainable attenuation is, of course, of

ed, and cut-off frequency adapted, with trimmer C_3. With component values as specified, the cut-off frequency may be set between 3 kHz and 10 kHz.

If it is necessary for the cut-off frequency to be set accurately, the IC may be provided with a stabilized clock signal via pin 8. The cut-off frequency is then equal to 1/10 of the clock frequency.

Synchronous system to measure μΩ

The circuit in the diagram on the next page uses a synchronous-detection scheme to measure low-level resistances. Other low-resistance-measuring circuits sometimes inject unacceptably large currents into the system on test. The present circuit synchronously demodulates the voltage drop across the system on test and can therefore use very low currents while measuring the resistance.

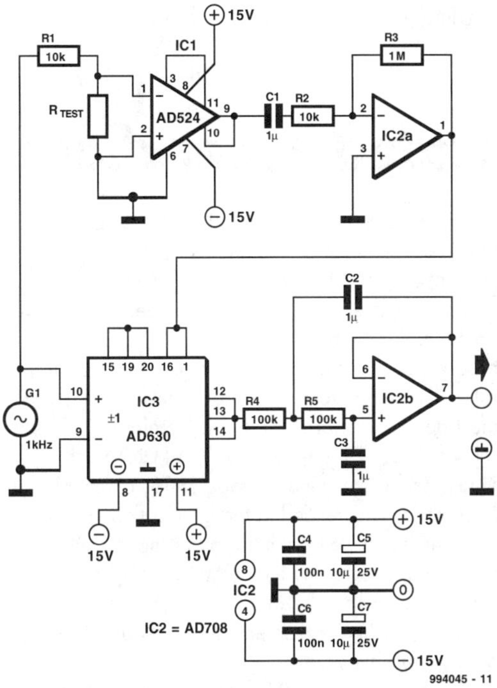

The generator, whose output is a 1 kHz signal at a peak level of 10 V, injects a 1 mA reference current into unknown resistor R_{TEST}. Instrument amplifier IC_1 and precision op amp IC_{2A} amplify the voltage across $R_{TEST} \times 10^5$. Synchronous detector IC_3 demodulates this voltage, which is then applied to low-pass filter IC_{2B}. The low-pass filtering attenuates all uncorrelated disturbances, such as noise, drifts, or offsets, while passing a direct voltage that is proportional to R_{TEST}.

The relationship between the output voltage and R_{TEST} is

$$V_{OUT} = 10 \times 2/\pi \times R_{TEST} \times 10^5/10^3,$$

so that

$$R_{TEST} = 0.0157 \times V_{OUT},$$

which is 15.7 mΩ V^{-1} at the output of the circuit.

994045 - 11

Temperature reference

It is often difficult to properly calibrate a temperature sensor since there is no suitable aid for doing so available. This article, which describes a temperature reference source, aims at putting this right. Since the source is made variable, the reference temperature may also be used for adjusting thermostats correctly. This may prove useful in the case of the recently published Titan 2000 audio power amplifier.

The diagram shows how a Type BDV64 power transistor, T_1, is used to provide a regulated-heat source and a calibrated sensor Type LM35 (IC_2) monitors the resulting temperature. The two devices are mounted on a common heat sink. At the same time, good thermal coupling between IC_2 and the sensor to be calibrated is of paramount importance.

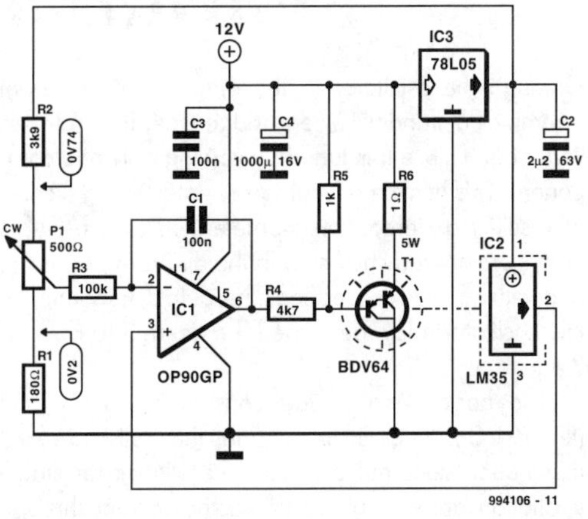

Circuit IC_1 functions as an on/off switch and actu-

994106 - 11

ates the power transistor (heater) when the temperature drops below the set value. The desired temperature is set with potentiometer P_1. The better the thermal coupling, the smaller the hysteresis of the system.

The circuit operates as follows. The output of IC_1 controls power transistor T_1. The specified values of resistors R_4 and R_5 ensure that the current through the transistor is not greater than 0.5 mA. This results in a dissipation of not greater than 6 W.

Sensor IC_2 is powered by a regulated 5 V supply. Its output is a direct voltage of 10 mV $°C^{-1}$. With component values as specified, the temperature may be set with P_1 between +20 °C and +74 °C. Given these data, it is fairly simple to construct a suitable scale for the potentiometer.

Almost any power transistor in a TO3P case and an amplification factor of ≥ 1000 may be used for T_1.

Temperature-compensated zener diode

Only zener diodes rated at 6 V have a negligible temperature coefficient (TC). At lower ratings, the coefficient becomes negative, and at higher ratings, positive. At a rating of 30 V, the coefficient is 0.01% K^{-1}, and remains constant at higher ratings.

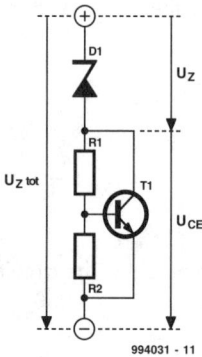

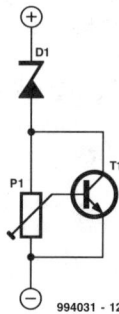

994031 - 11

The present transistor circuit enables a positive TC to be compensated by making use of the negative TC (-2.2 mV K^{-1}) of the base-emitter junction

For instance, an 18 V zener diode has a TC of 16 mV K^{-1}, which is ×7.3 as large as the TC of the base-emitter junction in T_1. This gives a guide to the ratio of the resistors in potential divider R_1-R_2: R_1 must be ×6.3 as large as R_2. If, therefore, R_2 is chosen as 1 kΩ, R_1 needs to be 6.3 kΩ. Therefore, to obtain an overall zener voltage of 18, a zener diode rated at 15 V is needed in the D_1 position.

If the zener voltage itself is not too critical, but it must be unaffected by temperature changes, variable compensation may be obtained as shown in the diagram at the right. The potentiometer should preferably be a 10-turn type.

The transistor in both cases may be general purpose n-p-n type such as the BC238.

Variable oscillator

Although the oscillator in the diagram at first sight resembles a standard Wien bridge type, it is in fact a variant of it since it is tuned by varying only one component. This has the advantage of not needing a carefully selected stereo potentiometer; instead, a single potentiometer may be used. In the diagram, this is P_1, and with values as specified, the output frequency of the oscillator may be varied between 340 Hz and 3.4 kHz.

The basic Wien bridge consists of R_1-C_1 and (R_2+P_1)-C_2. Since in this variant the well-known ×3 attenuation does not occur, the conditions for stable oscillation are met by including the current through

R_2+P_1 in the positive feedback loop. This means that the circuit cannot be based on a single operational amplifier, which is the reason that IC_{1b} has been added to it. Diodes D_1 and D_2 ensure reasonable stability of the signal.

The design requires that the resistance of R_4 is more or less equal to the impedance of network R_5-R_6-R_7P_2-D_1-D_2. Potentiometer P_2 is set to a position which ensures that the level of the output signal is just below that of the supply voltage: the distortion is then a minimum. For best results, it may be worthwhile to experiment with the values of R_5, R_6, and P_2.

Frequency control P_1 may have a linear or logarithmic characteristic, although the latter will normally give more linear control.

The frequency, f, is, at least theoretically, determined by

$$f = 1/2\pi R_1 C_1 \sqrt{a},$$

where $a = (R_2 + P_1)/R_1$.

Note the conditions that $R_1 = R_3$ and $C_1 = C_2$.

The circuit has a drawback in that the frequency is dependent to some degree on the peak value of the signal, which cannot be nullified with the present design. It can, however, by replacing the parallel network D_1-D_2 by a proper stabilization circuit.

The circuit draws a current of about 4 mA without load. When a supply voltage of ± 15 V is used, the peak output signal is 9.4 V r.m.s. With the use of a Type TL072 as specified, the circuit can work from supply voltages as low as ± 5 V.

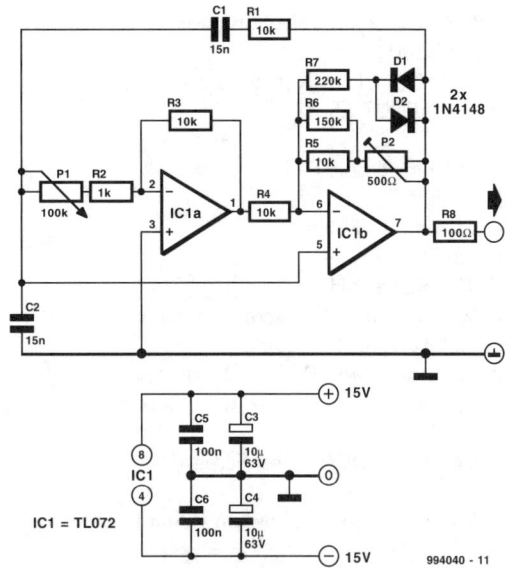

994040 - 11

1-chip LCD interface

The EDE702 Serial LCD Interface IC has been designed to provide a cost-effective LCD control solution for a wide variety of embedded designs. The chip, a firmware-programmed PIC16C54A, allows nearly any text-based LCD module to be controlled via a simple one-wire data link, freeing an additional 6 to 10 I/O lines on your microcontroller or microprocessor system.

Besides full LCD control, the EDE702 also allows the creation of custom characters. Another useful feature is the serial-controlled digital output pin for lighting an indicator LED, driving a sounder, etc.

With a 2400 or 9600 baud rate and selectable serial data polarity, the EDE702 can easily communicate with any device capable of sending asynchronous serial data (including the BASIC Stamp!). Connection to a PC's serial port (RS232) requires only one 33-kW resistor.

The application diagram shows how the EDE702 may be employed as the 'glue' between a PC and an LCD (based on the HD44780 controller). A 4-MHz resonator is used to clock the EDE702. If an external (TTL) oscillator

is used, its output should be connected to only OSC1 (pin 16), while OSC2 (pin 15) is left unconnected. The LCD contrast is set in the usual way with a 10–20 kW preset. The schematic may be easily tested with the aid of a small QBASIC program as shown in the listing. More interesting E-Lab products and datasheets in pdf format may be found on their website at http://www.elabinc.com. E-Lab's UK distributor is Dannell Electronics, http://www.dannell.co.uk.

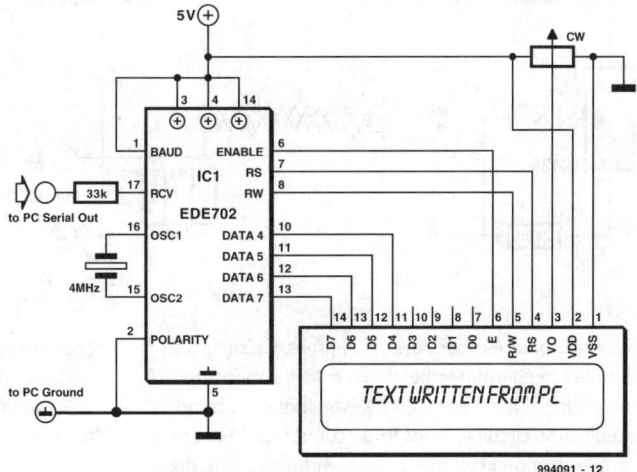

TEXT WRITTEN FROM PC

994091 - 12

REM Open communication channel to COM1 at 9600 Baud
OPEN "com1:9600,n,8,1,cd0,cs0,ds0,op0,rs" FOR OUTPUT AS #1

REM Clear Display
GOSUB 999
OUT &H3F8, &HFE
GOSUB 999
OUT &H3F8, &H1
REM Pause for LCD screen clear command to complete on LCD module
FOR delay=1 to 5000: NEXT delay
REM Write first row of text to LCD screen
GOSUB 999
PRINT #1, "EDE702 Test Screen";

REM Jump to second row on 2 line LCD
GOSUB 999
OUT &H3F8, &HFE
GOSUB 999
OUT &H3F8, &HC0

REM Write second row of text to LCD screen
GOSUB 999
PRINT #1, "Time is: "; TIME$;
END

REM Hold until Transmit Buffer is empty
999 IF (INP(&H3FD) AND &H40) = 0 THEN GOTO 999
RETURN

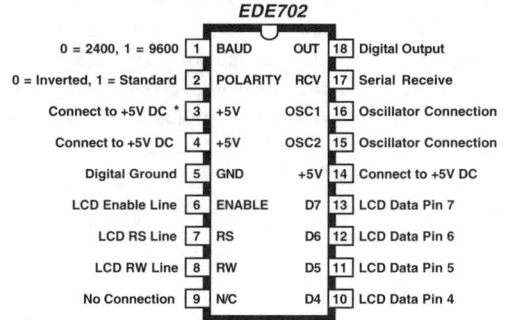

EDE702

0 = 2400, 1 = 9600	1	BAUD	OUT 18	Digital Output
0 = Inverted, 1 = Standard	2	POLARITY	RCV 17	Serial Receive
Connect to +5V DC *	3	+5V	OSC1 16	Oscillator Connection
Connect to +5V DC	4	+5V	OSC2 15	Oscillator Connection
Digital Ground	5	GND	+5V 14	Connect to +5V DC
LCD Enable Line	6	ENABLE	D7 13	LCD Data Pin 7
LCD RS Line	7	RS	D6 12	LCD Data Pin 6
LCD RW Line	8	RW	D5 11	LCD Data Pin 5
No Connection	9	N/C	D4 10	LCD Data Pin 4

994091 - 11

2-wire temperature sensor II

The Type LM35 temperature sensor from National Semiconductor is very popular for two reasons: it produces an output voltage that is directly proportional to

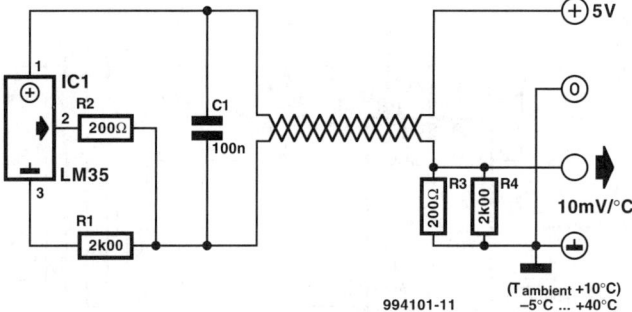

994101-11

($T_{ambient}$ +10°C)
−5°C ... +40°C

10mV/°C

the measured temperature in degrees Celsius, and it enables temperatures below zero to be measured. A drawback of the device is, however, that in its standard application circuit it needs to be connected to the actual measuring circuit via a three-wire link. This draw-

back is neatly negated by the present circuit.

When the LM35 is connected as shown, a two-wire link for the measurement range of −5 °C to +40 °C becomes possible. Actually, the circuit shown is a temperature-dependent current source, since it uses the variation of the quiescent current with changes in temperature. The values of resistors R_3 and R_4 are calculated to give an output voltage of 10 mV °C^{-1}. Where good accuracy is desirable or necessary, 1% resistors should be used. In this context, note that a loss resistance in the link between sensor and measuring circuit may cause a measurement error of about 1 °C for every 5 ohms of resistance. Capacitor C_1 eliminates undesired interference and noise signals.

At an ambient temperature of 25 °C, the circuit draws a current of about 2 mA.

3-frequency oscillator

The output frequency of the oscillator shown in the diagram may be derived via two control inputs, A and B, and may, therefore have three different values. If the logic level at both inputs is low, the oscillator is disabled.

The oscillator proper is formed by gate IC_{1c}. Depending on whether a high logic level is applied to IC_{1a} or IC_{1b}, either network R_1-C_2 or network R_2-C_3 determines the output frequency. If both inputs are high, the output frequency is somewhere between the other two. With values as indicated, the output frequencies are 1300 Hz, 200 Hz, and 2700 Hz.

Branches R_3-D_1 and R_4-D_2 ensure that the pulse duty ratio of the output signal is 1:1. If the oscillator is to be used in applications where this ratio is irrelevant, the two branches may be omitted.

The oscillator is particularly suitable for use in frequency shift keying modulators.

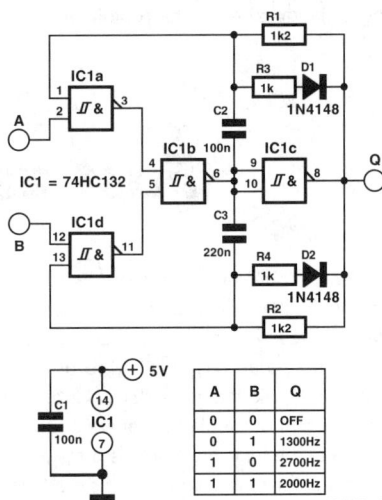

A	B	Q
0	0	OFF
0	1	1300Hz
1	0	2700Hz
1	1	2000Hz

994023 - 11

Constant-current source

The simplest version of a constant-current source, often used for that reason, consists of only a FET with the source and gate connected together. This utilizes the zero-gate-voltage drain current (I_{DSS}) of the FET. Sometimes a source resistor is added to allow the current level to be set more exactly.

A disadvantage of such a very simple current source is that the maximum drain-source voltage of most standard FETs is no more than 30 to 40 V. If we look at bipolar transistors instead, the range of available voltages is significantly larger. This is adequate justification for developing an alternative current source, using bipolar transistors, that is comparably simple.

In the example shown here, we use a BC547 and a small Darlington transistor (BC517). For convenience, we have chosen a current of 1 mA for calculating component values. Transistor T_2 controls the current, and resistor R_1 determines the base-emitter voltage of T_2. R_1 also

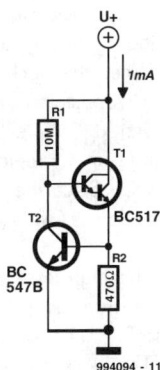

provides the base current for T_1, but the value of R_1 can be made high since T_1 is a high-gain Darlington transistor. The advantage of this is that the resulting error in the current setting is very small. If the value of R_1 is 10 MΩ, the base-emitter voltage of T_2 is less than 0.5 V, so that a current of around 1 mA flows through the current source when R_2 is 470 Ω.

The current regulation that a constant-current source must provide comes from the fact that T_2 controls the base current of T_1. This means that if the current through R_2 should increase, the base current of T_1 would be reduced since the collector current of T_2 would increase. Assuming that the amplification factor of T_2 is at least 10 000, the value of R_1 must be at least 10 MΩ to produce a voltage drop of 1 V. The error current through R_1 is thus negligible in comparison to the total current.

Since the current through R_1 varies with the applied voltage, the base-emitter voltage of T_2 will also vary. A disadvantage of this is that the internal resistance of the current source is reduced. In addition, the temperature dependence of T_2 shows up fully in the output current. For a number of applications, this is however not that important. In fact, this characteristic could be used intentionally, for example to provide temperature compensation or for a particular measurement or control circuit.

In spite of its simplicity, the circuit is capable of delivering a steady current. The prototype provided a measured current at room temperature of 0.91 mA with an input voltage of 5 V; which increased to 0.99 mA at 15 V, and 1.04 mA with a 30 V input.

Fast zener diode

Standard zener diodes are often too slow for application in signal-limiting circuits. If a fast zener is not available, the circuit in the diagram may help.

The standard zener, D_1, is linked to a direct voltage of 10 V via R_1, so that it conducts. Capacitors C_1 and C_2 buffer and decouple the zener voltage.

Diode D_2 at the junction R_1-D_1 is a fast, standard type. If the potential at the anode of this diode is higher than the zener voltage plus U_{D2}, C_1 and C_2 will compensate the inertia of the zener. So, the arrangement simulates a fast zener diode.

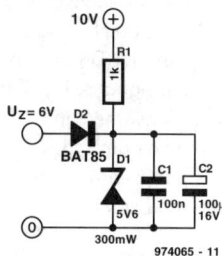

Discrete voltage inverter

The circuit in the diagram enables a negative voltage to be derived without the use of integrated circuits.

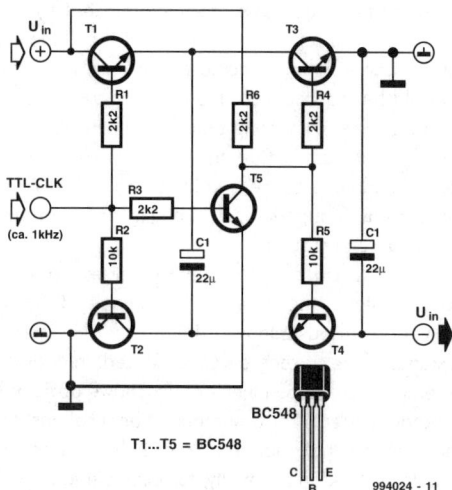

Instead, it uses five n-p-n transistors that are driven by a 1 kHz (approx) TTL clock.

When the clock input is high, transistors T_1 and T_2 link capacitor C_1 to the supply voltage, U_{IN}, which typically is 5 V. During this process, transistor T_5 conducts so that T_3 and T_4 are off.

When the clock input is low, T_5 is cut off, whereupon transistors T_3 and T_4 are switched on via pull-up resistor R_6 and either R_4 or R_5. This results in the charge on C_1 being shared between this capacitor and C_2 Since the +ve terminal of C_2 is at ground potential, its −ve terminal must become negative w.r.t. earth.

The high level at the clock input must be of the same order as the positive input voltage, U_{IN}, otherwise T_1 cannot be switched on.

The clock frequency should be around 1 kHz to ensure a duty cycle ratio of 1:1. Altering the ratio results in a different level of negative output voltage, but this is always smaller than that with a ratio of 1:1.

Fixed-gain line driver OPA3682

The OPA3682 provides an easy-to-use, broadband fixed gain, triple buffer amplifier. Depending on the external connections, the internal resistor network may be used to provide either a fixed gain of +2 video buffer or a gain of ±1 voltage buffer. Operating on a low 6 mA/channel supply current, the device offers a slew rate and output power normally associated with a much higher supply current. The output stage architecture provides high output current with minimal headroom and crossover distortion to give excellent single-supply operation. Operating from a single +5 V supply, the OPA3682 can deliver a 1–4 V output swing with over 100 mA drive current and a bandwidth of 200 MHz. This combination makes the OPA3682 ideal for use as an RGB line driver or a single-supply, triple ADC input driver.

Each amplifier has a dedicated disable pin (3, 6, 16). When the disable function is not used, a decoupling capacitor, C_1–C_3, links the relevant pin to earth. Correct decoupling, as well as faithful adherence to the circuit layout, is important. A good guide is the DEM-OPA368xE evaluation fixture sheet available from the Burr-Brown Corporation or its dealers or at http://www.burr-brown.com/ This sheet also outlines the rea-

sons why all components should be in surface mount technology (SMT).

Resistors R_1–R_3 determine the input impedance, and resistors R_7–R_9, the output impedance.

The quiescent current with all amplifiers enabled is about 18 mA and with the amplifiers disabled, about 900 µA.

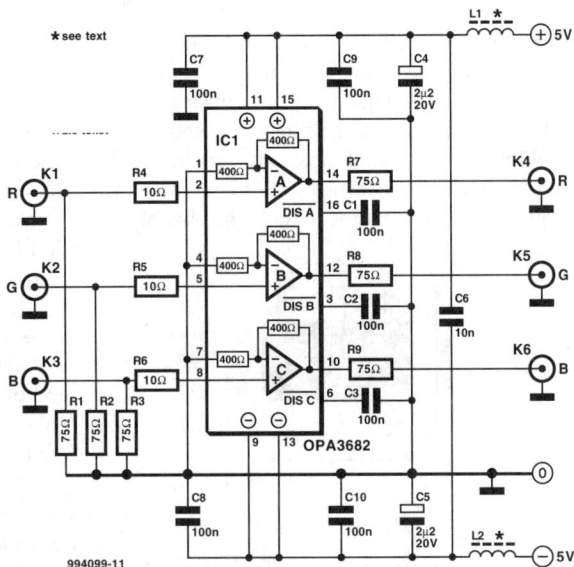

994099-11

Impedance matching network

When r.f. signals are transferred directly from a cable or other output terminal with an impedance Z_1 to a signal input terminal or cable with an impedance Z_2, reflections ensue that cause standing waves. Reflected signals then collide with incoming signals. The consequent superimposition of the two signals causes the resulting signal to be weak at certain points in the cable or network and very strong at others.

The matching network shown in the diagram matches two unequal impedances, provided that Z_1 is greater than Z_2. The table shows a number of frequently encountered values of Z_1 and Z_2 and the requisite resistors, as well as the resulting attenuation. The resistor values are the nearest

standard values in the E-96 series to the computed ones.

The matching of impedances in this manner is wideband and is often used in the test and measurement operations when 75 Ω ad 50 Ω appliances are used.

The resistor values are calculated from

$$R_1 = Z_1 - Z_2 R_2 / (Z_2 + R_2)$$

$$R_2 = Z_2 \sqrt{Z_1} / (Z_1 - Z_2)$$

where Z_1 and Z_2 are as described earlier and their, and the resistor, values are in ohms.

Z1	Z2	R1	R2	Attenuation
75Ω	50Ω	43Ω2	82Ω5	5,7 dB
150Ω	50Ω	121Ω	61Ω9	9,9 dB
300Ω	50Ω	274Ω	51Ω1	13,4 dB
150Ω	75Ω	110Ω	110Ω	7,6 dB
300Ω	75Ω	243Ω	82Ω5	11,4 dB

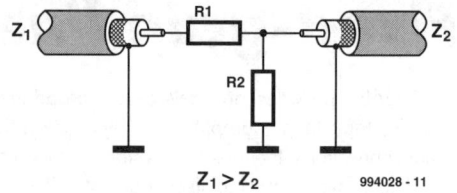

$Z_1 > Z_2$

994028 - 11

Microgate logic

A few years ago, attention was drawn to the AHC (advanced high-speed CMOS) series. Logic circuits in this series are three times as fast as similar circuits in the HC (high-speed CMOS) series. An important benefit of the series is that devices can work from 3.3 V and 5 V supplies. Subsequnetly, Texas Instruments and Philips Semiconductors brought out a number of new,

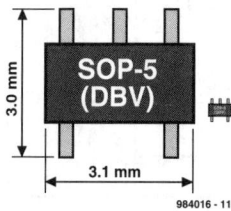

SOP-5
(DBV)

3.0 mm

3.1 mm

984016 - 11

single-gate logic, devices in the AHC series. As the name indicates, these devices provide only one gate and not four or six as had become usual. Single-gate logic circuits are housed in SOP-5 5-pin cases, which measure only 3×3.1 mm (quad-gate circuits housed in TSSOP 14-pin cases occupy 33.66 mm^2).

The AHC devices are marketed under the name Microgate Logic. Although they are not revolutionary, they offer several real advantages. Designers no longer need to worry about gates not being used. Also, it becomes possible to locate the gate at a more suitable space on a board than quad gates. All this makes the devices more functional and more compact, which improves their EMC properties. The placing of quad-gate or six-gate devices on a board is almost invariably a compromise which makes signal lines unnecessarily long.

Microgate Logic devices are distinguished from normal AGC chips by the addition of 1G to the type number. For instance, a Type AHC1G00 is a single-gate logic 74AHC00, that is, a single NAND gate.

The AHC series is further extended by much larger devices to meet the requirements of 16-bit or 32-bit wide buses. These chips have more gates than before, and are marketed as Widebus™ circuits. For instance, a Type 74AHC16244 is a dual 244, that is, a 16-bit data buffer.

More information on these new circuits may be found on Internet page www.ti.com/-sc/docs/as1/lit.htm.

Memory change-over tip

When the contents of two existing memory address have to be interchanged for one reason or an other, there is usually a need for an additional address or variable:

MOV dummy,var1
MOV var1,var2
MOV var2,dummy

This dummy variable is not always necessary:
XOR var1,.var2

XOR var2,var1
XOR var1,var2

This tip may well be of use when the memory space is limited. It may also be used with higher programming languages to save having to declare an additional variable.

Oven-controlled temperature stability

Accurate measurements that are not affected by ambient temperature may be taken when the part or circuit being tested is placed in an oven whose inside temperature is held constant after a short warming-up period. This works very well, indeed, when the temperature inside the oven is held higher than the maximum ambient temperature. This is because the inside of the oven may be heated but cannot be cooled. This type of control is frequently used to stabilize a crystal oscillator or a surface acoustic wave (SAW) filter.

The circuit consists of a heating element, R_{14}, which is thermally coupled to temperature sensor IC_3. The output of the sensor at 75 °C is +3.48 V, which rises linearly at a rate of 10 mV °C^{-1}.

Integrated circuit IC_1 comprises an operational amplifier (at pins 5, 6 and 7) and a comparator (at pins 1, 2 and 3). The op amp is arranged as a ×100 amplifier and delivers an error voltage that depends on the difference between the actual temperature and that set with P_1. The preset can set the wanted temperature between +55 °C and +105 °C. The stable voltage across potential divider R_5-P_1-R_6 is 4.096 V, which is provided by reference voltage source IC_2.

The error voltage across R_7 is applied to the oscillator which is based on the comparator in IC_1. It alters the duty factor of the oscillator output in such a way that when the temperature drops, transistor T_1, and thus heating element R_{14}, remains on a little longer than when the temperature is stable.

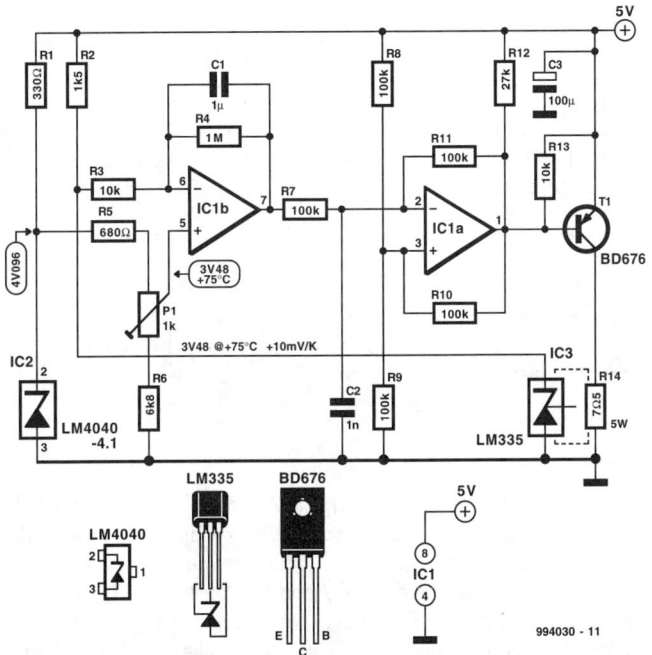

994030 - 11

PAL timing I

In the PAL television system, the CCIR B and G standards specify that the colour carrier is directly coupled to the line rate, with a 25 Hz offset. The frequency ratio and offset are chosen to suppress interference patterns, according to the formula

$$f_{colour}=283.75\,f_{line}+25\ Hz$$

At a line rate of 15 625 Hz, this means that the PAL colour carrier frequency is 4.43361875 MHz. Single-sideband modulation is frequently used to obtain the correct relationship with the line rate. For example, the frequency of a crystal oscillator can be offset by 25 Hz, divided by 1135 and then multiplied by 8 to obtain twice the actual line rate. This is a rather complicated procedure, which we think could be made a lot simpler.

There is a fixed ratio between the 25 Hz frame rate and four times the colour carrier frequency: four times the colour carrier frequency is exactly equal to 709 379 times the frame rate. An obvious approach is to use a crystal oscillator running at four times the colour carrier frequency and divide its output by 709 379 to obtain the frame rate. The line rate can then be derived from the frame rate with the help of a

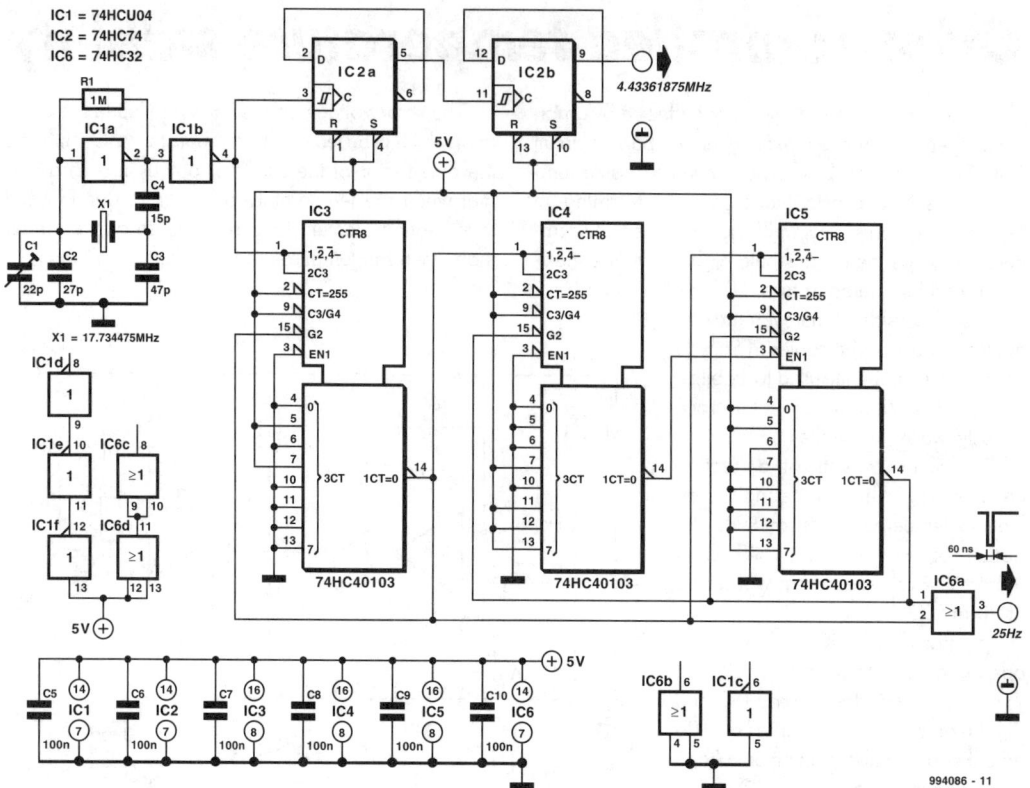

IC1 = 74HCU04
IC2 = 74HC74
IC6 = 74HC32

4.43361875MHz

X1 = 17.734475MHz

60 ns

25Hz

994086 - 11

PLL circuit (see PAL timing II on page 182.

The crystal oscillator is a standard Pierce configuration with a trimmer capacitor, built around a 74HCU04 (IC1). The values of C_2 and C_3 must be chosen to obtain the specified load capacitance for the crystal. An incorrect value of C_{load} can make it impossible to adjust the oscillator to the exact frequency. Two D-type bistables wired as binary scalers are used to obtain the colour carrier frequency.

Four ICs are used for the division needed to obtain the frame rate signal. IC_3, IC_4 and IC_5 are presettable synchronous down-counters (type 74HC40103) that are very well suited for timing and frequency-division applications. The requisite divisor is split into two factors: 11 and 64 489. The 74HC40103 works as an

$(1 + n)$ divider, so a value 10 is applied to the preset inputs of IC_3 for the first factor. The second factor is obtained by wiring IC_4 and IC_5 as a synchronous 16-bit divider, with the output of IC_5 fed back to both synchronous preset inputs. The preset value is again 1 lower than the divisor.

A disadvantage of the 74HC40103 is that glitches can occur, due to differences in internal delay times. These glitches are eliminated by clocking an OR gate (IC_{6a}) at the divider output with the divider input signal. The 25 Hz output signal has an active low pulse approximately 60 ns long, which is essentially equal to one period of the crystal oscillator.

The circuit draws a current of just over 12 mA, primarily due to IC_1.

PAL timing II

This design is complementary to that described in the preceding article (PAL timing I). It is intended to derive a line rate signal from a television frame rate signal, using a PLL. Naturally, this technique can also be used in situations where the line synchronization pulses are corrupted.

In the PAL television system, there are 625 lines per frame. In the PLL circuit, a nominal frequency of 15 625 Hz is divided by 625 and the result is compared to the 25 Hz input signal. A 74HC4040 (IC$_2$) is used for the divider. The correct divisor is obtained with the help of an AND circuit formed by several diodes, which produces the counter reset signal (625 decimal = 1001110001 binary).

The well-known HC version of the 4046 IC has been chosen for the PLL. HC logic must be used here to keep up with the fast pulse from the output of the 'PAL timing (1)' circuit. Since phase comparator 2 is used, the inputs are edge triggered, and no further requirements need be placed on the input signals.

As can be seen, the internal oscillator of the PLL IC is also used (pin 9). The necessary low-pass filter R$_3$-C$_2$ is not exactly in accord with the requisite formulas, but this version proved to yield the least jitter in practical tests. This points to a weakness of this cir-

cuit. With a normal RC oscillator, as used here, it is not possible to reduce the jitter of the 15 625 Hz signal to below about 200 ns. For most applications, this is not good enough, in which case there is no way of avoiding the use of an external crystal oscillator for the VCO in combination with a suitable divider.

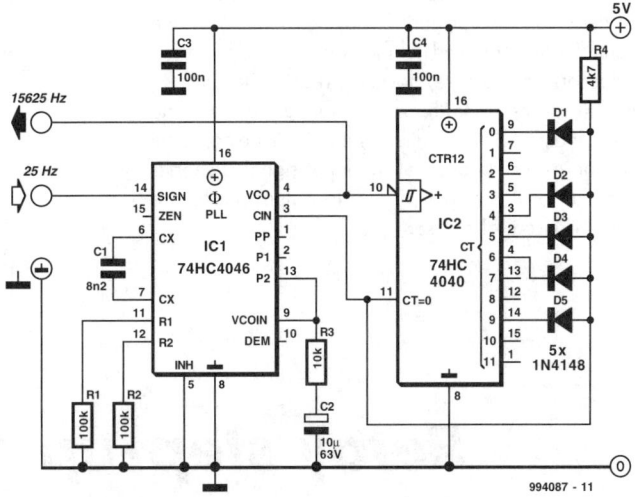

994087 - 11

Pulse doubler

The design of frequency and pulse doubling circuits is normally complex and critical. The circuit in Figure 1 is a pleasant exception to this. It is based on a standard monostable multivibrator (MMV) and produces an output pulse for both the leading and the trailing edge of the input pulse. The duration of the output pulses is determined by the time constant R$_3$-C$_3$.

The TTL input signal is linked to both the +T (positive pulse) and the –T (negative pulse) inputs IC$_{1a}$ via capacitors C$_1$ and C$_2$ respectively. The two inputs cannot be active simultaneously. This means that at the leading edge of the input pulse, the negative input must be high, and at the trailing edge, the positive input must be low.

Since IC$_{1a}$ is a retriggerable MMV, each output pulse is stretched by the time constant R$_3$-C$_3$. The linking of both outputs with the relevant input via resistor R$_1$ and R$_2$ respectively ensures that a quiescent trigger input is at a non-active level. But, because the other trigger input is then at an active level for the duration of the pulse, the design ensures that the MMV is not

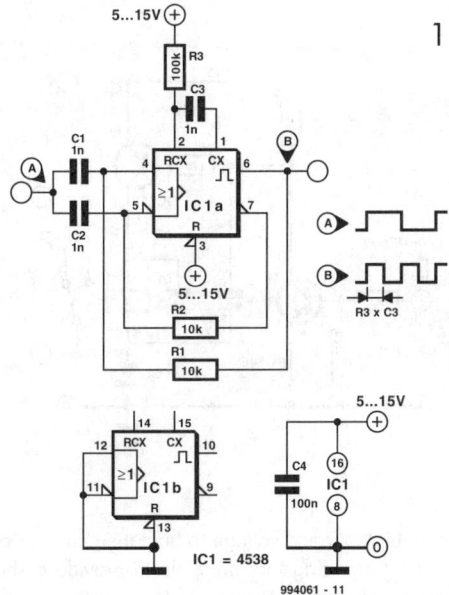

IC1 = 4538

994061 - 11

retriggerable. If the width of the input pulse is shorter than the width of the output pulse (determined by R_3-C_3), only one output pulse will be produced, since the MMV can be retriggered only when the output pulse has terminated. It should therefore be ensured that the time constants R_1-C_1 and R_2-C_2 are shorter than R_3-C_3 at all times.

If a retriggerable version of the circuit is desired, the design shown in Figure 2 may be used. Again, the input signal is applied to the two trigger inputs via capacitors. In this case, however, resistors R_1 and R_2 are linked directly to the supply voltage to make the inputs inactive. This results in the outputs being active as soon as the duration of the input signal is shorter than that of the output pulse. No output pulses are then produced, of course.

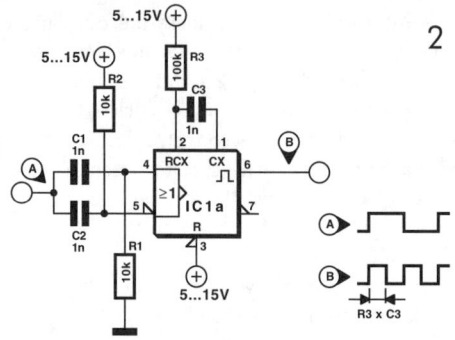

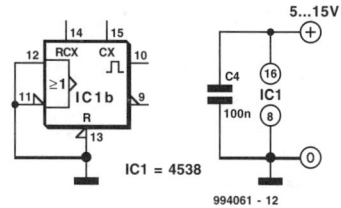

Relay step-up circuits

The circuits in the diagrams allow 12-volt relays to be operated from 6 V or 9 V, or 24-volt relays from 12 V. While most normal relays require the manufacturer-specified coil voltage to reliably pull the contacts together, once the contacts are together you only need

about half that rated voltage to hold them in. This circuit works by using that principle to provide a short burst of twice the supply voltage to move the contacts

and then applies the available 6 or 9 volts to the relay to lock the contacts in place.

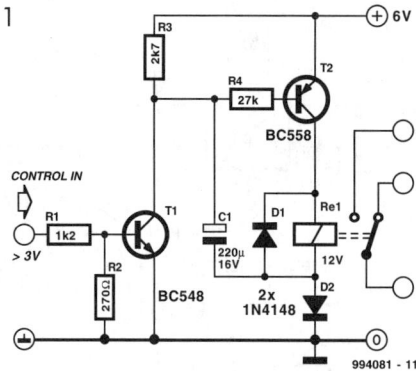

With reference to Figure 1, when the main supply is applied to the circuit, the $220\,\mu F$ capacitor, C_1, charges quickly to +6 V via resistor R_3. The circuit is now awaiting voltage on the control input. When a control voltage (which may be as low as 3 V) is applied to the control input, transistor T_1 switches on. The other transistor, a BC558, is also switched on. This allows connection of the relay coil to the main supply rail while T_1 short-circuits the positive terminal of the C_1 to ground. The negative terminal of the capacitor is then at a potential of –6 V, which is applied to the other side of the relay coil. The relay coil potential is then briefly 12 V — enough to actuate the contact(s). However, the coil voltage drops to the supply voltage fairly quickly. The period is determined by the RC time constant of the relay coil resistance and C_1.

While this circuit is simple and works well in many situations, it has a few weaknesses in its current form. The relay may remain energized for as long as one second after the control input has fallen. Also, if the control input goes high before the capacitor has fully recharged, it may not have enough energy to control the relay reliably. Also, the voltage drop across the diode limits the voltage to about 10.8 volts. The more

2

994081 - 12

complex version of the circuit shown in Figure 2 fixes these problems by using an extra transistor and diode. In this arrangement, the BC558 is now isolated from the recharge current of the capacitor. The new transistor provides fast charging for the capacitor. Charging is completed within the mechanical response time of the relay.

When using these circuits it should be noted that the contact pressure of the relay contacts may be a little lower than with the nominal coil voltage. It is therefore advisable to keep contact currents well below the maximum specified value.

Stroboscope filter

To drive a stroboscope from an audio signal, the signal must first be reduced to its low-frequency component. This can be done with the circuit presented here. Coupling capacitors C_4 and C_5, for the left- and right-hand channels respectively, prevent any direct voltages in the audio signals from reaching transistor buffer stage T_1. The buffered audio signal is applied to an active second-order low-pass filter, whose upper frequency limit can be adjusted over the range of approximately 80–170 Hz using the stereo potentiometer R_{14}-R_{15}. Good tracking between the two halves of this potentiometer is a basic requirement for proper operation of the filter. Minimum distortion of the low-frequency component of the input signal is not that important, since all all that is needed is a signal that is suitable for triggering the stroboscope circuit. This is achieved by applying the signal to a comparator with an adjustable reference potential. The trigger level of the circuit can be varied with R_4. When the signal amplitude is high enough, a pulse signal appears at the output of IC_{2a}.

An optoisolator is essential to provide electric isolation between the filter and the stroboscope. Normaly, this is present at the input of the stroboscope, so all that is necessary is to reduce the output voltage by 3 V (with D_1) and provide a current-limiting resistor (R_{13}) at the filter output.

Standard inexpensive op amps, such as the TL082, can be used for the filter. With a simple +12 V supply and component values as specified, the current drawn by the filter circuit is around 4.3 mA.

994008-11

111

Thermostat I

The MAX6501–MAX6504 integrated circuits contain a thermostat with fixed temperature thresholds. Available are thresholds of –15 °C, +5 °C, +45 °C, +55 °C, +65 °C, +75 °C, +85 °C, and +95 °C. The MAX6501 and MAX6502 are for use over the temperature range +35 °C to +115 °C, while the MAX6503 and MAX6504 can be programmed from –45 °C to +15 °C.

The tiny circuit, terminated in a 5-pin SOT-23-SMD case is ideally suited for building into an existing appliance. The supply voltage may be 2.7–5.5 V. The HYST input allows a hysteresis of 2 °C (HYST=GND) or 10 °C (HYST=V_{CC}) to be selected.

The difference between the four ICs is only in their output configuration. The MAX6501 and MAX6503 have open-drain outputs that need a pull-up resistor, whereas the MAX6502 and MAX6504 have a push-pull output that can gate GND or V_{CC}. At high temperatures, the MAX6501 and MAX6504 gate to ground, whereas the other two link the output to V_{CC}.

Further information about these useful circuits may be had from http://www.maxim-ic.com

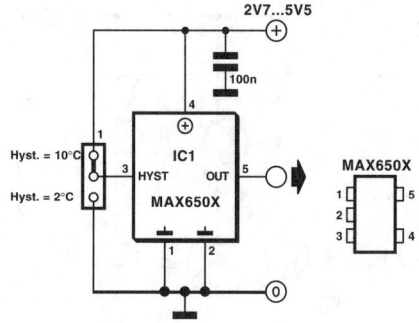

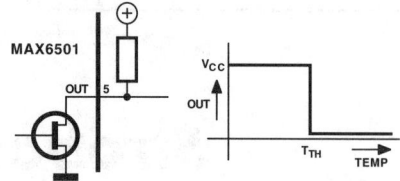

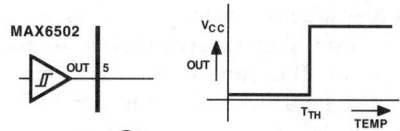

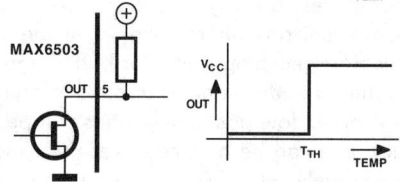

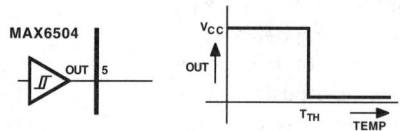

Thermostat II

The Type AD22105 from Analog Devices is an integrated circuit that contains a temperature sensor, a threshold comparator with hysteresis and an output stage. A single external resistor, R_{SET} allows setting the tripping threshold accurately anywhere between –40 °C and +150 °C. The value of R_{SET} is calculated with

$$R_{SET}=39\times10^{6}/(t_{SET}+281.6)–90.3\times10^{3},$$

where t_{SET} is the numerical value of the trip tempera-

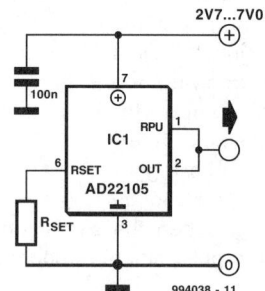

ture and R_{SET} is in ohms. This gives values of R_{SET} of, for instance, 47.5 kΩ for a trip temperature of 0 °C, 36 kΩ for 25 °C, and 12 kΩ for 100 °C.

The internal comparator trips when the ambient temperature measured by the sensor exceeds the set limit. The maximum error is ±2 °C at 25 °C and ±3 °C over the entire temperature range. The hysteresis, which prevents rapid operation of the comparator, is set at 4 °C during production.

The AD22105 needs a supply voltage of 2.7–7.0 V. Since the dissipation at 3.3 V is only 230 μW, the error caused by this is negligible. The low dissipation makes the device ideally suited for battery operation.

The output stage is an open-collector n-p-n transistor with the emitter at ground potential. A pull-up resistor of 200 kΩ may be connected between pin 1 and pin 7. The transistor comes on when the ambient temperature exceeds the set limit. The output may be connected directly to low-current LEDs and CMOS inputs.

The AD22105 is housed in a SO8 case.

Further information may be had from http://www.analog.com

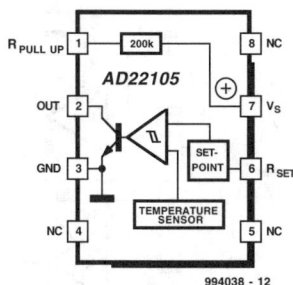

994038 - 12

Supply voltage monitor

A circuit for monitoring supply voltages of ±5 V and ±12 V is readily constructed as shown in the diagram. It is appreciably simpler than the usual monitors that use comparators, and AND gates. The circuit is not intended to indicate the level of the inputs.

In normal operation, transistors T_1 and T_3 must be seen as current sources. The drop across resistors R_1 and R_2 is 6.3 V (12–5–0.7). This means that the current is 6.3 mA and this flows through diode D_1 when all four voltages are present. However, if for instance, the –5 V line fails, transistor T_3 remains on but the base-emitter junction of T_2 is no longer biased, so that this transistor is cut off. When this happens, there is no current through D_1 which then goes out.

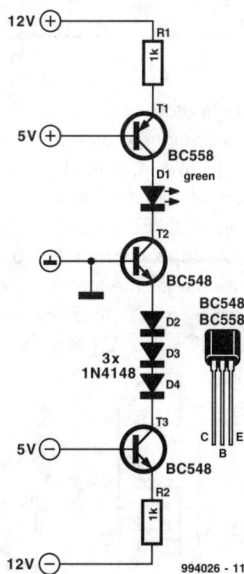

994026 - 11

Voltage boosting with inverters

A 'tree' of inverters is highly suitable for boosting a voltage. That in the diagram provides voltages that are whole multiples of the input voltage by clock-driven charging of capacitors. The voltage across the capacitors is added stage by stage to the input voltage. Here, integrated circuits, IC_2–IC_7, each containing six inverters are used. In each IC, except IC_2, one of the inverters is connected in series with the parallel

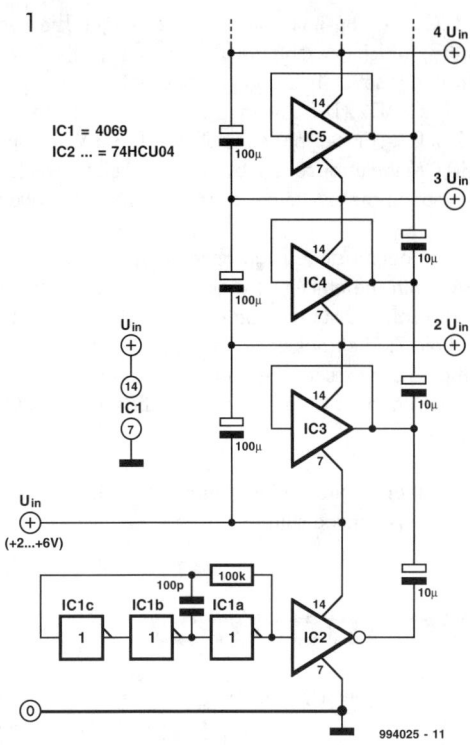

1

IC1 = 4069
IC2 ... = 74HCU04

994025 - 11

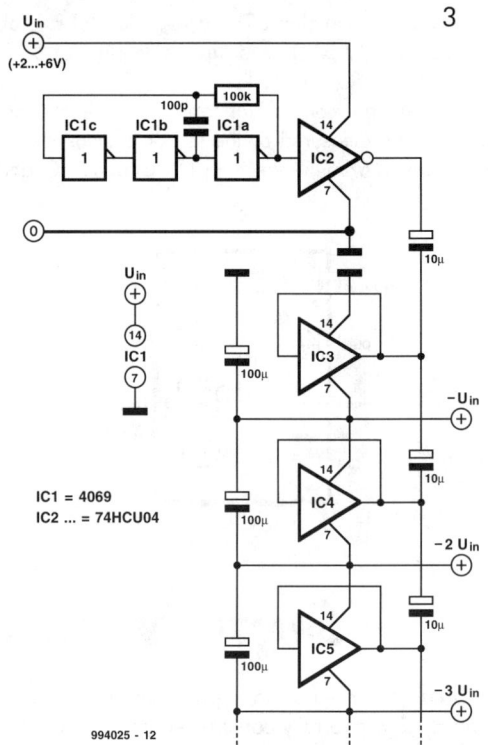

3

IC1 = 4069
IC2 ... = 74HCU04

994025 - 12

combination of the other five.

Circuit IC$_1$ is configured as a 50 Hz oscillator that controls inverting driver IC$_2$. A 10 μF capacitor interlinks the outputs of IC$_2$ and IC$_3$. The bidirectional properties of the MOSFET output of IC$_3$ ensure that the voltage across supply terminals 7 and 14 is identical to the input voltage. The increased voltage, equal to 2V$_{IN}$, is filtered by a 100 μF capacitor.

The outputs of IC$_3$ and IC$_4$ are also interlinked by

clock frequency is lower than 50 Hz, but then the available output current drops. If a current of 5 mA is drawn from the terminal at which 3V$_{IN}$ appears, the efficiency is about 90%. However, it drops to around 75% when the current is increased to 15 mA.

The circuit can also be arranged as a voltage-inverting booster. The ICs should then be arranged as shown in Figure 3.

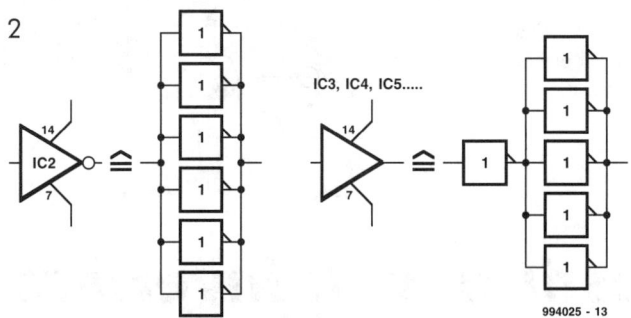

2

IC3, IC4, IC5.....

994025 - 13

a 10 μF capacitor, and a similar process as just described takes place. This continues up to the last IC on the tree.

The efficiency of the booster increases when the

3
Computers &
Microprocessors

ADC for Centronics port

The conversion of analogue signals, such as the output of a temperature sensor, into a digital code remains a challenge for many computer users. The analogue-to-digital converter (ADC) shown may be of help. It uses only a few components and a simple program in BASIC It is intended for use with PCs only.

The 8-bit converter IC is a Type TLC549 from Texas Instruments (IC$_2$).

A REF02 (IC$_1$) is used as reference source and supply regulator. It operates from 8–30 V and provides 5 V.

The signal to be quantized may have a level of 0–5 V and is applied to pin 2 of IC$_2$. A trailing edge applied to pin 5 of the IC then results in the signal being digitized. The MSB appears at pin 6.

Next, eight clock pulses are applied to the I/O-CLK input to shift all eight bits out of the converter. After the eighth clock pulse, the next cycle is started by setting a trailing edge at the /CS pin. To ensure correct conversion this line must have been at least 1.7 μs high. If BASIC is used, this condition is normally met automatically.

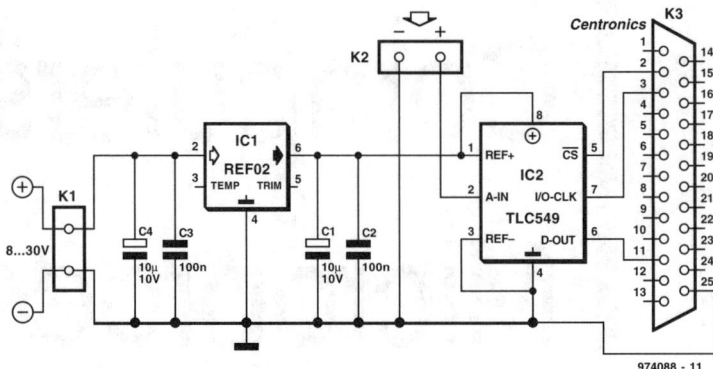

974088 - 11

The BASIC program below shows how this language may be used to quantize an analogue signal. Note that with a trick on line 240 the ever-present noise may be limited.

The converter draws a current of about 5 mA average.

```
10 Basis = 888:          REM Basis LPT1 (for LPT2: 632)
20 Delay = 1             REM conversion delay time
30 Average = 10          REM average cycles
40 CLS
50 Value = 0
60 FOR t = 1 TO Average
70 OUT (Basis), 0:       REM CS, I/O-CLK low
80 OUT (Basis), 1:       REM CS high, start conversion
90 FOR q = 1 TO Delay    REM wait-state for conversion time
100   NEXT q
120   OUT (Basis), 0:    REM CS+CLK low
130   OUT (Basis), 0:    REM CLK high
140   OUT (Basis), 0:    REM CLK low
150   FOR i = 1 TO 7:    REM write bit 7-0
160   x = INP (Basis + 1) AND 128:   REM read and discriminate input bit
170   IF X = 128 THEN a = 0
180   IF x = 0 THEN a = 1
190   Value = Value + a * 2 ^ (7 – i)   REM constitute number
200   OUT (Basis), 2:    REM CLK high
210   OUT (Basis), 0:    REM CLK low
220   NEXT i
230   NEXT t
```

116

```
240   Value = Value * 5 / (255 * Average):      REM mean value of "Average" numbers and
      conversion
250   LOCATE 10, 10:                            REM to measuring range (0–5 V)
260   PRINT USING "#.### Volt"; Value
270   GOTO 50
```

Inexpensive isolator for RS232

The isolator is intended to provide electrical isolation between a computer and the equipment connected to its serial port. For instance, users of the BASIC Stamp want to link the microcontroller to electrical loads only if there is no risk of damage to the PC. In such cases, the isolator described in this article may be of help.

Connector K_1 is linked to the serial port of the PC and derives from one of the lines, here the TxD line, a symmetrical supply voltage. The DTR and RTS lines may also be used, provided they are regularly switched between a positive and a negative voltage.

The other side of the isolator carries TTL levels. This side is powered by a low voltage. Since most micro-controller circuits have their own power supply, power-ing a few more gates should not present any diffi-culties.

The potential at the TxD line is converted into a sym-metrical direct voltage of ±6.8 V by D_5 and D_6. This voltage is used to supply IC_1.

The TxD signal is also applied to the LED in IC_2. Diode D_7 prevents the LED being damaged by a nega-tive input voltage.

The LED in the optoiso-lator will flash in rhythm with the applied data, while the digital code appears at pin 6 of the IC.

Buffers IC_{3b} and IC_{3c} magnify the digital signal to full TTL level.

The send signal of the microcontroller system is applied to optoisolator IC_1 via IC_{3a} and, after optical transfer, also appears at pin 6. There, it switches between ±6.8 V, a swing large enough to drive a stan-dard RS232 link.

RS232-driven shift register

The shift register is eminently suitable for driving several outputs via a two-wire RS232 connection. For instance, it may prove useful when all the gates of a microcontroller are occupied.

The RS232 interface must be set to 9600 baud, no parity, eight data bits, one stop bit. So, to send a bit via the RS232 bus, a data block of ten bits (eight data bits plus one stop bit plus one start bit) must be sent.

A logic 1 is sent as FF_{HEX}, which is eight ones and a logic low is sent as 00_{HEX}, that is eight zeros.

In the quiescent state, the output of the RS232 interface is −12 V. A logic 1 is represented by −12 V and a logic 0 by +12 V.

The internal protection diodes in IC_1, in conjunction with R_1, limit the input voltage to about 600 mV.

The DATA line (pin 2 of IC_2) carries the same signal as the RS232 bus, but converted to 0 V (logic low) and +5 V (logic 1).

A leading edge at the input of the circuit, such as that of the start bit of a new data block, causes a positive pulse at the input of IC_{1b}, which enables the Schmitt trigger. Capacitor C_3 is then discharged via D_1. During the discharge time, the output of IC_{1c} is high. As soon as C_3 is discharged, the output of IC_{1b} goes high, whereupon C_3 is charged again via R_3. After about 530 μs, the potential across C_3 is high enough to trigger IC_{1c}, whereupon the output changes from 1 to 0. The consequent trailing edge causes the input of IC_{1e}, which is normally held

high by R_5, to become low. This in turn results in a leading edge at the CLK input of IC_2, enabling the information on the DATA line to be written.

The lower branch in the diagram, IC_{1d} and IC_{1f}, functions in a similar manner, but the time constant

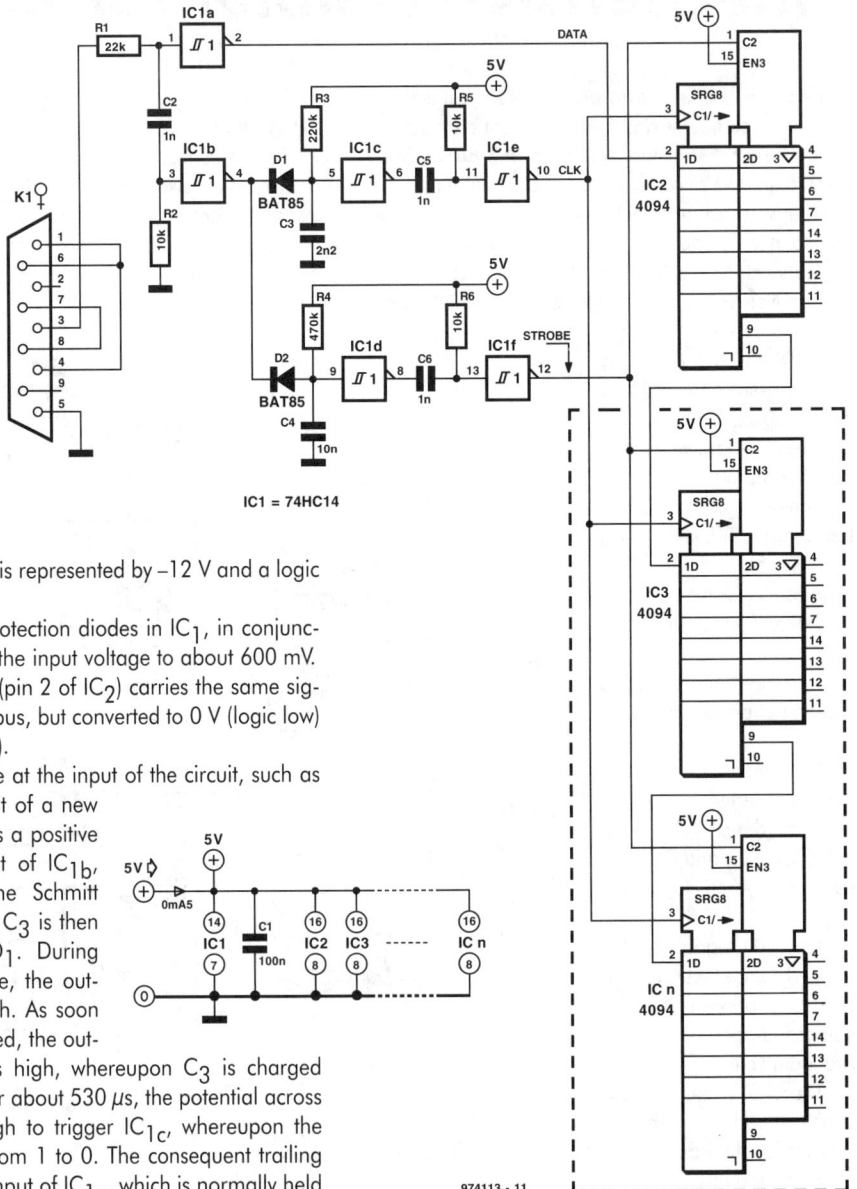

IC1 = 74HC14

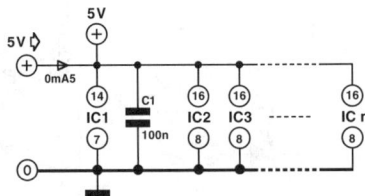

974113 - 11

118

R_4-C_4 is about ten times as long. When no signal is sent over the RS232 line for 5.16 ms, the STROBE signal becomes active, whereupon the data in the shift register is latched to the output.

As shown in the diagram on the following page, the circuit may be expanded by linking the carry out of IC_2 to a second (and subsequent) register (IC_3, IC_4, ...).

Speedometer for CPU ventilator

This circuit is based on an unusual concept, although it employs a 'classic' component, the 555, which is used in monostable mode (also called one-shot).

Both the fan motor and the 555 are powered via a 10-Ω series resistor, R_3. Note that pin 1 of the 555 is open-circuited instead of tied to ground as usual.

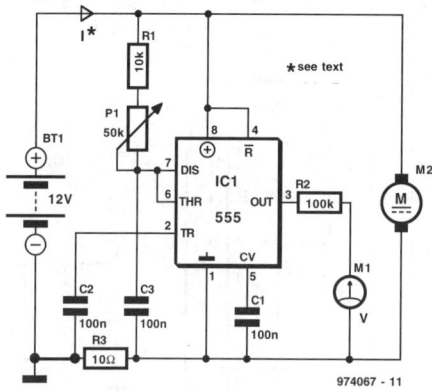

974067 - 11

In any d.c. motor, commutation (or current transfer) is required by electromagnets in order to build an alternating or rotating electrical field. The commutation process may be mechanical (brush and collector) as in many small motors used in toys, or electronic, based on a small circuit inside the motor housing. The latter solution is frequently used in CPU fans and cassette recorder motors. In most cases, commutation should occur at least every 180°, that is, twice during every full rotation of the motor.

The data sheets of the 555 tell us that the trigger input of the chip, pin 2, is at one-third of the supply voltage.

The commutation by the electromagnets in the motor produces a negative voltage peak of a few tenths of a volt across R_3. Whereas the internal reference voltage of the 555 will faithfully track such a negative pulse, the voltage at the trigger input remains virtually constant (with respect to capacitor C_2 which is connected to the 'true' ground). In case a larger current flows through R_3, the reference voltage is temporarily 'lifted', which means that the voltage at the trigger input (=C_2) drops briefly below the half-supply level presented by the CV (control voltage) input. The result is that the 555 is triggered.

Because of the pulse train produced by the monostable, the output voltage of the 555 will be proportional to the motor speed. The 'speedometer' readout is a small moving-coil meter with an f.s.d. current of 120 μA and an internal resistance of 750 Ω. If a meter with different specifications is used, the value of R_2 needs to be altered as appropriate.

It should be noted that this circuit will not work with motors having an internal noise suppression capacitor. If it is necessary to use such a capacitor, it should be fitted across the electrical unit formed by the circuit and the motor, i.e., not just across the motor. In that case, keep the wires connecting the motor to the circuit as short as possible, and use an electrolytic capacitor of, say, 100 μF/25 V, connected across the moving coil meter.

Switch box for PCs

Although the personal computer was developed for office automation, it is an undeniable fact that a large proportion of the 200 million or so in use worldwide are employed for games and other entertainment only. In these applications, the joystick plays a particularly important role.

The 15-pole D-connector in the diagram on the opposite page has provision for connecting two joysticks and a MIDI system (but this only if the connector is available on the sound card).

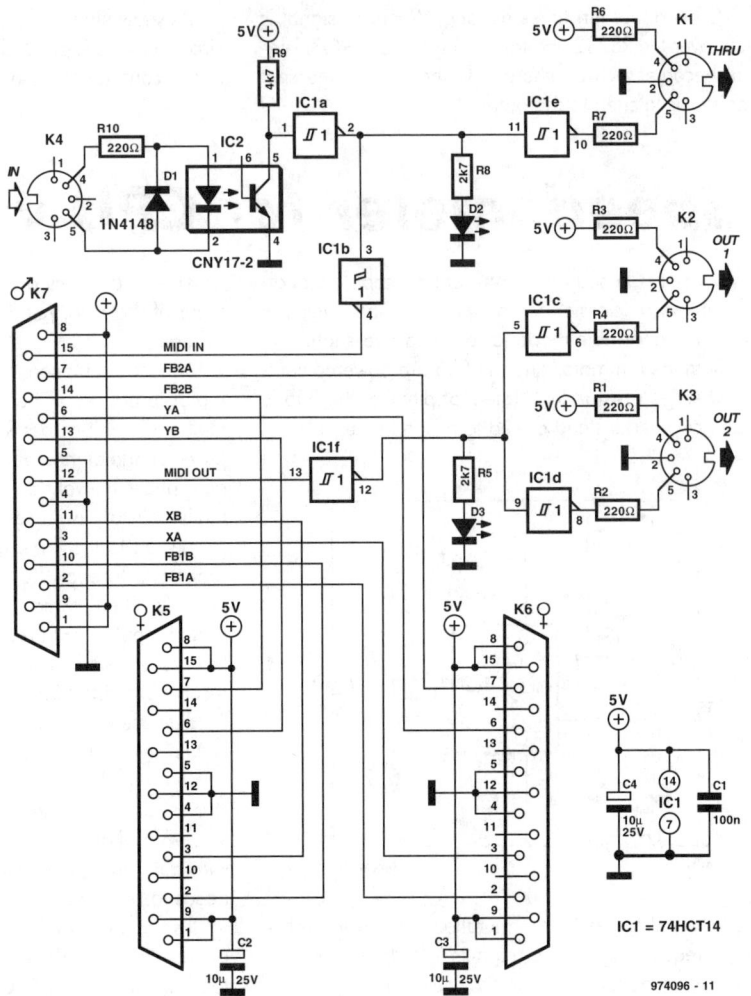

974096 - 11

Adaptor board for 18-pin PICs

Microchip's PICSTART-16B1 kit comes with two PIC controllers: an 18-pin PIC16C71 and a 28-pin PIC16C57. Both controllers have an internal EPROM, with the obvious disadvantage that the time needed to erase them is rather long.

The adapter allows a 16C71 to be simulated with a larger PIC device, the 28-pin PIC16C57. In this way, developing a PIC code can be continued without too much of a delay before the UV eraser has done the necessary.

The circuit is connected to a target system by a flat-cable and an 18-pin DIL plug which is inserted in the socket normally occupied by a PIC16C71. Although the present board also offers a crystal and ancillaries to operate the PIC's internal oscillator, these parts will seldom be required as they will be available on the target system board in most cases. For oscillator configurations, consult the PIC's datasheets.

If it is desired to fit 18-pin 16C71's on the present board, two options are available. The first is to fit an

18-pin narrow-DIL ZIF socket, which is not only expensive, but also hard to come by. A low-cost alternative is to mount an 18-pin socket with turned pins on the board in which another, inexpensive, socket is inserted, and fit the PIC in to this.

The board is connected to the target system by a flatcable with an 18-pin IDC-style DIL plug at one side, and a 20-way IDC socket at the other. The latter is connected to box-header K$_1$. Note that pins 19 and 20 are not used. The board may be made with the aid of the track layout.

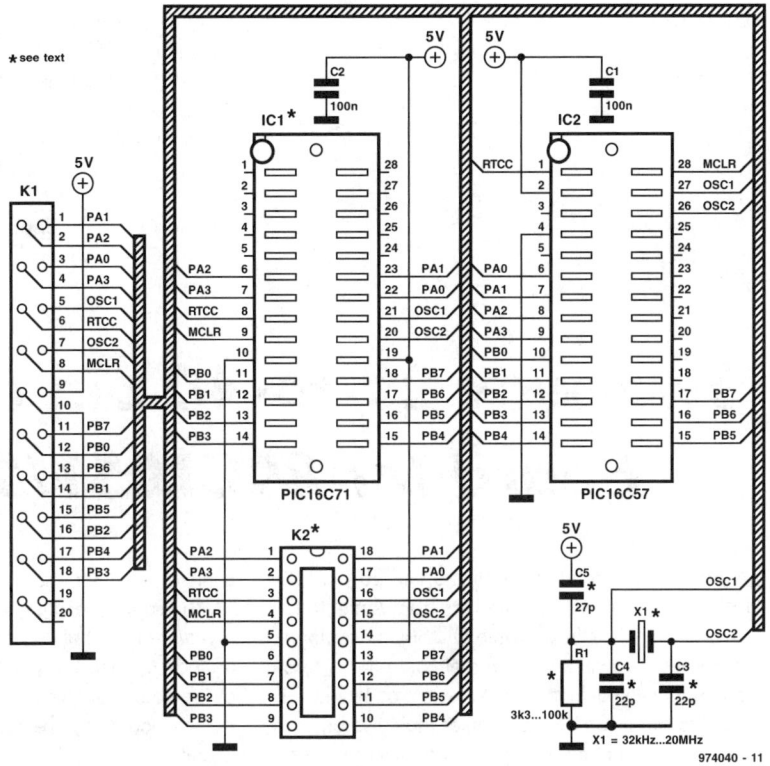

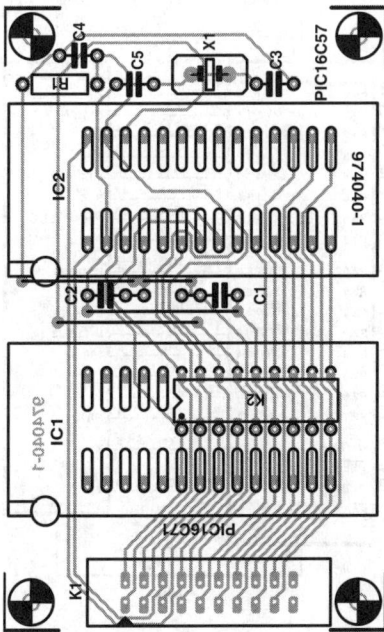

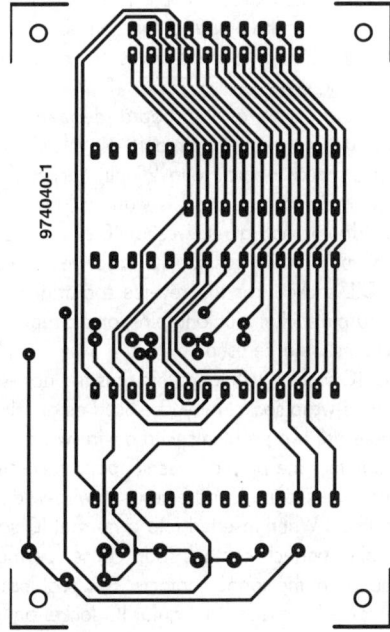

974040 - 11

Parts list

Resistors:
R_1 = 3.3–100 kΩ *

Capacitors:
C_1, C_2 = 0.1 μF
C_3, C_4 = 22 pF *
C_5 = $\geq$ 27 pF *

Miscellaneous:
IC_1, IC_2 = 28-pin ZIF socket, see text
K_1 = 20-pin boxheader or pinheader
K_2 = see text
X_1 = quartz crystal, 32 kHz–20 MHz *
18-pin DIL plug, IDC style (Eurodis 35633010T)

* optional, value depends on application, consult PIC datasheets

Two-way 20-to-40 pin adapter board for 89C1051/2051

The adaptor board maps the pins of the 20-pin PDIP AT89C1051/2051 microcontrollers from Atmel to the corresponding pins in a 40-pin DIL 80C51 footprint. The board is essential if an Atmel controller is to be used or emulated in a circuit having a 40-pin DIL socket (for an 80C51). The adapter may also be used the other way around in case ity is desired to use an 80C51 in a circuit which is physically designed to accept an Atmel controller. Note, however, that the 80C51 has no comparator on pinsP11 and P12, while the 89C2051 has the output of this comparator located internally on P36, in other words, the output is not bonded out to a pin.

The construction of the board depends on its intended use. If a change from a 40-pin DIL socket (on the target system board) to a 20-pin socket (for an Atmel controller or emulator) is wanted, position IC_1 on the adaptor board receives a 40-pin DIL socket with long pins, while position IC_2 receives a regular 20-pin DIL socket. For the reverse footprint conversion, a 20-pin socket with long pins, and a regular 40-pin socket, should be used.

Since IC sockets with long, thin pins are not easy to find, a wire-wrap socket mayh be used as an alternative. There are two problems with a wire-wrap socket. The first is that the plastic cross supports have the be removed. The second is more serious: wire-wrap pins are too thick! When inserted into a normal IC socket, they make perfect contact, but cause permanent deformation of the spring contacts. This problem may be overcome by stacking a regular IC socket on to the wire-wrap socket, and using the thinner pins for the connection with the target system socket.

The supply decoupling capacitor, C_1, should be a miniature type that can be dropped inside the 20-pin IC socket. If necessary, solder the cap at the track side of the board, or, even better, use an SMA device.

The board may be made with the aid of the track layout.

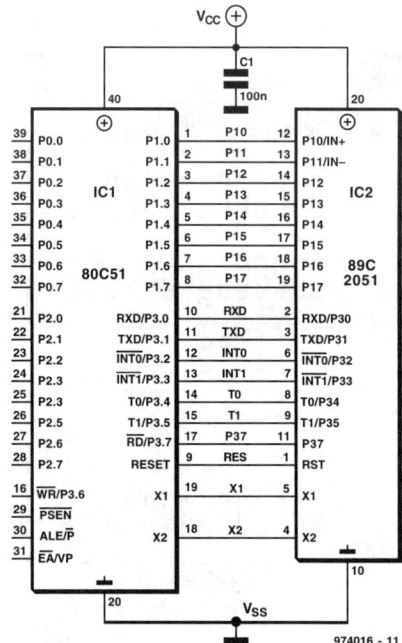

974016 - 11

974016-1

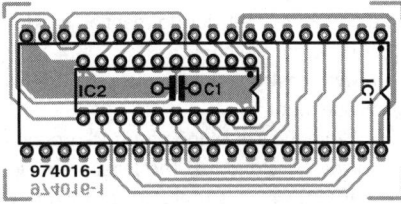

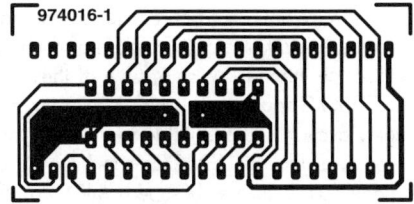

Yamaha DB50XG stand-alone soundcard

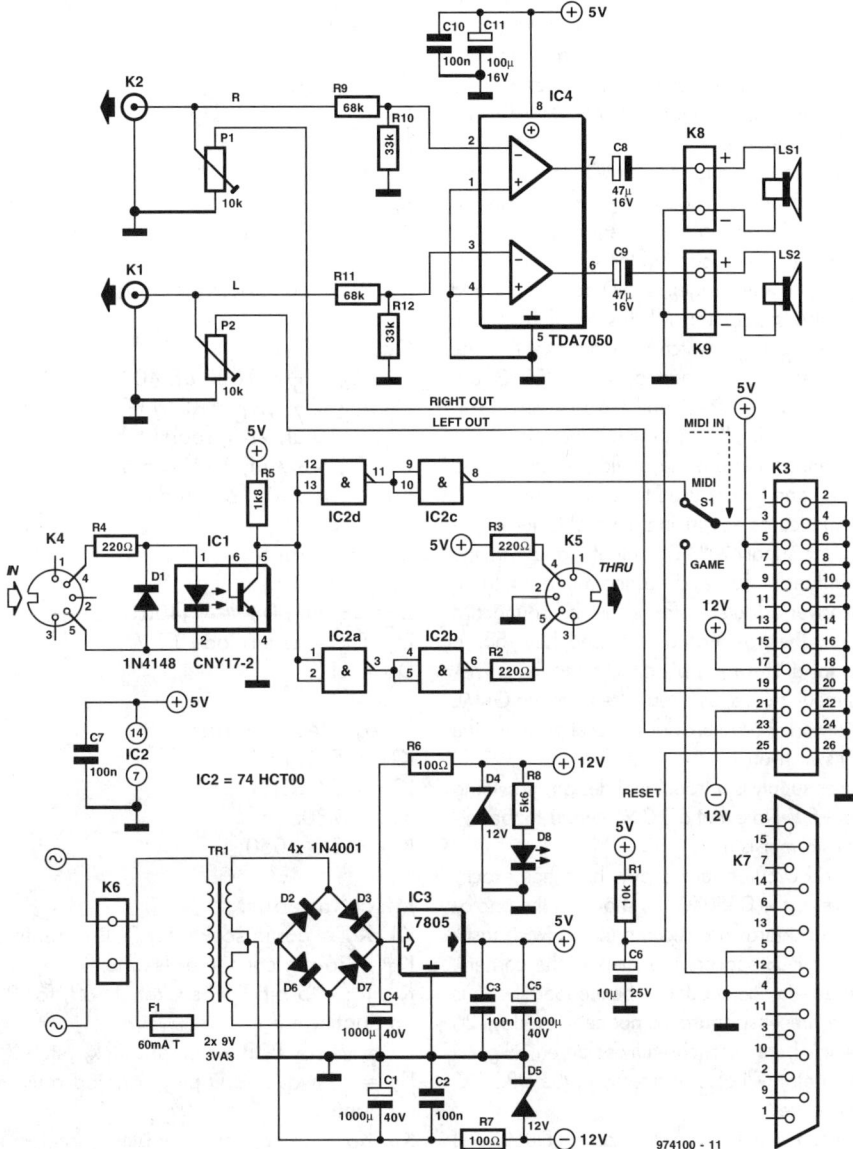

974100 - 11

There are several daughter boards that can be added to SoundBlaster (clone) cards to produce a much better sound than the internal FM chip. One of these is the Yamaha DB50XG sound card, which is relatively cheap and widely available. If a suitable sound card for the PC is not available, the DB50XG can be used as a stand-alone sound card with superb MIDI wavetable sound quality, simply by adding some hardware and a suitable power supply as shown here. The additional hardware allows the DB50XG to be driven by any MIDI source, whether computer or musical keyboard.

The DB50XG offers vastly improved sound quality over OPL and similar FM synthesizers which are used on low-cost soundcards to imitate MIDI wave samples. Thanks to the 18-bit DACs used on the DB50XG, the board even surpasses soundcards that do have an internal MIDI wavetable.

The circuit diagram shows that MIDI signals arrive via a standard 5-way DIN socket, J_4. Next, optoisolator IC_1 converts the MIDI 5 mA current loop into a TTL compatible signal which is fed to the MIDI THRU connector, K_5, via Schmitt trigger NAND gates IC_{2a} and IC_{2b}. It is also fed to the MIDIIN terminal of the DB50XG via IC_{2d}, IC_{2c}, switch S_1 and socket K_3. The stereo sound signals returned by the DB50XG are taken from the same socket and fed to a small on-board power amplifier, IC_4. The signals are also available to active loudspeakers via audio sockets K_2 and K_3. Presets P_1 and P_2 are used to set the sound level.

Many soundcards not having a waveblaster extension connector supply a MIDI output signal via the gameport connector. To avoid wasting money on (generally expensive) adapters offered by the soundcard manufacturer, the host board discussed here has a direct input for a 15-way cable attached to the gameport. If this input is used, S_1 should be set to the GAME position. Network R_5-C_1 provides a reset pulse for the DB50XG at switch-on.

The power supply is a traditional design, providing ±12 V and +5 V. Note that a 3.3 VA centre-tapped 9-0-9 volt transformer is used.

The printed circuit board shown here has exactly the same size as the DB50XG. As shown in the photograph, the two boards are secured in sandwich fashion using four PCB spacers mounted in the corners. Although a 26-way pinheader or boxheader shape is indicated on the host board component overlay, 26 pieces of solid, bare wire (carefully soldered, aligned and cut to length) will also fit directly in the DB50XG waveblaster socket.

Finally, lots of useful information on the DB50XG is available on the Internet. Just to mention two links:
http://www.yamaha.co.uk
http://www.castrop-rauxel.netsurf. de/home-pages/michael.banz
The official Yamaha site is good for background information on the XG standard. The latter web site contains an FAQ list which is useful for anyone having a DB50XG, or considering the purchase of one, as it contains much information on setting up with SoundBlaster cards.

Parts list

Resistors:
R_1 = 10 kΩ
R_2, R_3, R_4 = 220 Ω
R_5 = 1.8 kΩ
R_6, R_7 = 100 Ω
R_8 = 5.6 kΩ
R_9, R_{11} = 68 kΩ
R_{10}, R_{12} = 33 kΩ
P_1, P_2 = 10 kΩ preset. horizontal

Capacitors:
C_1, C_4, C_5 = 1000 µF, 40 V, radial
C_2, C_3, C_7, C_{10} = 0.1 µF
C_6 = 10 µF, 25 V, radial
C_8, C_9 = 47 µF, 16 V, radial
C_{11} = 100 µF, 16 V, radial

Semiconductors:
D_1 = 1N4148
D_2, D_3, D_6, D_7 = 1N4001
D_4, D_5 = zenerdiode 12 V
D_8 = LED

Integrated circuits:
IC_1 = CNY17-2
IC_2 = 74HCT00
IC_3 = 7805
IC_4 = TDA7050

Miscellaneous:
K_1, K_2 = audio socket for PCB mounting
K_3 = 26-pin connector (see text)
K_4, K_5 = 5-pin DIN socket, 180°, for PCB mounting
K_6 = 2-way PCB terminal block, pitch 7.5 mm
K_7 = 15-way sub-D plug, angled pins, for PCB mounting
K_8, K_9 = 2-way terminal block, pitch 5 mm

S₁ = 3-way jumper or single-pole changeover switch
TR₁ = 2x9V 3.3 VA

F₁ = fuse 60 mAT with holder for PCB mounting
Yamaha daughter board DB50XG

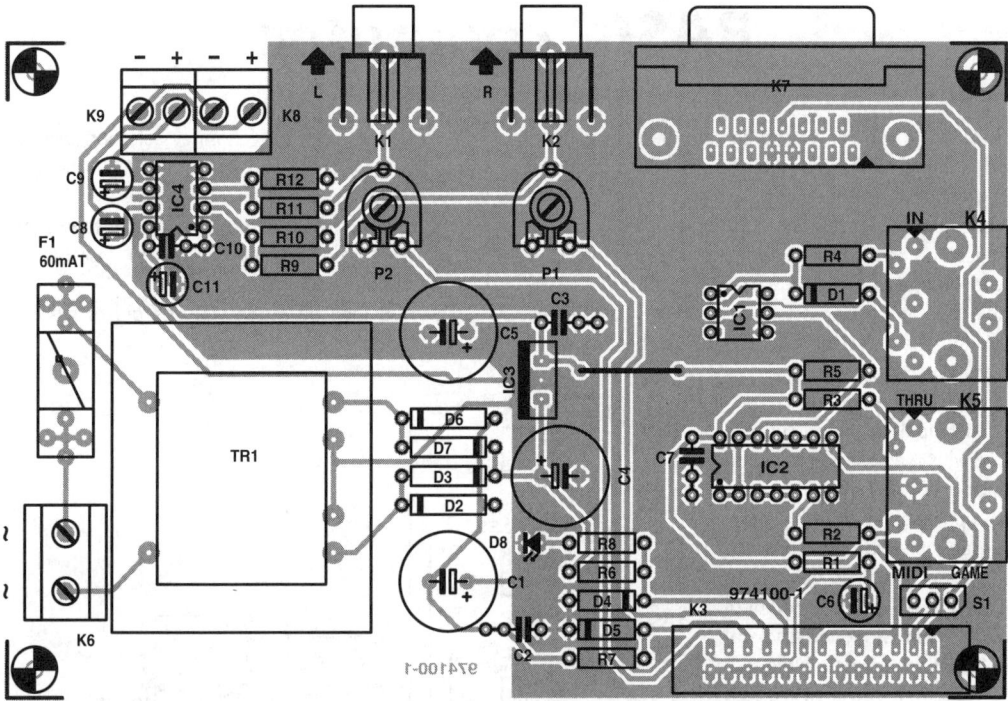

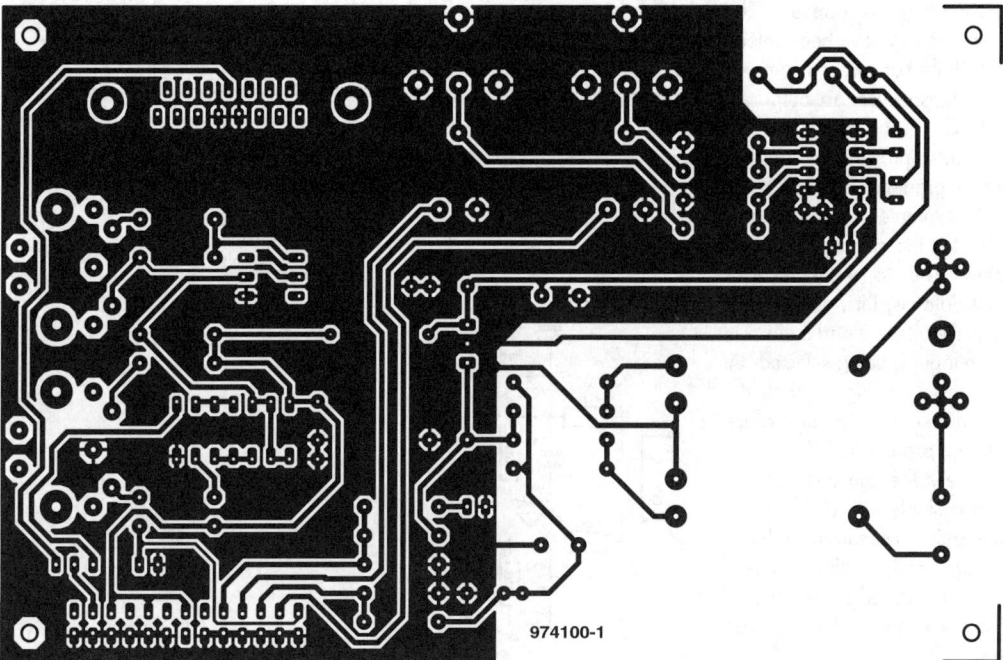

974100-1

125

A-D converter for MatchBox BASIC computer

The MAX186 is a converter with eight analogue inputs and is linked to the computer board via a length of 10 way flatcable. Although K_4 would appear to be the right connector for this link, K_1 was eventually chosen because bit operations are not possible on port P_2. A disadvantage of using K_1 is, however, that the 1-way cable has to be connected to a 20-way pinheader. Note that the converter may, in principle, be connected to any port as long as the supply voltage is at the right pins.

The inputs of the circuit are fitted with overvoltage protection resistors (R_1, R_3, etc.) as well as pull-up resistors (R_2, R_4 etc.). Consequently, inputs which are left 'open' are still held at a defined level, while additional ESD (electrostatic discharge) protection is provided. Note that the resistors cause a certain amount of attenuation if an input voltage is applied. The resistor values have been selected such that an external temperature sensor with an output of $1\,\mu A\,°C^{-1}$ (for instance, AD590 or LM334) provides the desired voltage gradient of 10 mV $°C^{-1}$.

An example of a control program for the converter is listed below. The converter receives eight databits; bit 7 is the startbit, while bits 4, 5 and 6 indicate which input is being selected. Bit 3 is used to signal that the measurement is to take place between ground and V_{REF}, while bit 2 tells the converter to perform a single-ended (i.e., non-differential) measurement. Bits 1 and 0, finally, initiate an A-D conversion based on the internal clock. Next, the 12-bit result can be read back.

A final remark: the channel selection bits are mixed up: bit 6 is the LSB, bit 5, the MSB, and bit 4, the middle bit.

The circuit draws a current not greater than 2 mA. For more information on the MAX186, visit Maxim's Internet site at www.maxim-ic.com

The PCB may be made with the aid of the track layout.

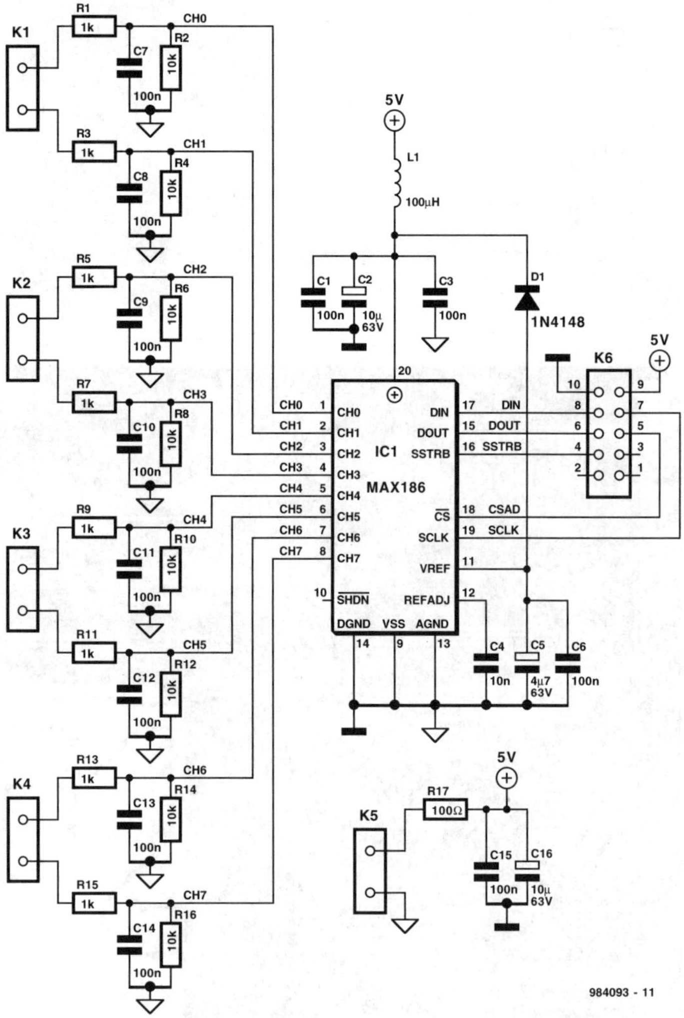

984093 - 11

126

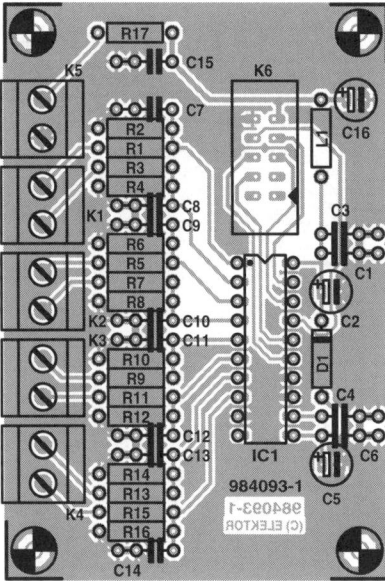

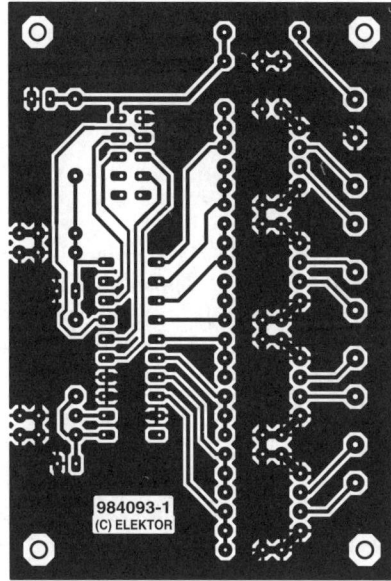

Parts list

Resistors:
$R_1, R_3, R_5, R_7, R_9, R_{11}, R_{13}, R_{15} = 1k\Omega$
$R_2, R_4, R_6, R_8, R_{10}, R_{12}, R_{14}, R_{16} = 10k\Omega$
$R_{17} = 100\ \Omega$

Capacitors:
$C_1, C_3, C_6{-}C_{15} = 0.1\ \mu F$
$C_2, C_{16} = 10\ \mu F$, 63 V, radial
$C_4 = 0.01\ \mu F$
$C_5 = 4\ \mu F$, 63 V, radial

Inductor:
$L_1 = 100\ \mu H$

Semiconductors:
$D_1 = 1N4148$

Integrated circuits:
IC_1 = MAX186DCPP or MAX186BEPP

Miscellaneous:
$K_1{-}K_5$ = 2-way PCB terminal block
K_6 = 10-way box header

READ_AD:

```
; This subroutine reads the MAX186 12-bit A-D converter.
; Before calling this routine the code for the desired
; channel has to loaded into integer variabele TEMP,
; as follows:
; TEMP:=1XYZ1110B where XYZ indicates the desired channel.
; The conversion result is then returned in TEMP.
;
; MAX187 connections:
; Data to 187     P1.0
; Serial clock    P1.1
; Data from 187   P1.2
; Chip select     P1.3
; Strobe          P1.4
;
```

```
INIT_AD:
  P1.0:=0               ;data in
  P1.1:=0               ;clock
  P1.3:=0               ;CS active

WRITE_AD:
  CNTR:=8
  WHILE CNTR>0 DO          ;send 8 bits of A-D command
    P1.0:=TEMP.7          ;msb bit to A-D
    TEMP:=TEMP SHL(1)      ;hold next bit ready
    P1.1:=1              ;clock-in data on pos. edge
    P1.1:=0
    CNTR:=CNTR-1
  WHEND

READ_AD:
  TEMP:=0               ;store result in this variable
  CNTR:=12
  P1.0:=0               ;read zeroes (else conversion starts)
  WHILE CNTR>0 DO          ;fetch 12 bit data
    P1.1:=1            ;supply clock pulse
    P1.1:=0            ;data valid after neg. edge
    TEMP:=(TEMP SHL 1)+P1.2 ;read data
    CNTR:=CNTR-1
  WHEND
  P1.3:=1               ;CS, turn off A-D
RETURN
```

Centronics in-system programmer

The programmer is based on an AT89S8252, which is an 8052-derived IC with a flash ROM of 8 Kbyte and a data EEPROM of 2 Kbyte. The ROM is to be programmed >×1000, and the EEPROM >×100 000.

Programming is carried out via only four lines available via the Centronics port. This is possible because the interface already uses the requisite TTL levels. The lines should be not longer than 1.5 m (5 ft) to ensure reliable operation.

To prevent a phantom supply arising between the IC and the Centronics port, the programmer should be connected to the PC only when both systems are on.

DOS software, called CISP, has been specially developed for programming the IC. The program is enabled via a batch file with parameters (see help CISP /? for the available options) or via its own menu (start up with CISP.EXE).

The software can program and read the internal memory of the processor. It can also be used to enable two protection bits. Both the input and output files are in Intel hex format.

The software is written in C and the source file, together with the EXE code, is available from the Publishers on a 3.5 in floppy disk (Order no. 986023-1).

The circuit as shown enables the software to be checked at an early stage. When the software is writing to a port or is verifying the content of one or more registers, a temporary address is set in the EEPROM during the test phase. The content of this address can be read and, if necessary, modified relatively easily.

Make sure when purchasing the IC to get a version with the suffix H (or higher letter), since older versions suffer from small errors in the programming protocol.

The IC shown in the diagram is a 44-pin PLCC version. It is also available in a 40-pin DIL case, but bear in mind that the pinouts of the two are different.

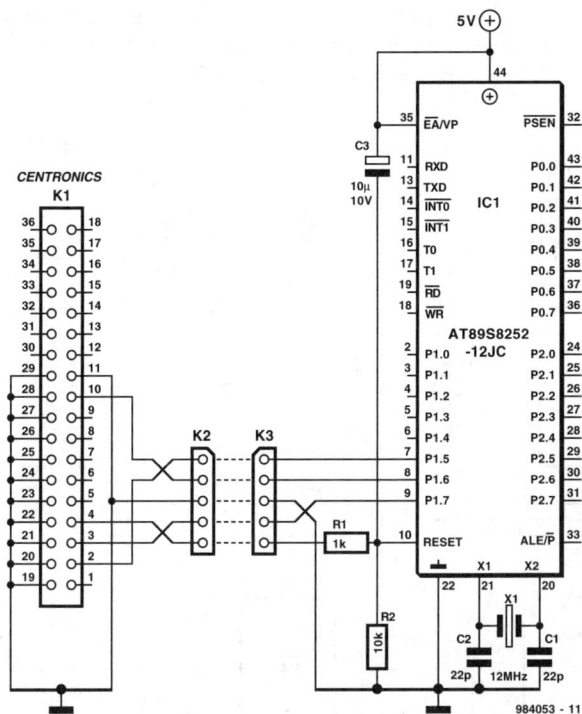

Extension board for MatchBox BASIC computer

The extension board contains the following building blocks: a memory extension, a real-time clock, an alphanumerical LC display and an 8-bit I/O port. Although the port lines are shown as connected to four LEDs and four push-buttons, the port may be used with many other configurations. Five connections have to be made between connector K_2 on the extension board and K_3 on the MatchBox computer board.

With the extension board linked to the main MatcbBox board, the memory structure of the computer is modified, mainly as a result of the clock being addressed in the memory range. The optimum memory allocation is then as follows:

- 256 bytes (PCF8582) on the MatchBox board, to which are added
- 512 bytes (PCF8594C-2, A2, A1, A0 = 010) and
- 1024 bytes (PCF8598C-2, A2, A1, A0 = 100) on the extension board

The available space for programs in the memory is then 1.5 kBytes.

If the Xicor 24C04 or 24C08 is used for the memory, pin 7 has to be strapped to ground via a short wire at the underside of the board. With the Philips versions, for which the board has been designed, the same pins are not connected.

The LC display is mounted on to the board with PCB spacers. The electrical connection is made with 14 short wires. Preset P_1 is adjusted for optimum legibility of the texts appearing on the display.

During testing it should be ensured that the directive 'LCD' is included in the 'FORMAT' statement. Also, the command LCDSET is required at the start of the program to enable the LCD to be initialised.

The GoldCap on the board enables the real-time clock to keep operating even if the supply voltage is absent for a few days.

The extension board draws a current of only a few milliamps, mainly on account of the LEDs.

The PCB may be made with the aid of the track layout.

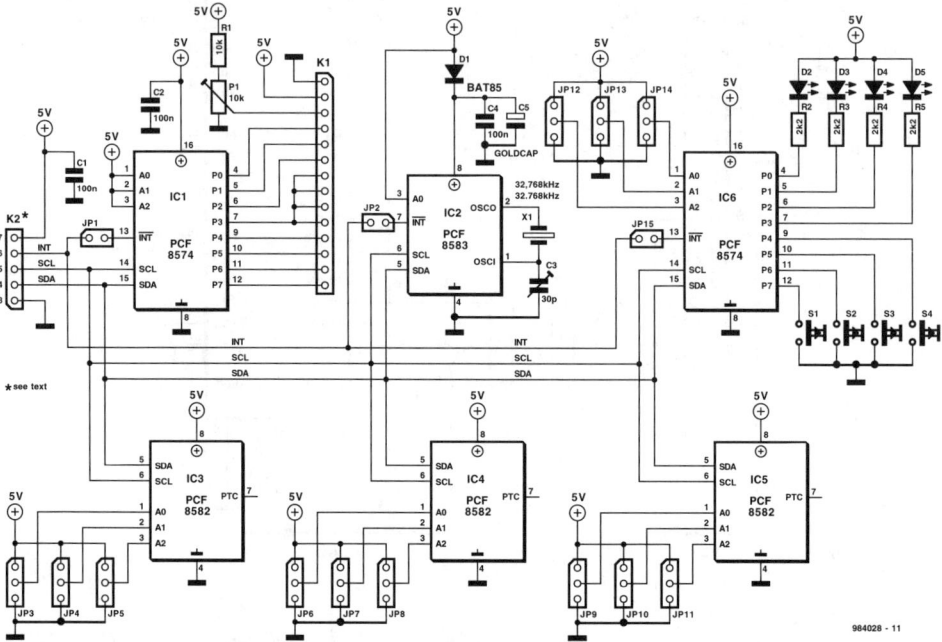

984028 - 11

Parts list

Resistors:
R_1 = 10 kΩ
R_2–R_5 = 2.2 kΩ
P_1 = 10 kΩ preset, horizontal

Capacitors:
C_1, C_2, C_4 = 0.1 μF
C_3 = 30 pF trimmer
C_5 = 0.47 μF or 1 μF; 5.5 V

Semiconductors:
D_1 = BAT85
D_2–D_5 = LED, high efficiency

Integrated circuits:
IC_1 = PCF8574P
IC_2 = PCF8583P
IC_3, IC_4, IC_5 = PCF8582P, PCF8594C-2 or PCF8598C-2
IC_6 = PCF8574P or PCF8574AP

Miscellaneous:
X_1 = 32.768 kHz quartz crystal
JP_1, JP_2, JP_{15} = 2-way SIL pinheader with jumper
JP_3–JP_{14} = 3-way SIL pinheader with jumper
S_1–S_4 = push-button switch
K_1 = 14-way SIL pinheader
K_2 = 5-way SIL pinheader

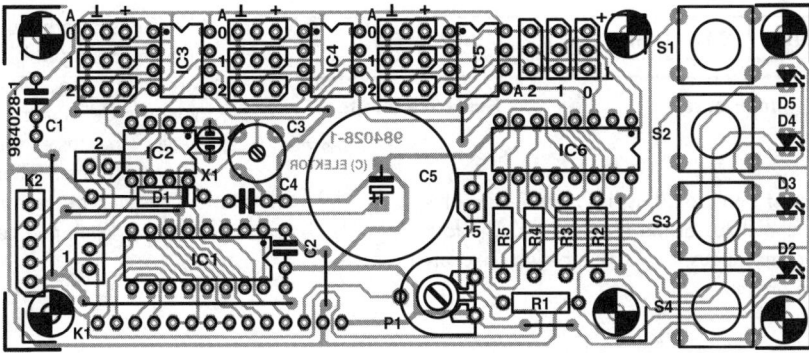

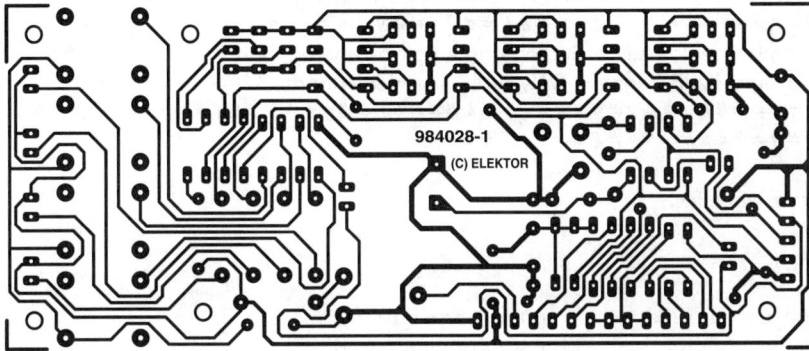

Game control adaptor

Most modern PCs are provided with a 15-way game port that contains four digital inputs for the buttons and four analogue inputs for the potentiometers in the analogue joysticks. However, there are also I/O cards on the market with only one game port.

The joystick inputs are based on a monostable multivibrator (MMV) whose on-time is set with a 100 kΩ external potentiometer as shown in Figure 1. This facility may be used for converting the game port for measuring analogue phenomena via, say, a resistor with negative temperature coefficient (NTC) or positive temperature coefficient (PTC) or a light-dependent resistor (LDR).

The software for reading the game port is relatively simple. A byte should be read at address 201_H (see Figure 2). The

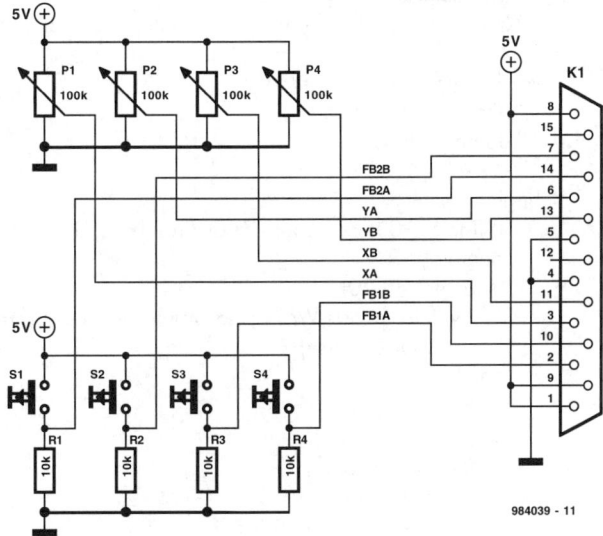

984039 - 11

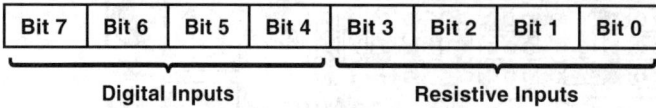

Bit 7	Bit 6	Bit 5	Bit 4	Bit 3	Bit 2	Bit 1	Bit 0

Digital Inputs　　　　　　　　**Resistive Inputs**

984039 - 12

four MSBs (most significant bits, i.e., bits 7–4) at this address give the status of the four buttons. The four LSBs (least significant bits, i.e., 3–0) are high only during the mono on-time.

The software must determine the time during which a bit is high via a fast loop. The analogue value is derived from the pulse width. The faster the loop, the more accurate the measurement. The listing shows a Pascal program with which up to four analogue levels can be measured.

```
{#################### Analog Game Port
##############################
# Example how to use an analog game port as analog input        #
# Copyright 1998 Segment B.V., Beek, The Netherlands             #
######################################################
####################}

program gametest;

uses crt;

const g_port = $201;        {game port's base address}
     max = 550;             {holds maximum value}
     offset = 50;           {holds minimum value}
     nr = 5;                {number of samples to average}
{#################### Measurement function
##########################}
function measure (var Value: integer; Input, Nr: integer):boolean;
{ 'Value' contains the result of the measurement
  'Input' selects the input channel
  'Nr' determines the number of samples used for averaging}

var i, counter, game : integer;
    bitgame : boolean;
    dummy : longint;

begin
    if ((nr > 100) or (input>4) or (input<1)) then
    begin
        gotoxy(5,20);
        writeln('Error!! Wrong parameter in measurement function');
        measure:=false;
    end
    else
    begin
        value:=0;
        dummy:=0;
        if input=4 then input:=8;
        if input=3 then input:=4;
```

```
            for i:=0 to (Nr-1) do
            begin
                counter:=0;
                bitgame:=true;
                port[g_port]:=0;
                while bitgame do
                begin
                    game:=port[g_port];
                    bitgame:=(((game and input)=input) and
                                (counter<((max*2)+offset)));

                    counter:=counter+1;
                end;
                counter:=counter-offset;
                dummy:=(dummy + counter);
                delay(1);
            end;
            value:=trunc (dummy/Nr);
            if ((value> (max * 2)) or (value<0)) then
            begin
                if value<0 then value:=-9999
                else value:=9999;
                measure:=false;
            end
            else measure:=true;
        end;
end; {function}


{###################### Input fire button status
######################}
procedure buttons(var key1, key2, key3, key4: boolean);

{returns boolean values to show fire button status}
{keyx := true if button pressed}
var game : integer;

begin
    game := port[g_port];
    key1:=((game and 128)<>128);
    key2:=((game and 64)<>64);
    key3:=((game and 32)<>32);
    key4:=((game and 16)<>16);
end; {procedure}


{###################### Main Program
##################################}
var connect, i : integer;
    value1 : integer;
    returnm, t1, t2, t3, t4 : boolean;
    e : char;
    value2 : real;
```

```pascal
begin
    ClrScr;
    gotoxy(5,3);
    writeln('Analog game port input');
    gotoxy(5,23);
    writeln('Press "e" to interrupt this program');
    gotoxy(1,8);
    writeln ('    INPUT 1:');
    writeln ('    INPUT 2:');
    writeln ('    INPUT 3:');
    writeln ('    INPUT 4:');
    gotoxy(5,15);
    write('digital inputs:');
    gotoxy(12,17);
    write('1:');
    gotoxy(22,17);
    write('2:');
    gotoxy(32,17);
    write('3:');
    gotoxy(42,17);
    write('4:');
    while e<>'e' do
    begin
        if keypressed then e:=readkey
        else
        begin
            for i:=0 to 3 do
            begin
                returnm := measure(value1, (i+1), Nr);
                if returnm then
                begin
                    gotoxy(30,(8+i));
                    write('        ');
                    gotoxy(30,(8+i));
                    write(value1);
                    gotoxy(35,(8+i));
                    write(' number of program loops ');
                end
                else
                begin
                    gotoxy(30,(8+i));
                    if value1<0 then
                        write('----  negative overflow             ')
                    else
                        write('++++  positive overflow            ');
                end;
            end;
            buttons(t1,t2,t3,t4);
            gotoxy(15,17);
            if t1 = true then
```

```
            write(' ON')
        else
            write('OFF');
        gotoxy(25,17);
        if t2 = true then
            write(' ON')
        else
            write('OFF');
        gotoxy(35,17);
        if t3 = true then
            write(' ON')
        else
            write('OFF');
        gotoxy(45,17);
        if t4 = true then
            write(' ON')
        else
            write('OFF');
        end; {else}
    end; {while}
    ClrScr;
end.
```

Improved power-down for the 8051

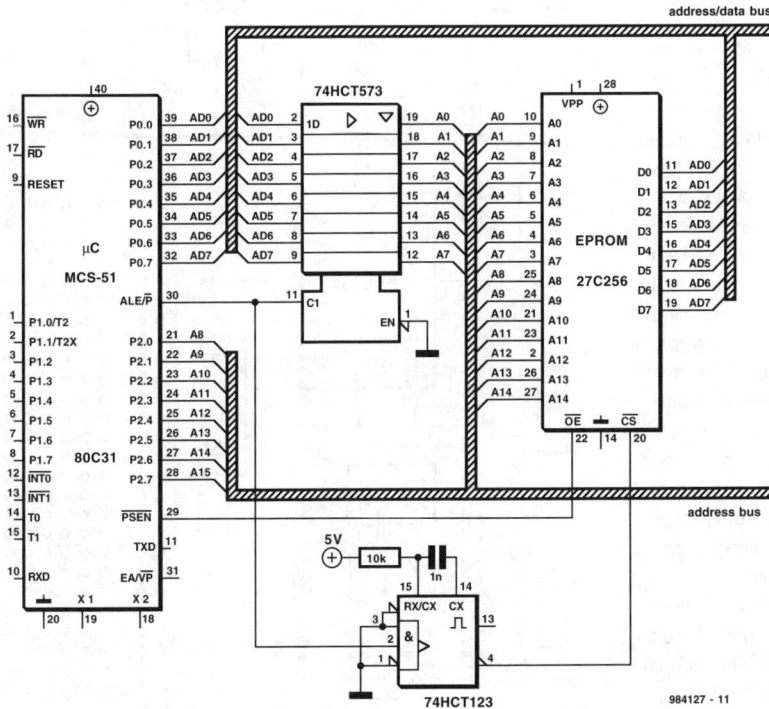

984127 - 11

Members of the 8051 family of microcontrollers (MCS51) are well-known and widely used. The controllers have a power-down mode in which the program processing is suspended by the clock oscillator and ended with a power-down instruction. To reduce the current drain, the supply voltage is reduced to a minimum of 2 V after the powered-down mode has been selected. This mode can only be disabled by a reset, for which the supply voltage needs to be returned to 5 V.

In simple applications of the 8051, the EPROM containing the program to be executed is enabled by making PSEN (program storage enable) active via its OE (output enable) terminal. There are also circuits where PSEN acts on the CS (chip select) terminal of the EPROM.

Use of the power-down mode has a drawback: line ALE (address latch enable), like PSEN, remains low during the power-down mode and so holds the EPROM active. It occupies the address/data bus with the accidentally same addressed byte.

This drawback can be removed by the circuit in the diagram. A retriggerable monostable evaluates the low and high edges of the ALE signal, which after a power-down and before a reset has a clock pulse. The output of the monostable sets a high on the CS input of the EPROM when the power-down mode is selected (and when, consequently, the disabled quartz oscillator can no longer generate an ALE pulse).This arrangement ensures that the EPROM can also be switched to the power-down mode.

Moreover, the monostable output may also be set on the address decoder of the system, or the combined CS lines of other peripheral equipment, so that these are also in the power-down mode.

Note that with component values as specified, the monostable has a time constant of about 4.5 μs.

Low-cost development system for PICs

The system allows testing the hardware operation of a PIC that has been programmed by the user. For such testing it is often necessary to connect simple input/output devices like LEDs and switches to various PIC port lines. This is possible on the prototyping area on the board used here. Provision is also made for connecting an external clock, a quartz-controlled clock, or an adjustable RC-controlled clock to the PIC. The first option is particularly useful if the program execution is to be slowed down to a speed at which it becomes possible to verify the operation of individual instructions.

The development system has an on-board 5-volt power supply stabilized by a 7805 regulator. The (unreg-

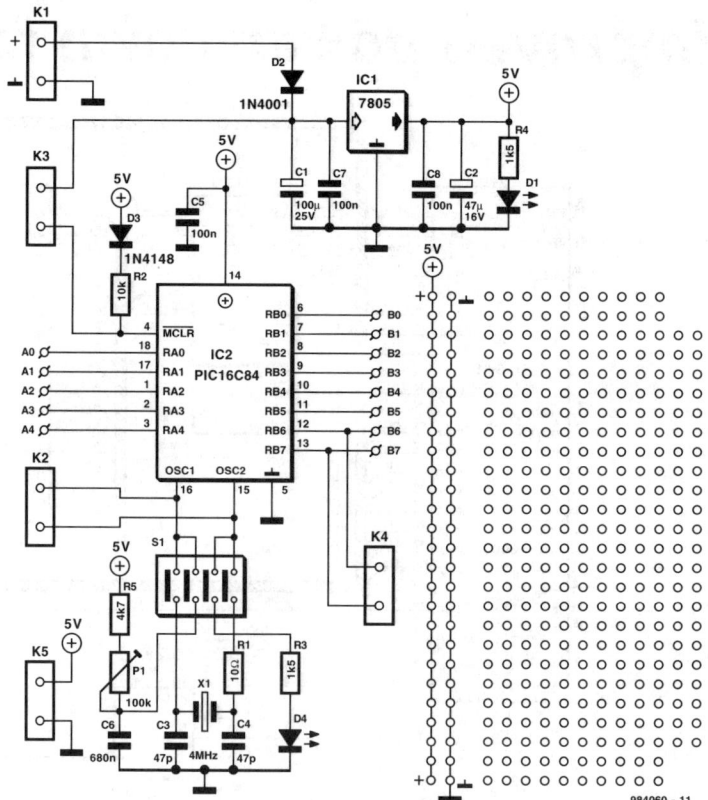

984060 - 11

ulated) input voltage (connected to K_1) should not exceed about 12 V, and the regulator may have to be fitted on a heat sink depending on the current drawn by devices connected to K_5 and any LED indicators etc. on the prototyping area.

DIP switch S_1 serves to select one of the PIC clock sources mentioned earlier. If so required, an external clock signal is connected to K_2. Diode D_4 serves to show up the activity of a very slow clock (single-stepping!).

Any or all of the port lines may be wired to the prototyping area via the solder pads around the PIC for connection to LEDs, switches, etc. used to simulate input/output devices.

Connectors K_3 and K_4 are intended for other projects and are therefore not used here. Connector K_5 makes the on-board 5-volt supply voltage available to external devices.

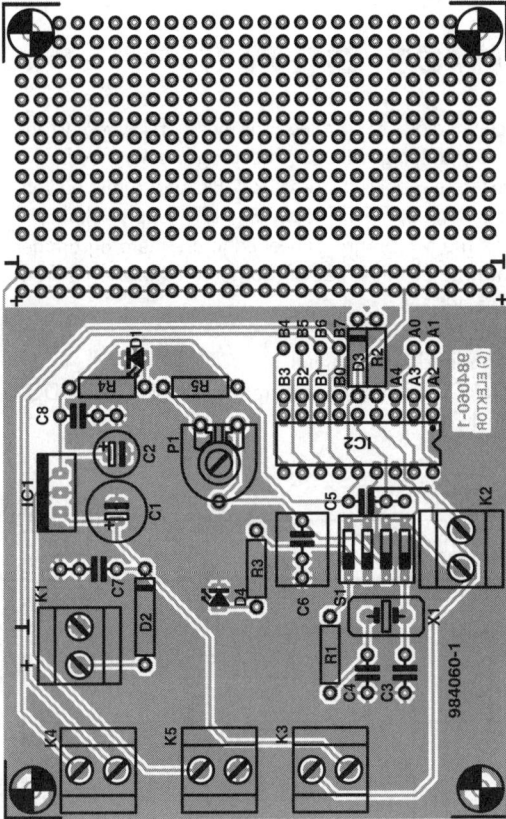

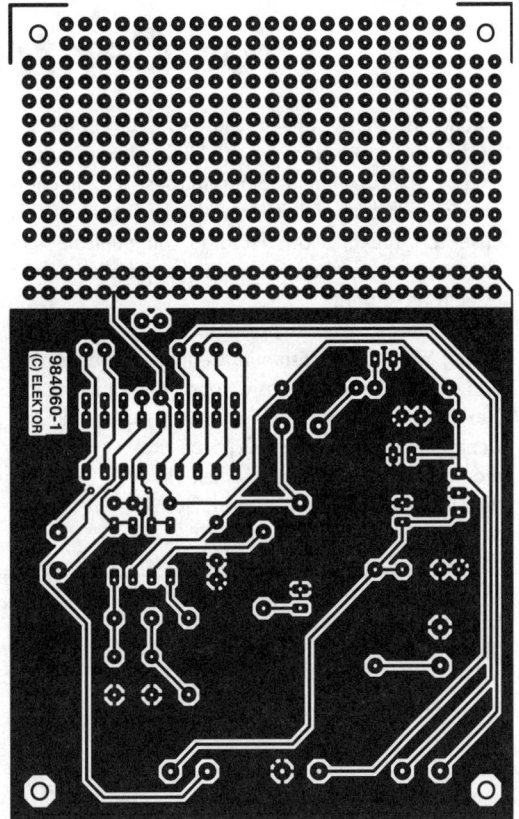

Parts list

Resistors:
$R_1 = 10\ \Omega$
$R_2 = 10\ k\Omega$
$R_3, R_4 = 1.5\ k\Omega$
$R_5 = 4.7\ k\Omega$
$P_1 = 100\ k\Omega$ preset, horizontal

Capacitors:
$C_1 = 100\ \mu F$, 25 V, radial
$C_2 = 47\ \mu F$, 16 V, radial
$C_3, C_4 = 47$ pF, ceramic
$C_5, C_7, C_8 = 0.1\ \mu F$
$C_6 = 0.68\ \mu F$

Semiconductors:
D_1, D_4 = low-current LED
D_2 = 1N4001
D_3 = 1N4148

Integrated circuits:
IC_1 = 7805
IC_2 = PIC16C84

Miscellaneous:
S_1 = 4-way DIP switch
K_1–K_5 = 2-way terminal block, pitch 5 mm
X_1 = 4 MHz quartz crystal
The PCB may be made with the aid of the track layout.

PIC16C84 programmer for Centronics port

Among the most popular shareware programs for PIC16C84 programming is PIPO2 from Silicon Studios. The circuit in the diagram uses PIPO2 in combination with a special driver written by Dave Tait. This driver, DTAIT.EXE, allows PIPO2 to communicate with PIC programmer hardware via the parallel printer ('Centronics') port.

The hardware does not amount to much and is cheap, too — the PCB-mount Centronics connector is probably the most expensive part! A single 74LS06 (hex open-collector inverter) is used to enable the Centronics port and the PIC (to be programmed) to talk to one another. Three data lines of the Centronics port, D0, D1 and D3 are first inverted and then applied to the PIC to be programmed. D0 supplies data, D1 clock pulses, and D3 programming pulses. Information returned by the PIC to the PC is first inverted by gate IC_{1b} and then applied to the BUSY line on the Centronics connector.

The PIC to be programmed is connected as follows:

K_3 DATA to RB7 (pin 13)
K_3 CLOCK to RB6 (pin 12)
K_4 MCLR to MCLR (pin 4)
K_2 +5 V to VDD(pin 14)
K_2 GROUND to VSS (pin 5)

The +Vin terminal of K_4 is connected to an external 12-volt supply for the programming voltage. Diode D_2 lights when programming pulses are applied.

Zener diode D_1 protects the base of T_1 against the programming voltage. The programmer needs two external supply voltages: 12 V (PIC programming voltage) and 5 V (74LS06 and the LED supply voltage).

The two programs needed to use the programmer may be obtained free of charge from the Internet from this url:
http://www.sistudio.com/
sistudio/download/html.
DTAIT.EXE is unzipped from the 'PINAPI Drivers DOS Pack 1'.

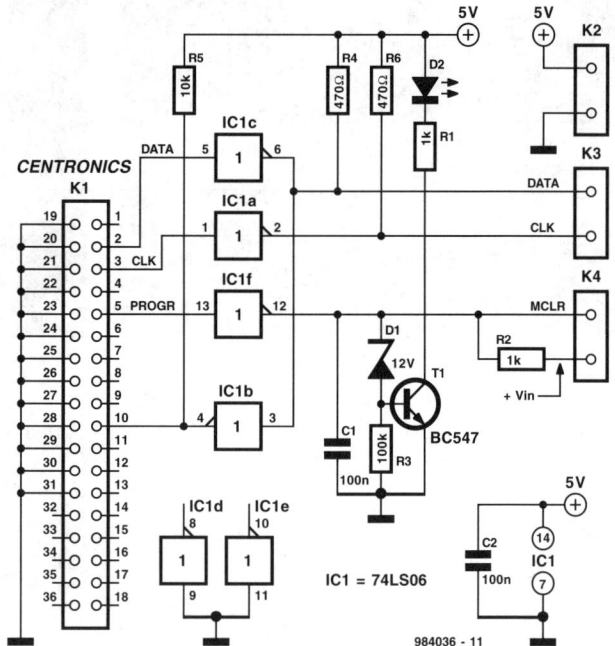

984036 - 11

When these programs are downloaded, all that needs to be done is writing the following batch file:

```
DTAIT.EXE 7406
PIP02.EXE
DTAIT.EXE REMOVE
```

and launching it from the DOS prompt. The printed circuit board may be made with the aid of the track layout.

Parts list
Resistors:
$R_1, R_2 = 1$ kΩ
$R_3 = 100$ kΩ
$R_4, R_6 = 470$ Ω
$R_5 = 10$ kΩ

Capacitors:
$C_1, C_2 = 0.1$ μF

Semiconductors:
D_1 = zener diode, 12 V, 400 mW
D_2 = LED
T_1 = BC547

Integrated circuits:
IC_1 = 74LS06

Miscellaneous:
K_1 = Centronics socket, PCB mount
K_2, K_3, K_4 = two-way terminal block, pitch 5 mm

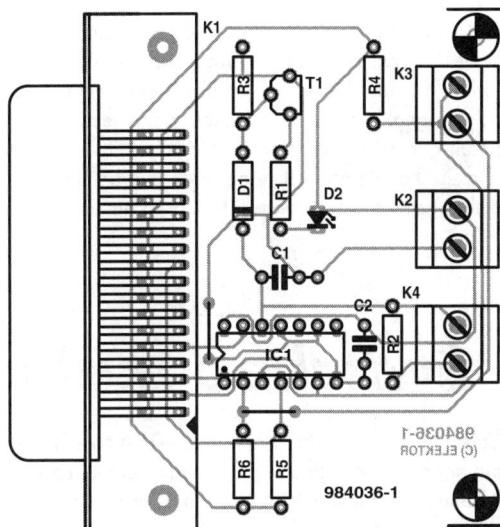

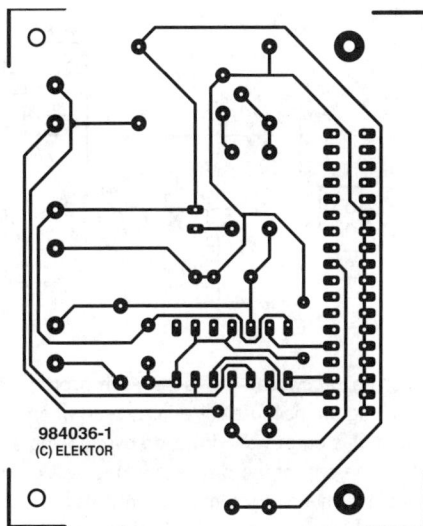

Printer port for BASIC Stamp

Users of the BASIC Stamp are generally satisfied with the product, but there are many who wish the device had a parallel printer port. Such a port is particularly helpful when the BASIC Stamp is used as a data logger.

Fortunately, it is possible to provide the Stamp with a printer port by adding just one IC. The relevant circuit is based on IC1, an EDE1400 from E-Lab Digital Engineering. This chip converts the serial data into a parallel port at TTL level. The IC works from a single +5-V power supply and needs an external 4 MHz crystal for its oscillator. The internal circuits (a programmed PIC!) process the Centronics protocol.

The requirements on the data to be converted are

not high: any serial signal at 2400 baud (no parity, eight data bits, and one stop bit) suffices. When the serial data have been received (the design is such that only a signal line and an earth return are needed), the circuit generates the requisite control signals for the printer. The internal watchdog/timer guarantees troublefree operation. A simple test program is shown below.

Switched mains output for ATX

Current personal computers with an ATX motherboard are not provided with a switched mains output socket. It is therefore no longer possible to achieve automatic switching on or off of peripheral units with the com-

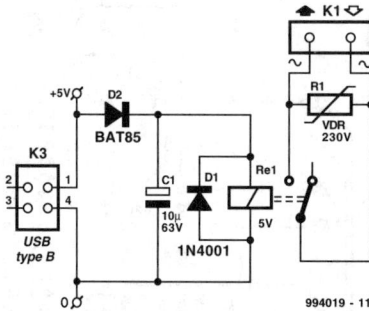

puter on/off switch. This is not such a problem when switching on, because it is quickly seen when a peripheral unit is on or not. However, overlooking switching off at the end of the day or working session leads to needless energy consumption an possible shortening of life of the peripheral unit. It is, fortunately, simple to do something about this.

An ATX motherboard contains all logic circuits for implementing a USB port. Peripheral equipment that

may be connected to this bus may draw a current of up to 100 mA from it, and a total of not more than 500 mA for all units together. This means that the USB port may be used without any difficulties to energise a relay. This relay in turn makes it possible to switch peripheral units on and off in tandem with the computer.

When there is a potential at the USB interface, relay Re1 is energised via diode D2. The relay contact closes and the peripheral units are powered. Capacitor C1 buffers any irregularities on the USB potential so that a brief dip in this does not immediately lead to the relay being de-energised. Diode D1 is a freewheeling device for the relay coil.

The design of the USB port is well thought out. There are two types of connector around. Downstream, that is, from computer to peripheral unit, a square one (Type A) is used, whereas upstream, that is, from peripheral unit to computer, a flat one (Type B) is used. By terminating each cable with a Type A and a Type B connector, connecting errors, such as those frequently encountered with RS-232 and DIN cables, are all but impossible. The cables are always terminated into plugs, whereas the equipment is fitted with sockets.

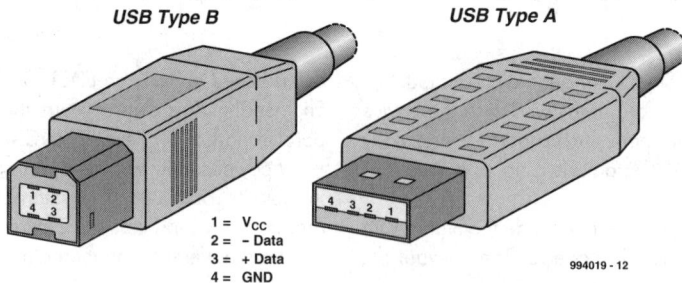

USB Type B   USB Type A

1 = V_{CC}
2 = – Data
3 = + Data
4 = GND

994019 - 12

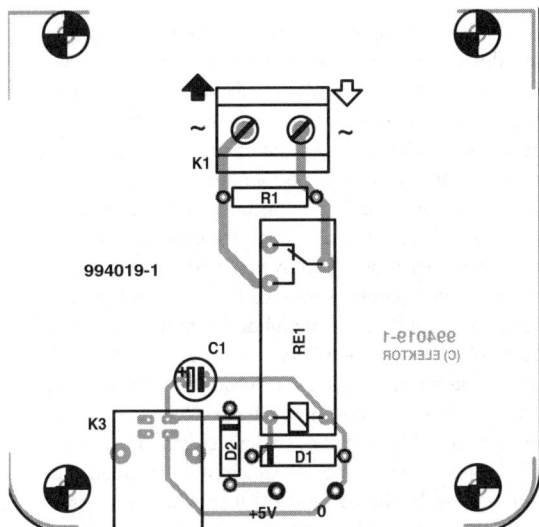

994019-1

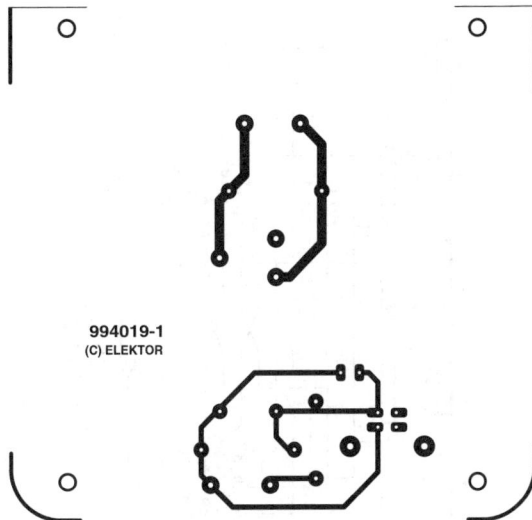

994019-1
(C) ELEKTOR

Parts list
Resistors:
R_1 = voltage-dependent resistor (VDR),
230/250V

Capacitors:
C_1 = 10 µF, 63 V

Semiconductors:
D_1 = 1N4001
D_2 = BAT85

Miscellaneous:
Re_1 = Siemens E-card relay, 5V, PCB mount
K_1 = 2-way PCB terminal block, pitch 7.5 mm
K_3 = USB socket, panel mounting, Type B

Line switch for PC sound card

With the equipment that is found in the domestic or office environment, the distinction between the various disciplines is becoming increasingly blurred. The stereo, television set and video recorder have long since been merged into a single audiovisual installation, and there are signs of the same sort of process with the computer. Formerly, this consisted of just a main enclosure, a monitor and a keyboard, but nowadays it is being surrounded by a steadily increasing number of peripheral devices, and it is in a manner of speaking growing in the direction of the audiovisual installation. The same effect can be seen in the application realm. In addition to the classical applications, the computer is being used more and more for digital audio.

Anyone who for example spends a lot of time with sampling and making his or her own CDs, soon finds

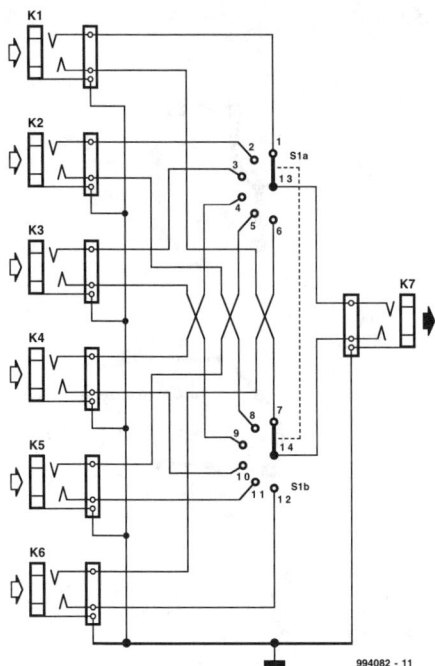

994082 - 11

that the single input connector of the sound card is insufficient to deal with a large number of audio sources. A cassette deck, MiniDisk player, phonograph, microphone — in principle, all of these can be connected, although the phonograph and the microphone naturally require special preamplifiers.

If you want to avoid the inconvenience of disconnecting and reconnecting cables, the only solution is to use a line switch box. Such a box is not at all complicated in the electronic sense, as can be seen from the schematic diagram. A handful of 3.5-mm jacks and a six-position, two-pole rotary switch are all that you need. The practical aspects, as usual, are rather more onerous. For this reason, we have developed a tidy printed circuit board for the line switch box, which eliminates hand wiring — which will no doubt be appreciated by the constructor! It should not be difficult to find a suitable enclosure for the line switch. Preferably use a metal enclosure and connect it to the circuit earth.

Parts list

K$_1$–K$_7$ = 3.5 mm stereo socket for PCB mounting
S$_1$ = rotary switch, 6 positions, 2 poles, for PCB mounting

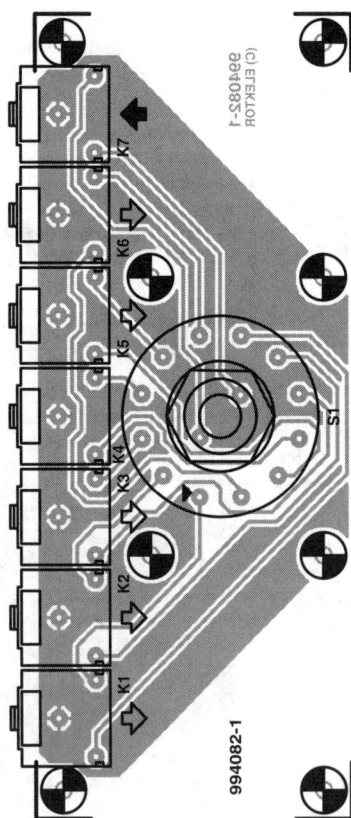

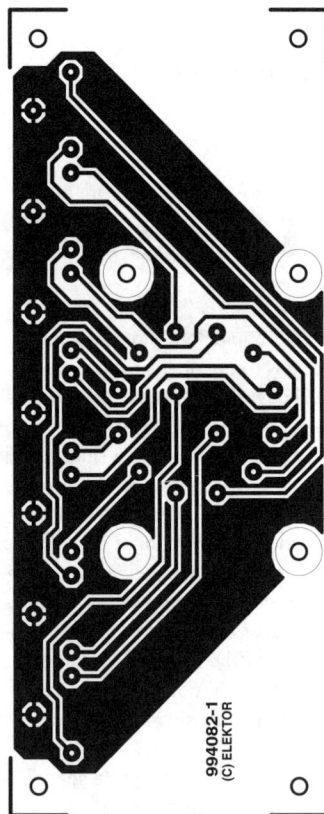

994082-1
(C) ELEKTOR

994082-1
(C) ELEKTOR

LOGO! interface

A special adapter cable is needed to connect a PLC of the Siemens LOGO! series to the serial interface of a PC. Of course, such a cable is commercially available but can also be made by a home constructor.

The interface circuit consists of a galvanic isolator and a level converter. The galvanic isolation is provided here by a dual optoisolator (Sharp PC827), although two single PC817 optoisolators may be used instead, or other types as long as their current transfer ratio (CTR) is at least 50% at a forward current I_F of 5-mA. Since the two optoisolators are built as inverters, a pair of inverter gates must be used to restore the signals — IC_{3a} (to the PC) and IC_{3b} (from the PC). Resistor R_3 acts as a current-limiting resistor, and R_4 is a pull-up resistor that holds the line securely high when the signal level is not definitely low.

The well-known MAX232 IC is used to convert the signal levels between 0 and +5 V (on the PLC side) and ±12 V (RS232) for the serial interface. Since only the RxD and TxD lines are needed, two drivers of the MAX232 can be connected in parallel in each direction.

As a rule, the LOGO! interface does not need its own power supply. The level converter, the phototransistor of IC_{2a} and the LED of IC_{2b} are powered from the RS232 interface. Zener diode D_2 limits the voltage to +5 V for this purpose. The LOGO! PLC provides the operating voltage for the inverters, the LED of IC_{2a} and the phototransistor of IC_{2b}.

The interface circuit draws a current of only around 10 mA from the RS232 interface. In certain rare cases this can overload the interface drivers of the PC, in which case an external +5 V supply will be necessary.

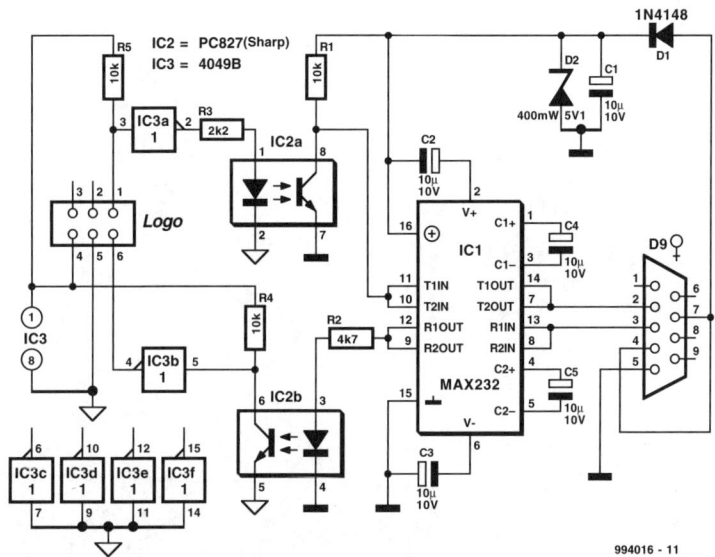

994016 - 11

143

4
General
Interest

Aquarium pump switch

The pump in an aquarium is normally switched off when the fish are being fed to prevent the food being dispersed too widely, which causes much of it to be lost. Such lost food remnants foul up the water. Unfortunately, it is quite easy to remember to switch the pump off when feeding is started, but all too often it is forgotten to switch it on again when feeding time is over.

The circuit shown here ensures that the pump is switched on again after a certain time lapse. It is based on a 555 timer arranged as a preset monostable multivibrator (MMV). The off time of the MMV can be set between 1 minute and 9 minutes with P_1.

Power is provided by transformer Tr_1, bridge rectifier D_4–D_7 and reservoir capacitor C_3. The rectified voltage is held at 12 V by regulator D_3-R_8.

The output of the timer (pin 3) controls a relay via T_2; one of the make contacts of the relay is connected in series with the supply line to the pump.

The state of the circuit is indicated by D_1, D_2 and inverter T_1. Diode D_1 is on when the MMV is inactive, whereas D_2 lights when the pump is switched off.

The pump may be disabled tem-porarily with two-way press-button switch S_1. This switch may also be used to switch the pump on again before the mono time has elapsed.

The circuit draws a current of only a few millamperes, excluding the current through the relay.

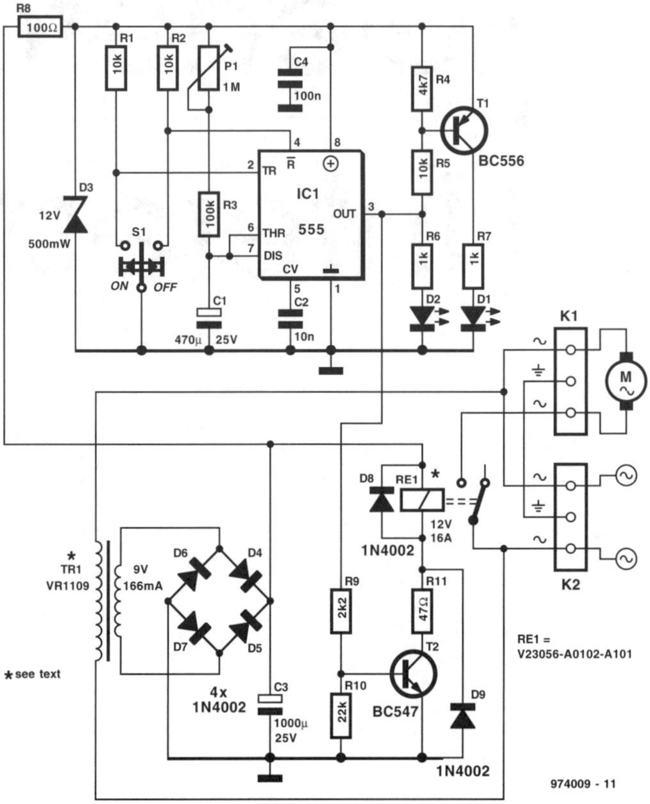

974009 - 11

Auto shuttle for model trains

The circuit on the next page is intended to make a model train shuttle continuously between two buffers. At the start and finish of each journey, one of the rails (it does not matter which) is interrupted and a resistor, R_x and R_x', placed in series with it at the break via K_4 and K_4' respectively. The value of these resistors is shown as 2.2 Ω, but in practice they should have a value that causes a potential difference, pd, of 1.5 V across them when the break in the rail is bridged by the train (that is, the value depends on the current drawn by the locomotive).

When the break is being bridged, the relevant optoisolator, IC_4 or IC_4', is actuated. Mind the polarity: when the locomotive hits the upper buffer in the diagram, IC_4' must be actuated and when it touches the lower buffer, IC_4. Almost any type of optoisolator may be used.

The outputs of the optoisolators at K_5 and K_5' are connected in parallel and applied to the input of the relay control circuit via K_3. Since it does not matter for the operation of this control circuit which of the buffers is touched, a large number of end-stops may be connected in parallel to K_3.

The circuit operates as follows. When the supply to the train is switched on, a power-up reset is effected by the circuit based on IC_{4b} before the train can move. As soon as one of the end-stop detectors is actuated, monostable multivibrator (MMV) IC_2 is triggered, whereupon the supply to the rails is broken via T_2 and Re_2: the train stops.

After about 2 seconds (time constant R_5-C_6), bistable IC_3 is triggered, whereupon the polarity of the rail voltage is reversed. When the mono time of IC_2 (which can be set with P_1) has elapsed, relay Re_2 is re-energized and the train starts to move again, but in the opposite direction.

The relay may be a 5 V or 6 V type. The power supply for the circuit is taken from that for the train via K_1: its polarity is irrelevant. The circuit draws a current of only a few milliamperes, to which must be added the current drawn by the relay. If the supply for the train is lower than 8 V, it may be that the drop across IC_1 becomes too small. In that case, it should be replaced by a low-drop type. It is also advisable in that case to replace the bridge rectifier by four Schottky diodes Type SB130 in a bridge configuration. If, on the other hand, the supply to the train is high, it may be advisable to mount IC_1 on a suitable heat sink.

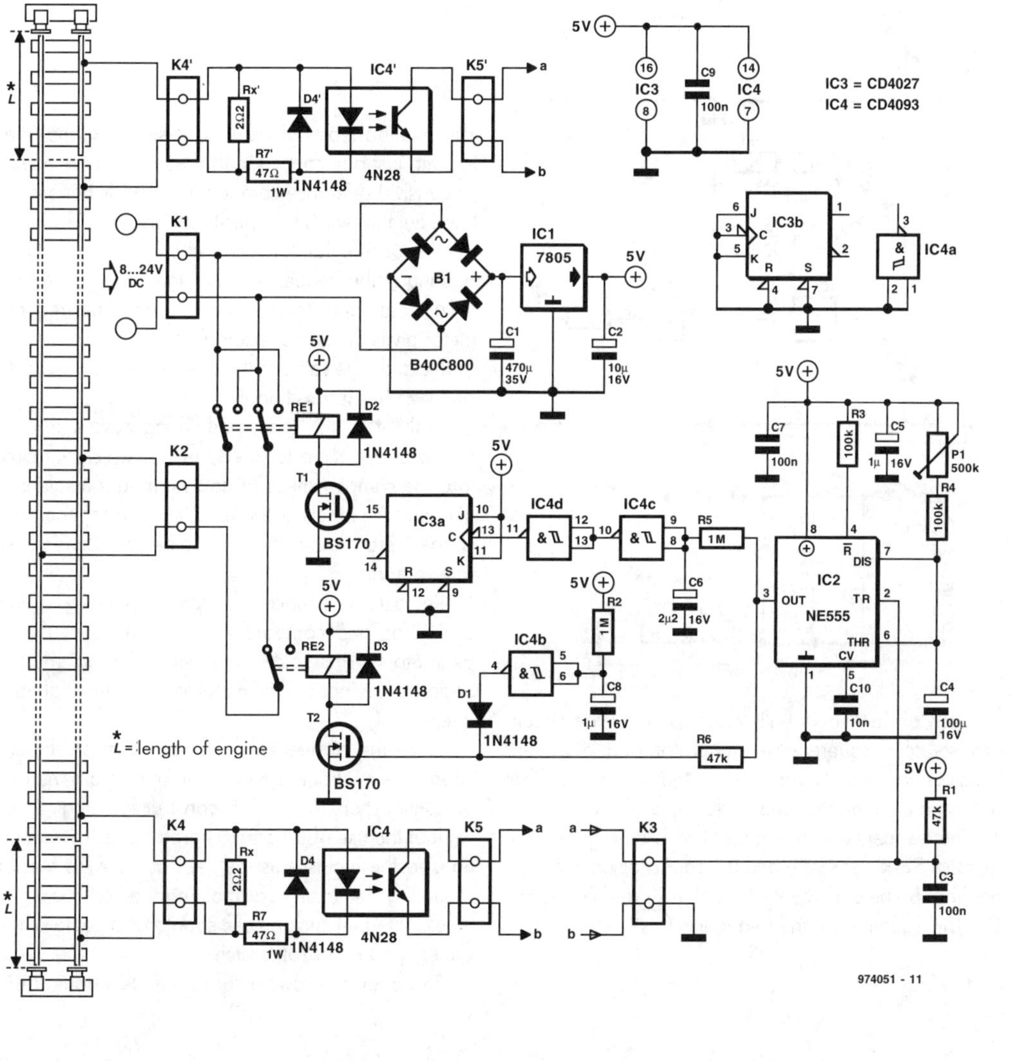

974051 - 11

147

Capacitor matching

Determining the value of a capacitor requires a fairly sophisticated meter. Often, however, all that is required is finding two capacitors that are roughly equal in value: their actual value is of less importance. The diagram shows an adaptor for matching a pair of capacitors with the aid of a frequency meter.

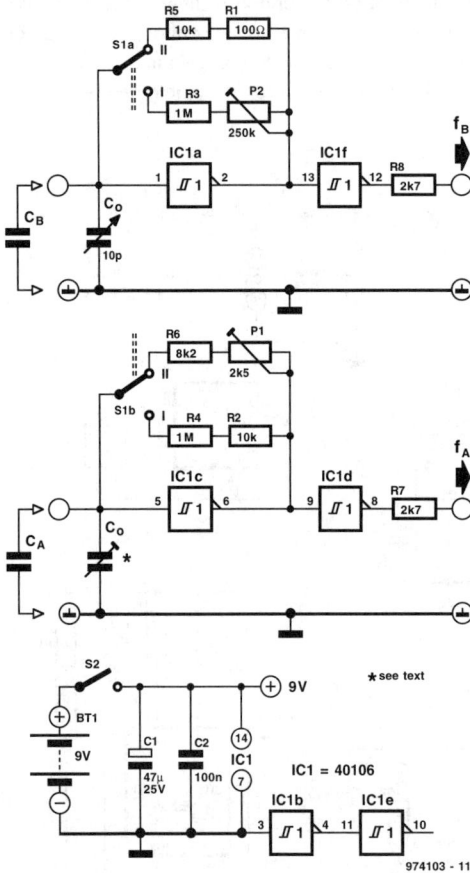

Each of the two virtually equal parts of the circuit consists of a square-wave generator and a buffer. Resistors R_7 and R_8 limit the output current of the buffer gates when the load impedance is low.

The frequency of one generator is determined by resistors R_5-R_1 or R_3-P_2 and the capacitor on test, C_B, and that of the other by R_6-P_1 or R_4-R_2 and capacitor C_A. The equation for the frequency, f, is

$$f = kRC,$$

where k is a constant, depending on the trigger thresholds of the gates and the level of the supply voltage.

Two meter ranges are available, depending on the position of S_1: $\leq 0.1\ \mu F$ and $\geq 0.1\ \mu F$.

The frequencies of the generators are related by

$$f_A/f_B = k(R_B/R_A)(C_B/C_A).$$

In both meter ranges, potentiometers P_1 and P_2 may be set to obtain

$$k(R_B/R_A) = 1.$$

This makes elimination of the constant possible, so that

$$f_A/f_B = C_B/C_A.$$

This relationship enables the adjustment of the adaptor with just one capacitor, that is, without the need of (expensive) reference components. This is done as follows. Set the switches to position I, use a 0.01 μF polyester capacitor for C_A, short-circuit C_B and note the reading of the frequency meter. Interchange the short-circuit and capacitor and turn P_2 until the frequency meter gives the same reading.

Next, do the same in the other meter range, that is, with switches in position II.

In this test, the gates and wiring have small stray capacitances of up to 20–30 pF, for which the circuit must be compensated. To do this, mount a solder pin at one of the inputs and wind a few turns of enamelled copper wire around this and solder both ends of the wire to earth.

Replace the capacitor on test by one with a value of 100 pF and compare the frequencies of the two generators. The solder pin with winding must be added to the input of the generator with the higher frequency.

Since the compensation is fairly critical, the stray capacitances should be borne in mind from the onset of construction. A floating construction, that is, one without the use of a board, is preferable. The connections to the inputs must be kept as short as feasible. Mounting the circuit in a small, earthed case is a good idea. Note that such a case should also contain a 9 V battery and an on/off switch.

The adaptor draws a current of just about 2 mA.

148

Car alarm

Every day a large number of private cars are stolen or broken into. One of the available deterrents to car thieves is a loud car alarm and that is why most new cars are now sold with one fitted as standard. If your car does not already have one, the alarm proposed here may be of interest.

The alarm is based on the fact that when a car is broken into, at least one of the front doors has to be opened. When that happens, the interior light comes on and this causes a slight temporary drop in the battery voltage. The present circuit detects such a drop and when it does so sounds an alarm.

The circuit consists of a number of distinct sections as shown in the diagram. When S_1 is closed, a trigger pulse, delayed by C_3, is applied to pin 4 of IC_{1a}. When this device has been triggered, its Q output goes high whereupon D_5 lights.

Monostable multivibrators (MMVs) IC_{3a} and IC_{3b} are reset by the low level at the Q output of IC_{1a}. When the mono time of the MMVs, which can be

set with P_3, has elapsed, the alarm is on. It follows that the driver (and his passengers) must have left the car and closed all doors before this happens.

The voltage drop detector is based on comparator IC_2. This op amp compares the voltages at its two inputs. That at pin 2 is a reference potential whose level is set with P_2. When the battery voltage applied to pin 3 drops, the output of IC_2 becomes low, which triggers IC_{3a}. The level at which the trigger pulse is generated depends, of course, on the setting of P_2.

To enable the driver of the vehicle to gain access without setting off the alarm, there has to be a certain delay on opening one of the front doors. This delay is provided by the circuit around IC_{3a}. A trailing edge on pin 5 of IC_{3a} results in a high level at the Q output and a low level at the Q output of this device. A yellow LED (D_6) then lights. The time needed to turn off S_1 (and thus disable the alarm) is set with P_1.

If the alarm is not disabled within the preset time, a trailing edge at pin 11 of IC_{3b} will cause the alarm

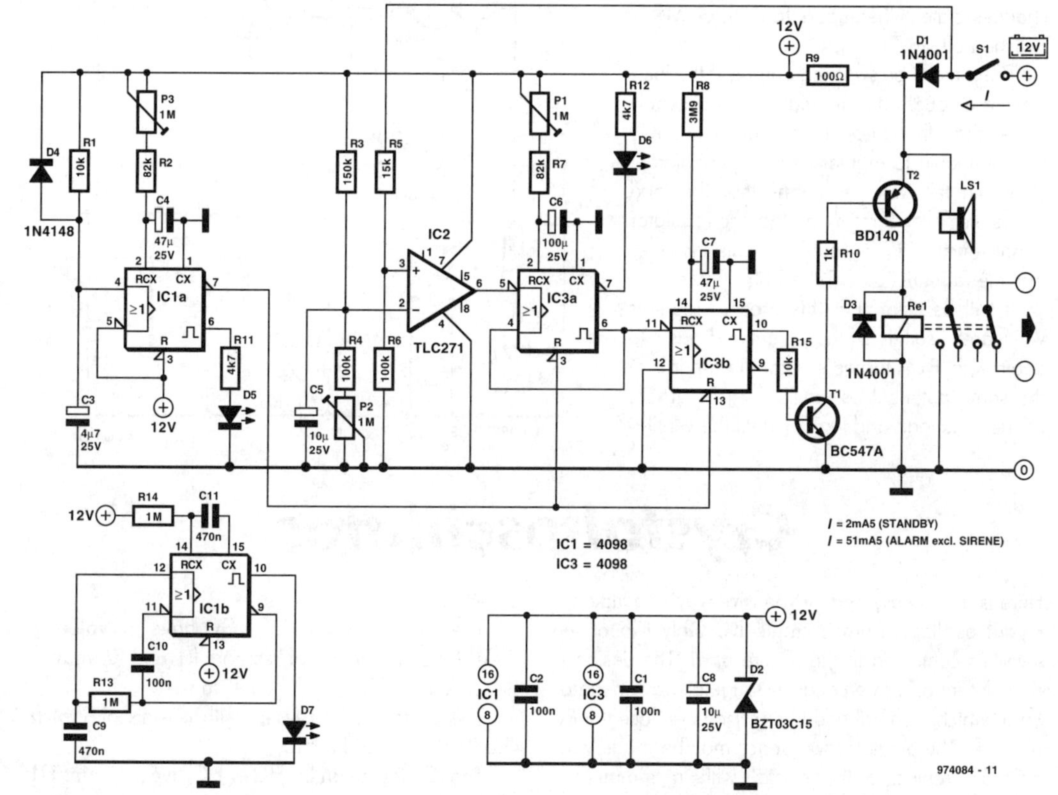

974084 - 11

149

to be sounded for 60 seconds (legally the maximum permissible time). During this period, the starter circuit is disabled, so that the engine cannot be started. After the alarm has ceased sounding, it is on again.

A visual indication of the state of the alarm as shown may be added. When S_1 is closed, IC_{1b} starts to oscillate and a red LED (D_7) begins to flash.

Cooling fan for regulator

This circuit is intended to convert some of the heat losses occurring in a voltage regulator into electrical energy and use this to operate a small fan.

The circuit is simply connected in parallel with the regulator, so that no modifications in the existing circuit are required. All that is needed are three wires from the regulator to the fan circuit.

The fan, current limiter R_2, and series transistor T_1 are connected in parallel with regulator IC_1. The transistor is driven by IC_2, which compares part of the output voltage, derived from potential divider R_4-R_5-P_1 with a reference voltage provided by zener diode D_1.

As soon as the output voltage of the regulator rises even slightly, the comparator changes state, whereupon fan motor M_1 is switched off via T_1.

The fan can be switched on only if the load current through the regulator is noticeably larger than that through M_1, otherwise IC_1 will be unable to regulate the output voltage. A rise in this voltage means that the load draws a smaller current and that the regulator is not functioning as required. This indicates that the fan must be switched off.

It will be clear that this process requires very careful setting of the potential divider. To begin with, P_1 is turned so that the wiper is at the same potential as junction R_4-P_1. Then, connect the load and turn P_1 until the ventila-

tor just begins to work. If a suitable load for this procedure is not available, use a 33 Ω, 1 W resistor.

The circuit is designed to operate with a transformer with 9 V secondary which, after rectification, provides a potential of about 11 V. If the input voltage is higher than this, increase the value of R_2 by 8.2 Ω per volt. Mind the dissipation of the resistor!

In the prototype, a brush-less fan, rated at 5 V, 600 mW, was used. These types produce rather less noise and interference than brush types.

974110 - 11

Crystal oscillator

Here is a very easy and inexpensive way of building a crystal oscillator from a single IC. Only two of the inverters contained in the IC are used. The design is reminiscent of a traditional rectangular-wave generator in which a crystal and two resistors replace the RC network. The oscillator frequency may be made variable by replacing the crystal with a trimmer of

22–68 pF.

The crystal frequency may have a value of 1–10 MHz. The value of resistors R_1 and R_2 may be 1–4.7 kΩ (but they must be the same).

The prototype was tested with crystals of 1 MHz, 3.579 MHz, and 8 MHz.

The IC may be an LS, HC or HCT type, but not TTL.

The oscillator draws a current of only a few milliamperes from a 5 V supply line.

The simplicity of the design has a drawback in that the frequency stability and the stability with temperature variations are not very good. This is because the oscillations depend to a large extent on the parallel capacitance of the crystal.

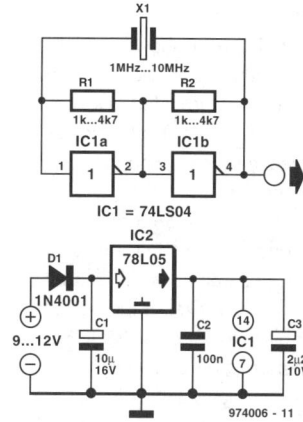

974006 - 11

Digital potentiometer

Xicor's digitally controlled E^2POT ICs provide ergonomic and long-lasting alternatives to mechanical potentiometers. The ICs in the X9CMME series have a 7-bit counter with reversible count direction and a decoder that enables one of the 100 analogue switches.

The outputs of the analogue switches serve as the wiper of a potentiometer, while the inputs are linked to a potential divider composed of 99 equal resistors. The counter state may be stored in a non-volatile EEPROM, so that it can serve as the output value at a subsequent start.

The X9CMM series is designed to operate from 5 V supply lines. The potential across the resistive divider must not exceed 10 V (only 4 V in case of the X9C102). The ON resistance of the analogue switch is about 40 Ω, so that the current through the wiper is limited to 1 mA.

E^2POT ICs have three inputs for the digital drive. The level at U/D determines whether a trailing edge at clock input INC lowers or raises the counter state. This action only takes place if chip select input CS is low. A leading edge at CS arranges for the counter state to be stored when INC is high. When CS is high, the IC

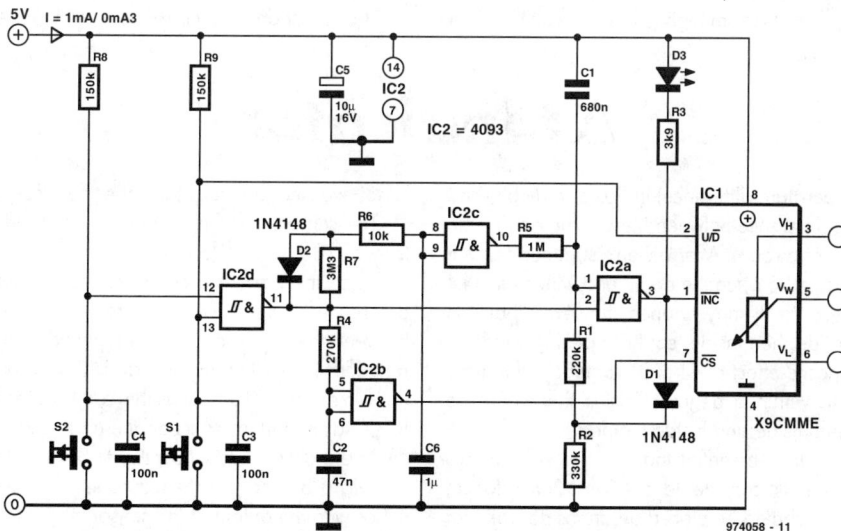

974058 - 11

is in the standby mode.

The circuit diagram shows a complete digital potentiometer based on a Type X9CMME. It is provided with two controls, S_1 and S_2, an optical indicator and a delayed frequency change-over of the clock generator.

When keys S_1 and S_2 are open, resistors R_8 and R_9 hold the inputs of IC_{2d}, a NAND, as well as the U/D input of IC_1 high. The low level at the output of IC_{2d} disables clock generator IC_{2a}. Frequency determining capacitor C_1 is discharged in the quiescent state.

When one of the keys is pressed (S_1 firmly, S_2 gently), the output of IC_{2d} changes state, so that the clock generator and IC_1 (via IC_{2b}) are enabled. Capacitor C_1 is then charged via R_1 and R_2 until the input level of IC_{2a} goes low, whereupon the gate output linked to the clock input of IC_1 changes state (from low to high). When this happens, C_1 is discharged via R_1 and D_1 until the upper trigger level of IC_{2a} is attained. The gate then changes state again and the above action repeats itself.

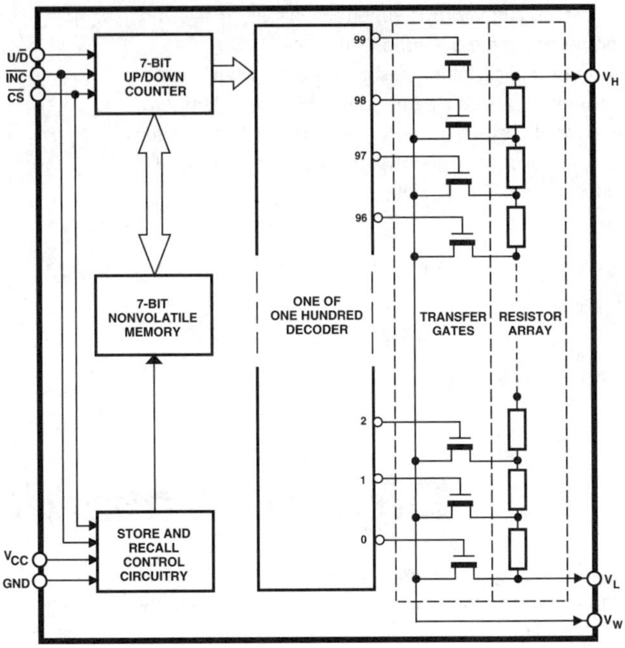

The clock signal is optically monitored by D_3.

When the output of IC_{2c} is high, the gate draws a portion of the charging current from C_1, which results in the clock frequency at INC being relatively low.

At the same time that the generator is enabled, C_6 begins to be charged gradually via R_6 and R_7 until IC_{2c} changes states (from high to low). Circuit IC_2 then

974058 - 12

contributes to the charging current to C_1, whereupon the clock frequency increases: in the prototype, the frequency rose from 1.3 Hz to 3.1 Hz in four seconds..

When the keys are released, the clock generator stops. At the same time, C_6 is discharged rapidly via R_6 and D_2, so that the frequency is low again when the keys are operated anew.

The switch-off delay owing to R_4-C_2 enables actual counter state to be stored by the internal logic.

The circuit draws a current of 0.3–1.0 mA.

Δ*t indicator*

The Δt (temperature difference) indicator may be used to show the difference between the outside and inside temperature of a room. After a warm summer's day, it is often pleasant to open the doors and windows to let in some fresh air. It may happen, however, that it is warmer outside then inside, so that opening windows has the opposite effect of what is wanted. The indicator shows to within a degree Celsius the difference between the outside and inside temperature.

It is, of course, essential that the two sensors are placed in carefully considered positions. For instance, the inside one should be placed on an inside wall that

is not exposed to sunlight or other heat sources.

The sensors are Type LM35CZ, which has a sensitivity of 10 mV $°C^{-1}$ (above 0 °C)

Since the present circuit needs to measure only a difference between two temperatures, the difference between the sensor outputs is amplified by differential amplifier IC_{3a}. Its amplification is ×10 to ensure that the offset (max. 10 mV) of the comparator following IC_{3a} has minimum effect on the performance.

To enable outside sensor IC_1 to be linked to the circuit via a balanced screened cable, the resistance in series with the output of the sensor is split into R_2 and

R_3. The inside sensor, IC_2 may be linked to IC_{3a} via a single resistor (R_4).

To hold the output of IC_{3a} within a suitable operating area commensurate with the supply voltage, it is set to 2.5 V via R_5–R_7-P_1-IC_{3b}. Window comparator IC_{3c}-IC_{3d} following it therefore operates symmetrically around a level of 2.5 V. Preset P_1 sets the output potential of IC_{3a} to the centre of the window.

The output of the comparator is applied to three LEDs: D_1 (green) shows when the outside temperature is lower than that inside; D_2 (yellow) lights when the inside and outside temperatures are equal; and the lighting of D_3 (red) indicates that the outside temperature is higher than the inside one. Any differences in brightness of the diodes may be obviated by giving R_{10}–R_{12} slightly different values from those shown.

The circuit is calibrated by closing P_2 completely and adjusting P_1 until the three LEDs just light. The window can then be set between 1 °C and 10 °C with P_2.

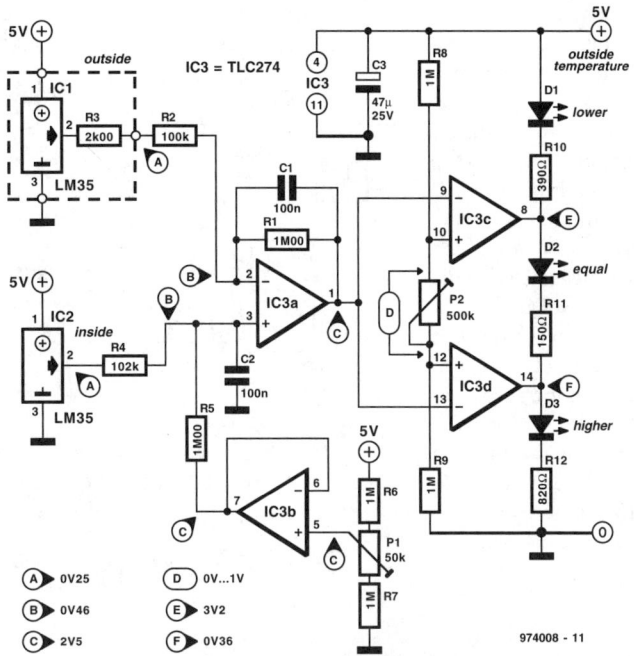

The measurands shown in the circuit refer to equal temperatures at the sensors and an ambient temperature of 25 °C.

The circuit draws a current of not more than 8 mA.

Intruder alarm

The alarm uses a pyrosensor to detect the presence of animals or human beings by changes in heat radiation.

The contact of the relay in the pyrosensor is linked to the input of the circuit and is closed in the quiescent state. If an animal or person approaches the sensor, the relay contact opens. The input of IC_{1b} then goes low and its output becomes high. Pin 8 of IC_{1c} goes low, which enables the MMV (monostable multivibrator) formed by IC_{1c} and IC_{1d}. Owing to the feedback to pin 9 of IC_{1c}, the MMV output remains low for about three minutes, even if the sensor is disabled. If on completion of the three-minute period the sensor is still actuated, a new period also three minutes long is started. Alarm pulses generated when the MMV is enabled are ignored.

When the MMV is quiescent, its output is high and counter IC_2 remains reset. The counter position is then zero and, since output Q_0 (pin 3) is not linked to T_1

and the buzzer via a diode, the buzzer remains inactive.

When the MMV is enabled by the pyrosensor, its output goes low. The counter is then no longer reset and begins to count the clock pulses from IC_{1a}. The buzzer is then actuated intermittently via diodes D_1–D_5 and T_1.

When the counter reaches its highest position, Q_9 (pin 11), the high level at this output impedes the clock at the enable input (pin 13). The counter stops counting and retains this position. After a short while, the mono time elapses and pin 11 of IC_{1d} goes high, whereupon the counter is reset. A 600 μs pulse is then passed to IC_{3b} via R_3 and C_2. This pulse briefly disables IC_{3a} which reenables the MMV, provided the sensor is still actuated.

The mono time may be changed by altering time constant R_4-C_1.

When the sensor contact has been closed almost

153

continuously for a lengthy period, the mono time may be a little shorter when the MMV is enabled for the first time after this period. If this is found inconvenient, the recovery time may be lengthened by increasing the value of R_2

The circuit draws a current of 1–2 mA, which increases to 13–14 mA (via R_1) when the relay contact is closed, and to 15–16 mA when the buzzer is actuated.

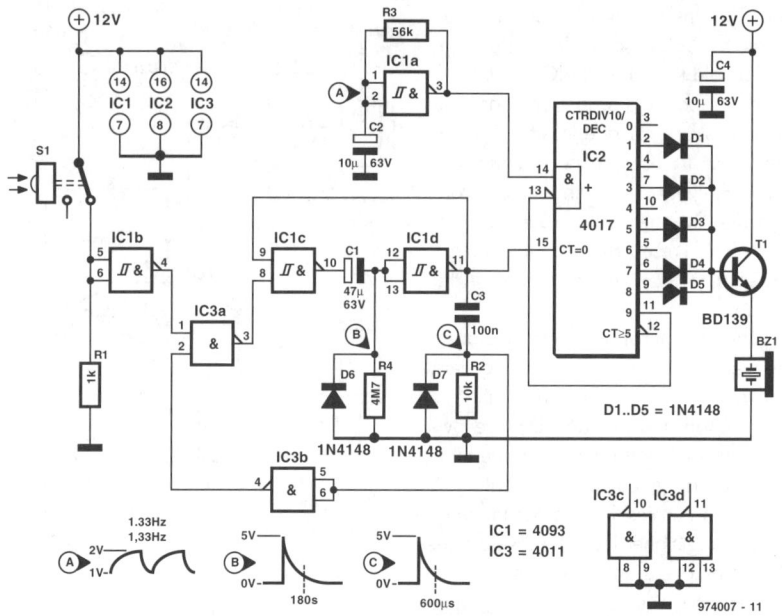

Object protection

Valuable items are best protected against theft by a customized alarm system. The protection system described here can not only protect that valuable vase in the hall, but also windows, doors or closed spaces, such as rooms or hallways. If anything or anybody approaches the protected item or space, an alarm sounds.

The circuit is based on a measuring bridge and proximity sensor which is controlled by a traditional Wien-Robinson oscillator. The measurand is rectified and applied to a differential amplifier, which drives a piezobuzzer.

A Wien-Robinson bridge as in the circuit diagram consists of a high-pass filter and low-pass filter in series, and the whole shunted by potential divider R-½R. The divider provides a signal at a level of $1/3U_e$ in the pass band. The output voltage at the resonance frequency $f_R=1/2\pi RC$ drops to a minimum (theoretically 0). The

bridge has a frequency response reminiscent to that of a notch filter and an instantaneous phase shift from $-90°$ to $+90°$ at the resonance frequency.

Such a phase shift is a prerequisite for the use of the filter in a stable oscillator. Unfortunately, the output

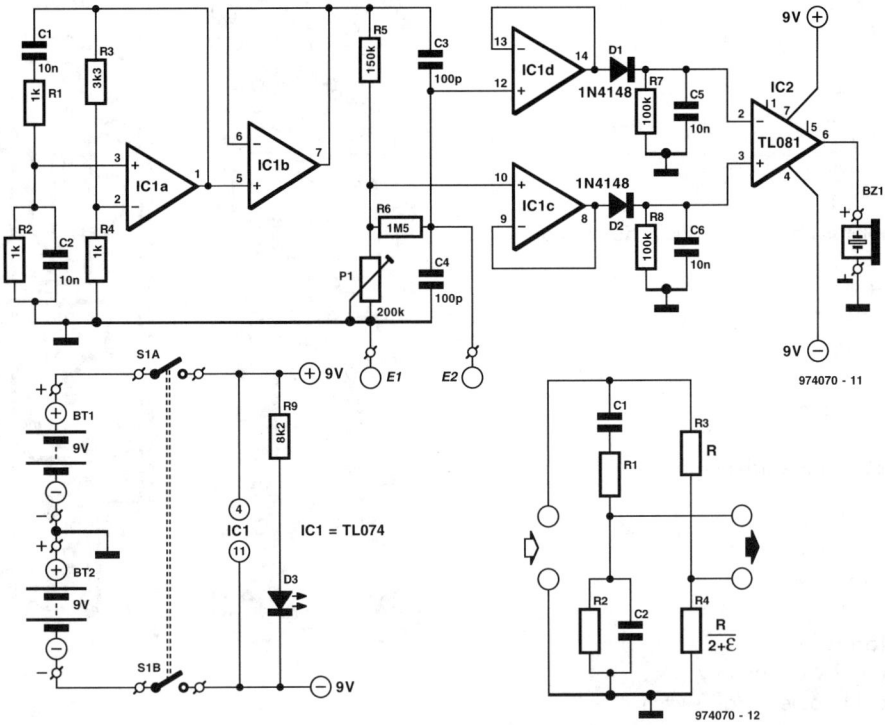

974070 - 11

974070 - 12

of the bridge at the resonance frequency is zero or very nearly so, which prevents any feedback by a factor ε unless the divider is altered slightly. The smaller ε, the more stable the oscillator is. However, a good frequency response requires a large ε and a high output voltage at the resonance frequency, so that the following amplifier can compensate for the losses caused by the bridge. So, if ε is too small, the oscillator does not work, and if ε is too large, the output becomes so large that it overdrives the system. A precisely arranged ε is not easily obtained, whence a Wien-Robinson oscillator is normally provided with an output control.

This is not needed, however, in the oscillator in the circuit in Figure 2. Here, ε=1.3, which ensures that the oscillator works reliably but overdrives so that it behaves as a square-wave oscillator. Its frequency is 15.9 Hz.

After the signal has been buffered in IC_{1b}, it is applied to a measurement bridge consisting of a resistive and a capacitive potential divider.

The proximity sensor connected across E_1 and E_2 is in parallel with C_4. It consists of two metal strips (aluminium foil or tin plate) that are arranged in parallel to form a kind of capacitor.

The sensor is located near the item to be protected. As long as nothing or nobody approaches the item, the capacitance of the sensor is small. Preset P_1 is then adjusted until the voltage across C_3-C_4 is slightly higher than that across R_5-P_1.

The situation changes when somebody, or an animal, or an object comes into the vicinity of the item and acts as the dielectric of the sensor capacitance. This increases the value of the capacitance, whereupon the potential at the junction of the capacitive divider drops, while that across the resistive divider remains constant.

The measurand is buffered by IC_{1c}-IC_{1d} and rectified and smoothed by D_1-C_5 or D_2-C_6 as the case may be. The output of the rectifier is applied to comparator IC_2 which actuates the piezobuzzer when the voltage from the capacitive divider (at its inverting input) drops below that of the resistive divider (at its non-inverting input).

The circuit, except the sensor, is best built on a printed-circuit board as shown, which may be made with the aid of the track layout.

155

Parts list
Resistors:
$R_1, R_2, R_4 = 1\ k\Omega$
$R_3 = 3.3\ k\Omega$
$R_5 = 150\ k\Omega$
$R_6 = 1.5\ M\Omega$
$R_7, R_8 = 100\ k\Omega$
$R_9 = 8.2\ k\Omega$
P_1 = multiturn preset, 200 kΩ, horizontal

Capacitors:
$C_1, C_2, C_5, C_6 = 0.01\ \mu F$
$C_3, C_4 = 0.1\ \mu F$

Semiconductors:
$D_1, D_2 = 1N4148$
$D_3 = LED$, high efficiency

Integrated circuits:
$IC_1 = TL074CN$
$IC_2 = TL081CP$

Miscellaneous:
$Bt_1, Bt_2 = 9\ V$ dry battery
S_1 = double-pole on/off switch
Bz_1 = piezobuzzer

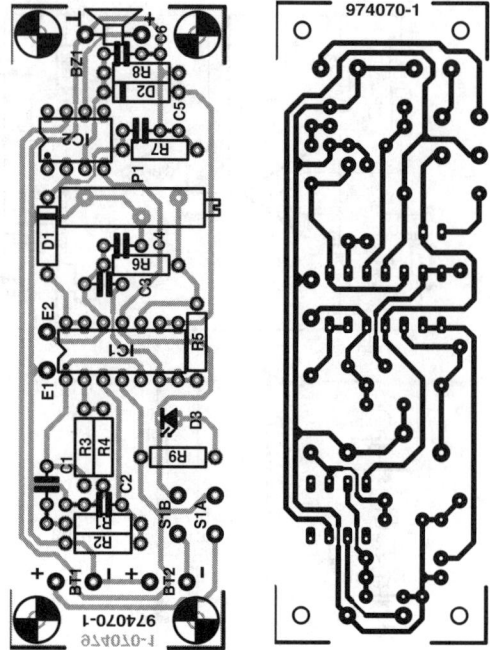

974070-1

PIC-controlled light barrier

The microcontroller-driven light barrier operates with two infra-red senders which, if set up correctly, provide a measure of directivity.

Senders D_5 and D_6 may be standard infra-red diodes whose beam is directed on to infra-red receiver IC_3. Since the senders operate with a 36 kHz carrier, the receiver may be an inexpensive, easily available model with integrated demodulator, filter and amplifier, such as used in IR remote control systems. LEDs D_7–D_{11} function as status, operation and position indicator.

The light barrier is controlled by a Type 16C54 microcontroller, IC_1, which is clocked by a 1 MHz crystal. The outputs of IC_1 are buffered by an integrated octal power driver, IC_2. This enables a buzzer, relay or lamp rated at up to 500 mA to be driven in addition to, or instead of, the LEDs.

The PIC contains two programs. After switch-on,

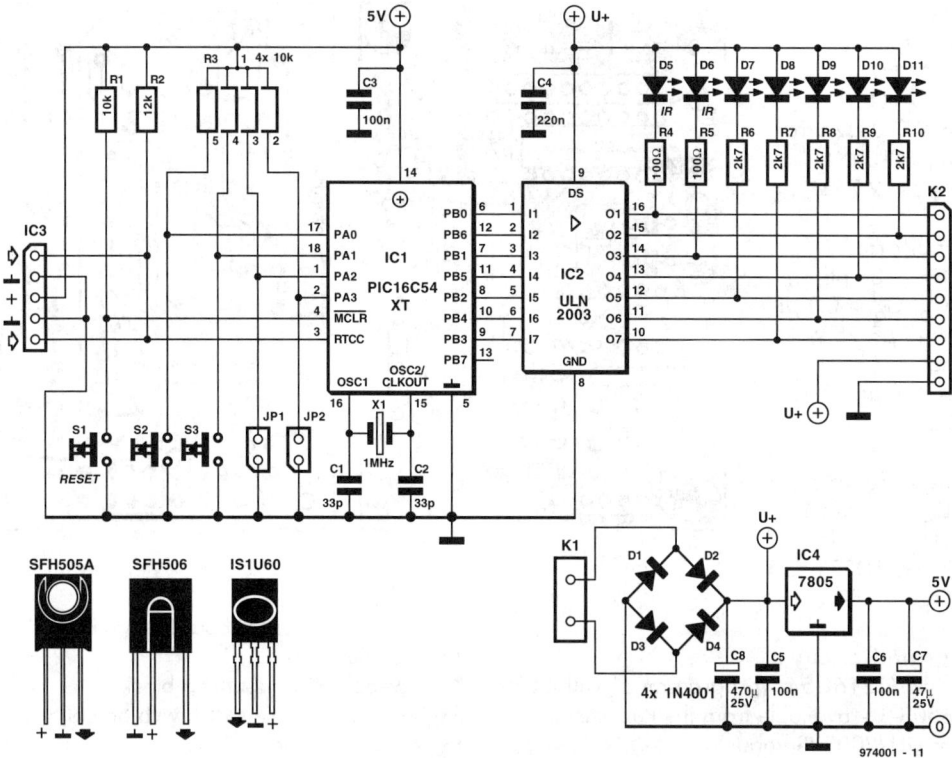

the operating program starts with a routine which initializes the port lines and the various registers. Thereupon, ports PB_0 and PB_1, to which the IR senders are linked, alternately generate four pulses each.

These pulses are detected by the receiver and read via RTCC. If fewer than four pulses are received, the PIC repeats the process twice to ensure that no pulses are lost or that no spurious pulses are received. If during none of the three processes four pulses are received, the program interprets this as an interruption (break) in the light beam.

Whether this arrangement ensures that the system does not react to spurious pulses can be checked by closing jumper JP_1 or JP_2, which disables the control process. The PIC then reacts to the second (JP_1) or first (JP_2) spurious pulse. This action is, however, useful only if the IR beam is set up accurately and screened cable is used.

When the light beam from D_5 is interrupted, D_8 and D_{11} light for 2 seconds, D_7 for about 8 s, and D_9 for around 20 s. In case of D_6, D_{10} lights for about 2 s and D_9 for around 20 s. With the use of an appropriate evaluation circuit, it may be determined in which sequence, or in which direction, the light barrier was interrupted.

The clearly different switching times permit various uses of the alarm outputs. After the last LED has gone out, the program starts anew, so that the light barrier cannot be disabled by covering the IR diodes.

The position program is an important aid during the setting up of the (invisible) infra-red beams. It starts when S_2 (for D_5) or S_3 (for D_6) is held down at the same time as S_1 is pressed for a reset. As long as there is no IR link established, D_{11} flashes (for D_5) or D_{10} (for D_6). When the link has been established, the relevant LED lights continuously. A return to the operating program is effected by another reset.

The circuit draws a current of about 25 mA plus that through the load connected to IC_2. The power supply may be a 9 V mains adaptor, which need not be regulated.

The light barrier may be constructed on the printed-circuit board shown, which is, however, not available ready-made.

157

Parts list:

Resistors:
R_1 = 47 kΩ
R_2 = 12 kΩ
R_3 = 4×10 kΩ array
R_4, R_5 = 100 Ω
R_6–R_{10} = 2.7 kΩ

Capacitors:
C_1, C_2 = 33 pF
C_3, C_5, C_6 = 0.1 μF
C_7 = 47 μF, 25 V, radial
C_8 = 470 μF, 25 V, radial

Semiconductors:
D_1–D_4 = 1N4001
D_5, D_6 = infra-red LED, e.g. LD271
D_7–D_{11} = LED

Integrated circuits:
IC_1 = PIC16C54 (programmed with EPS 976503-1 – available from the Publishers)
IC_2 = ULN2003 (Sprague)
IC_3 = SFH505 or SFH506-36 or ISU60 (36 kHz)
IC_4 = 7805

Miscellaneous:
JP_1, JP_2 = jumper for board mounting
K_1 = soldering pin
K_2 = 9-pole PCB mounting block
S_1–S_3 = single-pole switch with on contact
X_1 = crystal, 1 MHz
PCB may be made with the aid of the track layout.

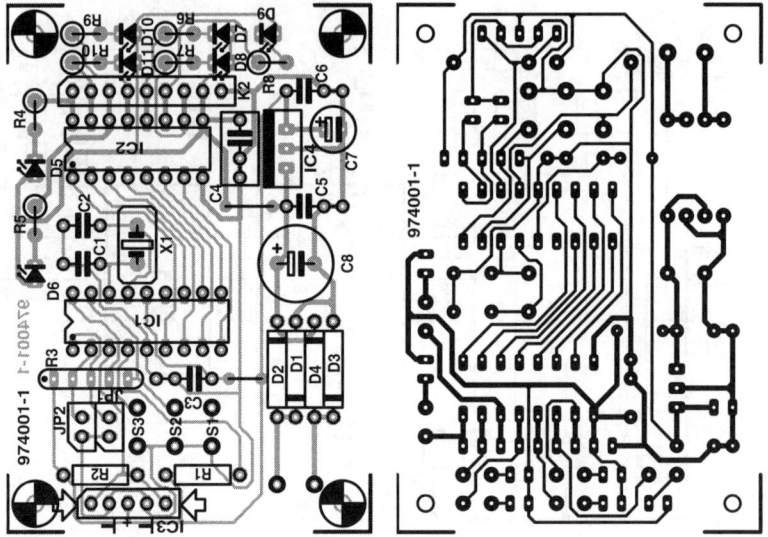

Pin-ball machine

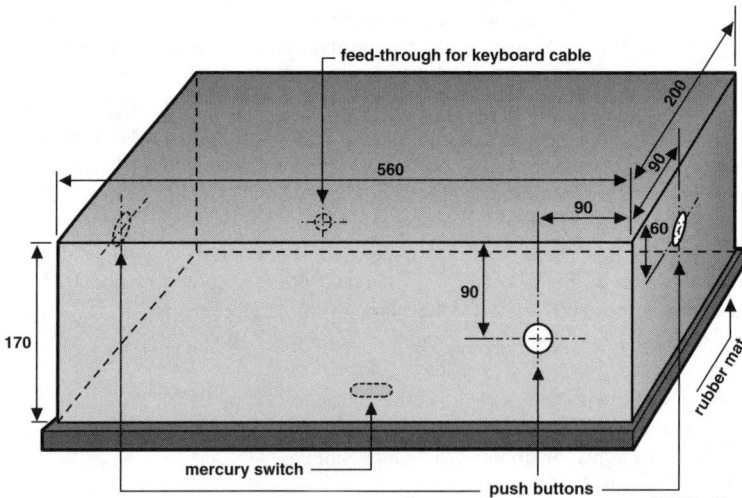

feed-through for keyboard cable

560

200

90

90

90

60

170

90

mercury switch

push buttons

rubber mat

974095 - 11

Commercially available electronic pin-ball machines, such as that from Microsoft, are controlled via a standard keyboard. Since this does not really resemble the traditional mechanical pin-ball machine, an alternative is presented here.

The machine is made from a few boards of wood, two switches, a push-button switch, a mercury switch and a discarded keyboard.

The crux of the design is a wooden box that is wide enough to resemble a 'real' pin-ball machine. The two switches are fitted at the sides of the box. and the start push-button at the front.

The mercury switch, which functions as the nudge detector, is fitted inside the box. Normally, this switch is allowed to react three times before the game is over.

In practice, this switch makes sense only if your program has allowed for a key to simulate pushing against the box. Usually, the space bar is used for this.

Glue a thick rubber mat or four rubber feet at the underside of the box to give a a feel of reality to movements of the box.

Link the switches to the relevant keys of the keyboard. If desired, this may be done semi permanently by fitting a mini DIN socket on the keyboard.

Real luxury is obtained if a disused keyboard is mounted inside the box. Only the cable from the keyboard then emerges from the interface. Connect this to a PC and play away!

The artist's impression shows what a practical construction may look like.

Rev(olution) counter

The proposed circuit is a general-purpose rev(olution) counter for cars. It is suitable for 4-, 6- or 8-cylinder engines.

In a traditional manner, a direct voltage is derived from the ignition pulses. The voltage level is directly proportional to the number of pulses per unit time. The direct voltage is converted by an ADC (analogue-to-digital converter) into a BCD (binary-coded decimal) signal, which is read with the aid of a decoder and a set of 7-segment displays.

The requisite pulses are taken directly from the circuit breaker (CB) in the engine compartment and applied to K_1. Any unwanted peaks are removed by low-pass filter R_3-C_6, while the level is held to a safe value by D_1.

The signal is subsequently amplified by T_1 and then applied to monostable multivibrator (MMV) IC_1. This stage converts the signal into a series of regular pulses, which are integrated by C_1. In other words, the potential across this capacitor is a measure of the number of pulses, that is, engine revo-

lutions. This voltage is measured by ADC IC_3. This circuit has four BCD outputs and three digit-drives and, in conjunction with IC_2, a BCD-to-7-segment decoder, drives displays LD_1–LD_3.

The number of engine cylinders is determined by correcting the potential across C_1 with the aid of divider R_4–R_9. Four-cylinder engines produce four pulses, six- cylinder engines, six pulses, and eight-cylinder engines, eight pulses, for every two

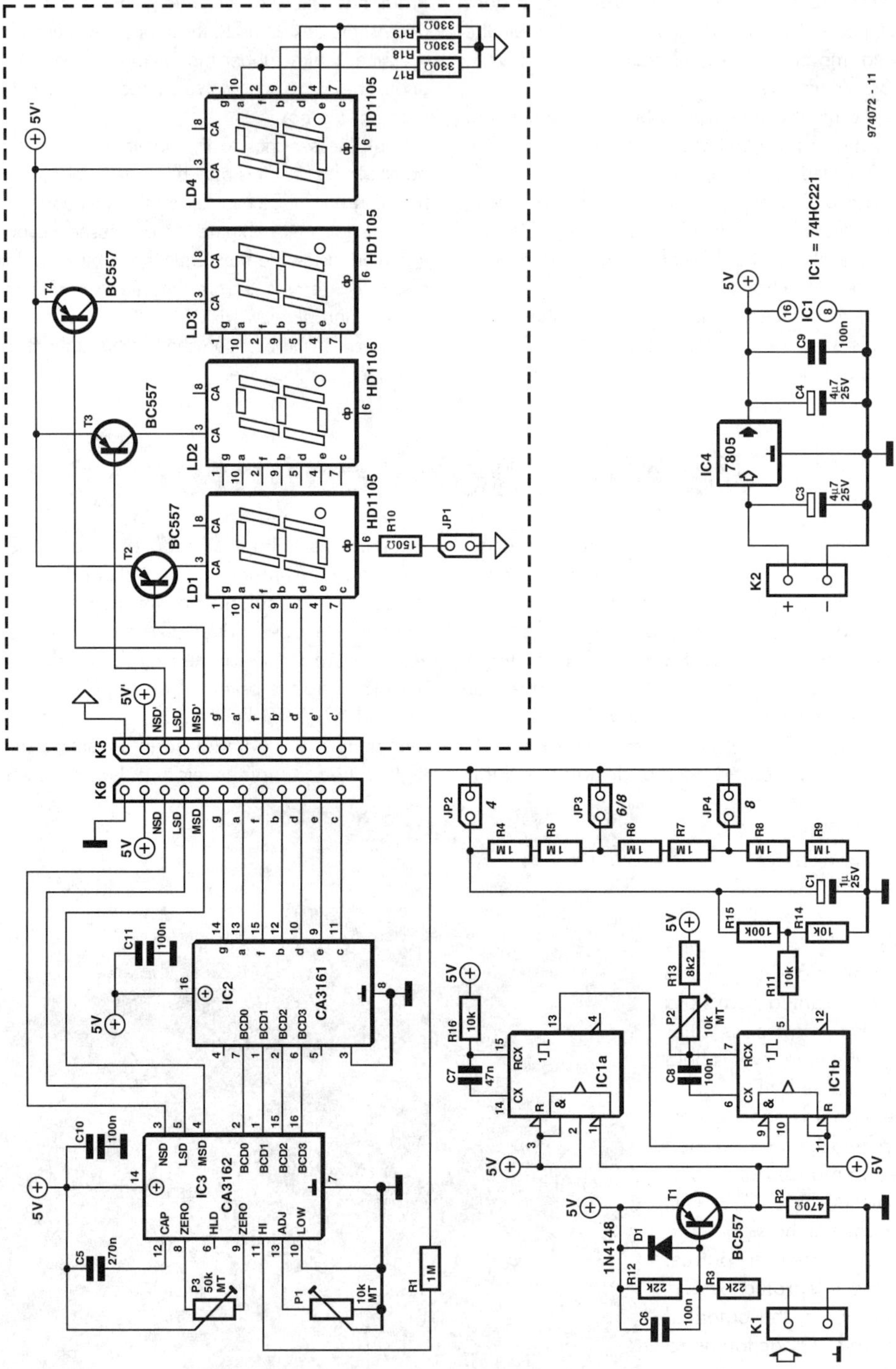
IC1 = 74HC221
974072 - 11
IC2
CA3161
IC3
CA3162
IC4
7805
HD1105
BC557
1N4148
LD1
LD2
LD3
LD4
K1
K2
K5
K6

revolutions. In the case of a four-cylinder engine, JP_2 is short-circuited and the potential across C_1 is applied to IC_3. With six-cylinder engines, JP_3 is short-circuited so that the voltage across C_1 is divided by 4/6, and with eight-cylinder engines, JP_3 and JP_4 are short-circuited which results in the potential across C_1 being divided by two.

To calibrate the circuit, remove any jumpers and short-circuit the input of IC_3 (R_1) to earth. Adjust the offset with P_3 until the display reads '000'. Next, apply a voltage varying from 0 V to 1 V to the input of IC_3, measure every step with a DVM (digital voltmeter) and adjust P_1 for a display of exactly the same voltage. Finally, use a good-quality rev counter as reference, or apply a suitable voltage from a function generator with digital display to K_1, and adjust P_2 until both readings are the same.

The printed-circuit board, which is not available ready made, may be cut into two to separate the display section from the remainder. The two parts should then be interconnected by a length of flatcable between K_5 and K_6.

In case it is desired to get a display of '3400' instead of '340' when the number of revolutions is 3400, add LD_4. When only three displays are used, place JP_1 as indicated. This causes the decimal point of LD_1 to light to show that the display reading must be multiplied by 1000.

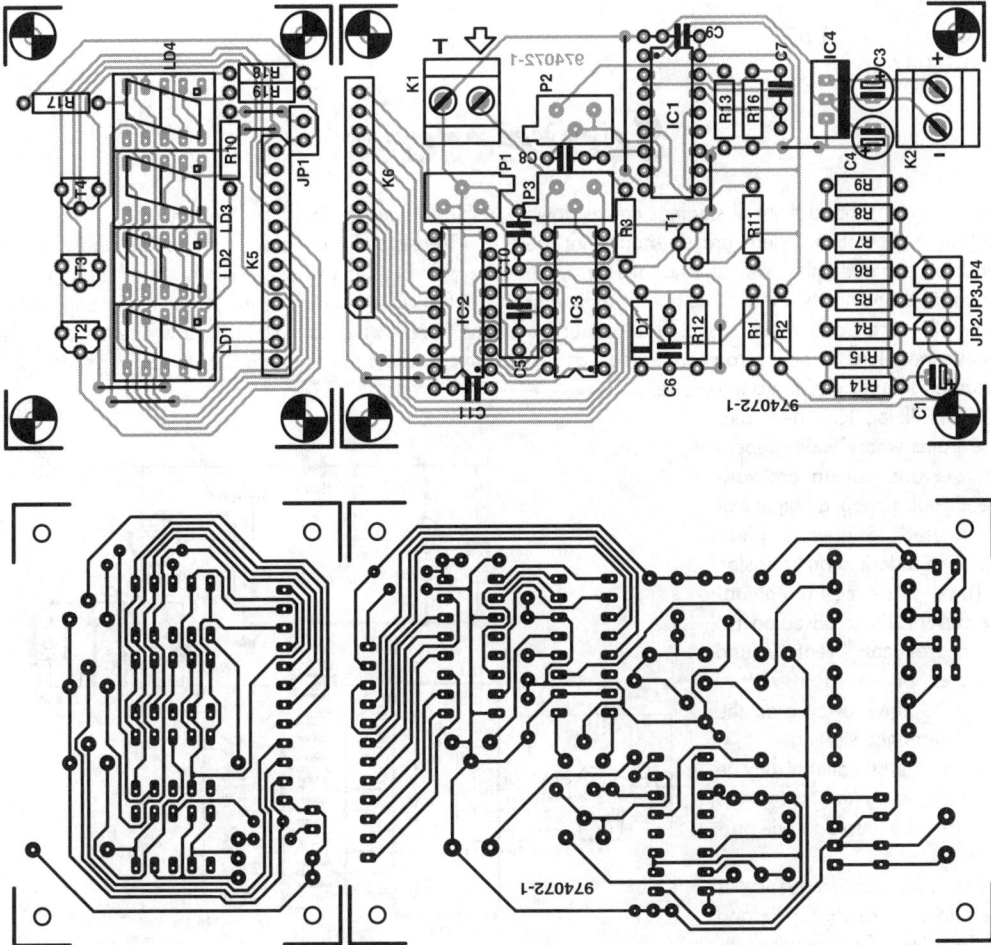

Parts list

Resistors:
R_1, R_4–R_9 = 1 MΩ
R_2 = 470 Ω
R_3, R_{12} = 22 kΩ
R_{10} = 150 Ω
R_{11}, R_{14}, R_{16} = 10 kΩ
R_{13} = 8.2 kΩ
R_{15} = 100 kΩ
R_{17}–R_{19} = 330 Ω
P_1, P_2 = 10 kΩ, multiturn preset for vertical mounting
P_3 = 50 kΩ, multiturn preset for vertical mounting

Capacitors:
C_1 = 1 μF, 25 V, radial
C_3, C_4 = 4.7 μF, 25 v, radial
C_5 = 0.27 μF
C_6, C_8–C_{11} = 0.1 μF
C_7 = 0.047 μF

Semiconductors:
D_1 = 1N4148
T_1–T_4 = BC557

Integrated circuits:
IC_1 = 74HC221
IC_2 = CA3161E
IC_3 = CA3162E
IC_4 = 7805

Miscellaneous:
LD_1–LD_4 = HD11050
JP_1–JP_4 = jumper
K_1, K_2 = 2-way terminal block for board mounting
K_5, K_6 = 12-pin SIL header, right-angled
PCB may be made with the aid of the track layout.

Scarecrow

It often happens that a flight of starlings comes down into a tree. Sometimes there are so many that the branches bend under their weight. And, of course, the amount of droppings below the tree does not do your garden (or your car) much good. The scarecrow requests these birds in an environment-friendly as well as starling-friendly way to go elsewhere.

An oscillator, IC_1, generates pulse trains whose width, repetition rate and pattern are variable. Its pulse-shaped output signal is used to drive a piezo buzzer via a darlington transistor, T_1. Power is derived from a variable current source based on T_2. The buzzer can create sound pressures of up to 100 dB – enough to drive away even the most determined starling.

Up to 256 bit patterns may be set with S_1 to make sure that the birds do not get used to one particular sound.

The oscillator frequency is determined by R_1-R_2-C_1-P_1 and may be set with P_1 to a pulse rep-

etition frequency (p.r.f.) of 0.5–5 Hz.

Transistor T_1 is on as soon as Q_3 and those of outputs Q_5–Q_{13} whose switch is closed are high.

The current drawn by the circuit is primarily that of the buzzer which is <150 mA. It is, therefore, advisable to use a mains adaptor as the power source.

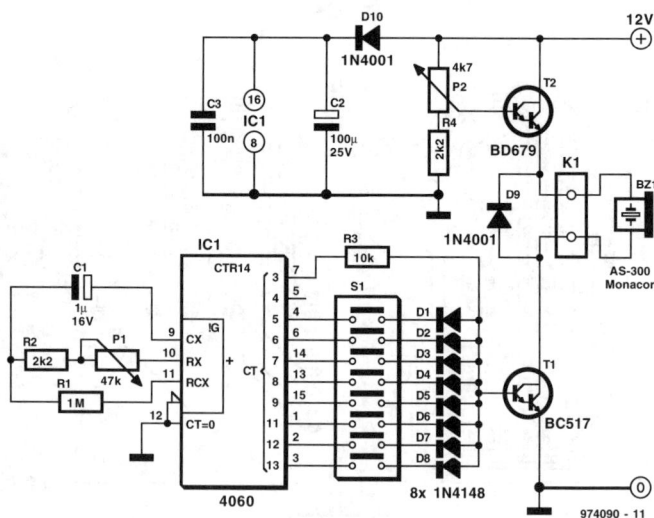

974090 - 11

162

Spark-plug monitor

As the name indicates, the monitor is intended for checking whether the spark-plugs of a car actually produce a good ignition spark. There are commercial monitors for this purpose on the market but these cost rather more than the one described here.

The circuit consists of two anti-parallel connected LEDs that are placed in series with the spark-plugs. Two diodes are used since the direction of the current is not always evident and also because oscillations occur in the ignition circuit. High-efficiency diodes that light brightly at a current of about 1 mA should be used.

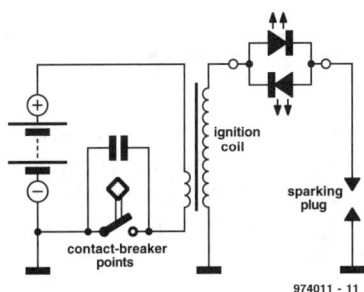

974011 - 11

The brightness of the diodes depends on the level of the current drawn by the spark plugs and this, of course, is a measure of the strength of the spark. If the spark-plug is short-circuited, which is highly unlikely (and could be ascertained with a ohmmeter), neither of the diodes lights.

The monitor must be provided at one side with a small piece of tin-plate that is inserted into the spark-plug socket and at the other with a short length of tough wire that is wound a couple of times around the spark-plug lead.

Normally, only one of the LEDs will light, but, owing to oscillations in the LC circuit formed by the coil and the capacitor in the ignition circuit, it may happen that both light.

When the engine is started, or when it runs at low constant speed, the diode flickers, but at higher engine speeds it lights uniformly.

In the case of an engine with point contacts as in the diagram, the brightness of the diode remains virtually constant when the engine speed increases. This is because although the ignition frequency rises, the pulse duration shortens.

If at higher engine speeds the brightness of the diode diminishes, either the gap in the contact breaker is set incorrectly or the contacts 'float'. Both of these events result in a shortening of the time that the contacts are closed. As a consequence, there is insufficient time for the ignition coil to recover its electric field. When the contacts are then opened, only a weak spark will be produced.

In multi-cylinder engines it is advisable to give each spark-plug its own monitor, which makes it possible for the four or six spark-plug currents to be compared.

Warning! Make sure that the ignition switch is off when connecting the monitor to the spark-plug socket since random ignition pulses may be lethal.

Starter for fluorescent tubes

Owing to the difficulty of obtaining certain parts, the starter will have more curiosity than construction value for many readers. Constructors should bear EMC Regulation EN50081-1 and EN60555 Parts 2 and 3) pertaining to low-frequency interference in mind.

Note that, owing to D_1, the starter operates with direct voltage. At switch-on, C_1 is discharged which prevents the thyristor, Thy_1, being triggered.

Just before the peak of the first period of the operating voltage is reached, Thy_1 is triggered via R_4, so that a positive current flows through the ballast coil and the filament of the tube.

At the next zero crossing, Thy_1 is cut off so that no current flows through the filament of the tube. At the same time, a voltage is induced in the ballast coil which triggers the tube to some extent.

Capacitor C_1 is charged by the voltage drop across D_2 via R_5 and R_6, until, after a few positive half cycles, the potential across it has risen to a level at which Thy_2 is triggered halfway during the positive half cycle and removes the voltage from the gate of Thy_1.

163

If the current through Thy1 is lower than the hold current, Thy1 is cut off, whereupon a counter-e.m.f. is generated that is strong enough to trigger the tube. Potential divider R_2-R_3 ensures that C_1 remains charged and Thy1 remains off. If the tube is not triggered, the foregoing process repeats itself.

Thyristor Thy2 is a special type from SGS Thomson which has a very sensitive gate. A current lower than 1 mA is enough to trigger the device.

Thyristor Thy1 is typified by a large hold current (>175 mA).

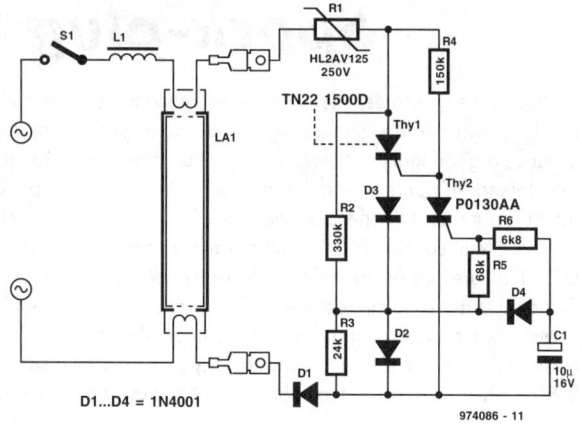

D1...D4 = 1N4001

974086 - 11

Telecom control box

The control box provides a bidirectional connection between two telephones, two fax machines, or a fax machine and a fax modem. It also allows a fax machine to be used as scanner or printer. It is built from relatively few standard, inexpensive components.

The function of the circuit is straightforward: the two units act as if a normal telephone communication is established. The data exchange is started manually at one of the units while the other is switched to the receive mode. Its power may be provided by a 9 V dry or rechargeable battery or a 12 V mains adaptor (which need not be regulated).

Field-effect transistors T_1 and T_2 are arranged as 20 mA constant-current sources and simulate the power supply of the trunk line. Resistors R_1 and R_2 represent the off-the-hook receiver. Capacitors C_1 and C_2 only pass alternating currents, such as data or speech signals.

The data traffic between the two units (the typical fax flutter or the dial tone in case of a telephone) may be monitored with a pair of headphones by tapping into the line between the two capacitors. The requisite 5:1 or 10:1 transformer can be obtained from an old telephone or purchased from an electronic retailer.

The circuit is so tiny (and even more so if the monitoring facility is omitted) that it may be constructed 'in the air' or on a terminal strip. It may also be built into the telephone without too much trouble. In the latter case, a miniature switch must be added for switching between the trunk line and the box.

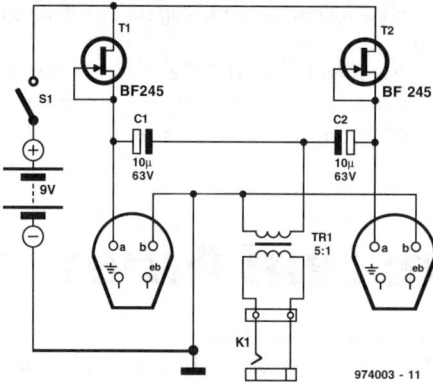

974003 - 11

Variable-pulse generator

There is often a requirement for a generator that provides variable-width pulses. Examples abound: stair lighting; time/interval switch in private and commercial vehicles; time switch for room lighting, radio or stereo equipment, or ventilators. The integrated timer frequently used is the 555. This has a drawback, however, in not being able to provide very long pulse-widths. Moreover, long pulse-widths cannot be set accurately with a potentiometer.

In the present circuit, the pulse-width can be set very accurately over a wide range. The circuit is controlled by an RS bistable consisting of gates IC_{3a} and IC_{3b}. It is set by operating switch S_1 at the Set input (pin 6).

The output of the bistable (pin 3) trips the output relay via T_1. Freewheeling diode D_1 protects the circuit against inductively induced surges. The status of the circuit is indicated by D_3.

The bistable is returned to its output state via the Reset input (pin 1). A reset also occurs on switch-on via R_4-C_2.

The Reset input is driven by the output of of a counter/ oscillator. In the quiescent state, decade counter IC_2 is disabled via its Enable input while IC_1 oscillates and counts. This does not matter, however, since a push on the button resets both counters to zero. The oscillator pulses then start being counted and when the 16348th pulse arrives at pin 3, the decimal counter is clocked.

Next, S_2, which may be a rotary switch, a DIP switch or a jumper, enables one of the next ten pulses to be selected and applied to the reset line of the multivibrator. Diode D_2 prevents any feedback to the outputs of IC_2.

The time elapsing until 16348 or 163480 pulses are generated depends, of course, on the oscillator frequency and, therefore, on the setting of P_1. A

1 kΩ preset makes switching times up to 2 s possible. Resistance and pulse duration are virtually directly proportional.

The circuit draws a current of 15 mA (mainly on account of the LED), excluding the relay current.

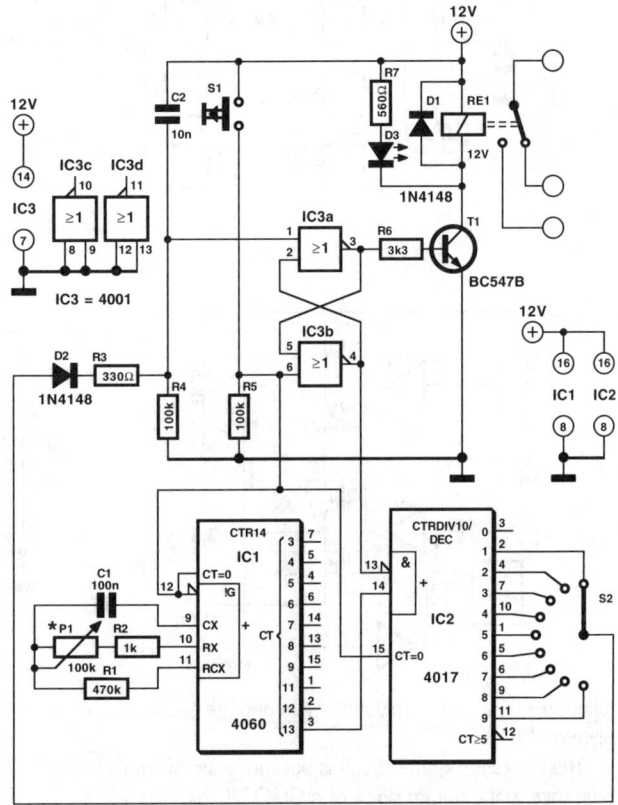

974017 - 11

Voltage inverter/doubler

The circuit in Figure 1 consists of a charge pump which can be switched for use as a voltage inverter or a voltage doubler. It requires two CMOS ICs Type 74HC14, but Type 40106 may also be used, although this provides a rather smaller current.

Circuit IC_{1a} is arranged as an oscillator operating, with components as specified, at a frequency of about 160 kHz.

Inverters IC_{1b}–IC_{1f} are in parallel to ensure sufficient output current.

The signal is also applied to a second group of

inverters in parallel, IC_{2b}–IC_{2f}, via C_2. Since this capacitor passes a pulse only at the edges of the oscillator signal, this group of inverters is augmented by positive feedback provided by IC_{2a}. Resistor R_2 ensures that the feedback circuit is triggered via C_2.

Capacitor C_3 is the charge pump that transfers the requisite energy. Because of the relatively high frequency, the value of this capacitor need not be high.

Adjacent to the diagram is shown how the circuit can be used as either a voltage in-verter or a voltage doubler. Note that the inverted or doubled voltage is

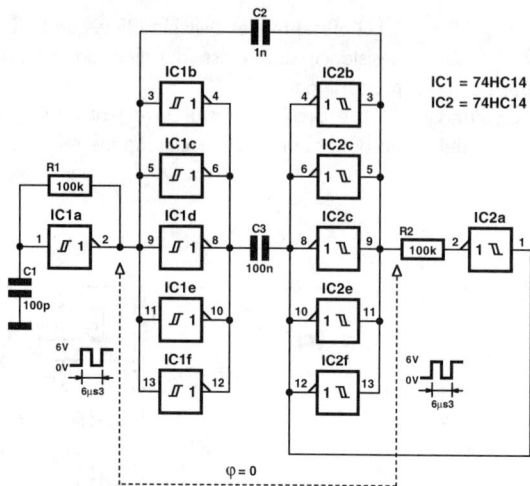

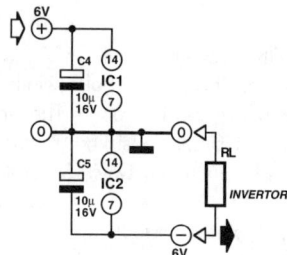

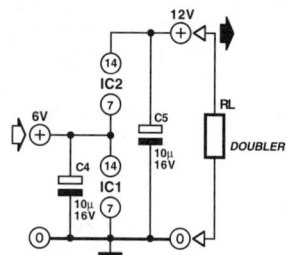

IC1 = 74HC14
IC2 = 74HC14

974010 - 11

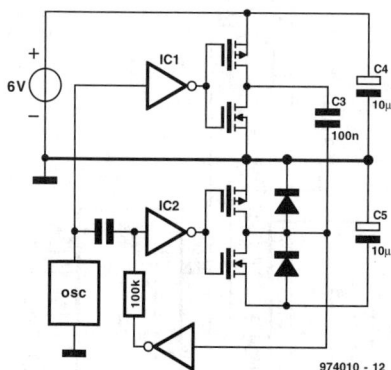

974010 - 12

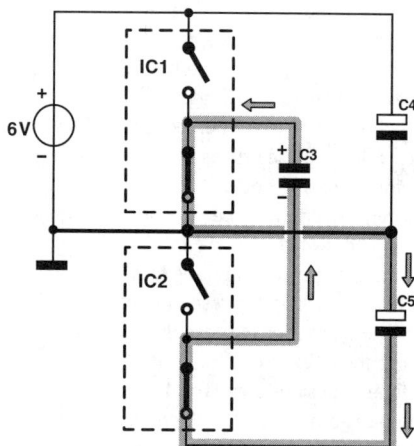

974010 - 14

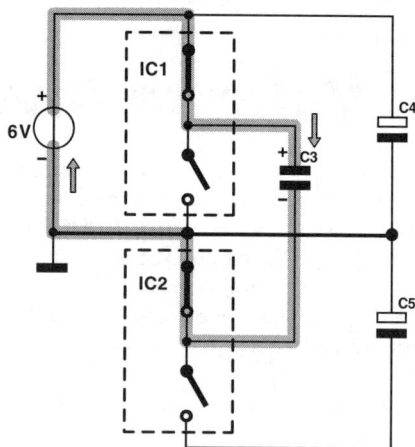

974010 - 13

taken from the +ve supply pin of IC_2 and the –ve supply pin of IC_1.

The operation of the circuit is best understood with reference to the output stage of a CMOS IC as shown in Figure 2.

Both groups of inverter switch in phase. In case of a voltage inverter this results in C_3 being charged during one half-period (in parallel with the supply line—see Figure 3a), whereas during the other half-period the capacitor is in parallel with C_5, so that this capacitor is also being charged—see Figure 3b. During voltage-doubling, virtually the same happens, but in this case the potential across C_2 is superimposed on the output of IC_1.

The unloaded output voltage is virtually double the supply voltage (when the circuit is acting as a doubler). When the output current is 10 mA and the supply voltage is 6 V, the output voltage drops to –5 V (inverter) or rises to +11 V (doubler). When the circuit is unloaded, it draws a current of only 1.5 mA.

150-watt lamp dimmer

The α1108APA from Alpha Electronics is a lamp dimmer circuit that requires few external parts. The IC provides soft-start and soft switch-off options to protect the filament of incandescent bulbs.

A resistor with negative temp[erature coefficient (NTC), R_1, is connected in series with the load to protect the IC against high inrush currents which may occur as the dimmer is turned on, or when a lamp is changed while the dimmer is on. Without the NTC, the circuit is suitable only for lamps with a power of 75watts or less.

phase blanking. These values correspond to minimum and maximum brightness of the lamp, respectively.

The choke-capacitor filter between the dimmer IC and the mains input is obligatory in applications where EMI suppression is a design requirement.

The α1108APA is suitable for use with a.c. voltages between $80V_{rms}$ and $276V_{rms}$.

The PCB may be made with the aid of the track layout.

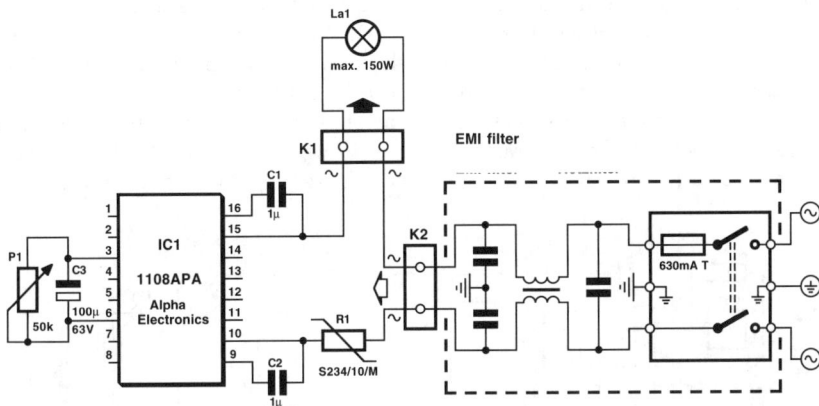

974023 - 11

The brightness of the lamp increases with the resistance presented by potentiometer P_1.

The α1108APA contains two anti-parallel silicon-controlled rectifiers (SCRs), which are connected in series with the load, and switched by the internal phase control circuit. The control circuit is run off the mains voltage, and also realizes the necessary protection functions including complete SCR shut-off in case of thermal overloading, turn-on of the SCSs in case of a voltage surge between anode and cathode, and disabling of the SCRs the first time the circuit is closed (as a result of a fast dv/dt peak or compensation currents caused by parasitic capacitance). The voltage on capacitors C_1 and C_2 is proportional to the voltage between the control inputs C+ and C-. These capacitors are discharged at each zero crossing of the mains input voltage, and they are charged again (depending on the control voltage) to the firing voltage of the SCRs. The range of the control voltage is about 6V, where 0V means total phase blanking, and 6V no

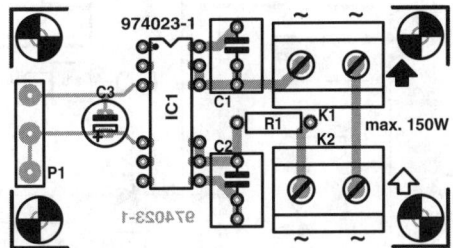

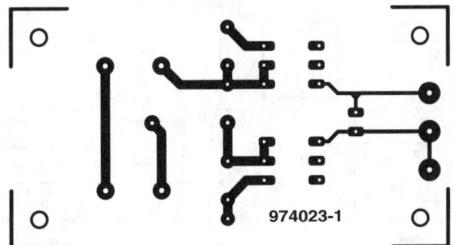

974023-1

974023-1

167

Parts list
Resistors:
R_1 = 10 Ω, negative temperature coefficient (NTC)
P_1 = 50 kΩ linear potentiometer, polythene shaft

Capacitors:
C_1, C_2 = 1 μF
C_3 = 100 μF, 63 V, radial

Integrated circuits:
IC_1 = α1108APA (Alpha Electronics)

Miscellaneous:
K_1, K_2 = 2-way terminal block, pitch 7.5 mm
EMI filter, e.g. Belling-Lee type L2777
Fuse, 630 mA

Adaptive windscreen wiper control

Although a very useful car accessory, the typical adjustable windscreen wiper interval control never seems to get it right; the delay between wiper activity is either too short or too long, and you seem to be forever busy tweaking the interval control pot to match the amount of rainfall. A more or less automatic adjustment to rainfall variation has an intuitive touch: measure the time between two wiper actions performed by the driver, store this delay, and then use it to control the wipers automatically.

The circuit is based on common-or-garden CMOS logic wired to form a clock/counter configuration. The unit is controlled by a single push-button, S_1, and on/off switch S_2. The push-button acts as a start/stop control which determines the length of the wiper interval. Regrettably, it will not be possible in many cases

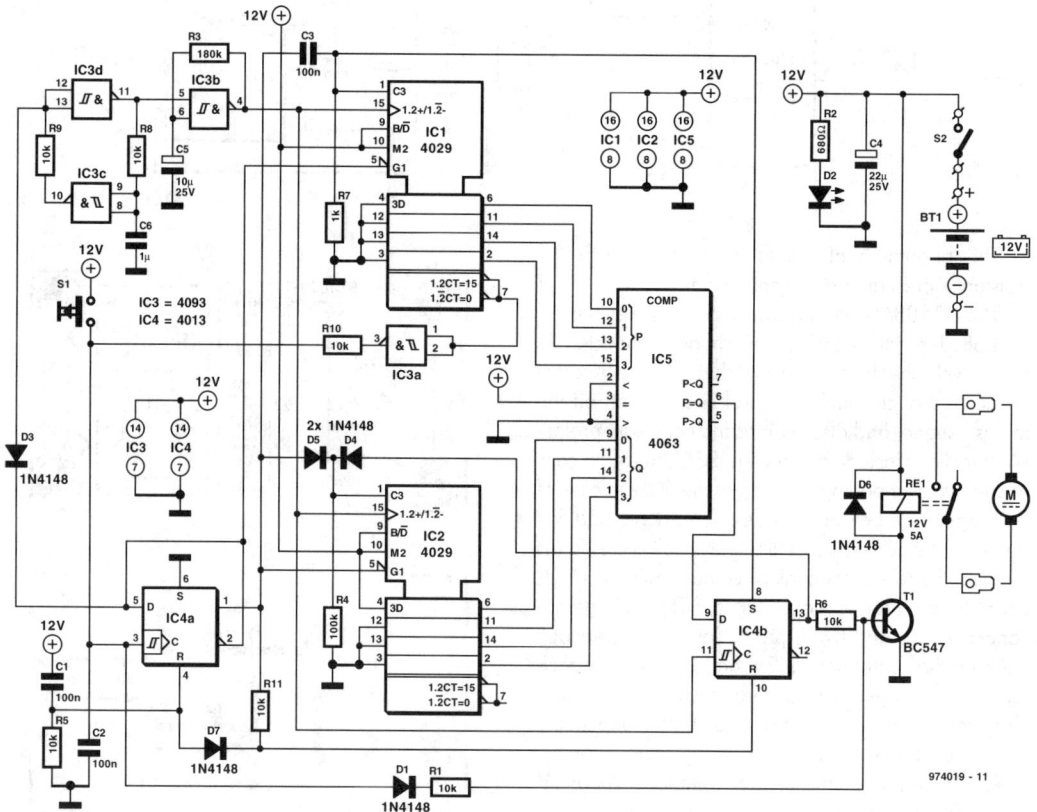

974019 - 11

to employ the wiper control lever already fitted on the steering column, and a suitable location will have to be found for the push-button on the dashboard. With luck, the existing wiper control simply switches the 12-V supply to the wiper motor relay. In that case, the switched voltage may be taken to the 'lower side' of S_1 (see circuit diagram), whereupon the circuit will work without the external push-button.

At the heart of the circuit are two 4094 counters, IC_1 and IC_2. The first of these sets the (variable) wiper delay, while the second operates in free-running mode. Both counters receive a 1 second clock signal from Schmitt trigger oscillator IC_{3b}, which is not enabled until the bistable around IC_{3d} and IC_{3c} is set. Because of 4-bit comparator IC_5, the wiper relay, Re_1, will be actuated only when the state (output value) of IC_2 equals that of IC_1.

The interval is the time elapsing between two actions on push-button S_1. On the first action, IC_1 starts to count, and continues counting until either S_1 is pressed again or IC_1 produces a carry-out pulse via inverter IC_{3a}. This happens when the counter has cycled through all its possible states (16). Consequently, the Q output of IC_{4a} drops low, causing IC_2 to start counting, and IC_1 to hold. Diode D_7 and resistor R_{11} prevent the wiper relay from being actuated during the interval adjustment period, because IC_1 and IC_2 then briefly produce the same output value, 1, which equals the preset value loaded by IC_2.

When the counters produce equal output states, the data input of D bistable IC_{4b} goes high. At the next clock pulse, the Q output will change to high, causing IC_2 to be preset with the value 1 again, and the relay driver, T_1, to be switched on. This again clears the 1 at the input of IC_{4b}, causing the PE (preset enable) signal to disappear and IC2 to start counting again.

The wiper motor(s) may be controlled directly by Re_1, or indirectly using the existing relay fitted in the car. In the latter case, the specification of Re1 may be lighter than indicated in the parts list.

Parts list

Resistors:
R_1, R_5, R_6, R_8–R_{11} = 10 kΩ
R_2 = 680 Ω
R_3 = 180 kΩ
R_4 = 100 kΩ
R_7 = 1 kΩ

Capacitors:
C_1, C_2, C_3 = 0.1 μF
C_4 = 22 μF, 25 V, radial
C_5 = 10 μF, 25 V, radial
C_6 = 1 μF, pitch 5 mm or 7.5 mm

Semiconductors:
D_1, D_3–D_7 = 1N4148

D_2 = red LED
T_1 = BC547

Integrated circuits:
IC_1, IC_2 = 4029
IC_3 = 4093
IC_5 = 4063

Miscellaneous:
Re_1 = relay, 12 VDC coil, 250 VAC/8A contact
2 off car-type spade terminals
S_1 = push-button, 1 make contact
S_2 = on/off switch
PCB may be made with the aid of the track layout.

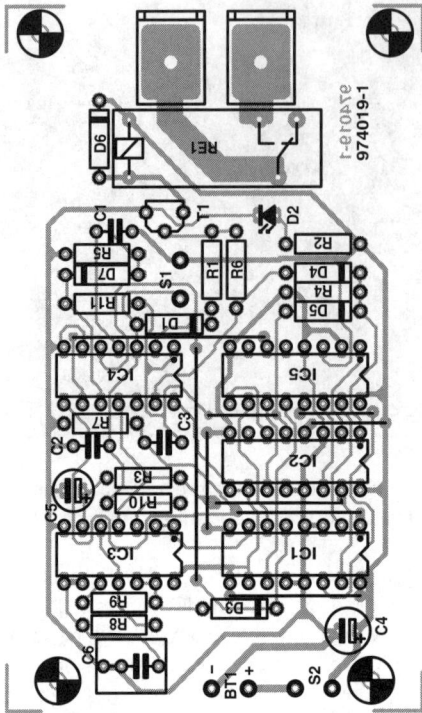

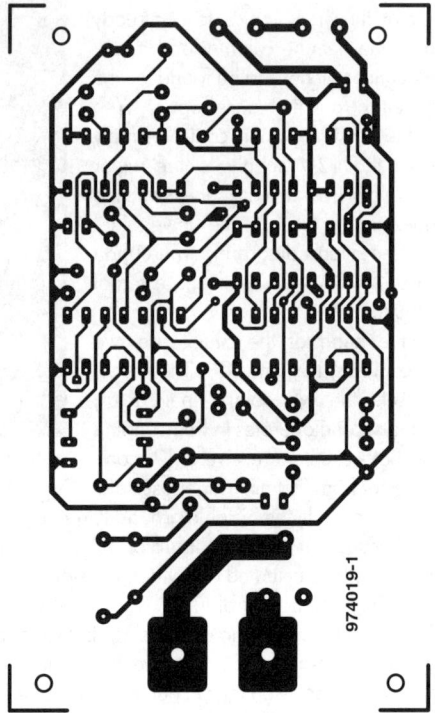

Battery saver

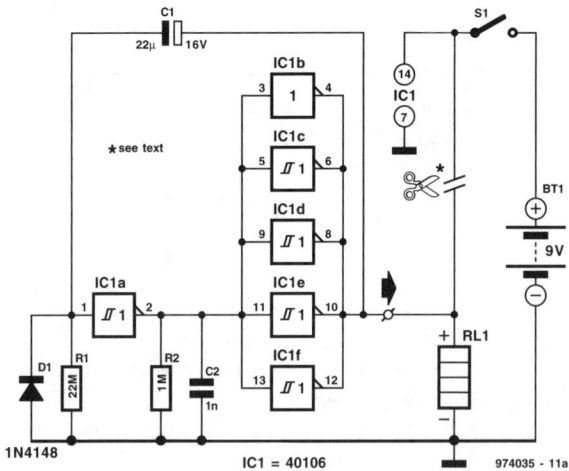

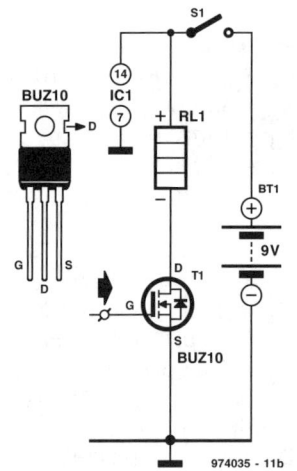

974019-1

974019-1

IC1 = 40106

974035 - 11a

1N4148

BUZ10

974035 - 11b

The saver is intended to prevent a battery-operated instrument such as a multimeter without an automatic off switch to be left on for days on end and so completely discharge the battery.

The circuit described is inserted in the +ve supply line and breaks the +ve supply to the instrument after this has not been used for about six minutes.

Gates N_1 and N_2 form a monostable multivibrator (MMV). When the supply is switched on, capacitor C_2 arranges for the input of N_2 to be grounded, so that the output level is about equal to the supply voltage. The load is then energized.

At the same time, the 9 V level at the output of N_2 is applied to the input of N_1, whereupon the output of this gate goes low. This is the initial state: no change. However, capacitor C_1 is gradually being discharged via R_1, which, owing to the high value of this resistor, is a very slow process. Nevertheless, after about six minutes the potential across C_1 will have dropped suf-

ficiently to cause a low level at the input of N_1. The output of this inverter changes state, so that the output of N_2 becomes low and this causes the supply to the load to be discontinued.

Restarting is effected by pressing on/off switch S_1. So as to make the current to be switched as large as possible, the remaining gates in IC_1 are linked in parallel with N_2. Each gate can provide a current of about 0.5 mA, so that the total output current is about 2.5 mA, which is quite sufficient for most test instruments. If a larger output current is needed, a FET (BUZ10) may be added in series with N_2: this raises the current to a couple of amperes. In that case, make sure that the +ve terminal of the load is fixed and the –ve line is switched.

If the delay of 6 minutes is found too long or too short, it may be altered by changing the value of R_1 empirically.

Linear optoisolator

The fact that the Texas Instruments TIL300 optoisolator contains two photodiodes is exploited here to endow the device with a virtually linear transfer characteristic. The trick is to include one of the photodiodes in the feedback circuit of the LED driver, while the other is used to drive an output buffer as usual. Provided that the two photodiodes are virtually identical, the feedback circuit irons out any non-linearity of the transmit diode and the photodiode.

Non-linearity of the circuit should be ≤ 2% or so, which is not bad for such a simple setup. The advantage of using a TLC271 is that its common-mode range goes down to 0 V, allowing small input and output voltage levels to be used, while the supply voltage may remain asymmetrical. A prototype of the circuit produced an output signal of 10 V_{pp} at 50 kHz, albeit at considerable distortion. For accurate operation, the frequency should be much reduced. In this respect, it is recommended to experiment with the

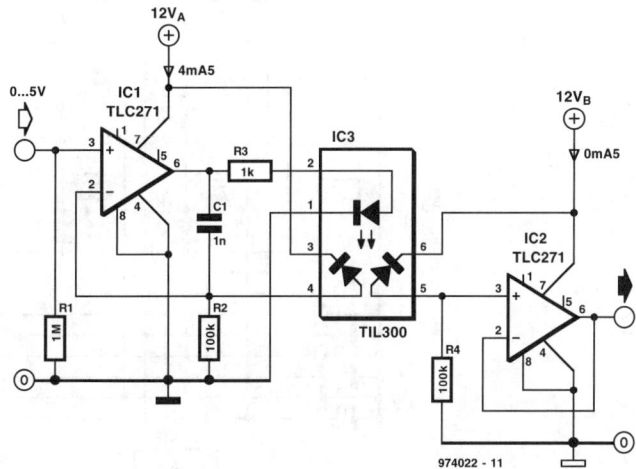

value of C_1, which may need to be fine-tuned to achieve the best possible frequency compensation (minimize overshoot in the output signal). In the circuit below, the TLC271 is used in high-bias mode here (pin 8 tied to ground). The use of faster and more accurate op amps will undoubtedly produce even better results.

Octopush

Octopush is a game for two players (or teams) to pit their wits against one another: they are identified as red or green. There are two rows of push-buttons, four red and four green, with adjacent light-emitting diodes (LEDs). These flash sequen-

tially between red and green at a speed that can be manually adjusted. The object is for a player to press his or her associated push-buttons during the brief time that the player's LEDs are alight. Each push-button enables an energy store to be tanked

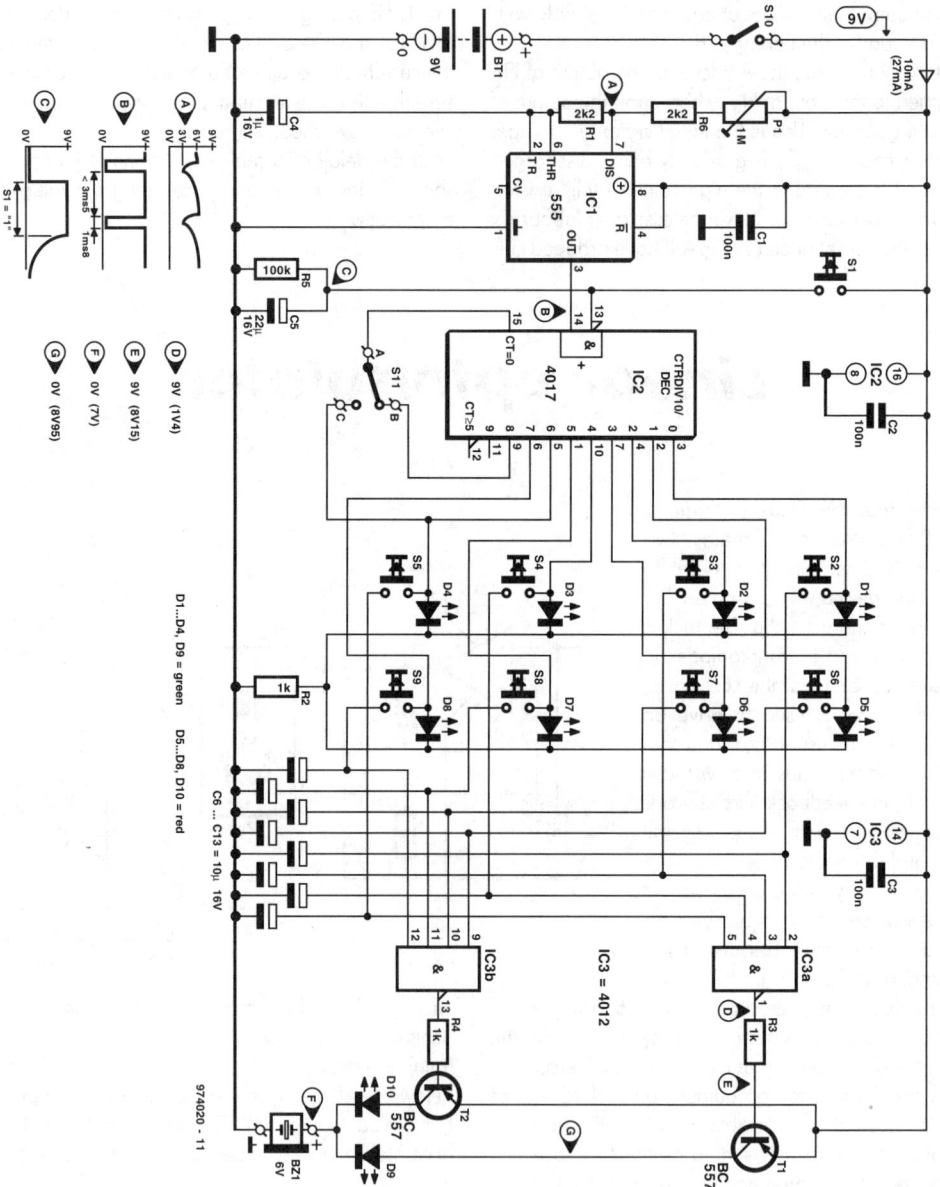

up during this period. However, any push-button pressed when its LED is not on will drain off energy. When all four energy states are tanked up, the output of a 4-input NAND gate switches on a transistor and a red/green jumbo LED in its collector circuit indicates the winner. A buzzer indicates when the game is over.

Control of the timer speed (by the referee only!) enables the game to be played with varying degrees of skill to satisfy both young and old.

A 'freeze' button, which pauses the display on one of the LEDs for about two seconds can provide a more leisurely method of play. Again, a slower timer speed will introduce an element of skill, players taking turns to choose which LED may be targeted for storage.

A dice facility, using six of the LEDs, also quasi-random or skill-dependent (slower speed), is provided for use with other games. For some games, a heads/tails facility can be employed by choice of red and greens. For four-a-side games, quizzes for example, all eight LEDs can be used, together with the freeze button.

In the circuit on the next page, counter clock pulses are provided by IC_1, a 555 timer wired in astable mode. Timing components P_1, R_1, R_6 and C_4 enable any pulse speed to be set between one and 180 per second (test points a and b).

At switch-on, the output of IC_1 cycles eight of the decoded outputs of IC_2 (a 4017 Johnson counter), as the reset pin, 15, is connected to pin 9 via S_{11}. The resulting high output on pin 9 provides a reset pulse after the eighth output. The positive output pulses energize D_1–D_8 in turn, causing them to cycle at a speed determined by the setting of P_1.

Red push-buttons S_6–S_9 each connect a counter output to inputs 9-12 of a 4-input NAND gate in IC_3. Similarly, green push-buttons S_2–S_5 link the counter output to the inputs of the other NAND gate. Electrolytic capacitors C_6–C_{13} (one at each input of IC_3) serve as the tank circuits. Although any of these can be charged instantly by pressing the associated push-button when the LED indicates that the output pulse is present, it will also be discharged instantly if the push-button is actuated when the pulse is absent.

If all 'red' inputs of IC_3 are high, output pin 13 goes low and switches on T_2. This in turn switches on D_9 and brings on solid-state buzzer Bz_1.

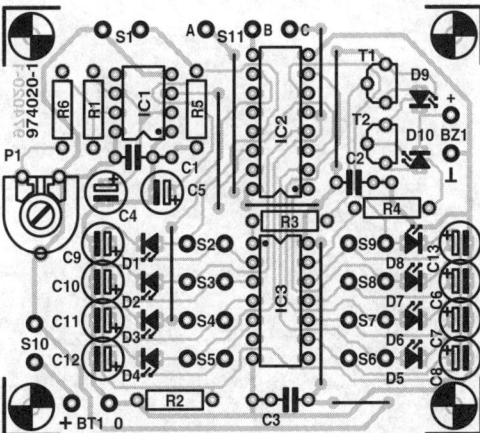

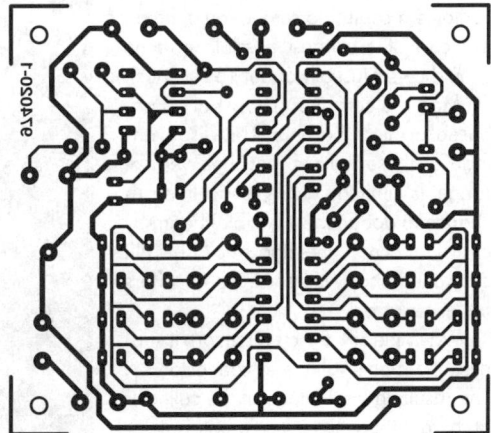

Conversely, if all green inputs are high, output pin 1 of IC_3 goes low and T_1 switches on to energize D_{10} (test point d).

Freeze button S_1 takes the clock inhibit input of IC_2 (pin 13) high, and the charge on C_5 holds the displayed LED momentarily (test point c). The stated value of C_5 gives a display of about two seconds, which seems adequate for dice or capture purposes, but this can be varied as desired.

Parts list
Resistors:
R_1, R_6 = 2.2 kΩ
R_2, R_3, R_4 = 1 kΩ
R_5 = 100 kΩ
P_1 = 1 MΩ linear potentiometer

Capacitors:
C_1, C_2, C_3 = 0.1 μF
C_4 = 1 μF, 16 V, radial
C_5 = 22 μF, 16 V, radial
C_6–C_{13} = 10 μF, 16 V, radial

Semiconductors:
D_1–D_4, D_9 = green LED
D_5–D_8, D_{10} = red LED
T_1, T_2 = BC557

Integrated circuits:
IC_1 = 555
IC_2 = 4017
IC_3 = 4012

Miscellaneous:
S_1 = push-button, 1 make contact
S_2–S_5 = red push-button, 1 make contact
S_6–S_9 = green push-button, 1 make contact
S_{10} = switch, 1 make contact
S_{11} = switch, 1 change-over contact
Bt_1 = 9 V (PP3) battery with holder and clip
Bz_1 = active buzzer, 6 V
PCB may be made with the aid of the track layout.

PIR controlled shop-bell

A passive infra-red (PIR) detector coupled with an electric light may be used for intruder protection. PIR detectors are available as stand-alone units which usually have a switched output for controlling external loads. The Argos 431/5595, for example, has a switching capacity of 2 000 watts.

This circuit will work with stand-alone PIR units as well as combined lamp units. In the latter case, you only use the PIR section, which usually contains a control to set the 'on' time. In this case, the shortest possible on-time should be set (usually about 15 seconds).

Most electronic shop-bells are based on light barriers. The disadvantage of these units is that their vertical range is limited, giving shop-lifters a chance to dodge the (invisible) beam by crawling across the doorstep. The PIR controlled shop bell shown here offers better security.

When the PIR detects a person, it supplies the mains voltage to connector K_1. The resulting low voltage at the collector of the phototransistor inside optoisolator

IC_2 is first 'cleaned' in a low-pass filter, R_2-C_2, to prevent interference and false detection. The resultant pulse at the CLK input of IC_{1b} causes a '1' to be clocked by this bistable. Because the Qoutput goes high, the bell relay is energized via darlington transistor T_1. When switch S_1 is closed timer IC_{3d} determines the delay before the bistable is reset and the bell is switched off. The length of the delay is

adjustable with preset P_1. This delay (max.8s) is also useful to discourage children from toying with the shop bell. If S_1 is open, the bell sounds until the shop-keeper presses the other switch (a push-button), S_2.

The circuit is powered by a 9 V mains adaptor.

Current drain is about 25 mA with the relay energized. The control input is designed for a drive voltage of 230 V, 50Hz.

Finally, to enable the PIR detector to work in daylight, the internal light/darkness sensor must be covered.

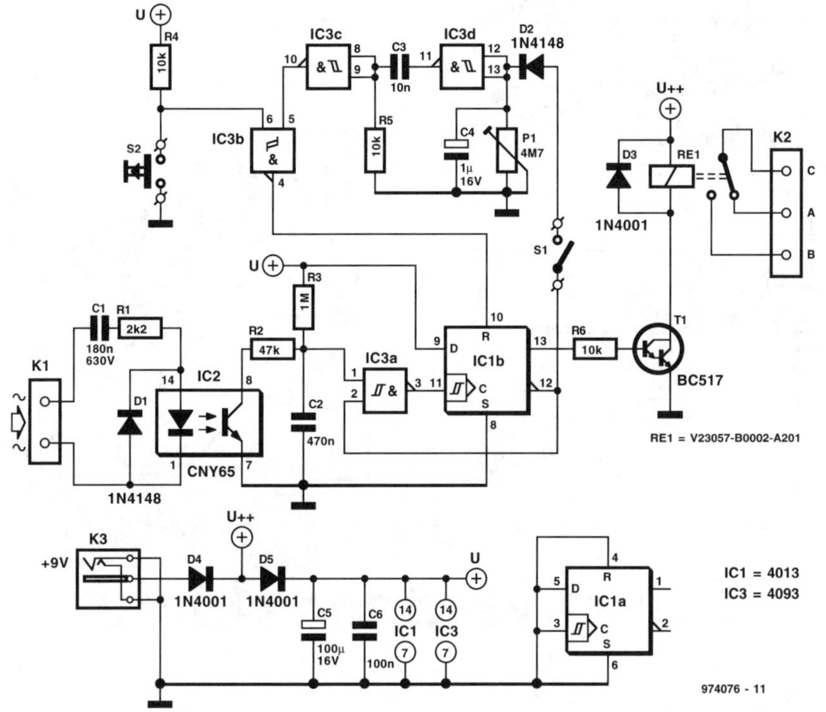

974076 - 11

Parts list

Resistors:
R_1 = 2.2 kΩ
R_2 = 47 kΩ
R_3 = 1 MΩ
R_4, R_5, R_6 = 10 kΩ
P_1 = 4.7 MΩ (5 MΩ) preset, horizontal

Capacitors:
C_1 = 0.18 μF, 630 V DC
C_2 = 0.47 μF
C_3 = 0.01 μF
C_4 = 1 μF, 16 V, radial
C_5 = 100 μF, 16 V, radial
C_6 = 0.1 μF

Semiconductors:
D_1, D_2 = 1N4148

D_3, D_4, D_5 = 1N4001
T_1 = BC517

Integrated circuits:
IC_1 = 4013
IC_2 = CNY65 (Temic)
IC_3 = 4093

Miscellaneous:
S_1 = on/off switch, miniature
S_2 = pushbutton, 1 make contact
Re_1 = V23057-B0002-A201 (Siemens)
K_1 = 2-way terminal block, pitch 7.5 mm
K_2 = 3-way terminal block, pitch 7.5 mm
K_3 = mains adaptor socket for PCB mounting
PCB may be made with the aid of the track layout.

175

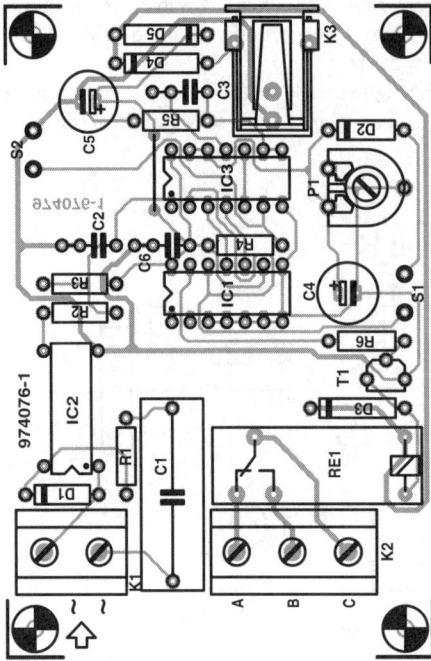

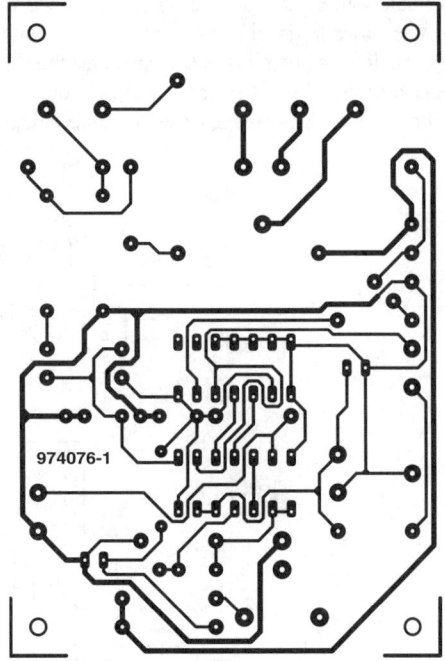

Selective door chime

The chime is for use in situations where there is bell or chime at the front door as well as at the back door. When loud bells or chimes are used, it may be difficult to determine which one is being rung.

This problem is solved by the present circuit which can be set to produce either two or three different sounds. When two of these circuits are built, the one at the front door is set to produce, say, two tones, and the one at the rear door, three tones. It is even possible to couple the two circuits via terminal 'C' on the boards. This may be useful in large premises because it enables both chimes to sound irrespective of which doorbell switch is pushed. In other words, the front door loudspeaker may produce a two-tone

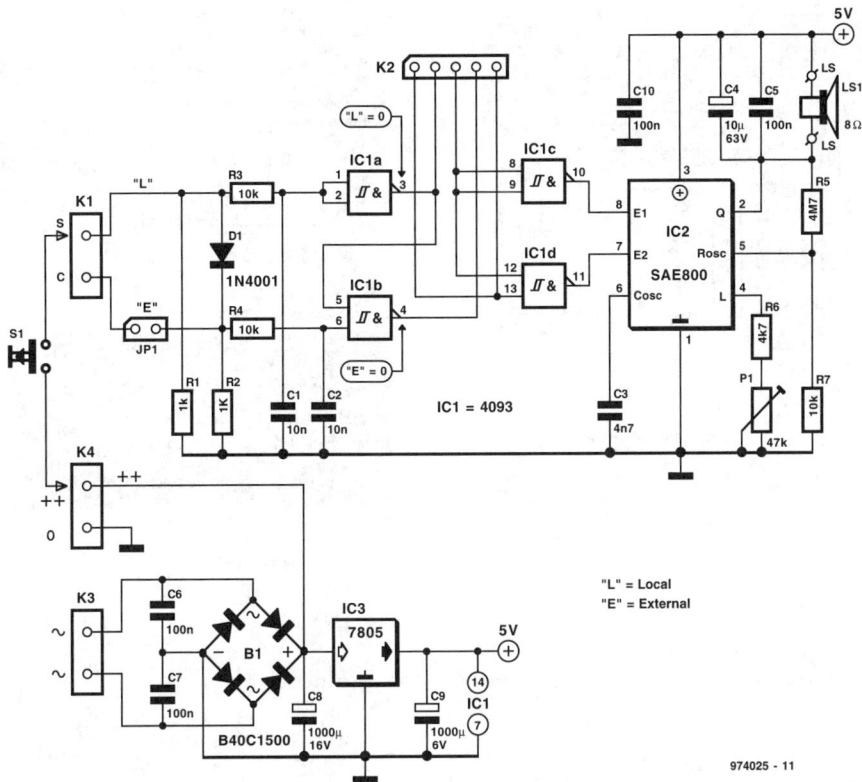

974025 - 11

chime sound to indicate that the rear door needs answering.

The circuit is based on the Siemens SAE0800, a three-tone chime IC with an on-chip integrated audio amplifier. The chime frequencies are determined by the oscillator frequency of the chip, which, in turn, depends on the value of the RC combination connected to pins 5 and 6. The loudspeaker volume is set with a combination of a preset and a fixed resistor, P_1 and R_6 respectively, at pin 4. The trigger inputs of the IC, E1 and E2, determine the number of chime notes generated by the IC, as follows:

E 1	E 2	mode
≠	≠	3 tones
GND/n.c	≠	2 tones
≠	GND/n.c	1 tone

where ≠ indicates a leading edge.

On one board, install links 1-2 and 3-4 on pinheader K_2 to select two tones. On the other board, install links 2-3 and 4-5 to select three tones.

Each board has a simple rectifier and regulator circuit which allows it to be powered from an inexpensive a.c. mains adaptor. The quiescent current drain is about 5 mA. At maximum sound level (set with P_1) this rises to about 400 mA.

If it is decided to link the front and rear door chimes, jumper JP_1 has to be placed on both boards. Also, a cable with at least two wires must interconnect the 'C' and ground ('0') terminals of the two units. Alternatively, use a three-core cable, and feed the '++' voltage of one unit to the '++' terminal on the other. In that case, the rectifier on the daughter board may be omitted, since both units can be powered by a single mains adapter.

The loudspeaker should be a weather-resistant (if necessary) miniature type rated at about 1 watt.

177

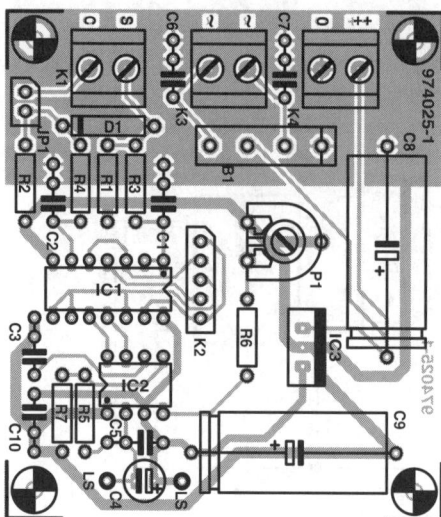

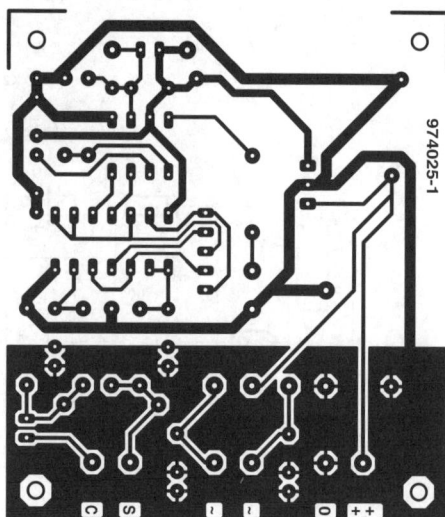

Parts list

Resistors:
R_1, R_2 = 1 kΩ
R_3, R_4, R_7 = 10 kΩ
R_5 = 4.7 MΩ
R_6 = 4.7 kΩ
P_1 = 47 kΩ preset, horizontal

Capacitors:
C_1, C_2 = 10 μF
C_3 = 4.7 μF
C_4 = 10 μF, 63 V, radial
C_5, C_6, C_7, C_{10} = 0.1 μF
C_8 = 1000 μF, 16 V
C_9 = 1000 μF, 6 V

Semiconductors:
B_1 = B40C1500
D_1 = 1N4001

Integrated circuits:
IC_1 = 4093
IC_2 = SAE0800 (Siemens)
IC_3 = 7805

Miscellaneous:
K_1, K_3, K_4 = 2-way terminal block, pitch 5 mm
K_2 = 5-way SIL pinheader
Ls_1 = miniature loudspeaker, 8 W
JP_1 = 2-way pinheader with jumper
S_1 = doorbell switch
PCB may be made with the aid of the track layout.

Ultrasonic intruder alarm

Sound becomes more directional as the frequency increases, a principle that is exploited here to create an inaudible ultrasonic sound beam which, when interrupted, sets off an acoustic alarm. The transmitter and receiver operate at a frequency of about 40 kHz, using special ultrasonic transducers from Murata.

The transmitter signal is generated and amplified by four gates in IC_1, which is run off a 12 volt power source. Preset P_1 enables he transmitter frequency to be matched to the transducer's resonance frequency, and so achieve maximum transmit power.

The receiver consists of a tuned preamplifier, T_1, a detector/rectifier, T_2-D_1, and an audible alarm tone generator, IC_2. The base bias of T_2 is adjusted with preset P_2 for maximum range of the system, which is 2 –3 metres. Obviously the transmitter and receiver have to be pointed at each other accurately.

The received ultrasonic sound causes a rectified voltage on the base of T_3. When the beam is interrupted, this voltage disappears, causing the collector of T_3 to swing to the supply potential. This, in turn, lifts the reset condition of IC_3, which responds by starting to oscillate and so producing an alarm tone in the loudspeaker.

Current drain of the transmitter is about 10 mA. In stand-by mode (beam received), the receiver draws about 20 mA from an unregulated 12 V d.c. source.

Finally, because walls reflect ultrasonic sound, the system is suitable for use out of doors only.

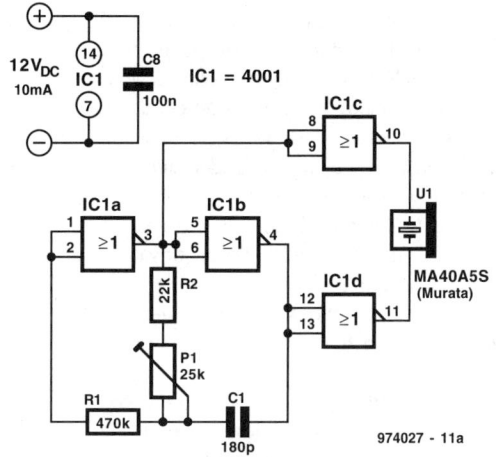

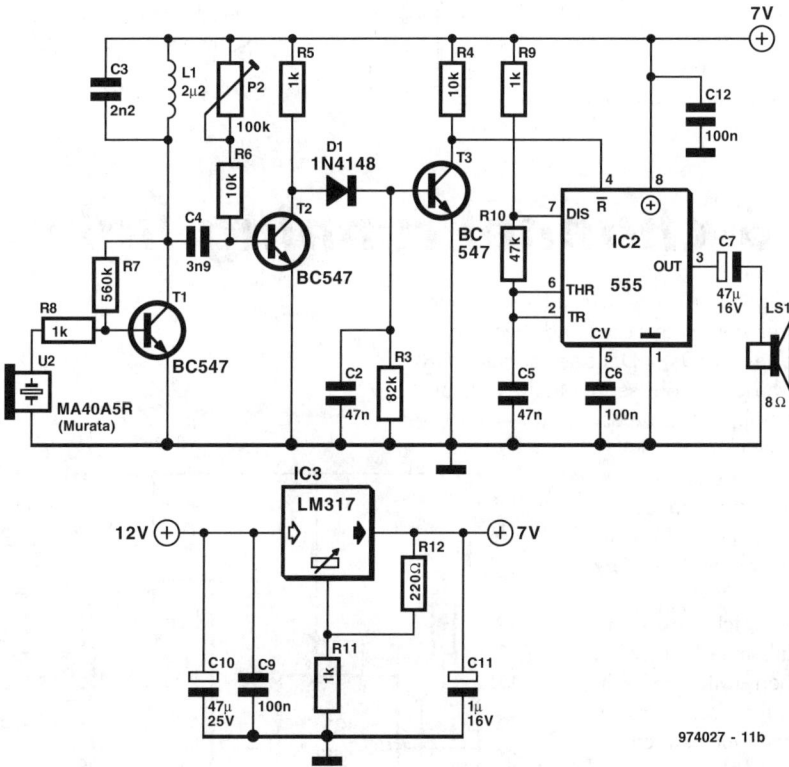

974027 - 11a

974027 - 11b

Water leakage detector

The detector produces an audible tone when its probes detect the presence of a conductive liquid, such as water. It may be used as a cellar flooding alarm, or near a washing machine to give a warning immediately there may be a leaking hose.

When water is detected, the probes pass a small

179

current which causes the base potential of darlington transistor T_1 to be pulled well below 0.6 V. As a result, the collector voltage swings high, lifting the reset condition of the oscillator, which will start to work. Conversely, if no water is detected, T_1 keeps IC_1 reset.

The well-known 555, IC_1, is used in astable (oscillator) mode, directly driving a small loudspeaker. The alarm tone is set to a frequency of about 700Hz by network R_1-R_3-C_1 and has a duty factor of about 0.5.

An optional push-button switch, S_1, enables the circuit to be tested. When the button is pressed, the loudspeaker should sound.

Current drain of the circuit is about 10 mA when no water is detected, and about 50 mA when the alarm sounds. These values were measured with a supply voltage of 6 V.

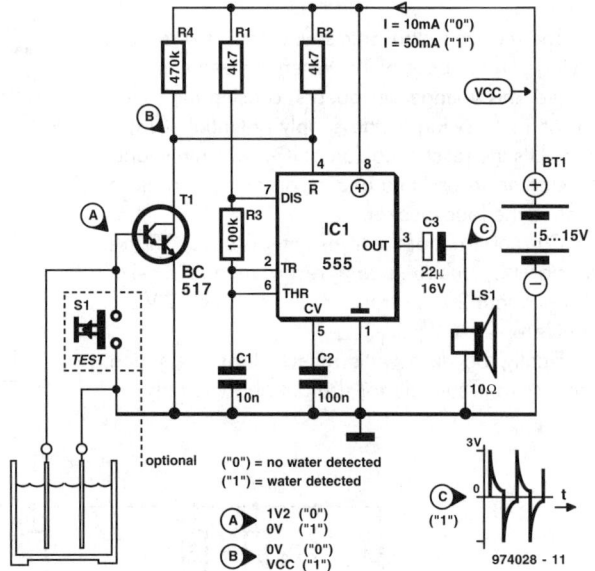

6-channel running light

The running light comprises two integrated circuits (ICs), a resistor, a capacitor and seven light-emitting diodes (LEDs), Decade scaler IC_2 ensures that the LEDs light sequentially. The rate at which this happens is determined by the clock at pin 14.

The clock is generated by IC_1, which is arranged as an astable multivibrator (AMV). Its frequency is determined by R_1-C_1.

The touch switch, consisting of two small metal disks is optional. When switch S_1 is in position 'off', the circuit may be actuated by the touch switch. By the way, this enables the circuit to be used as an electronic die (in which case the LEDs have to be numbered from 1 to 6).

The running light is powered by a 9 V battery or mains adaptor and draws a current not exceeding 20 mA.

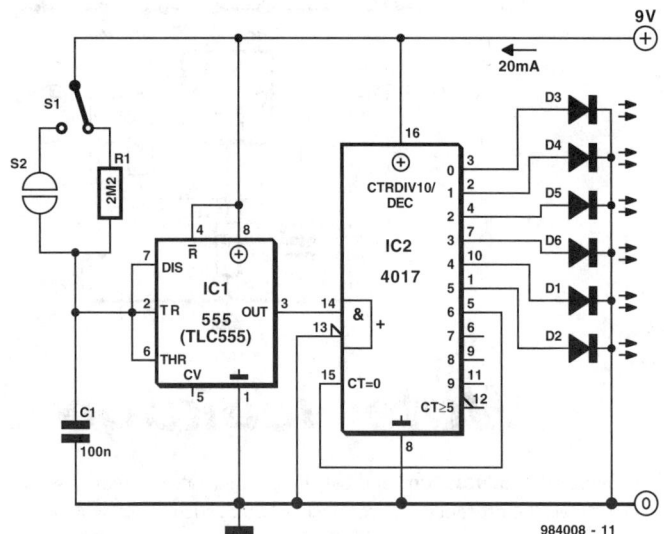

Ambient-noise monitor

Excessive noise is bad for our health and bad for our surroundings. It cannot be said too often: too many young people go prematurely deaf because of prolonged exposure to loud sounds. There cannot be any pleasure in excessively loud music: it hurts and, like skin cancer, the terrible effects do not immediately become noticeable.

The monitor in the diagram gives a visible warning or actuates a relay when the ambient noise is at a dangerous level.

The noise sensor is a two-terminal electret microphone that is powered via R_1. The audio signal is applied to op amp IC_1, whose input resistance is fixed at 47 kΩ by R_2. The signal amplification can be set from unity to ×250 with P_1.

Operational amplifier IC_2 functions as a comparator which likens the amplified signal with a reference voltage of 3.3 V. If the signal at the non-inverting input of the op amp exceeds the reference voltage, the output of IC_2 changes state (goes high), whereupon T_1 is switched on. When this happens, the relay is energized or the LED lights. The relay contacts may be used to operate a warning light or buzzer, or to switch the noise source off. In the latter case, C_2 prevents the circuit returning to its original state (which would cause the noise source to come on again). The capacitor is charged to the peak value of the signal. Owing to the presence of D_1, it cannot be discharged via the output of IC_1, but only, and very slowly, via the high-resistance input of IC_2.

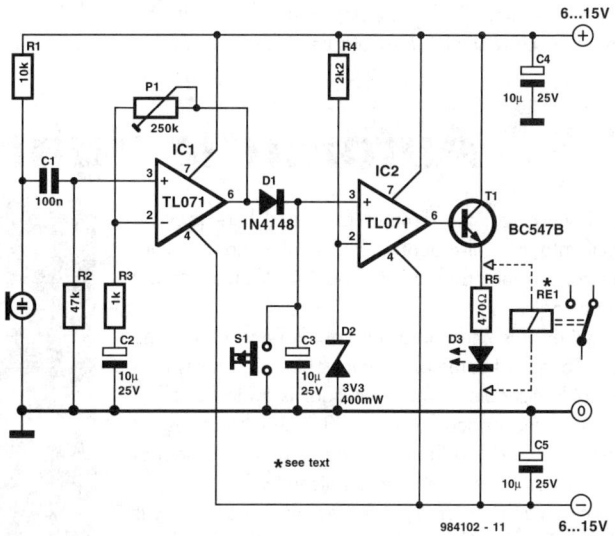

The monitor is reset with S_1.

Auto power off

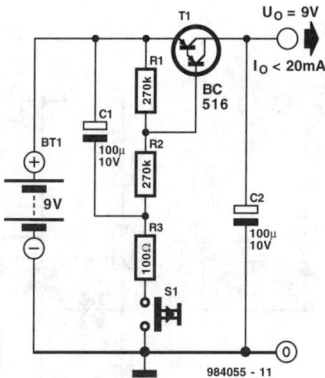

We are surrounded by battery-operated equipment of all kinds, and this array is growing still. Manufacturers and designers lean over backwards to make sure that their equipment draws a small current and can thus be operated by a battery.

This has a drawback, however, because even if the equipment in question draws only a small current, when it is not switched off, the battery is flat after a few days or weeks. The circuit presented here can prevent this happening.

It may be added to all kinds of equipment operating from a 9 V battery and switches this off automatically one minute after a preset time has elapsed. The peak switching current is 20 mA, which is more than enough for most applications.

The switch is formed by a p-n-p darlington, T_1,

181

which is actuated by push-button switch S_1. The very high amplification of the darlington enables it to be kept on fairly long with the aid of a relatively small-value capacitor, C_1 (= 100 μF). Resistor R_3 limits the charging current of C_1 to ensure a long life of S_1.

Resistors R_1 and R_2, in conjunction with C_1, determine the switch-on time. When this time has elapsed, R_1 ensures that T_1 is switched off.

Since the darlington can handle a U_{BE} of –10 V, a polarity protection diode is not needed.

Automatic light dimmer

In many cases, the dimmer presented here may be built into a wall-mounted box containing the light switch. It is intended for use with 240 V incandescent lamps only.

When it is fitted, and the light is switched on, the lamp does not come on fully for about 400 ms (which is not noticeable). When the light is switched off, it stays on unchanged for about 20 s, and then goes out gradually. This has the advantage that it is not immediately dark when the light is switched off.

When light switch S_1 is turned on, capacitor C_2 is charged via R_1, C_1 and bridge rectifier D_1–D_4. Zener diode D_5 limits the potential across C_2 to about 15 V. After a short while, diode D_6 lights, whereupon a potential difference ensues across light dependent resistor (LDR) R_3, which is sufficient to trigger triac Tr_1. The light then comes on.

When the light switch is turned off, C_2 is discharged via P_1, R_2 and D_6. When the potential across C_2 drops, the brightness of the LED diminishes, so that the p.d. across R_3 also drops. The increasing resistance of R_3 effects phase angle control of the triac so that the light is dimmed gradually. The dimming time may be altered with P_1 within the time range determined by network R_2-C_2.

The circuit operates correctly only, of course, when the LDR is not exposed to light other than that from the LED.

The type of LDR is not particularly important, as long as it is not too long: in the prototype, a model with a length of 5 mm was used.

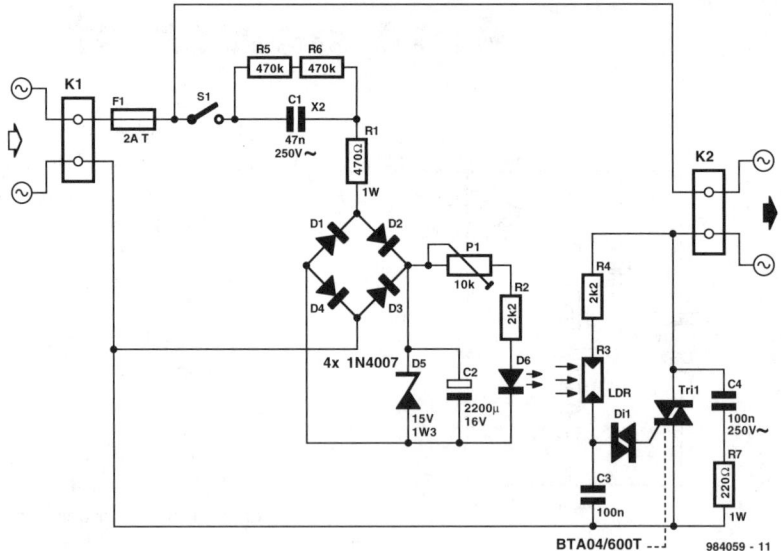

984059 - 11

Automatic air (de)humidifier

The (de)humidifier circuit is based on a special humidity sensor Type NH-3 from Figaro. Depending on the sensor output, the circuit drives a ventilator that is part of an air humidifying installation. The ventilator is

capacitor C_1, resistor R_1 and zener diode D_1. The pulsating voltage is used to drive the sensor. It is also transformed to a 7.5 V supply voltage by D_2 and C_2.

The sensor needs an alternating drive voltage at a

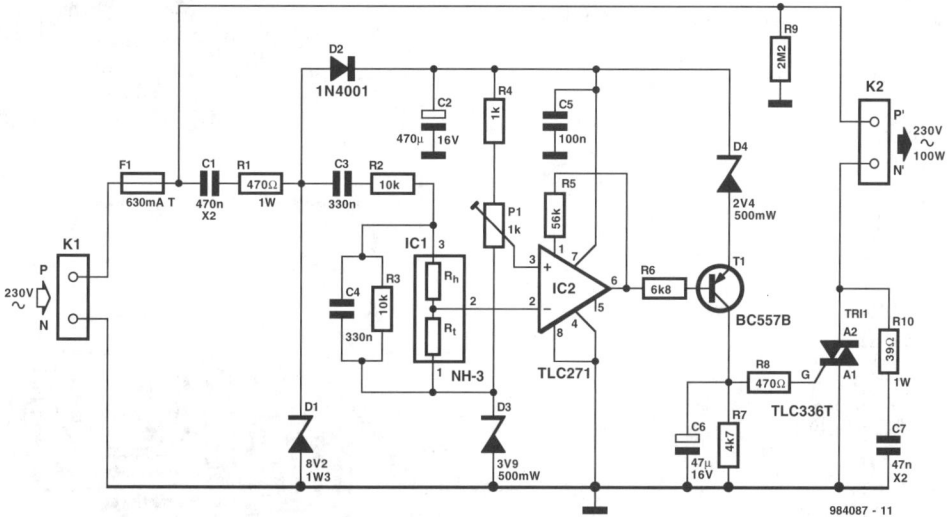

984087 - 11

switched on and off by a triac.

So as to keep the circuit as simple as possible, the supply voltage and the test voltage are drawn directly from the mains supply. The 240 V mains voltage is converted into an 8.9 V pulsating direct potential by

level not higher than 1.5 V. This potential is obtained from the pulsating direct voltage by network R_2-R_3-C_3-C_4, which removes the direct voltage component and lowers the level to 1.4 V. At the same time, the network functions as a 50 Hz bandpass filter.

To ensure that the drive voltage for the sensor does not fall outside the common-mode range of op amp IC_2, an offset potential of 3.9 V is applied to the sensor as well as to the voltage reference source of the op amp. This potential is provided by zener diode D_3. The reference level is set with P_1. The op amp is given some hysteresis by R_5.

When the humidity of the ambient air rises above that corresponding to the level with P_1, the output voltage of IC_2 is about 6 V. This results in T_1 being cut off by D_4, whereupon the triac is also disabled. When the humidity drops below that corresponding to the level set with P_1, a pulsating potential appears at the output of IC_2. This voltage is used to charge capacitor C_6. The charged capacitor thereupon provides a steady current to the triac.

When T_1 is cut off for some time, capacitor C_6 is discharged via resistor R_7. Capacitors C_1 and C_7 are discharged via R_9, so that after the mains has been switched off, no dangerous po-tential remains at the pins of the mains connector (K_1).

The (de)humidifier is best built on the PCB shown, which may be made with the aid of the track layout. Bear in mind that parts of the board will carry mains voltage, which makes careful working and the enclosing of the board in a plastic case imperative.

The unit may be converted into a dehumidifier by interchanging connections 1 and 3 to sensor IC_1.

Parts list

Resistors:
R_1 = 470 Ω, 1 W
R_2, R_3 = 10 kΩ
R_4 = 1 kΩ
R_5 = 56 kΩ
R_6 = 6.8 kΩ
R_7 = 4.7 kΩ
R_8 = 470 Ω
R_9 = 2.2 MΩ
R_{10} = 39 Ω, 1 W
P_1 = 1 kΩ preset

Capacitors:
C_1 = 0.47 µF, 250 V a.c.
C_2 = 470 µF, 16 V, radial
C_3, C_4 = 0.33 µF, metallized polyester, 5%
C_5 = 0.1 µF, high stability
C_6 = 47 µF, 16 V, radial
C_7 = 0.047 µF, 250 V a.c.

Semiconductors:
D_1 = zener diode, 8.2 V, 1.3 W
D_2 = 1N4001
D_3 = zener diode, 3.9 V, 500 mW
D_4 = zener diode, 2.4 V, 500 mW
T_1 = BC557B

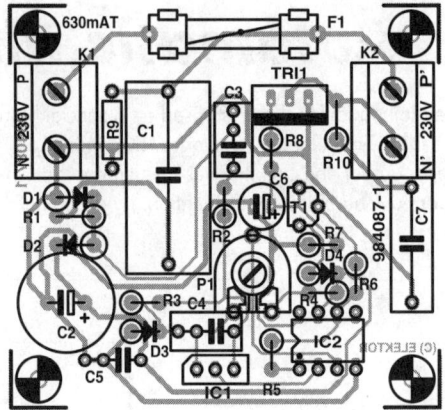

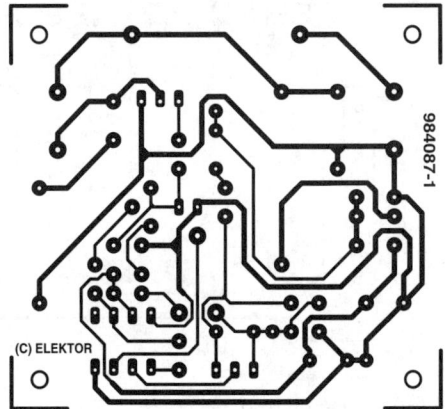

Integrated circuits:
IC_1 = NH-3 (Figaro)
IC_2 = TLC271CP
Tri_1 = TLC336T (SGS)

Miscellaneous:
K_1, K_2 = 2-way terminal block for board mounting, pitch 7.5 mm
F_1 = fuseholder with 630 mA slow fuse

Cable analyser

Many constructors have various cables lying around and after a time do not know any longer what they are for or what their pin connections are. It is not always possible to check this with a multimeter. The analyser may be of help in such a situation. In most cases, an analyser for checking cables with D9 and D25 connectors will suffice.

The shape of the analyser will depend to a large extent on the type of cable to be checked. It may be made as a connector, as a bus, or as a feed-through cable. Since only standard

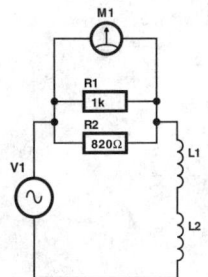

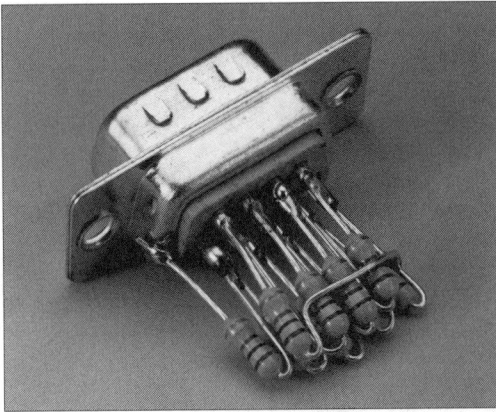

cable to be tested and measure the resistance between pin 1 and the case. The value so obtained in kilohms is the number of the pin at the other end of the cable.

The arrangement is shown in the diagram. If at all possible, use resistors in the E96 series, since these give best accuracy.

In some cases, a simple improvement may be possible, whereby the nine resistors are linked in series instead of in parallel. The advantage of this is that all nine resistors have the same value: 1 Ω or 1 kΩ. The test method remains the same: the value in Ω or kΩ measured on the multimeter is the nuber of the pin at the other end of the cable.

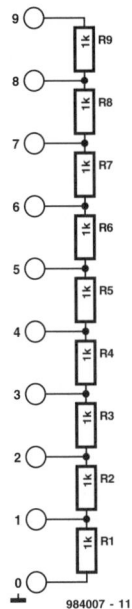

984007 - 11

components are needed, the cost is low.

Solder a resistor of 1 kΩ between pin 1 of the analyser and the case; one of 2 kΩ between pin 2 and the case, and so on, increasing the value of the resistor by 1 kΩ for each successive pin.

When this is completed, connect the analyser to the

Car immobilizer

A starter motor immobilizer is an effective (but not certain) means of protecting your car against theft. It has the drawback that a would-be thief will try to render it inoperative and in the process damages your car. The present circuit is a simple version of car immobilizer and tends to confuse the thief. This is because the car appears to function normally, but it does not start. Has it broken down or is there some sort of protection circuit active?

The circuit does not need additional controls, indicators, switches or keypads to be fitted in the car. The setup is 'invisible'. The only external sensor is the brake pedal. After the ignition has been switched on, the brake pedal has to be pressed for at least five seconds before voltage is applied to the coil. Since the thief does not know this, he/she will try everything to get the car started. Since it is only the coil to which voltage is not applied, all other electrical functions will work normally, but it is just impossible to get the car started.

When the ignition is switched on, the circuit is powered by the voltage at terminal

PC_2. Until the brake pedal is pressed, the potential at terminal PC_1 remains low, so that the relay remains unenergized. When the brake pedal is pressed, capacitor C_6 is charged via resistor R_3. The time, t, it takes for the capacitor to become fully charged is determined by network R_8-C_6. When this time has elapsed, the output of IC_1a goes low, whereupon voltage is applied to the base of T_2 via IC_1b. When T_2 is

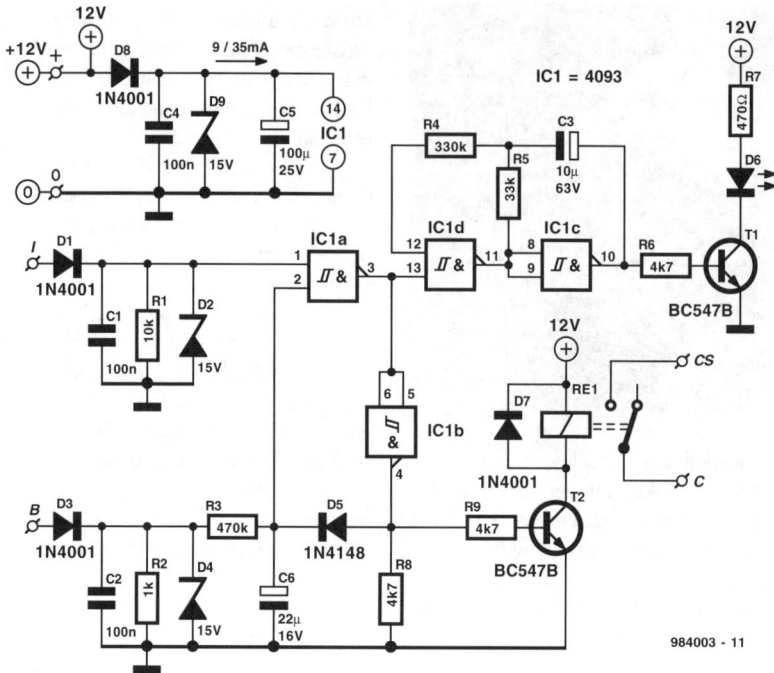

984003 - 11

on, the relay is energized , whereupon its contact changes over and voltage is applied to the coil. After the pressure on the brake pedal is released, diode D_5 ensures that the voltage remains applied.

Gates IC_{1c} and IC_{1d} form an oscillator, which causes diode D_6 to flash when the starter is immobilized. This has the disadvantage, of course, that it discloses the protection circuit.

Finding the right points to which to connect the circuit should not be a problem in most cars. The ignition voltage is normally available at the radio/cassette terminals, while the potential coupled to the brake pedal is usually available at the brake lights.

The PCB may be made with the aid of the track layout.

Parts list

Resistors:
R_1 = 10 kΩ
R_2 = 1 kΩ
R_3 = 470 kΩ
R_4 = 330 kΩ
R_5 = 33 kΩ
R_6, R_8, R_9 = 4.7 kΩ
R_7 = 470 Ω

Capacitors:
C_1, C_2, C_4 = 0.1 μF
C_3 = 10 μF, 63 V, radial
C_5 = 100 μF, 25 V
C_6 = 22 μF, 16 V, radial

Semiconductors:
D_1, D_3, D_7, D_8 = 1N4001
D_2, D_4, D_9 = zener diode, 15 V, 400 mW
D_5 = 1N4148
D_6 = LED
T_1, T_2 = BC547B

Integrated circuits:
IC_1 = 4093

Miscellaneous:
PC_1–PC_6 = PCB terminal (pin)
Re_1 = 12 V car-type relay, 1 change-over contact
PCB may be made with the aid of the track layout.

984003-1
984003-1 (C) ELEKTOR

984003-1
(C) ELEKTOR

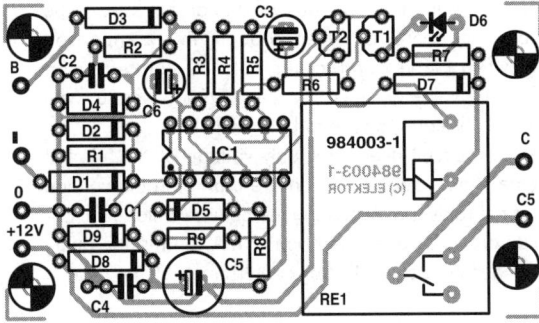

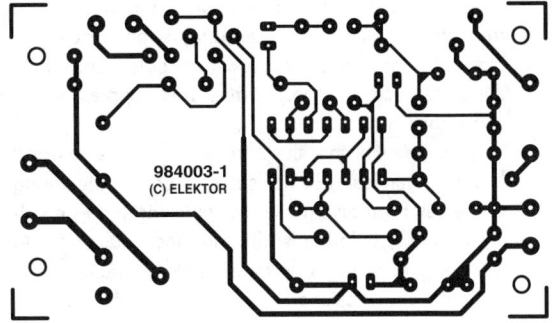

Car interior lights delay

Most cars do not have delayed interior lights. The circuit presented can put this right. It switches the interior lights of a car on and off gradually. This makes it a lot easier, for instance, to find the ignition keyhole when the lights have gone off after the car door has been closed.

Since the circuit must be operated by the door switch, a slight intervention in the wiring of this switch is unavoidable.

984065 - 11

When the car door is opened, the door switch closes the lights circuit to earth. When the door is closed (and the switch is open), transistor T_1, whose base is linked to the switch, cuts off T_2, so that the interior light remains off.

When the switch closes (when the door is opened), the base of T_1 is at earth level and the transistor is off. Capacitor C_1 is charged fairly rapidly via R_3 and D_1, whereupon T_2 comes on so that the interior light is switched on.

When the door is closed again, T_1 conducts and stops the charging of C_1. However, the capacitor is discharged fairly slowly via R_5, so that T_2 is not turned off immediately. This ensures that the interior light remains on for a little while and then goes out slowly.

The time delays may be varied quite substantially by altering the values of R_3, R_5, and C_1.

Circuit IC_2 may be one of many types of n-channel power MOSFET, but it should be able to handle drain-source voltages greater than 50 V. In the prototype, a BUZ74 is used which can handle D–S voltages of up to 500 V.

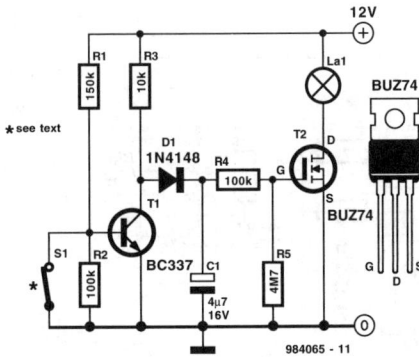

'Chip card' as security key

The 'chip card' consists of a small board on which an integrated circuit, IC_1, forms the key. The logic circuit for the key is hard-wired. The inputs and outputs of this circuit are terminated in a socket. It is advisable to use a surface-mount device (SMD) for IC_1 (as well as for the other ICs in the circuit) and a micro-miniature or miniature type of socket (and matching plug), to

keep the safety key as small as possible.

An identical circuit, based on IC_2, must be built for the lock into which the key fits. Added to this are 4-bit comparator IC_3, and 14-stage binary counter IC_4.

The comparator likens the logic circuit in the lock with that in the key. To that end, both circuits are fed with a 4-bit signal by the binary counter. If the two

187

match, the output of IC3 goes high and switches on transistor T1, whereupon the relay is energized and diode D2 lights. Any short, low-level pulses caused by the difference in transit times are suppressed by capacitor C2.

It is essential that the hard-wiring in the key and lock are identical. It is advisable to protect the wiring in the key from prying eyes by removing the typecoding from IC1 before it is wired in, and additionally to use opaque casting resin for embedding the key.

In the circuit as shown, there is a remote possibili-

ty that, owing to a spurious voltage across R4-C2, the relay is energized even though the comparison between the key and lock is not 100 per cent right. This may be obviated by shunting R4 with a Type 1N4148 diode (anode to base of T1).

The circuit needs a supply voltage of 5–15 V, but care should be taken to ensure that the rated voltage for the relay is available. In the quiescent state, the circuit draws a current of about 1 mA, and in the operating state a current wholly determined by the relay.

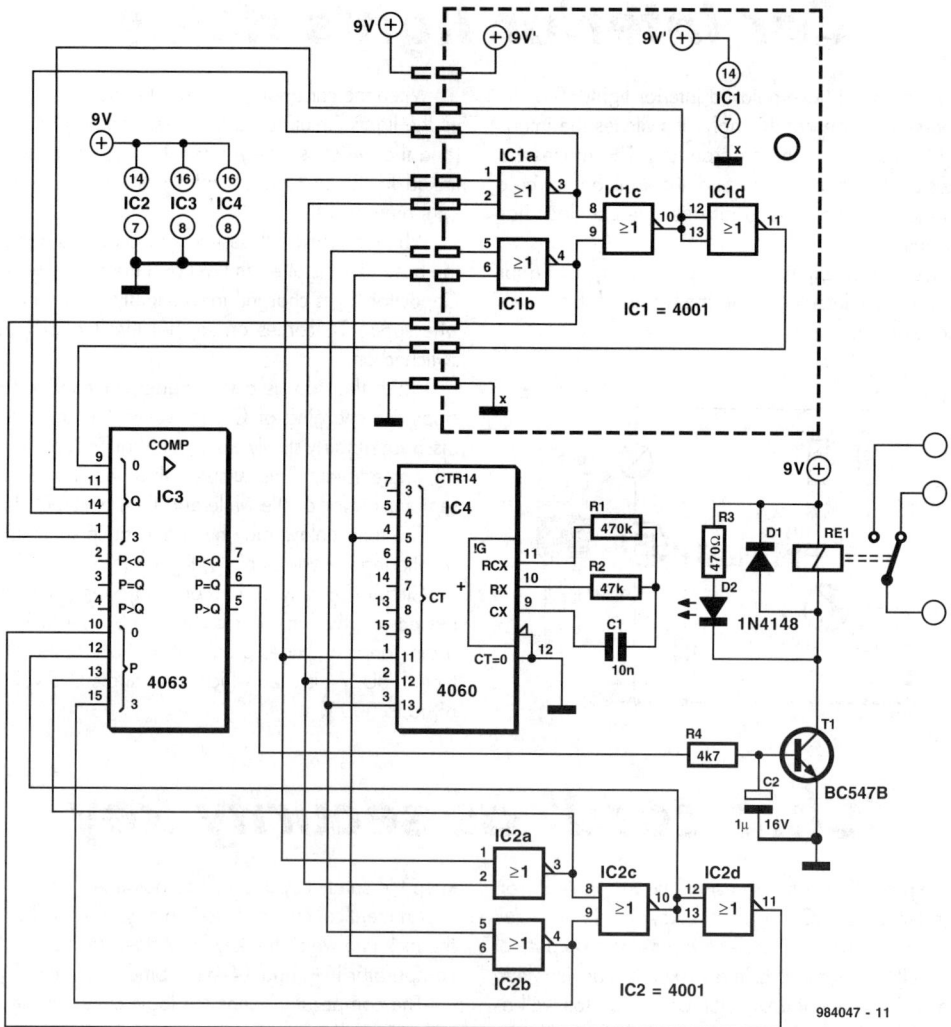

984047 - 11

Digital output with sink/source driver

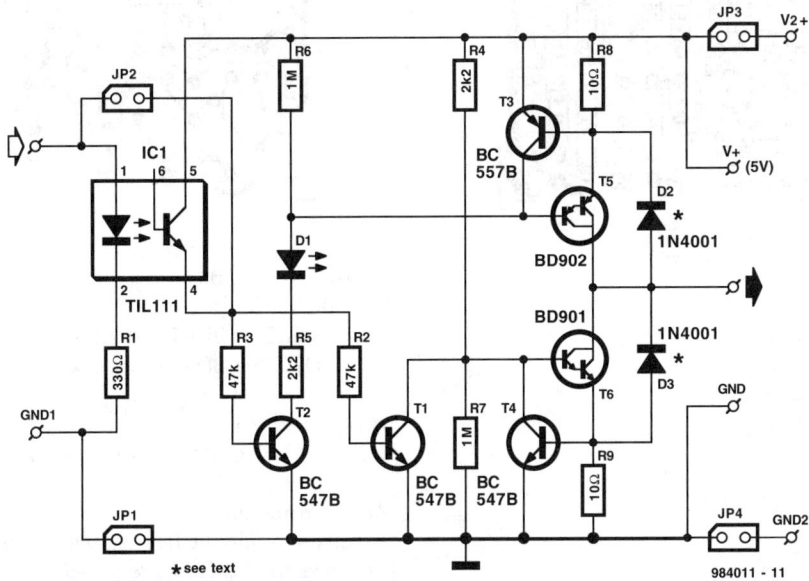

★ see text

984011 - 11

A PC or a microprocessor system may be used to control loads like lamps, relays, and motors, either by software or hardware. This article describes one of the latter. The circuit diagram shows a one-channel power driver with an (optional) electrically isolated input and a power output capable of sinking as well as sourcing current.

If electrical isolation is not required at the input, omit optoisolator IC_1 and fit the two jumpers. In that case, the circuit is driven by a TTL-compatible logic signal. If IC_1 is used, the driver responds to a current-loop signal with strength of between 10 mA and 20 mA.

Diode D1 in the collector line of amplifier stage T_2 provides a 'channel active' indication. The balanced sink/source power driver consists of a pair of complementary darlington transistors with associated current limiting resistors, R_8–R_9. Resistor R_8 determines the maximum source current, and R_9, the maximum sink current. Both currents are calculated from $I = 0.65\ U/R$. The driver board itself draws a current of only a few milliamps.

Diodes D_2 and D_3 are required only if inductive loads like relay coils are to be controlled, and different darlington pairs are employed. As opposed to the specified types, a BD911/912 complementary pair, for instance, does not have internal anti-surge diodes across the collector-emitter path. As a matter of course, the darlington transistors have to be cooled depending on the currents they sink or source.

Jumpers JP_3 and JP_4 have to be fitted if the controlled load(s) are not already connected to a supply line. When they are fitted, the load current is drawn from the driver board.

This driver is pretty fast: it can handle switching frequencies up to about 3 kHz without problems if the specified optoisolator is used. Higher frequencies may be achieved if a faster optoisolator is used. If more than one channel is to be controlled, build as many driver boards as needed.

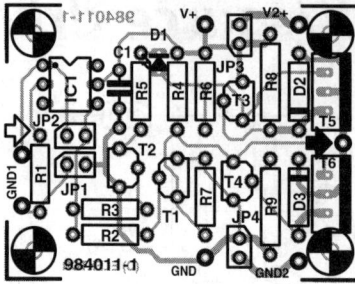

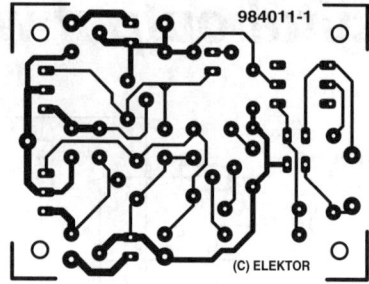

Parts list

Resistors:
$R_1 = 330\ \Omega$
$R_2, R_3 = 47\ k\Omega$
$R_4, R_5 = 2.2\ k\Omega$
$R_6, R_7 = 1\ M\Omega$

Capacitors:
$C_1 = 0.1\ \mu F$

Semiconductors:
D_1 = LED
D_2, D_3 = 1N4001 (optional, see text)

T_1, T_2, T_4 = BC547B
T_3 = BC557B
T_5 = BD902 or BD912 (see text)
T_6 = BD901 or BD911 (see text)

Integrated circuits:
IC_1 = TIL111 or 4N35 or CNY17-2

Miscellaneous:
JP_1–JP_4 = 2-pin SIL header with jumper
Heatsinks for T_5/T_6, as required
PCB may be made with the aid of the track layout.

Electronic spirit-level

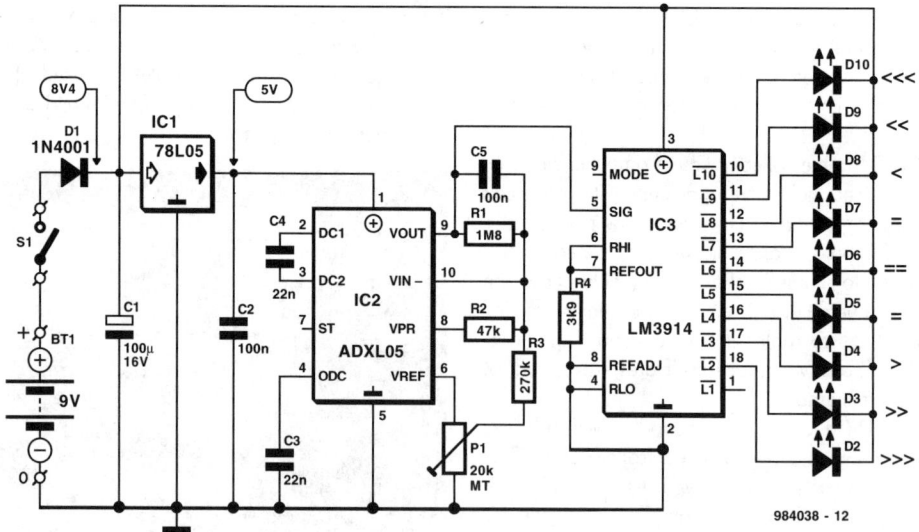

984038 - 12

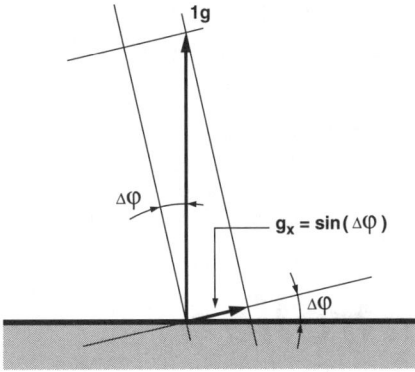

$$g_x = \sin(\Delta\varphi)$$

984038 - 11

The circuit, an electronic equivalent of a traditional spirit-level, is based on IC_2, a gravity force (g) sensor from Analog Devices. This sensor will detect a relative gravitational force of 0 g when it is positioned horizontally. It has a sensitivity of about 200 mV/g. Its internal output buffer is set to supply a gain equal to (R_1/R_2) or about ×38.3, so that the sensitivity is increased to 7.66 V/g.

The output of IC_2, pin 9, drives the signal input of an analogue-to-digital converter (ADC) cum LED-bar driver, IC_3. The single resistor connected to the driver,

R_4, arranges for full-scale diode D_{10} to light at an input voltage of 1.25 V. Consequently, the LED bar has a step size of 0.125 V, while the sensitivity of the spirit-level works out at 16.32 mg per LED, or one LED for each degree of angle.

It should be noted that the operation of the circuit is affected by temperature variations. With a drift of 0.4 mV $°C^{-1}$ at unity gain, the rate of change (temperature drift gradient) at the output equals 15 mV $°C^{-1}$ for a gain of ×38.3. This results in a gradient of 8 °C per LED, so the circuit has to be calibrated before use.

Preset P_1 and resistor R_3 double as a network for offset compensation (approx. ±0.3 V) as well as for ensuring that diode D6 lights when the sensor is held exactly horizontal (0-g potential). Preset P_1 is to be adjusted until D_6 lights. Assuming that the sensor is on a fairly horizontal surface, turn the instrument 180°. Divide the number of active LEDs which is then shifted to the left or to the right by two, and carefully adjust P_1 until this number is shifted around D_6. Check by turning the circuit 180 degrees again — the error with respect to D_6 should be equal but in the opposite direction. The spirit-level draws a current of about 20 mA from a 9-volt (PP3) battery.

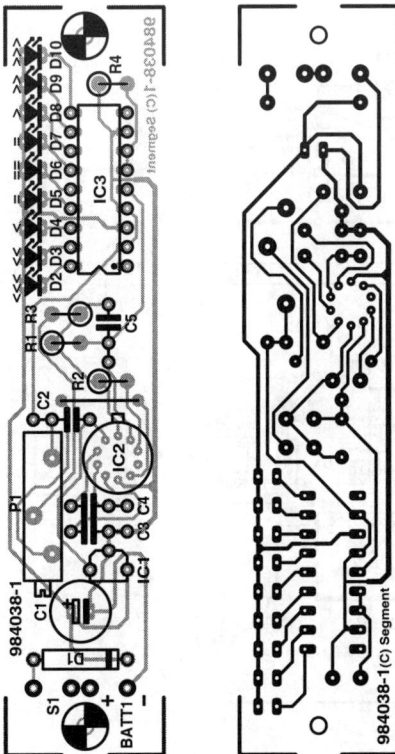

Parts list

Resistors:
R_1 = 1.1 MΩ
R_2 = 47 kΩ
R_3 = 270 kΩ
R_4 = 3.9 kΩ
P_1 = 20 kΩ multiturn preset, horizontal

Capacitors:
C_1 = 100 μF, 16 V, radial
C_2 = 0.1 μF, high stability
C_3, C_4 = 0.022 μF MKT (metallized polyester)
C_5 = 0.1 μF MKT (metallized polyester)

Semiconductors:
D_1 = 1N4001
D_2, D_{10} = LED, red, high efficiency
D_3, D_4, D_8, D_9 = LED, yellow, high efficiency
D_5, D_6, D_7 = LED, green, high efficiency

Integrated circuits:
IC_1 = 78L05
IC_2 = ADXL05JH (Analog Devices)
IC_3 = LM3914N

191

Miscellaneous:
S_1 = on/off switch, 1 make contact.
9 V (PP3) battery with clip-on connector and wires.

Some parameters
$G_{sensor} = 200$ mV g^{-1}
$A_{amp} = R_1/R_2 = 1.8/0.047 = \times 38.3$

$G_{LM3914} = 8x$ V^{-1}
$G = G_{sensor} \times A_{amp} \times G_{LM3914} = 0.2 \times 38$
$.3 \times 8 = 61.3x$ g^{-1}
$1/G = 16.3$ mg x^{-1}
$\Delta\varphi$ $x^{-1} = \arcsin(1/G) \approx 1° x^{-1}$

where x is the number of LEDs that light

Doorbell-controlled burglar-deterrent light

Burglars often ring the doorbell to see if there is anybody at home! The present circuit may trick them into believing that someone is about to answer the door by turning on a lamp after a short delay. The length of the delay is adjustable between about 5 seconds and 125 seconds. Likewise, the 'on' time of the lamp is adjustable between 25 s and 600 s.

The circuit is designed for easy connection to an existing doorbell switch, provided this is part of a standard 8-volt AC or DC doorbell circuit. The input of the circuit is simply connected in parallel with the doorbell switch. A relay is used at the output of the circuit so that a lamp, exterior or in the hallway, is easily connected up.

When the doorbell is pressed, capacitor C_1 is charged to a level exceeding the 2.5 V switching

984029 - 11

threshold set up for comparator IC_1. As a result, the comparator output swings high and triggers monostable multivibrator (MMV) IC_{2a}. This MMV determines the time (delay) it takes before the lamp is switched on. This delay is adjusted with preset P_1. The next MMV, IC_{2b}, is used to create the 'lamp-on' period, which is also adjustable (P_2). An indicator diode, D_4, is provided to check the relay activity.

The only unconventional thing about the power supply is that it exploits the fact that the small 9-volt transformer can readily supply 15 volts. This property should allow a low-power 12-volt relay to be used without problems, the unregulated voltage dropping to about 10 V when the relay trips.

In case the doorbell voltage is lower than about 8 volts, the value of R_1 may have to be reduced. If the doorbell is DC-controlled, the + connection of the doorbell switch should go to the anode (+ side) of D_1. Here, too, the value of R_1 may be reduced if necessary.

The circuit board is designed for incorporation into a mains adaptor enclosure. Due attention should be given to the connection of the wiring between terminal block K_2 and the mains socket in the case. Use only mains-rated wires, and properly secure the ends on the PCB terminal block and the mains plug/socket combination. Also with electrical safety in mind, the LED should not protrude any further from the case than strictly required.

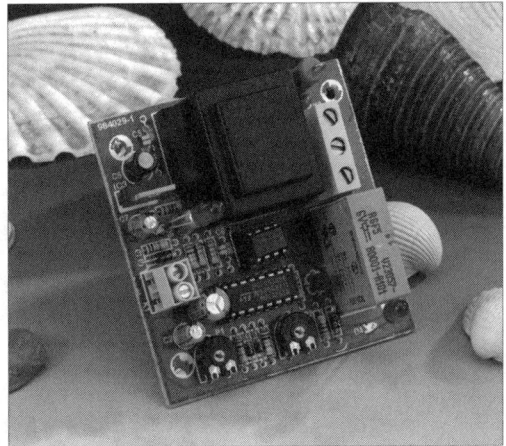

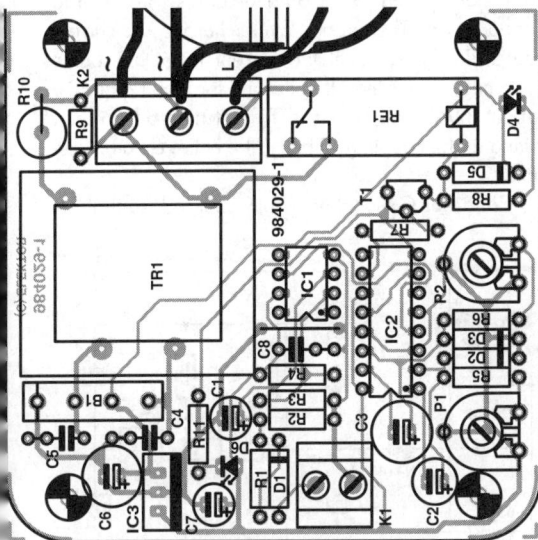

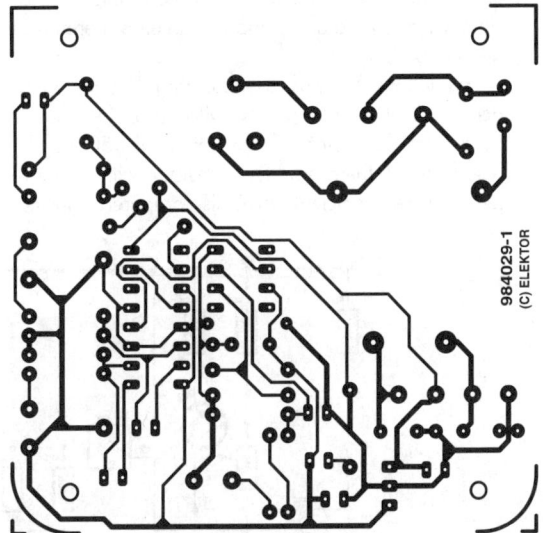

Parts list
Resistors:
$R_1, R_2, R_5, R_6 = 100\ k\Omega$
$R_3, R_4 = 10\ k\Omega$
$R_7, R_8 = 4.7\ k\Omega$
R_9 = VDR 230V (UK: 240V), small model
R_{10} = 47 Ω, 1W
$R_{11} = 2.2\ k\Omega$

P_1, P_2 = 2.5 MΩ, preset, horizontal

Capacitors:
$C_1, C_7 = 10\ \mu F$, 63 V, radial
$C_2 = 47\ \mu F$. 10 V, radial
$C_3 = 220\ \mu F$, 10 V, radial
$C_4, C_5, C_8 = 0.1\ \mu F$
$C_6 = 100\ \mu F$, 25 V, radial

Semiconductors:
B_1 = B80C1500
D_1, D_5 = 1N4001
D_2, D_3 = 1N4148
D_4, D_6 = LED 5 mm
T_1 = BC547B

Integrated circuits:
IC_1 = TLC271CP
IC_2 = 4538
IC_3 = 4805

Miscellaneous:
K_1 = 2-way terminal block, pitch 5 mm
RE_1 = Relay, 12 V, 1 make contact, for board mounting
K_2 = 3-way terminal block, pitch 7.5 mm
TR_1 = mains transformer, 9 V, 1.5 VA
PCB may be made with the aid of the track layout.

General-purpose alarm

The alarm may be used for a variety of applications, such as frost monitor, room temperature monitor, and so on.

In the quiescent state, the circuit draws a current of only a few microamperes, so that, in theory at least, a 9 V dry battery (PP3, 6AM6, MN1604, 6LR61) should last for up to ten years. Such a tiny current is not possible when ICs are used, and the circuit is therefore a discrete design.

Every four seconds a measuring bridge, which actuates a Schmitt trigger, is switched on for 150 ms by a clock generator. In that period of 150 ms, the resistance of thermistor R_{11} (resistor with negative temperature coefficient—NTC) is compared with that

of a fixed resistor. If the former is less than the latter, the alarm is set off.

When the circuit is switched on, capacitor C_1 is not charged and transistors T_1–T_3 are off. After switch-on, C_1 is charged gradually via R_1, R_7, and R_8, until the base voltage of T_1 exceeds the threshold bias. Transistor T_1 then comes on and causes T_2 and T_3 to conduct also. Thereupon, C_1 is charged via current source T_1-T_2-D_1, until the current from the source becomes smaller than that flowing through R_3 and T_3 (about 3 μA). This results in T_1 switching off, so that, owing to the coupling with C_1, the entire circuit is disabled.

Capacitor C_1 is (almost) fully charged, so that the

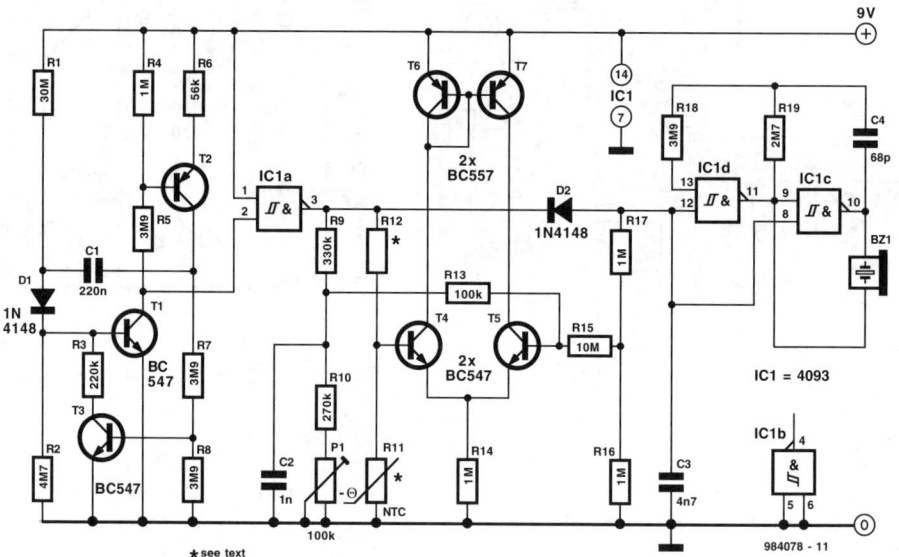

*see text

984078 - 11

194

anode potential of D_1 drops well below 0 V. Only when C_1 is charged again can a new cycle begin.

It is obvious that the larger part of the current is used for charging C_1.

Gate IC_{1a} functions as impedance inverter and feedback stage, and regularly switches on measurement bridge R_9–R_{12}-C_2-P_1 briefly. The bridge is terminated in a differential amplifier, which, in spite of the tiny current (and the consequent small transconductance of the transistors) provides a large amplification and, therefore, a high sensitivity.

Resistors R_{13} and R_{15} provide through a kind of hysteresis a Schmitt trigger input for the differential amplifier, which results in unambiguous and fast measurement results.

Capacitor C_2 compensates for the capacitive effect of long cables between sensor and circuit and so prevents false alarms.

If the sensor (R_{11}) is built in the same enclosure as the remainder of the circuit (as, for instance, in a room temperature monitor), C_2 and R_{13} may be omitted. In that case, C_3 will absorb any interference signals and so prevent false alarms.

To prevent any residual charge in C_3 causing a false alarm when the bridge is in equilibrium, the capacitor is discharged rapidly via D_2 when this happens.

Gates IC_{1c} and IC_{1d} form an oscillator to drive the buzzer (an a.c. type).

Owing to the very high impedance of the clock, an epoxy resin (not pertinax) board must be used for building the alarm. For the same reason, C_1 should be a type with very low leakage current.

If operation of the alarm is required when the resistance of R_{11} is higher than that of the fixed resistor, reverse the connections of the elements of the bridge and thus effectively the inverting and non-inverting inputs of the differential amplifier.

A thermistor such as R_{11} has a resistance at $-18\,°C$ that is about ten times as high as that at room temperature. It is, therefore, advisable, if not a must, when precise operation is required, to consult the data sheet of the device or take a number of test readings.

For the present circuit, the resistance at $-18\,°C$ must be 300–400 kΩ. The value of R_{12} should be the same. Preset P_1 provides fine adjustment of the response threshold.

Note that although the prototype uses a thermistor, a different kind of sensor may also be used, provided its electrical specification is known and suits the present circuit.

General purpose oscillator

The oscillator shown in Figure 1 is frequently used in digital circuits, but unfortunately it suffers from a nasty drawback caused by noise. When the amplitude of the noise is higher than the hysteresis of the gates used for the oscillator, spurious switching pulses are generated near the zero crossings. This problem can be cured only by ensuring that the rise time of the input signal is shorter than the reaction time of the relevant gate.

When the oscillator is built with fast logic gates, such as those in the HC-series, the likelihood of the problem occurring is great. As long as the positive feedback is fast enough, nothing untoward will happen, but when delays occur owing to the transit time of the components used, the problem may rear its head.

In the configuration of Figure 1 (left), the signal passes through two inverters and thus experiences twice the transit time of a single gate. The upper signal in the oscilloscope trace in Figure 2

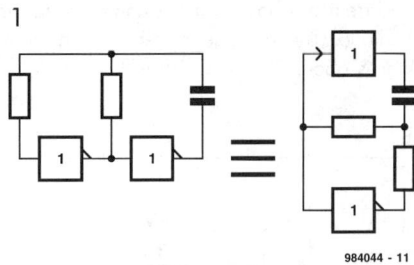

984044 - 11

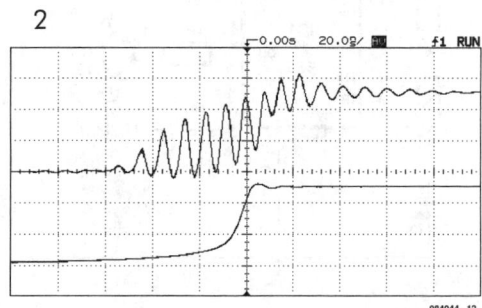

984044 - 12

shows the result of this: the gates used are simply too fast for this type of oscillator. If one of the inverters is replaced by a buffer, and the oscillator is modified as shown in Figure 1 (right), the transit time is limited to that of one gate: the lower trace in Figure 2 shows that the oscillator then works correctly.

The practical circuit diagram of the general-purpose oscillator is shown in Figure 3. Note that two XOR gates are used to ensure that the transit time of the buffer is equal to that of the inverter.

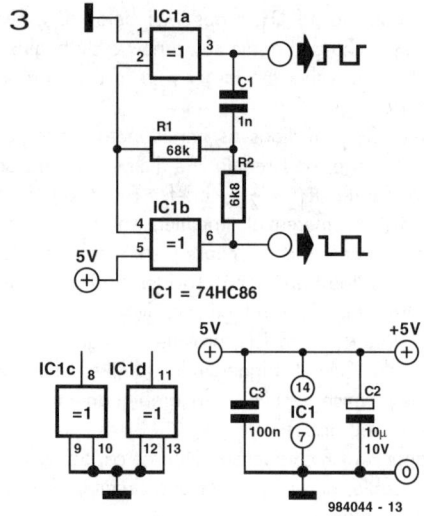

984044 - 13

Infra-red intruder alarm

The alarm circuit uses infra-red light beams to bridge distances between 3 m and 5 m (10 ft to 16 ft), but if the transmit diode is given a reflector, larger distances are possible. When the beam is interrupted, a buzzer sounds.

The transmitter is based on IC_1, which generates 10 μs wide pulses at a rate of 20 kHz. During the pulse, a current of about 100 mA flows through the transmit diode. The average current drawn by the transmitter is about 12 mA, which will normally preclude a battery-operated supply.

The receive diode is normally cut off, but comes on when it is exposed to infra-red light. The more intense the infra-red light, the larger the photo current. The received pulses cause an alternating voltage across resistor R_1.

The a.c.-coupled amplifier based on transistors T_1–T_4 provides an amplification of ×200 at a frequency of 20 kHz.

The bandwidth of the receiver is purposely limited to enhance the stability of the circuit.

The pulses arriving from the transmitter are intercepted by tone decoder IC_1. Provided the pulse rate is correct, the output of the decoder is logic low. This holds bistable IC_3 in the reset state, so that the buzzer remains inactive.

When the infra-red signal fails (because the beam is broken), the bistable is set, whereupon the buzzer is actuated. The sounding of the buzzer cannot be interrupted with switch S_1. When, however, the beam is restored, pressing S_1 causes the bistable to be reset, whereupon the buzzer is switched off.

The receiver draws a current of about 30 mA in the quiescent state, which rises to about 50 mA when the buzzer sounds.

The relatively large currents make battery operation uneconomical; it is far better (and safer) to use an appropriate mains adaptor.

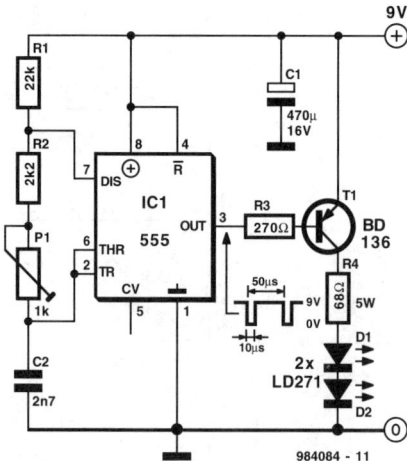

984084 - 11

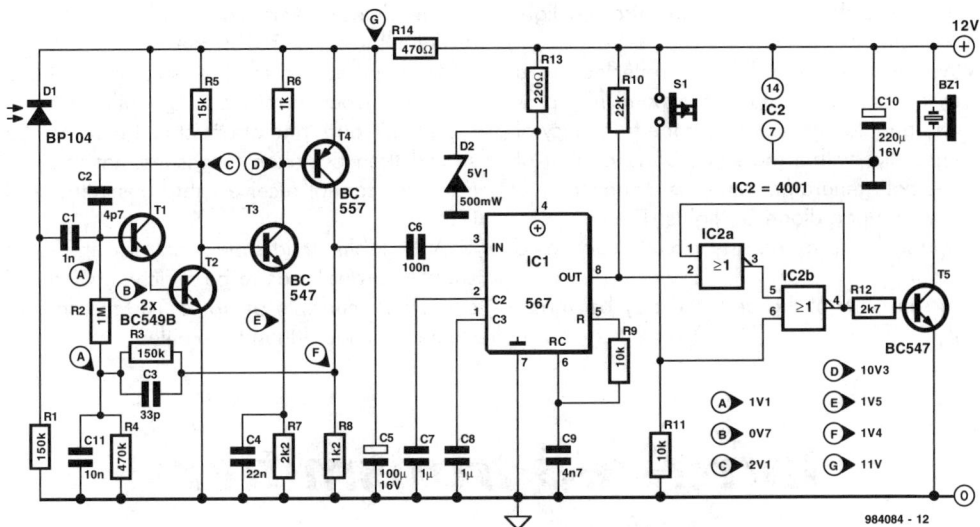

984084 - 12

Infra-red proximity detector

The detector is intended for sensing obstructions at distances of a few millimetres to a few centimetres. Similar detectors are used in the industry and health services, for instance, to open a water tap via a magnetic valve.

The sensor, IC_2, is a Type SFH900 optoisolator from Siemens or similar. A phase-locked loop (PLL) in decoder IC_1 compares the frequency of the input signal from IC_2 with that of an internally generated signal. When the two signals fall within the same band, the output at pin 8 of IC_1 changes state (from high to low).

The internal oscillator generates a signal at a frequency of about 4.5 kHz (determined by time constant R_1-C_1). The rectangular signal at pin 5 switches on the light-emitting diode in IC_2 via

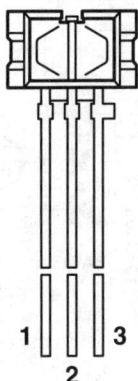

SFH900

984058 - 12

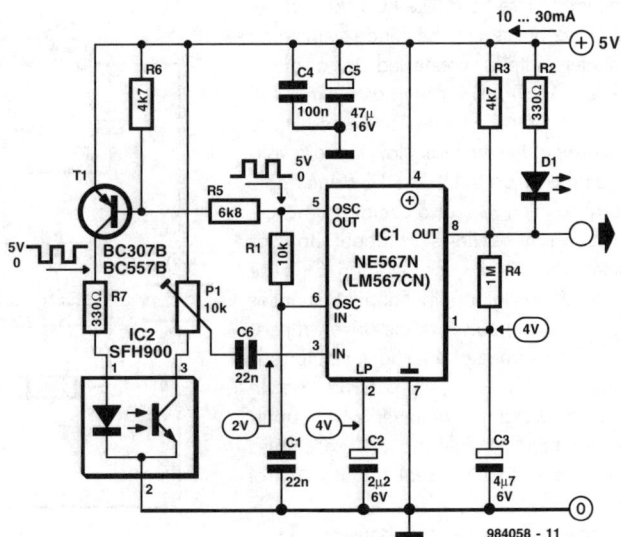

984058 - 11

197

T_1. The diode then transmits an infra-red light signal pulsing at 4.5 kHz.

When the infra-red light is reflected by a nearby object, the photo transistor in IC_2 provides a signal to pin 3 of IC_1. If the frequency of this signal lies within the same band as that of the internal generator, pin 8 is connected to earth, whereupon diode D_1 lights. The comparison by the PLL prevents the circuit reacting to stray light.

The sensitivity of the detector may be varied with P_1.

The detector with components as specified draws a current of 10–30 mA.

As stated earlier, the optoisolator may be one of several types. It may also be built from a discrete LED and phototransistor, but great care should then be taken to ensure that the photo transistor cannot receive light transmitted by the LED.

A suitable solid-state relay at the output enables larger loads to be switched. Circuit IC_1 can switch currents of up to 100 mA to earth. Diode D_1 should then be omitted.

Infra-red transmitter

The transmitter is intended primarily for the testing of infrared receivers. In most cases, such a test is possible without having the associated remote control available. Of course, the test will then be limited to just the infra-red sensor device in the receiver, but it is very useful to begin with when it comes to faultfinding in an infra-red controlled system like a TV set or a VCR. The combination of this test transmitter and the simple infrared receiver described elsewhere in this chapter forms a simple one-channel on/off remote control.

Like a 'real' remote control unit, the transmitter emits pulses modulated on to a carrier signal of about 36 kHz. Because it is not necessary for the testing purposes outlined above to have a true encoded pulse train (like RC5, RECS80 et al.), the bursts can be made with three counter outputs connected to a diode triplet, D_1, D_2, D_3, acting as a wired-OR gate. Schottky diodes are used here because of their inherent low forward voltage drop of just 0.4 V. The 14-stage ripple carry counter, IC_1, also contains an oscillator which oscillates at about 36 kHz determined by R_2, P_1, R_1 and C_1. The buffered oscillator output at pin 9 supplies carrier (clock) pulses to the driver transistor via R_3. If one of the counter outputs is logic high, however, T_1 is turned off. In this way, a signal is obtained which, apart from encoding information, resembles that produced by a real remote control unit.

Fast p-n-p switching transistor T_1 is used in common-emitter mode to make sure the circuit keeps transmitting even if the battery is almost 'flat'. In fact, the test transmitter will keep working reliably at 36 kHz at battery voltages down to 1.7 V. Because of the small duty factor of the 60-mA peak-pulse through sender diode D_4, the average current drawn from the 3-volt Li-ion battery is only about 6 mA when the battery is full, and 2.5 mA when the battery is discharged to 2.5 V. A test signal is emitted when push-button S_1 is pressed.

The whole circuit including the battery is conveniently built on a compact PCB designed to fit in a special key-fob case available from certain retailers.

984049 - 11

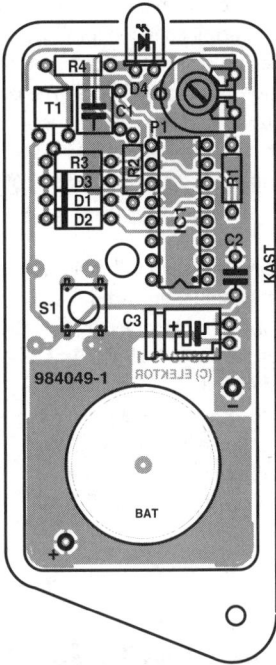

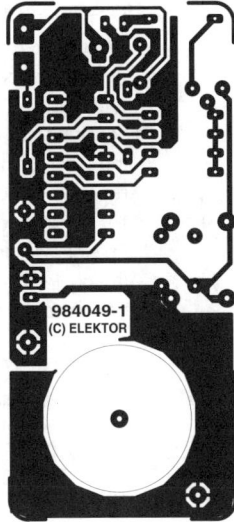

Parts list

Resistors:
R_1 = 5.6 kΩ
R_2 = 270 kΩ
R_3 = 3.3 kΩ
R_4 = 22 Ω
P_1 = 2.5 kΩ preset, horizontal

Capacitors:
C_1 = 0.0012 μF, pitch 5 mm
C_2 = 0.1 μF, rpitch 5 mm
C_3 = 100 μF, 10 V, radial

Semiconductors:
D_1, D_2, D_3 = BAT85

D_4 = LD271 or LD274
T_1 = BC327

Integrated circuits:
IC_1 = 74HC4060 (Philips, do not use Texas Instruments)

Miscellaneous:
S_1 = miniature push-button, e.g., Conrad o/n 70 04 79-44
Bt_1 = 3 V Li-ion battery, 20 mm diameter
Case: as appropriate
PCB may be made with the aid of the track layout.

Infra-red receiver

The infra-red receiver is intended to form an infra-red remote control system with the infra-red transmitter described in the previous section.

The system does not use any kind of coding or decoding, but the carrier of the transmitter is modified in a simple manner to provide a constant switching signal. Since the receive module, IC_1, switches from low to high (in the quiescent state, the output is high)

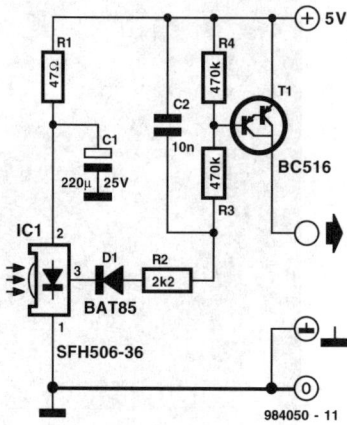

984050 - 11

when the carrier is received for more than 200 milliseconds, the carrier is transmitted in the form of short pulse trains. This results in a pulse at the output of the receiver with a duty factor a little over 12.5%. The carrier frequency is 36 kHz, so that the output frequency of IC_1 is 281.25 Hz. This signal is rectified with a time constant that is long enough to ensure good smoothing, so that darlington T_1 is open for as long as the received signal lasts.

A drawback of this simple system is that it may pick up signals transmitted by another infra-red (RC5) controller. In this case, only the envelopes of the pulse trains would appear at the output of T_1. This effect may, of course, be used intentionally. For instance, the receiver may be used to drive an SLB0587 dimmer. Practice has shown that the setting of the SLB0587 is not affected by the RC5 pulses.

The receiver draws a current of about 0.5 mA.

555 in memory mode = latch

It is not generally known that the familiar Type 555, which can switch currents up to 200 mA, can be used as a latch with control input. When input pins 2 (trigger) and 6 (threshold) are linked and connected to half the supply voltage, the output can be switched as follows. When the potential at pins 2 and 6 is at supply voltage level, the output is switched to ground; when the pins are at ground potential, the output assumes supply voltage level.

The circuit uses this mode of operation of the 555 to realize a two-wire on/off switch. The combination of S_1 (closed), R_2 and R_1 provides half the supply voltage to the input (pins 2, 6) of IC_1. When S_2 is closed, the output, pin 3, goes high so that D_2 (on) lights. When S_1 is opened, the input at pins 2, 6 of IC_1 rises to more that 2/3 of the supply voltage, whereupon IC_1 is disabled and the output goes low. Diode D_1 (off) then lights.

Network R_3-C_1 at the reset input, pin 4, forces the latch to come up in the off state when power is first applied.

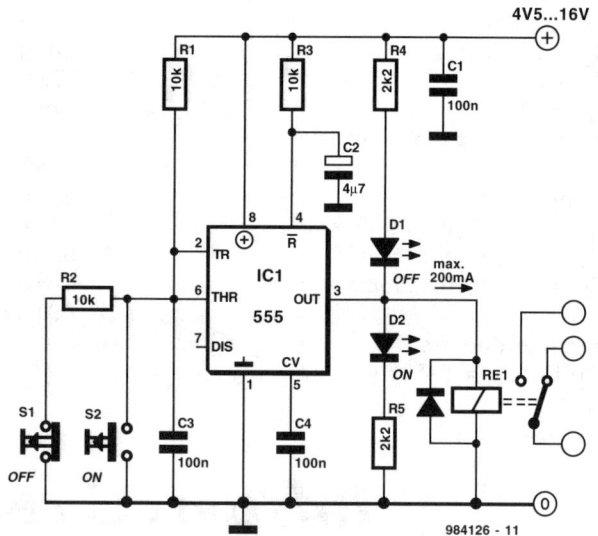

984126 - 11

LED lighting for consumer unit cupboard

The consumer unit (or 'electricity meter') cupboard in some houses is often a badly lit place as far as natural light is concerned. If the bell transformer is also located in this cupboard, it may be used to provide emergency lighting by two high-current LEDs. These diodes are powered via a small circuit that switches over to four NiCd batteries when the mains fails.

The output voltage of the bell transformer is rectified by bridge B_1 and buffered by capacitor C_1. The batteries are charged continuously with a current of about 7.5 mA via diode D_1 and resistor R_2. The base of transistor T_1 is high via R_3, so that the transistor is cut off.

When the mains voltage fails, C_1 is discharged via R_1; when the potential across the capacitor has dropped to a given value, the battery voltage switches on T_1 via R_3 and R_1, provided switch S_1 is closed. When T_1 is on, a cur-

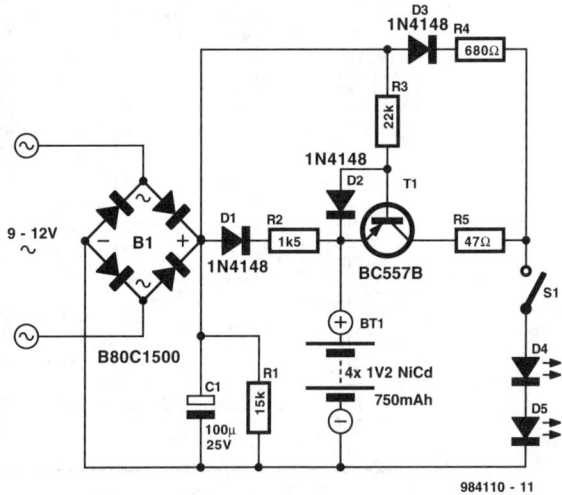

984110 - 11

rent of some 20 mA flows through diodes D_4 and D_5. The light from these LEDs is sufficient to enable the defect fuse or the tripped circuit breaker to be located.

LED-bar off indicator

The indicator may be combined, in principle, with any circuit that contains an LED bar display driven by a Type LM3914 IC. It ensures that an LED will light when all LEDs driven by the LM3914 are out. This prevents one drawing the erroneous conclusion that, since all the LEDs are out, the circuit is switched off. The circuit then continues to draw current, which, especially if it is battery powered, costs unnecessary money, apart from other considerations.

The LED in the monitor draws a current of only 1 mA.

When the LEDs forming the bar, D_1–D_{10}, are all out, there is no potential difference across R_3, so that T_1 is off and T_2 is on. This results in T_3, in conjunction with R_5 and the internal reference

voltage of IC_1, to form a current source that supplies a constant current through D_{11} so that the diode lights.

When one of diodes D_1–D_{10} lights, a potential difference ensues across R_3, which causes T_1 to come on. This results in T_2 being switched off so that there is no collector current through T_3. Consequently, there is no feedback at the emitter of T_3, so that the current through R_2 rises appreciably.

The current through R_2 determines the current through the LEDs in the bar. Therefore, when T_3 is enabled, the current through R_2, and thus the total current in the circuit, is reduced considerably.

201

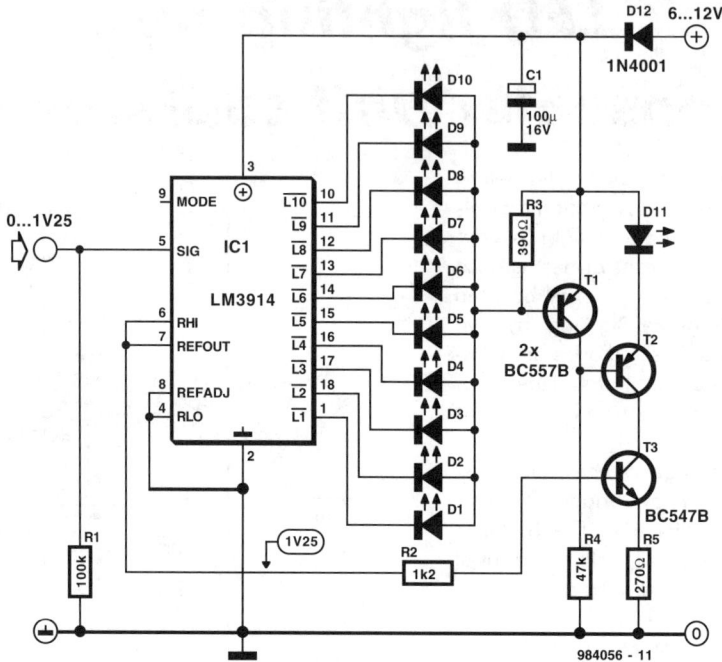

984056 - 11

Light from flat batteries

Button or coin cells that appear to be flat in normal use may often be discharged further. This is because in many cases, for instance, a quartz watch stops to function correctly when the battery voltage drops to 1.2 V, although it can be discharged to 0.8 V.

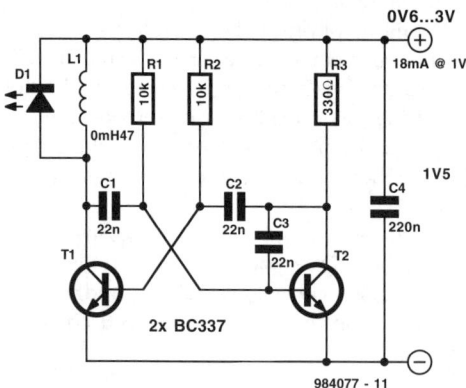

984077 - 11

Normally, however, not much can be done with a single cell. In the present circuit, a superbright LED is made to work from voltages between 1 V and 1.2 V. This may be used for map-reading lights, a keyhole light, or warning light when jogging in the dark. When a yellow, superbright LED is used with a fresh battery, it may be used as an emergency reading light or to read a front door nameplate in the dark or to find an non-illuminated doorbell.

Normally, LEDs light at voltages under 1.5 V (red) or 1.6–2.2 V (other colours) only dimly or not at all.

The present circuit uses a multivibrator of discrete design that oscillates at about 14 kHz. The collector resistor of one of the transistors has been replaced by a fixed inductor, which is shunted by the LED. Because of the self-inductance, the voltage across the LED is raised, so that the diode lights dimly at voltages as low as 0.6 V and becomes bright at voltages from about 0.8 V up.

The circuit requires a supply voltage of 0.6–3 V and draws a current of about 18 mA at 1 V.

202

Low-impact muscle stimulator

Used with care, this circuit can provide a small degree of muscle stimulation. Two electrodes, about 1–5 cm apart, are fixed on the skin covering the muscle area. The circuit generates voltage pulses at a variable rate of between 0.6 Hz and 4 Hz. The output voltage level is variable from 0 V to 250 V. The highest pulse energy supplied by the circuit is limited to a value of about 0.4 mJ, which is below the accepted safe level.

Circuit IC_1 is configured as an astable multivibrator (AMV). Its output signal frequency is adjustable with preset P_1. Transistor T_1 and preset P_2 form an adjustable voltage source which charges capacitor C_4 (via resistor R_3) to the voltage level set with P_2. Using the capacitor energy expression $[0.5CU^2]$, it is found that C_4 contains an amount of energy which is smaller than or equal to 0.4 mJ.

By charging C_3 via R_3, a simple and safe means is available of limiting the maximum power to be transferred to the muscle. As regards the on-time of the pulses, the output reactance of IC_1, in combination with that of C_3, causes transistor T_2 to conduct for just 0.5 ms. During that time, T_1 connects the low-voltage winding of Tr_1 in parallel with C_4, enabling the energy stored in the capacitor to be transferred to the electrodes, and from there to the muscle tissue. Diode D_3 indicates the pulse activity.

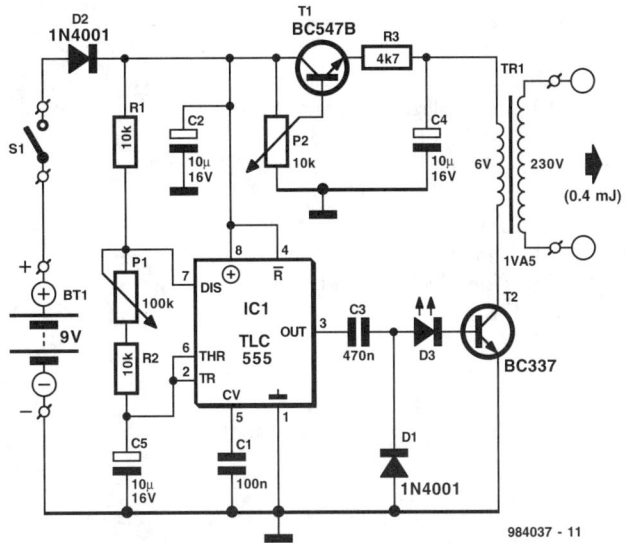

984037 - 11

WARNING. Not medically approved to any standard, this circuit should not be used by people suffering from heart ailments or epilepsy. The circuit is generally safe for use on arm and leg muscles, but not for heart stimulation. In case of doubt, see your GP.

Parts list

Resistors:
R_1, R_2 = 10 kΩ
R_3 = 4.7 kΩ
P_1 = 100 kΩ preset, horizontal

Capacitors:
C_1 = 0.1 μF metallized polyester (MKT)
C_2, C_4, C_5 = 10 μF, 16 V, radial
C_3 = 0.47 μF metallized polyester (MKT)

Semiconductors:
D_1, D_2 = 1N4001
D_3 = LED, high efficiency, red
T_1 = BC547B
T_2 = BC337

Integrated circuits:
IC_1 = TLC555CP (must be CMOS type)

Miscellaneous:
Bt_1 = 9V (PP3) battery, with clip-on leads
S_1 = on/off switch
Tr_1 = mains transformer, 6 V, 1.5 VA
PCB may be made with the aid of the track layout.

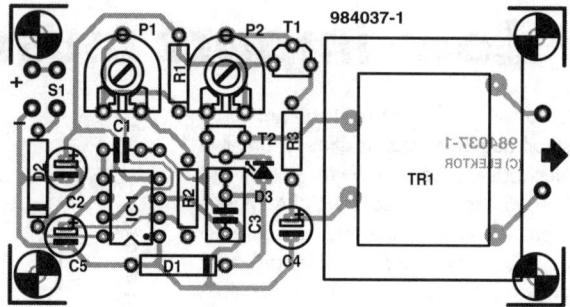

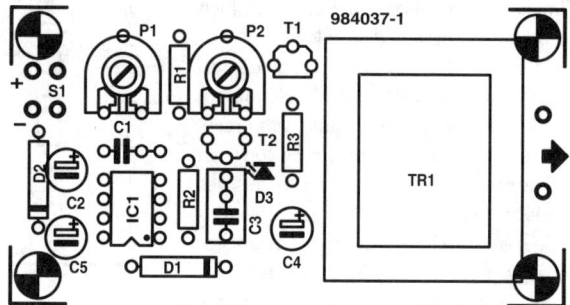

Mains master/slave control Mk2

This circuit allows one or several mains-operated apparatus ('slaves') to be automatically switched on when a 'master' apparatus is switched on. The design is complete with a PCB and uses a current transformer made from 1.5 mm dia. enamelled copper wire, which means that high master currents are allowed. In principle, up to 10 A is permissible, but, since the terminal block connectors on the printed circuit boards are rated at 5 A, this is the peak current the master device is allowed to draw from the mains. Consequently, 5 A fuses (or smaller, depending on the application) are needed in the 'master' and 'slave' supply lines, so that up to 1 kW of power may be switched in each of these 'channels' (assuming a mains voltage of 230-240 V, 50 Hz).

Transformer Tr_1 is home made. The primary winding which carries the 'master' current consists of 12 turns (1 layer) of 1.5-mm (approx. 16 SWG) enamelled copper wire (ECW) on a type ETD29 core. The secondary winding consists of 700 turns of 0.2-mm

(36 SWG) diameter ECW. Wind carefully to ensure its fitting on the core in its entirety. If this is not done, the secondary winding will touch the core material. The primary and secondary windings must be isolated from with a layer of insulating tape. Each of the wire ends of the primary is to be connected to three terminals of the core base. The primary winding drops less than 90 mV at a 'master' power consumption of 100 watts.

The 12:700 turns ratio of the transformer ensures a maximum sensitivity of about 42 mA, or just over 10 watts of 'master' power consumption. If the 'slave(s)' need to be switched on at a higher 'master' wattage, preset P_1 must be adjusted accordingly

Resistors R_1 and R_2 limit the relay current to 1 A (peak), and capacitor C_1 should be a type rated for operation at mains voltages. Note the use of two series-connected resistors in positions R_1-R_2 and R_3-R_4. Neither of these combinations should be replaced by a single resistor because the maximum voltage rating of this is then easily exceeded.

The target value for the relay voltage is about 28 V, which is derived from the mains voltage in combination with the aid of C_1. Although the relay shunt capacitor, C_6, 'wastes' about 6 watts of mains power, it is smaller and cheaper than a 1.5-VA mains transformer!

Apart from adjusting the sensitivity, P_1 also determines (to some extent) the time the 'slave' remains on after the 'master' has been switched off. Normally this delay will be about 3 seconds. For reliable operation of the relay, the U_{GS} of T_1 should be greater than about 4 V. The mains master/slave control draws a current $\leq$ 27 mA with the relay switched on.

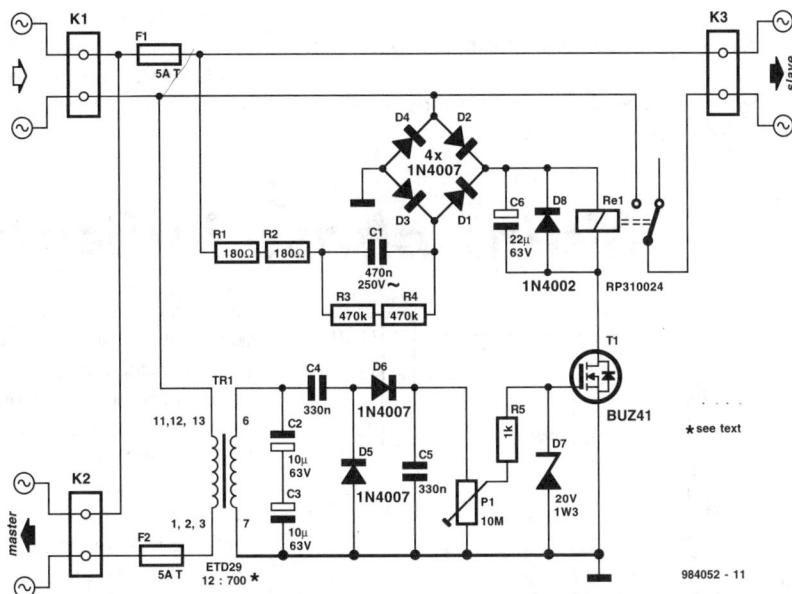

984052 - 11

WARNING. Great care should be taken when working with this circuit since it is connected directly to the mains.

Parts list

Resistors:
R_1, R_2 = 180 Ω
R_3, R_4 = 470 kΩ
R_5 = 1 kΩ
P_1 = 10 MΩ, preset, horizontal

Capacitors:
C_1 = 0.33 μF, 250 V a.c., high stability
C_2, C_3 = 10 μF, 63 V, radial
C_4, C_5 = 0.33 μF
C_6 = 22 μF, 63 V, radial

Semiconductors:
D_1–D_6 = 1N4007
D_7 = zener diode 20 V, 1.3W
D_8 = 1N4002
T_1 = BUZ41A

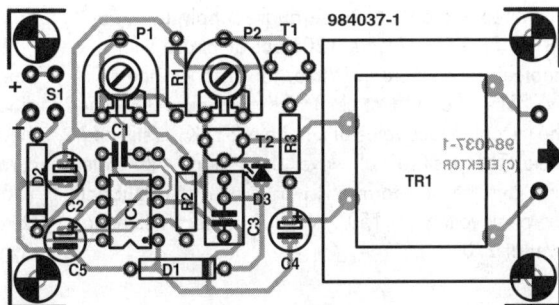

984037-1

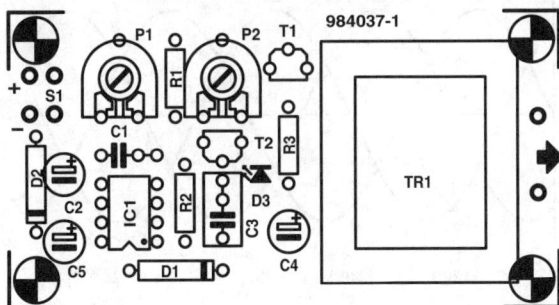

984037-1

Miscellaneous:
K$_1$, K$_2$, K$_3$ = 2-way terminal block, pitch 7.5 mm
F$_1$, F$_2$ = fuse, 5 A slow, with PCB mount holder
Re$_1$ = 24V, 250V a.c., 16A

Tr$_1$ = ETD29 core (Philips); primary: 12 turns 1.5 mm dia. enamelled copper wire; secondary: 700 turns 0.2 mm dia. enamelled copper wire.
PCB may be made with the aid of the track layout.

Mains phase indicator

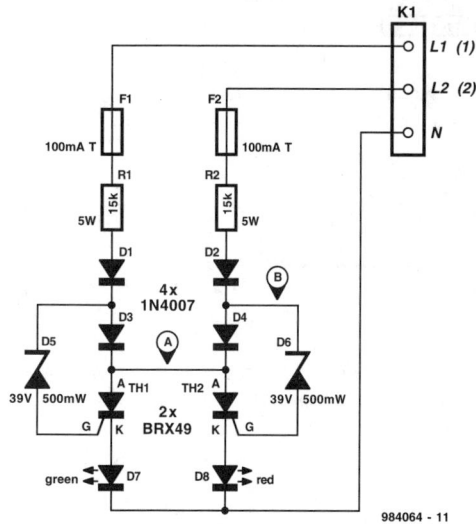

984064 - 11

The three voltages of a three-phase supply, L$_1$, L$_2$ and L$_3$ (or R, G and B) are 120° out of phase with one another—see Figure 1. When, for instance, the positive half-wave of L$_1$ at pin 1 of K$_1$ in Figure 3 begins, the instantaneous value of L$_2$ at pin 2 of K$_1$ is still negative. The positive half-wave of L$_2$ starts 120° later and cuts the waveform of L$_1$ at a level of about half the peak voltage at 150°. At 180° L$_1$ becomes negative; at 270°, L$_2$.

When two connections are interchanged as shown schematically in Figure 2, L$_1$ appears at pin 2 and L$_2$ at pin 1. It is, therefore, necessary only to establish in what order the half-waves arrive at two given terminals to determine the phase. The third phase is not discussed here, but this does not alter the argument

This requirement is met by the circuit in Figure 3. It uses a pair of thyristors, which are arranged so that the first one to be triggered cuts off the other. It should be noted that the circuit is completely symmetrical. Diodes D$_1$ and D$_2$ ensure that only positive half-waves are taken into account. The current is limited by R$_1$ and R$_2$. The two phases are combined by D$_3$ and D$_4$. The higher of the two positive voltages is always at A, and its phase is between 0° and 270°. The potential at A rises until the breakdown voltage (39 V) of zener diode D$_5$ is reached, whereupon thyristor Th$_1$ comes on and D$_7$ (green) lights. The potential at A then drops to a level equal to the sum of the breakdown voltage of Th$_1$ and the drop across D$_7$.

When a positive half-wave appears at pin 2, the potential at B can be higher than that at A only by the diode voltage of D$_4$ and this cannot be as high as the zener voltage of D$_6$. Instead, diode D$_7$ draws current from terminal 2 in the time interval between 150° and 270°. Thyristor Th$_1$ is cut off at 270° when L$_2$ drops below zero and the hold current of the thyristor ceases.

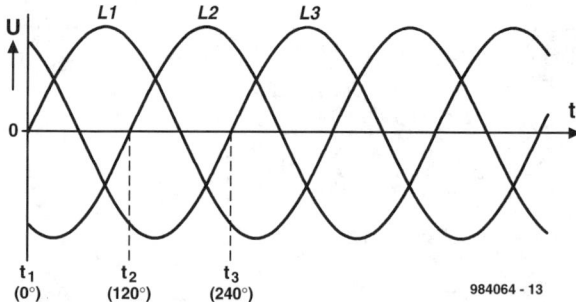

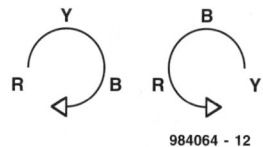

984064 - 12

984064 - 13

When, however, both terminals are interchanged, Th$_2$ is triggered first and draws current from terminal 1 at 150°, so that only the red LED (D$_8$) lights. The 20 ms interval between 270° and 360° cannot be discerned by the human eye.

Since the circuit operates with and from the mains supply, appropriate safety measures must be observed during the construction. It is imperative that the enclosure is strapped to the mains protective earth. Plugs and sockets used must, of course, be of the appropriate standard, and cable inlets must be provided with a strain relief. Do not use inferior materials!

Modem off indicator

The modem off indicator is intended especially for serious Internet surfers. It will be seen that the circuit of the indicator cannot be much simpler, or there might be nothing left. In spite of its simplicity, it may prove to be a cost-saving device, since it shows at a glance whether the telephone line is free again after the modem has been used. This obviates high telephone charges in case for some reason the modem continues to operate.

The circuit depends on the fact that there is a potential of about 40 V on the telephone line when it is not busy. This voltage drops sharply when a telephone call is being made. If, therefore, the circuit is linked to telephone terminals a and b, the lighting of the green LED shows that the line is not busy in error.

The bridge rectifier ensures that the polarity of the line voltage is of no consequence. This has the additional benefit that polarity protection for the LED is not necessary.

To make sure that the telephone line is not loaded

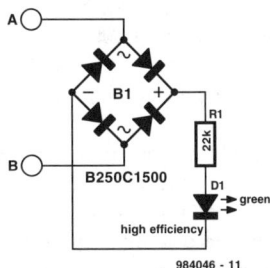

984046 - 11

unnecessarily, the LED is a high efficiency type. This type lights at a current as low as 2 mA, and this is, therefore the current arranged through it by resistor R$_1$.

WARNING. In spite of the liberal age we live in, it is highly probable that in many countries it is not allowed to connect the indicator across the telephone lines. Seek advice of your local telephone company that owns or operates the telephone network.

Humidity control

The humidity control is based on a light-dependent resistor, LDR, or a resistor with negative temperature coefficient, NTC—see diagram. There are other devices as well: the main requirement is that they can be driven by an alternating voltage, that is, that they are non-polarized.

An LDR is usually connected in series with a fixed resistor whose resistance should be equal to that of the LDR when it is not exposed to light. In the diagram (b), the network is connected as a twilight switch, that is, the circuit switches on the mains when the LDR is in darkness. When the two components are interchanged, the circuit switches the mains on when the LDR is exposed to light.

In network (c), the fixed resistor should also have

the same value as that with an NTC (at 20 °C). This network is suitable for use as thermostat in a greenhouse. When the temperature in there drops below a value set with P$_1$, the network switches on a heater.

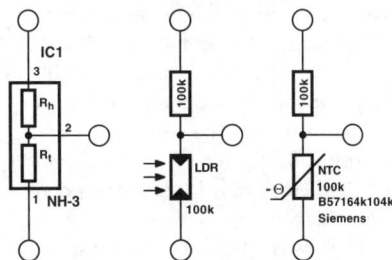

984095 - 11

Moisture detector

The function of the detector is to sound a buzzer, or, optionally, actuate a relay, when moisture sensed between a pair of probes.

The circuit has a 'memory' in the form of a bistable (US: flip-flop), IC_{1a}-IC_{1b}, which enables or disables a tone oscillator, IC_{1c}. The bistable is reset either by C_1 and C_2 when the supply voltage appears, or by push-button S_1. This may not reset the alarm, however, which will sound again until the probes are 'dry'.

The (passive) buzzer may be replaced by a relay actuating an externally connected alarm, lamp or other signalling device. Because the duty factor of the coil voltage is about 0.5, the rating of the relay coil voltage must be lower than the supply voltage.

The circuit draws a stand-by current consumption of 4–5 mA. This rises to about 40 mA when the relay is actuated. The supply voltage is not critical and may be anywhere between 3 V and 15 V. Note, however, that it may not be possible to use a relay if a supply voltage lower than about 8 V is used. If the circuit is found to be too sensitive, the value of resistor R_2 may be lowered.

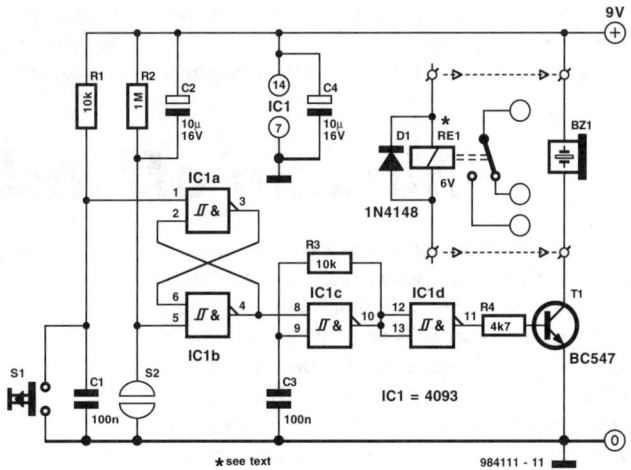

Multi-colour LED

A two-colour LED consists of two light-emitting diodes, usually red and green, encapsulated in the same case. It has three pins: two for the anodes, and one for the common cathode. In this way, each diode can be activated separately. Various colours may be obtained by varying the current through one or both of the diodes. For example, four colours may be obtained when the currents are arranged as follows: red, green, orange ($I_R \approx 2I_G$) and yellow ($I_G \approx 2I_R$).

In the present circuit, the diodes are driven by CMOS three-state buffers Type 4503, which, unlike most CMOS ICs from the 4000 series, are capable of supplying an output current of up to 10 mA. The currents are limited by resistors R_1–R_6, whose values should be ascertained empirically for brightness and colours according to requirements.

The circuit was originally developed to indicate the state of three inputs, a, b, and c

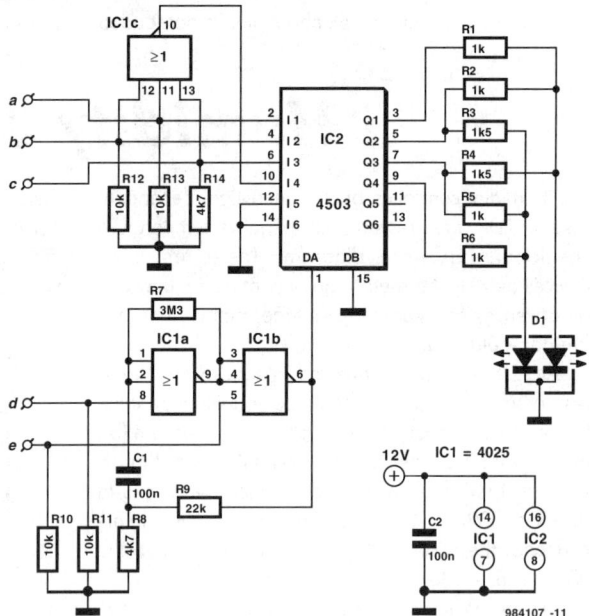

208

(non-binary, i. e., only one of these is at 1 at any time), with the configuration ($a=b=c=0$) representing the fourth state. The latter is decoded by NAND gate IC_1. An additional effect is produced by gates IC_{1a} and IC_{1b}, which form an oscillator circuit that outputs about two pulses per second. These pulses are used to control the common-enable input, DA (pin 1) of IC_2 so as to produce a flickering effect. The oscillator is controlled via inputs d and e. Pulling both

of these logic high disables the oscillator and the LED driver. With $e=0$ and $d=1$ the outputs of IC_2 are switched to three-state, and the circuit is in power-down standby mode.

Although designed for a 12-V supply voltage, the circuit will work with supply voltages between 5 V and 16 V. Non-used inputs of CMOS ICs must, of course, be tied to ground via 10–100 kΩ resistors.

Op amp with hysteresis

At first glance, the circuit in the diagram on the next page does not look out of the ordinary, and yet, it is. This is because it combines two characteristics that are

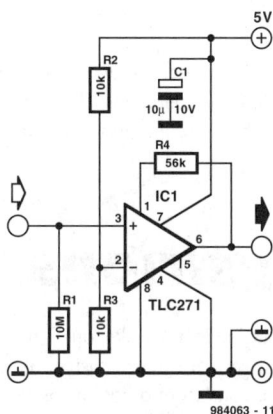

984063 - 11

usually assumed to be incompatible: hysteresis and a high input impedance. In a standard op amp circuit, this is, indeed, true, because the creation of hysteresis is normally achieved by positive feedback to the +ve input of the op amp. Unfortunately, the requisite resistance network causes a drastic deterioration of the original high input impedance of the op amp.

So, when a high input impedance and hysteresis are wanted, the solution is to obtain the needed positive feedback by coupling the resistor network not to the +ve input but to the offset correction pin. When this done, the hysteresis so obtained is calculated from

$$U_h = 1.2/R_4 U_o,$$

where U_h is the hysteresis voltage and U_o is the output voltage of the op amp, both in volts. The value of R_4 must be in kΩ. The level of U_o depends, of course, on the load.

Pulse rate monitor

The pulse rate monitor enables you to listen to your heartbeat, for instance, while you are exercising.

The transducer used for detecting the pulse is an electret microphone, X_1 in the diagram. The model used has two (polarized) terminals. As usual with this type of microphone, it functions via a series resistor, R_1. The potential drop across this resistor is applied to op amp IC_{1a} via C_1. The amplification of the op amp is set to between ×40 and ×1000 with preset P_1.

Network R_4-C_3 in the feedback loop of IC_{1a} is a low-pass filter with a cut-off frequency of 34 Hz. Higher frequencies are not needed for the present application. A pulse rate of 180* is equivalent to a fre-

quency of 3 Hz. So as to cater for a wide range of pulse rates, the cut-off frequency is made just over 11 times as high as that representing the highest pulse rate.

Operational amplifier IC_{1c}, in conjunction with push-pull am-plifier T_1-T_2, creates a headphone amplifier, whose output resistance is equivalent to the value of R_9, that is, 47 Ω. This makes the circuit usable for virtually any kind of headset. The output is short-circuit-proof. In case of certain headphones, such as that used with Sony Walkman™ sets, it is best to connect the two earphones in series.

Operational amplifier IC_{1b} is used as an active

potential divider. The voltage across the actual divider, R_5-R_6, is half the supply voltage. This voltage is buffered by IC_{1b}, taken from the low-resistance output, pin 7, of this op amp and used as reference for IC_{1a}, and as operating voltage for the electret microphone. The voltage is decoupled by C_4 to remove any interference signals from it.

The supply voltage for the pulse rate monitor is decoupled by capacitor C_7, immediately after polarity protection diode D_1.

Owing to the use of CMOS op amps, the current drain does not exceed 10 mA, so that operation from a 9 V battery is perfectly feasible. A dry alkaline-manganese battery will have a life of about 50 hours.

984013-11

* Unless you are a young superfit top-class athlete, you should see your GP immediately when you find you have a pulse rate of 180. As a general guide, the absolute maximum pulse rate for a young, very fit person is 180, for a fit middle-aged person, 160, and for a fit elderly person, 140. When exercising, the pulse rate of a not very fit person should not exceed 60% of these maxima.

Reflector for pedestrians

Pedestrians run grave dangers when, in badly-lit areas, they cross a road at night in dark clothing, because this means that car drivers may see them too late There are special armbands available that reflect the light of oncoming traffic, telling the driver that there is someone on the road.

case for electronics and battery
clothes peg (screwed on)
reflector disk
light bulb

984014 - 11

It is also possible to make an active means for warning car drivers or motorcyclists that you're crossing the road. It consists of a flat enclosure that houses two PP3 (6AM6; MN1604; 6LR61) 9 V batteries, and the small circuit shown in Figure 2.

Glue a piece of self-adhesive, reflecting tape (available in the motor trade) on the lid of the enclosure and drill a 10 mm hole in the centre of it. Fit a 6 V bulb in a suitable holder in the 10 mm hole. Fit a suitable clip and, optionally an on/off switch, at the rear of the enclosure. See Figure 1.

The electronic circuit consists of a light-sensitive switch, composed of P_1, a photoconductive cell, LDR, and IC_{1a}; an inverter, IC_{1b}; an oscillator, IC_{1c} with R_1 and C_1; a buffer, IC_{1d} with R_2 and T_1; and a 6–9 V bulb that draws a current of not more than 50 mA.

The photoconductive cell (or light-sensitive resistor) should be exposed to ambient light, but not to the light bulb, of course. Its sensitivity is set with P_1.

When the ambient light causes a potential drop across the LDR that is below the level set with P_1, IC_{1a}

changes state, so that its output goes low, whereupon the output of IC_{1b} goes high, which actuates the oscillator. The buffer then switches the light bulb on and off in the rhythm of the oscillator.

Optionally, the light bulb may be replaced by a light-emitting diode rated at 1 cd or higher, and bias resistor. Make sure, however, that the current through the transistor does not exceed its rating of 50 mA.

The two 9 V batteries should be connected in parallel. The circuit needs a supply of 3–12 V.

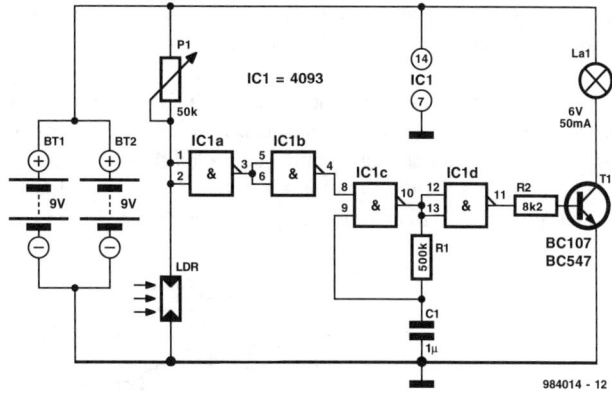

Sounds from the Old West

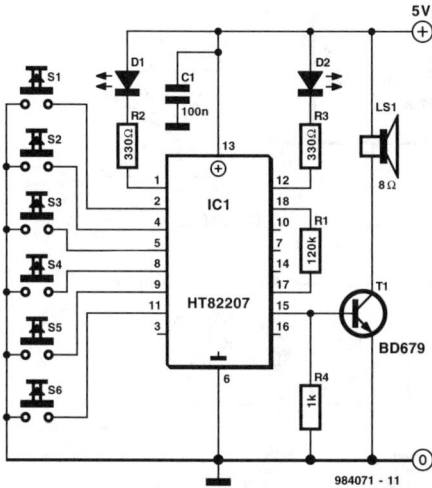

The standard 18-pin integrated sound generator Type HT82207 from Holtek produces a variety of sounds typical of the Old West. Apart from the device, only a (small) loudspeaker and the necessary selectors need to be added. The various sounds are selected by S_1–S_6 as listed below— see diagram on the next page.

In the quiescent state, the circuit draws a current not exceeding 1 μA.

S_1 – bugle
S_2 – neighing
S_3 – sound of hooves
S_4 – pistol shot
S_5 – crack of a rifle shot
S_6 – cannon fire

Symmetrical full-wave rectifier

Active rectifiers often consist of an op amp circuit with diodes in the feedback loop. Such traditional arrangements work very well in many applications, but the diodes introduce, by definition, non-linearity. However, the flaw so introduced is compensated adequately if the open-loop amplification of the op amps is sufficiently high. This is well and good at low frequencies, but the open-loop gain of op amps drops appreciably at high frequencies. This results in insufficient compensation of the the non-linearity introduced

by the diodes, which gives rise to distortion.

The rectifier presented in this article is free of such problems, since it has no diodes in the signal paths: only linear components.

The circuit is based on analogue multiplexer IC_3 and comparator IC_2. The signal to be rectified is initially buffered by op amp IC_{1a}. From there, it is applied to two inputs of the multiplexer: one in the original state of the signal and the other in inverted form by IC_{1b}. The input signal is also applied to comparator IC_2. The output of this stage will toggle between high and low in the rhythm of the input signal.

On command of the output signal from the comparator applied to its pin 11, the multiplexer switches between the original signal at X_1 and the inverted signal at X_0. This results in a signal at the output of the multiplexer whose polarity no longer varies, but remains constant: the signal is rectified. The user can even choose between the positive voltage at pin 14 and the negative signal at pin 15. The rectified signals are buffered by IC_{1c} and IC_{1d}.

The rectifier operates over a band extending from 0 Hz to 100 kHz.

The rectifier needs a power supply of ± 5 V, from which it draws a current of about 6 mA.

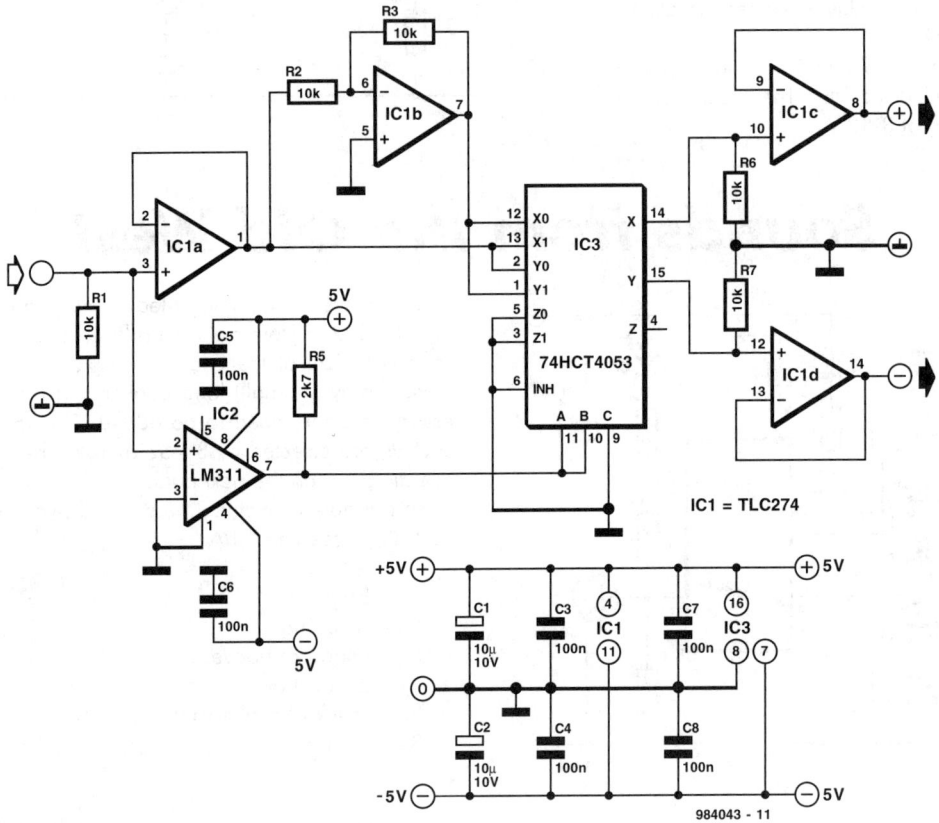

984043 - 11

Torchlight dimmer

This circuit was originally designed to control the brightness of an electric torchlight, but may find many other applications because of its high efficiency, ease of operation and ability to control (lamp) loads drawing several amps. The dimmer offers brightness control from nil to maximum in 16 steps by means of a small push-button. When the push-button is released, the selected brightness is retained. One of the most remarkable things about this circuit is that it hardly

adds to the battery load, its own current drain amounting to no more than about 4 mA (at a battery voltage of 3.5 V).

The 16 discrete brightness values are obtained by comparing two counter states. One of these actually determines the lamp brightness, while the other performs a cyclic count from 0 to 15. The lamp current is then only switched on if the second value is smaller than or equal to the first. To make sure the switching

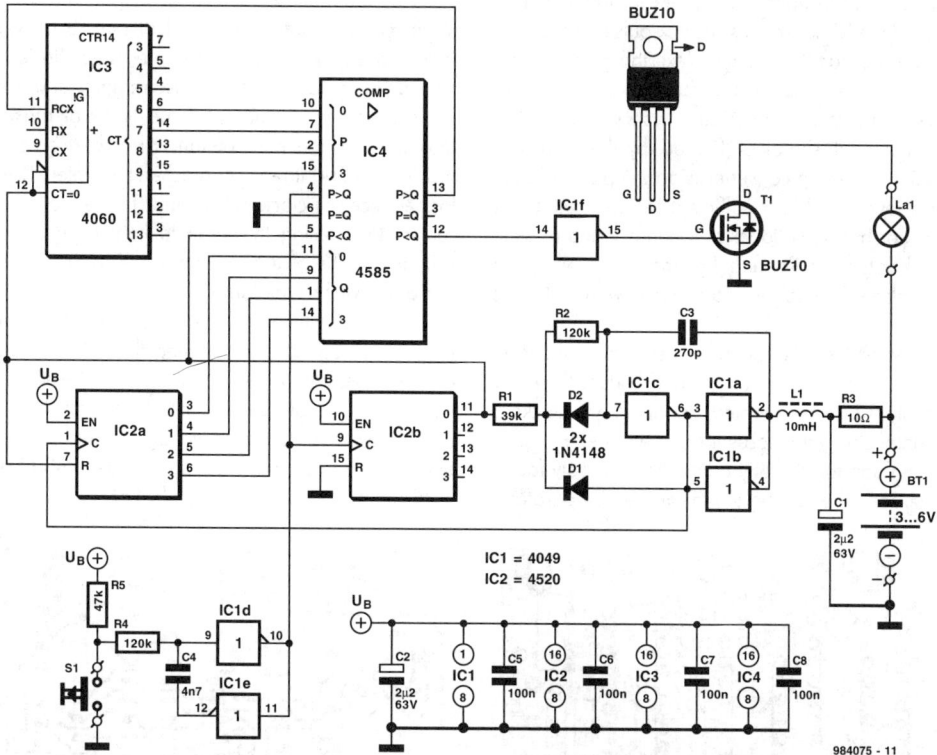

984075 - 11

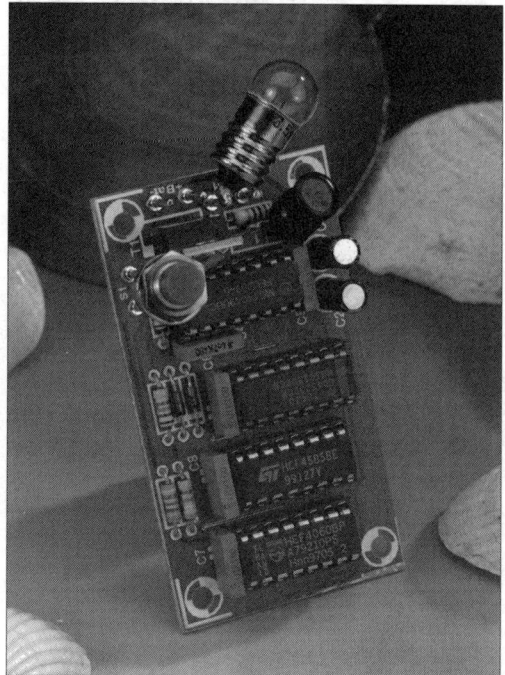

losses remain as small as possible, a power MOSFET with a very low on-resistance is used. The BUZ10 used here does, however, call for a drive voltage of at least 6 V, so that an additional voltage step-up converter is required.

Counter IC_{2b} only acts as a bistable to allow the circuit to be switched on by means of the lamp brightness push-button, S_1. The circuit is switched off (current drain ≤ 5 mA) if output Q_0 of IC_{2b} (pin 11) supplies a logic high level. The oscillator (f = about 27 kHz) formed by gates IC_{1a}, IC_{1b} and IC_{1c} is then disabled, so that the outputs of IC_{1a} and IC_{1b} are logic high. The ICs in the circuit are then powered via choke L_1 and the output transistors of IC_{1a} and IC_{1b}. Although unusual, this is possible because these transistors can also pass a voltage level at the IC outputs to the supply connection, instead of the other way around (which is far more usual). Because of the logic-high level at the reset inputs of counters IC_3 and IC_{2a}, comparator IC_4 receives input data which causes it to pull its P<Q output (pin 12) logic high. The result is that inverter IC_{1f} pinches off T_1, and the lamp remains off.

213

When the push-button is pressed for the first time, the bistable in IC_{2b} receives a clock pulse from switch debouncing circuit IC_{1c}-IC_{1d}. Next, the counters and the oscillator are enabled. The duty factor of the oscillator signal is determined by resistors R_1 and R_2. The oscillator output signal is filtered by R_3 and C_1. Although the step-up converter is only capable of supplying a few mA, that is sufficient for the CMOS ICs and T_1. For battery voltages between 3 V and 6 V, the specified values of R_1 and R_2 enable a voltage of 8.5 V to about 16 V to be created for powering the ICs and driving T_1.

As long as the push-button is held down, the level at the cascading input of IC_4, pin 4, causes the P>Q output, pin 13, to be enabled, so that IC_3 is clocked. The counter slowly increases the value at the 'P' inputs of the comparator, thereby controlling the duty factor (mark/space ratio) of the signal at the comparator's P<Q output, pin 12. As soon as IC_3 reaches its maximum counter state, the signal at pin 13 of IC_4 no longer changes, so that the counter is not started at 0 again. The P<Q output then also remains at 0, so that T_1 is driven hard and the lamp lights at maximum brightness. If the push-button is released before the maximum brightness is reached, counter IC_3 no longer receives clock pulses and 'freezes' at the current state, causing the lamp to light at the selected brightness. The next action on S_1 resets the entire circuit and switches the lamp off

If so desired, the brightness control rate may be reduced by doubling or trebling the value of C_3. To compensate for the resultant drop in the IC supply lines, the value of choke L_1 has to be increased proportionally. The IC supply voltage should always be between 8 V and 16 V (maximum value of 4000 series CMOS ICs).

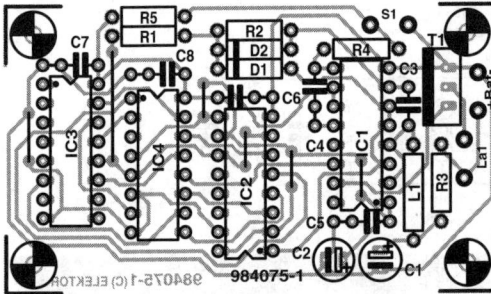

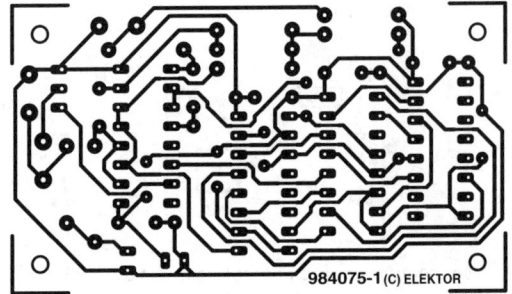

Parts list

Resistors:
$R_1 = 39\ k\Omega$
$R_2, R_4 = 120\ k\Omega$
$R_3 = 10\ \Omega$
$R_5 = 47\ k\Omega$

Capacitors:
$C_1, C_2 = 2.2\ \mu F$, 63 V, radial
$C_3 = 270\ pF$, ceramic
$C_4 = 0.0047\ \mu F$
C_5–$C_8 = 0.1\ \mu F$

Inductors:
$L_1 = 10\ mH$ choke

Semiconductors:
$D_1, D_2 = 1N4148$
$T_1 = BUZ10$

Integrated circuits:
$IC_1 = 4049$
$IC_2 = 4520$
$IC_3 = 4060$
$IC_4 = 4585$

Miscellaneous:
$S_1 = $ push-button, 1 make contact
$Bt_1 = $ torchlight battery, 3–6V
$La_1 = $ torchlight lamp
PCB may be made with the aid of the track layout.

Stepper motor control

The control is a compact design for driving bipolar stepper motors. Driver stages IC_2 and IC_3 are modern types from ST Microelectronics. Because of the combination of CMOS logic with D-MOS power transistors, these devices need few external components. Also, compared with the previous generation of bipolar devices such as the L298, the D-MOS transistors drop lower voltages so that the internal dissipation is smaller.

used when several L297s are driven in tandem and should be left open in the present design. Pin 3 (home) is an output that indicates when outputs A, B, C, and D, assume the binary code 0101, and is not used in the present design. The other pins are of less importance and will in most cases not be used at all – further information may be obtained from Internet address http://www.us.st.com

Since the current through the motor coils must not

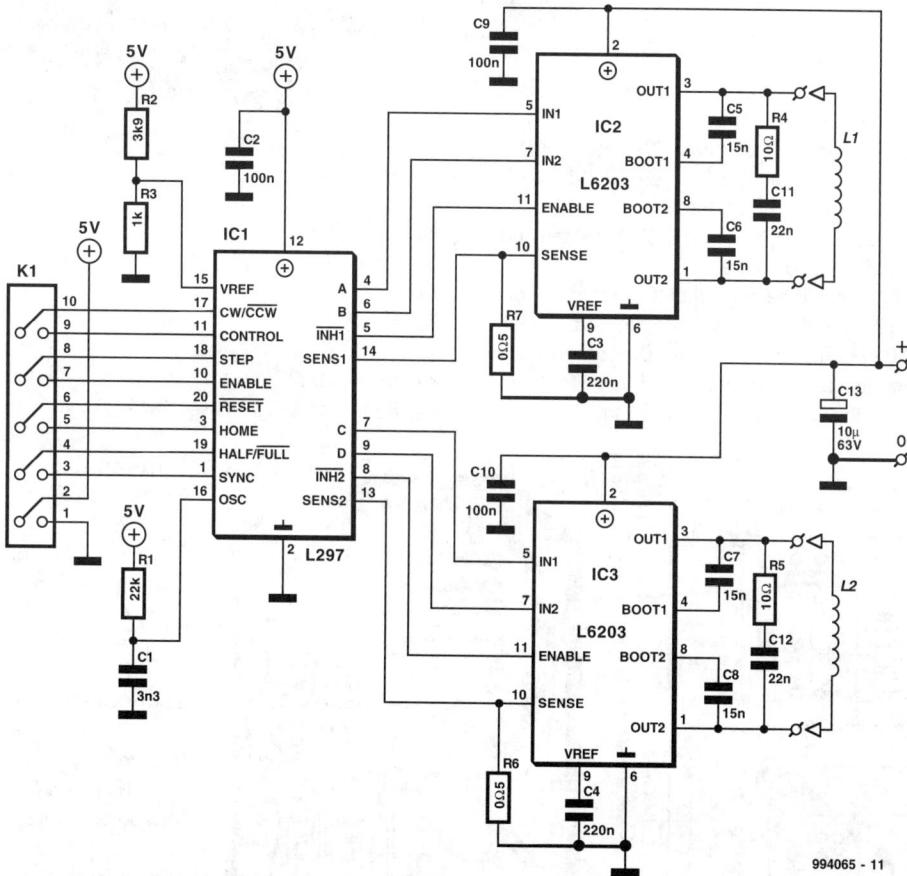

994065 - 11

The input stage, IC_1, enables the motor to proceed one step for each pulse at its input pin 18. The level at pin 17 (CW/CCW) determines whether the motor rotates clockwise or anticlockwise. The level at pin 19 decides whether the motor moves whole or half steps for each pulse at pin 18. In normal operation, pin 20 (reset), pin 11 (control), and pin 10 (enable) should be linked to the +5 V supply. Pin 1 (sync) is an output

only be switched on and off, but also be reversed, the driver ICs contain a complete bridge formed by four D-MOSFETs. The upper two need to be driven by a potential that is higher than the supply voltage, and this is obtained with the aid of a bootstrap circuit (C_5, C_6). Network R_4-C_{11} suppress voltage peaks across the motor terminals. Most of the other capacitors in the diagram are decoupling (bypass) elements.

The driver ICs can handle currents of up to 4 A at voltages up to 42 V. For safety's sake, it is better for the voltage to remain well below 42 V; the current is internally limited to 4 A. Any tendency of the current to rise above this level is sensed by resistors R_6 and R_7, whereupon the IC is disabled. The value, R, of these resistors must, therefore, be in line with I_m, the motor current: $R = 1/I_m$.

The driver ICs are provided with internal thermal protection, but when the dissipation is large, it is advisable to mount them on a suitable heat sink. They do not get damaged by heat, but they do switch off the motor when the temperature rises above the maximum specified temperature.

Finally, it should be noted that IC_1 operates from a +5 V supply (via K_1) from which it draws a current of about 50 mA. The voltage at pins 0 and + is intended for the stepper motor and should be equal to, or a little higher than, the rated motor voltage.

Parts list

Resistors:
R_1 = 22 kΩ
R_2 = 3.9 kΩ
R_3 = 1 kΩ
R_4, R_5 = 10 Ω
R_6, R_7 = 0.5 Ω, 3 W (see text)

Capacitors:
C_1 = 3.3 nF
C_2, C_9, C_{10} = 100 nF
C_3, C_4 = 220 nF
C_5–C_8 = 15 nF
C_{11}, C_{12} = 22 nF
C_{13} = 10 μF, 63 V, rad.

Semiconductors:
IC_1 = L297 (ST Microelectronics)
IC_2, IC_3 = L6203 (ST Microelectronics)

Miscellaneous:
K_1 = 10-way header
PC_1–PC_6 = PCB pins
L_1, L_2 = bipolar stepper motor

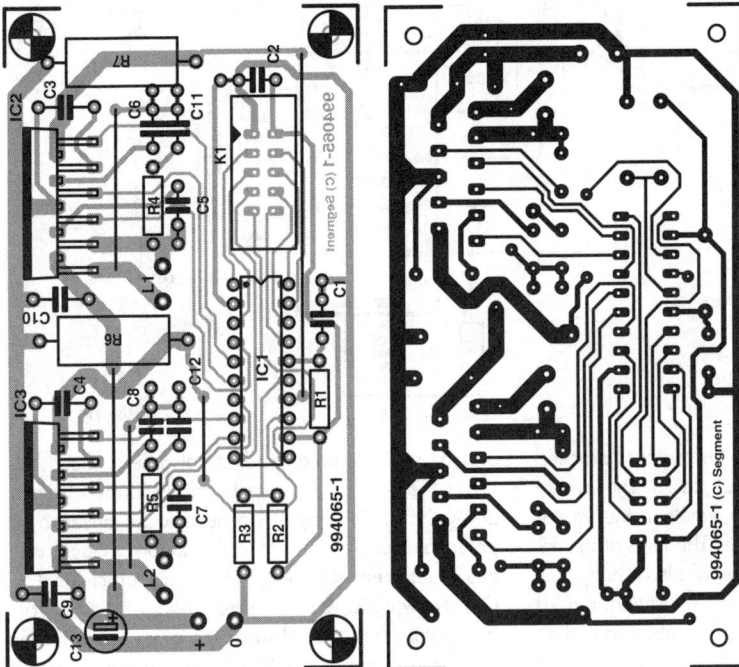

Speed controller for model train

The Type ZN419CE servo control IC from Ferranti is ideal for use in model railway direct-current engines. The internal circuit diagram of this device is shown in Figure 1. The IC varies the pulse width of input pulses in direct proportion to the displacement of a potentiometer. It operates with positive-going input pulses which can be coupled directly or via a capacitor to input pin 14. The advantage of a.c. coupling is that should a fault occur which causes the input signal to become a continuous level, the servo will remain in its last quiescent position, whereas with direct coupling the servo output arm will rotate continuously.

The active input is a Schmitt trigger input (pin 14), which allows the servo to operate consistently with slow input edges and supplies the fast edge required by the trigger monostable independent of input edge speed. The input pulse is compared with the monostable pulse in a comparison circuit and one output is used to enable the correct phase of an on-chip power amplifier. Additional, a deadband, that is the period during which the motor must not move, is effected by capacitor C_3 between pin 13 and earth.

Any difference between the input and monostable pulses is used to drive the motor in such a direction as to reduce this difference so that the servo takes up a position that corresponds to the position of the potentiometer in the controller.

The relation between the setting of the joystick on the controller and the duty factor of the output signal (0–100%), which determines the speed of the engine, depends on the potential at pin 12. This potential is set with P_2, while the rest position of the joystick and the motor is set with P_1.

The potential at pins 5 and 9 are applied to the inputs of NAND gate IC_{2b}. The pulse-width modulated (PWM) output of this gate is applied to TTL-compatible

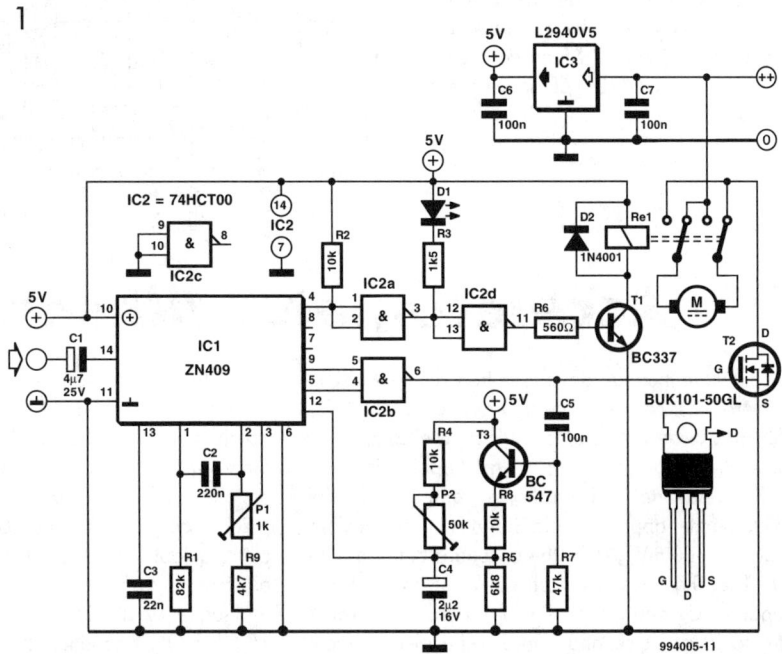

power MOSFET T_2, which switches the motor. The PWM signal drives the pulse expansion circuit via C_5 and T_3. In this way, pulses that are only just outside the deadband are expanded, which improves the control over the motor.

The potential at pin 4, which decides the direction of travel, is applied to relay Re_1 and Re_2 via gates IC_{2a} and IC_{2d}, and buffer T_1. The relays serve to reverse the polarity of the motor. The direction of travel is indicated by diode D_1. When the relays are in the rest position, the direction of travel should be forward, since their rating of 12–16 A applies only to short periods of time.

The supply voltage for the servo is held steady by regulator IC_3. This should be a low-drop type if the input to $K_2 \leq 8$ V; with higher voltages, a standard 7805 may be used. The supply voltage for the motor is, of course, not regulated.

Transistor T_2 can handle currents of up to 25 A. If a higher current is needed, two of these transistors can be connected in parallel. A somewhat less expensive transistor is the BUK100 which can handle currents of up to 13 A.

If a printed-circuit board is used for the construction, it should be capable of working with the high currents mentioned earlier (separation of tracks!).

The motor should be decoupled for noise and interference by ceramic capacitors rated ≥ 50 V: 0.01 μF between casing and earth and 0.01–0.1 μF across the terminals.

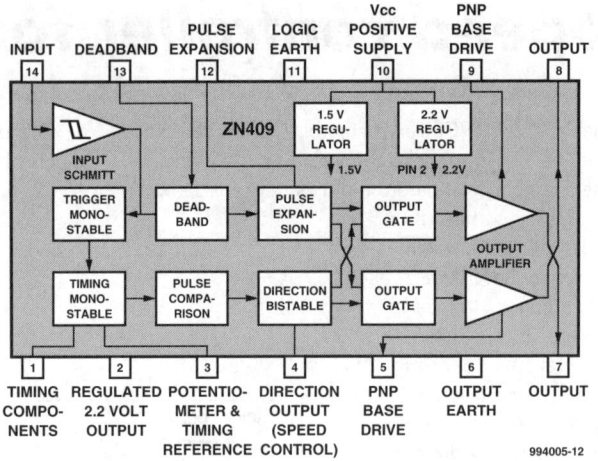

		PULSE	LOGIC	Vcc POSITIVE	PNP BASE	
INPUT	DEADBAND	EXPANSION	EARTH	SUPPLY	DRIVE	OUTPUT
14	13	12	11	10	9	8

ZN409

INPUT SCHMITT

1.5 V REGU-LATOR — 1.5V
2.2 V REGU-LATOR — PIN 2 2.2V

TRIGGER MONO-STABLE — DEAD-BAND — PULSE EXPAN-SION — OUTPUT GATE

OUTPUT AMPLIFIER

TIMING MONO-STABLE — PULSE COMPA-RISON — DIRECTION BISTABLE — OUTPUT GATE

1	2	3	4	5	6	7
TIMING COMPO-NENTS	REGULATED 2.2 VOLT OUTPUT	POTENTIO-METER & TIMING REFERENCE	DIRECTION OUTPUT (SPEED CONTROL)	PNP BASE DRIVE	OUTPUT EARTH	OUTPUT

994005-12

Accelerometer tilt sensor

The circuit in the diagram shows how a Type ADXL05 accelerometer can be connected to a low-cost CMOS 555 to provide a frequency output. The component values indicated apply for a ±1 g tilt application.

The nominal 200 mV g^{-1} output of the accelerometer appears at pin 8 and is amplified x2 to a level of 400 mV g^{-1} by the on-board buffer amplifier. The 0 g bias level at pin 9 is about 1.8 V. Capacitor C_4 and resistor R_3 form a 16 Hz low-pass filter to lower noise and improve the measurement resolution.

The 555 operates as a voltage-controlled oscillator where R_5, R_6 and C_5 set the nominal operating frequency. The values of R_5 and R_6 give a duty cycle of about 50% when a +1.8 V (0 g) input signal is applied to pin 5 of the 555. To prevent any change in frequency owing to supply variations, the 555 operates from the accelerometer's +3.4 V reference rather than directly off the +5 V supply line.

The output frequency of the circuit is determined by the charging and discharge times set by R_5, R_6, and C_5.

With the circuit and component values shown in the diagram, the output scale factor at pin 9 of the accelerometer will be ±400 mV g^{-1}, so the voltage output will be ±1.8 V ±0.4 V. The output scale factor at pin 3 of the 555 will be about 16,500 Hz ±2.600 Hz g^{-1}. The characteristic shown is the circuit's output frequency vs the voltage occurring at pin 5 of the 555.

Frequency stability of the circuit is good. With a 15.5 kHz 0 g frequency, the measured 0 g frequency drift over the range 0–70 °C temperature range was 5 Hz °C^{-1}, which is 0.03% °C^{-1}. The change in frequency vs supply voltage is less than 10 Hz with a supply voltage of 5–9 V.

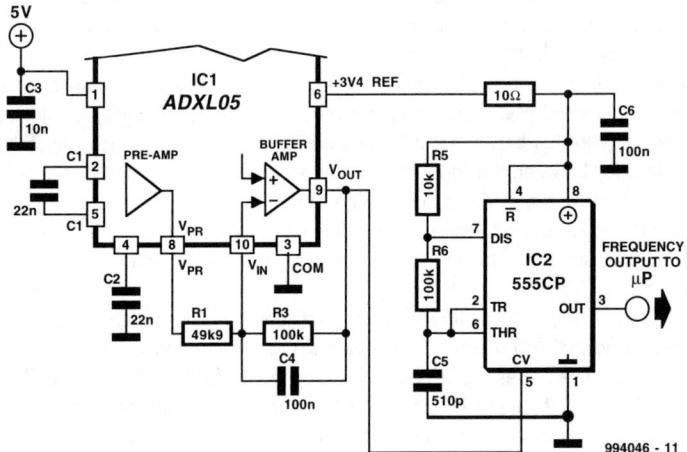

5V

C3
10n

IC1
ADXL05

+3V4 REF

10Ω

C6
100n

C1
22n

PRE-AMP

BUFFER AMP

V$_{OUT}$

R5
10k

C1

V_{PR}
V_{PR}

V_{IN}
COM

R6
100k

R
DIS

IC2
555CP

FREQUENCY OUTPUT TO μP

C2
22n

R1
49k9

R3
100k

TR
THR

OUT

C4
100n

C5
510p

CV

994046 - 11

218

AVC logic family

Driven by customer demand, Philips Semiconductor's AVC Logic family is the fastest logic on the market today and offers ultra low noise and low voltage for applications such as DRAM modules, personal computer, workstations, network servers, telecommunication switching equipment and base stations. (AVC is an acronym of Advanced Very-low-voltage CMOS).

Devices in the AVC logic family are intended for use in systems that operate from a supply voltage of 1.2–3.6 V. The manufacturer claims that this is the first family that offers a delay of not more than 2 ns at such a low supply voltage.

As an example, with a supply voltage of 3.3 V, the 74AVC16244 (a three-state buffer) has a typical delay of 1 ns. At 2.5 V this increases to 1.1 ns, and at 1.8 V, to 1.5 ns. This is about 40 per cent faster than attainable with current logic families.

The devices in the new family are provided with protection that makes hot insertion possible. This is an important aspect when, for instance, expansion cards are to be added to communication systems that should not be switched in any circumstances.

3 Volt Products Positioning

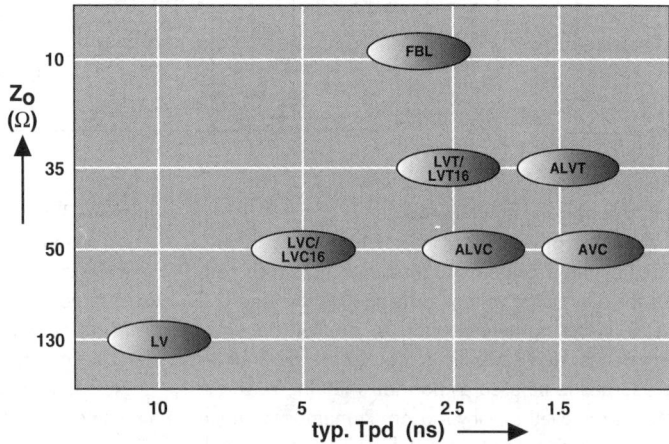

Bell transformer supply for wave-file player

If the wave-file player described in the February 1999 issue of *Elektor Electronics* is used as a programmable doorbell, it is, of course, handy if its supply is derived from the doorbell transformer. In the design of the circuit in the diagram, it was borne in mind that as little as possible should be changed in the existing wiring of the doorbell and that it would be convenient to be able to switch the doorbell back on when the wave-file player is removed to program it with a new sound.

The alternating voltage at the secondary of the doorbell transformer is rectified by diode bridge D_2–D_5, and smoothed by capacitor C_2. Zener diode D_6, a fast type, serves to suppress transients. The resulting direct voltage is stabilized by regulator IC_1 which is set for 6.8 V. When the drop across the protection diode in the wave-file player is deducted from this, a direct voltage of about 6 V remains, which is

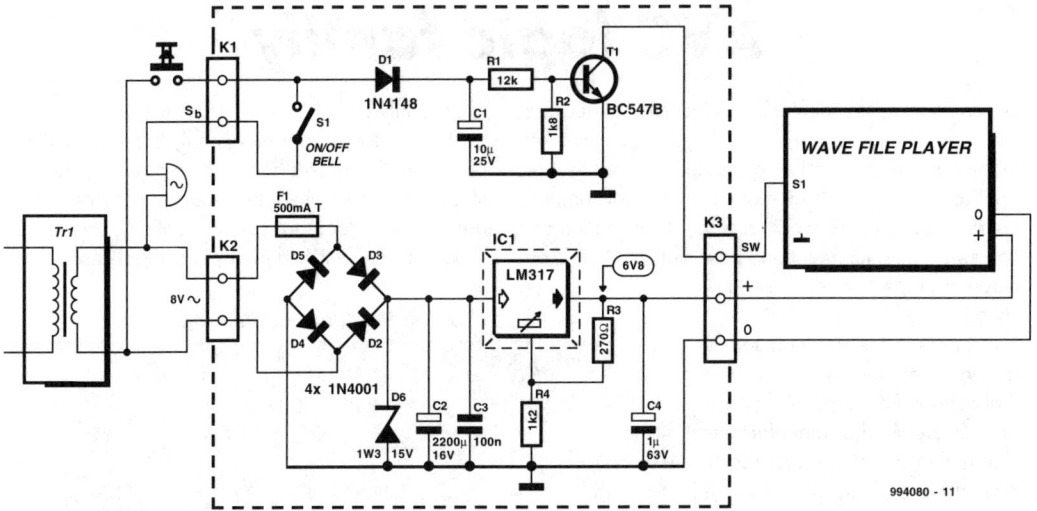

994080 - 11

used to supply the amplifier IC

The need of modifying the existing wiring is precluded by the circuit around transistor T_1, which converts the alternating voltage arriving from the doorbell into a switching signal that is applied to, and processed by, the wave-file player. Switch S_1 enables the doorbell to be used as normal when the wave-file player is removed for any reason. Remember to open the switch again when the wave-file player is replaced, otherwise the player does not work and it is set to the wrong speed of 9600 baud via the S_1 input.

If desired, the unit can be coupled to a computer via a screened cable to make it possible for the program of the wave-file player to be altered without it being necessary to remove the unit.

Note that the voltage regulator should be mounted on a small appropriate heat sink of 24 K W^{-1}.

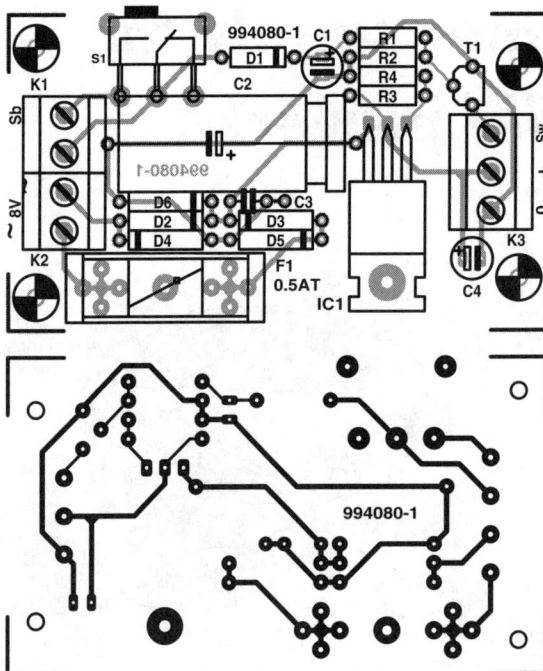

Parts list

Resistors:
R_1 = 12 kΩ
R_2 = 1.8 kΩ
R_3 = 270 Ω
R_4 = 1.2 kΩ

Capacitors:
C_1 = 10 µF, 25 V, radial
C_2 = 2200 µF, 16 V
C_3 = 0.1 µF, ceramic
C_4 = 1 µF, 63 V, radial

Semiconductors:
D_1 = 1N4148
D_2–D_5 = 1N4001
D_6 = BZT03, 15 V, 1.3 W
T_1 = BC547B

Integrated circuits:
IC_1 = LM317T

Miscellaneous:
K_1, K_2 = 2-way terminal strip, pitch 5 mm
K_3 = 3-way terminal strip, pitch 5 mm
S_1 = slide switch

F_1 = fuse holder with 500 mAT fuse
Enclosure, e.g. Bopla E410
Heat sink for IC_1: 24 K W^{-1}
PCB may be made with the aid of the track layout.

Bipolar relay with single supply

When a bipolar relay is to operate from a single supply voltage, normally two coils are needed. When, in Figure 1, the push-button RESET switch briefly connects the second terminal of coil 1 with $+U_b$, the relay is in one position. Pressing the SET switch briefly causes the relay to switch over to the other position. The relay can, however, change over only if the currents through the coils flow in opposite directions.

There are bipolar relays available that make operation with a single coil possible, but these are quite expensive. These relays are operated with voltages of opposite polarity or with a voltage whose polarity can be reversed.

An alternative, suggested by Robert Friberg in *Model Railroad Electronics*, 4 ('Adaptor for bipolar switches', p. 81), is shown in Figure 2. As in the case of a bipolar relay with two coil, the terminals of the solenoid are alternatively linked to the positive supply voltage, U_b. At the same time as one terminal is connected to $+U_b$, the other is linked to earth by a transistors whose base is connected via a series resistor to the first terminal, that is, $+U_b$.

Low-power bipolar relays with a coil resistance of 2–5 kΩ may be combined with small-signal transistors, such as the BC547. These transistors need a base resistor of about 10 kΩ. Higher power relays need power transistors and base resistors appropriate to the relay coil resistance.

The transistors should be protected against voltage spikes by free-wheeling zener diodes with a zener voltage slightly higher than the operating voltage.

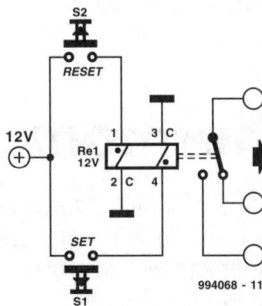

994068 - 11

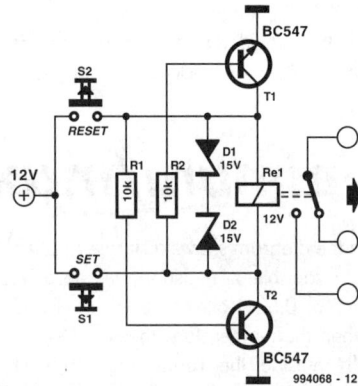

994068 - 12

Code lock

The combination of a couple of thyristors, key switches and a relay as shown in the diagram forms a suitable, reasonably tamper-proof code lock for use in, say, a car. To open the lock, a number of keys must be pressed in a prescribed sequence to energize a relay, whereupon the battery voltage is applied to the ignition switch.

The first key to be pressed is S_4, whereupon capacitor C_1 is charged via resistor R_1. The charge on this capacitor ensures that transistor T_1 is on for about 15 seconds, during which relay Re_1 is energized. However, within these 15 seconds, key S_5 must be pressed to switch on thyristor THR_1. Then S_6 must be pressed to switch on THR_2. Finally, S_7 is to be pressed

221

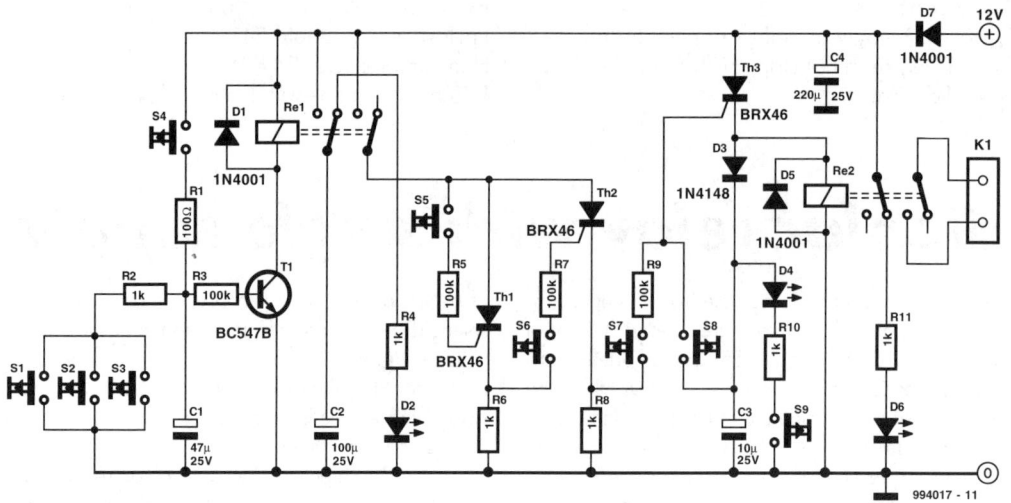

994017 - 11

to switch on THR3. When that is done, relay Re2 is energized, and the lock is open.

If within the initial 15 seconds the wrong key is pressed, C_1 is discharged, whereupon T_1 is cut off and Re_1 is deactuated. When this happens, D_2 lights briefly. The lock can then be open only by starting with pressing S_4 again. This applies also if S_5 is not pressed within the initial 15 seconds.

The lock is provided with a rapid access facility. In the actuated state, capacitor C_3 is charged via THR3.

When the ignition is switched off for only a brief period (for instance, when the driver stops to post a letter [sic]), the capacitor is discharged only slowly, so that THR3 may be switched on again by pressing S_8. Since this makes the lock less secure, the rapid access facility may be disabled by pressing S_9. Capacitor C_3 is then discharged rapidly and D_4 lights briefly.

The circuit draws a quiescent current of about 12 mA, which rises to some 80 mA when both relays are actuated.

Minimum/maximum thermometer

Although the diagram shows a simple circuit, the thermometer is capable of measuring temperatures with a resolution of 0.5 °C over a range of 30 °C. It also shows when the temperature measured falls outside the range and retains minimum and maximum measured temperatures. All this is made possible by microcontroller IC_1, a Type PIC16F84, and assembler programme THER15. The device is available ready programmed from the publishers (Order no. 996514-1). Readers who wish to program the device themselves can obtain the source code on a diskette.

The temperature sensor is a Type SMT160, available from Smartec (www.hy-line.de/Sensor/), which does not, as is usual, output a voltage commensurate with the temperature, but a

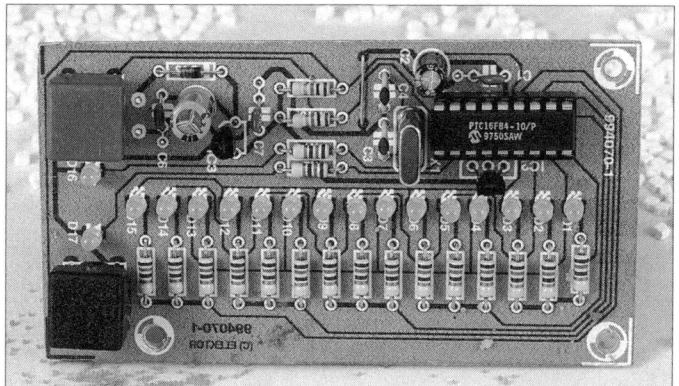

pulse-width modulated signal. This makes a separate analogue-to-digital converter unnecessary.

Only a PIC microcontroller with integral EEPROM can be used if minimum and maximum temperatures are to be retained. In the present design, a 15-way LED display, (D_1–D_{15}) is used to show temperatures in the ranges 0–15 °C and 16–30 °C. Which of these ranges is indicated by D_{16} and D_{17}. The simultaneous lighting of two adjacent LEDs indicates a half degree between their two values.

The LEDs are multiplexed in five groups of three each, so that for the display of the temperatures only one port is needed. The change from one temperature to the next takes place at a frequency of 67 Hz. When a temperature below the range is measured, only D_2 lights; when the temperature is above the range, only D_1 lights.

The stored minimum and maximum temperatures are displayed by the relevant LED when push-button switches S_1 and S_2 respectively are pressed. At the same time the relevant range LED lights. The minimum and maximum values can be erased by pressing (and holding) S_1 and then S_2 or S_2 and then S_1 respectively.

The circuit draws a current of not more than 25 mA (in case four LEDs light simultaneously). A 100 mA regulator is therefore perfectly all right. Power may be obtained from a suitable 8–12 V mains adaptor, but if low-current LEDs are used, a 9 V battery may also be considered. Note that the PIC cannot be used in the SLEEP mode since this would disable the minimum and maximum temperature memories.

The construction of the thermometer on the printed-circuit board in Figure 2 is straightforward. When purchasing the LEDs, make sure that their brightnesses are uniform; their colour does not really matter and is to individual taste. It is advisable to solder them in place last when the distance between the lid of the enclosure and the board has been established (since the diodes should just protrude through the lid).

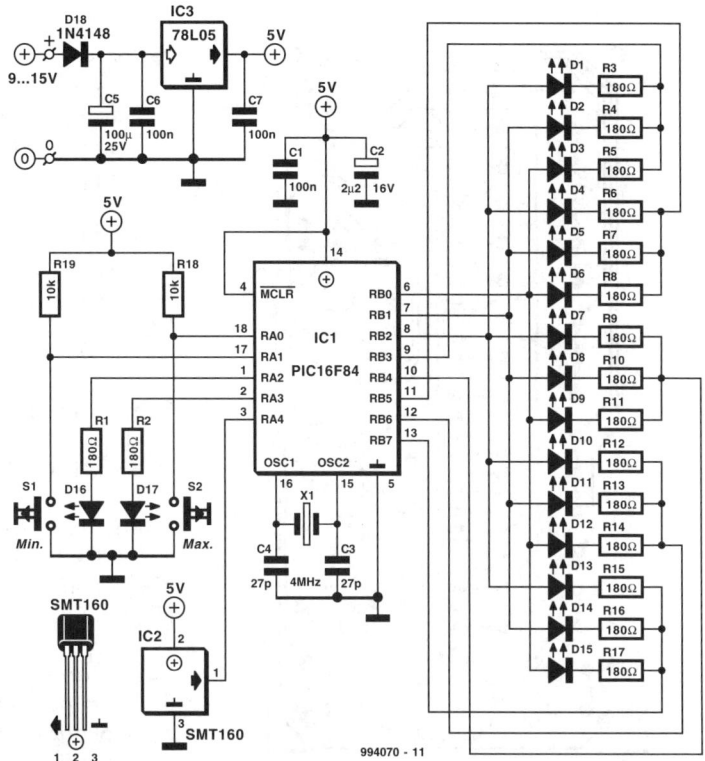

994070 - 11

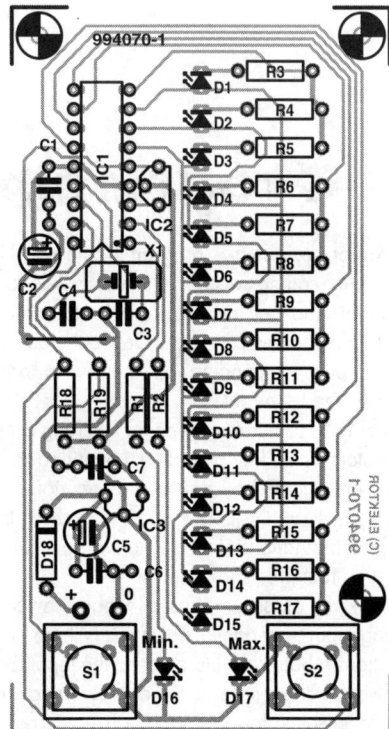

994070-1
(c) ELEKTOR-1

223

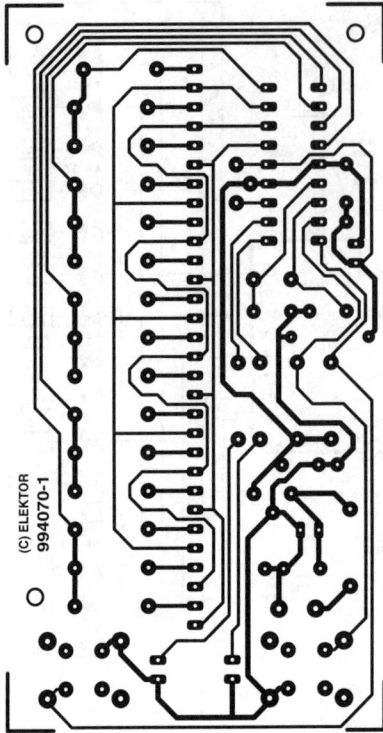

(C) ELEKTOR
994070-1

Parts list

Resistors:

R_1–R_{17} = 180 Ω (if standard LEDs are used; value should be adapted when low-current LEDs are used)

R_{18}, R_{19} = 10 kΩ

Capacitors:

C_1, C_6, C_7 = 0.1 µF
C_2 = 2.2 µF, 16 V
C_3, C_4 = 27 pF
C_5 = 100 µF, 25 V

Semiconductors:

D_1–D_{17} = LED (see text)
D_{18} = 1N4148

Integrated circuits:

IC_1 = PIC16F84-10P (996514-1)
IC_2 = SMT160 (Smartec)
IC_3 = 7805

Miscellaneous:

S_1, S_2 = push-button switch with make contact
X_1 = quartz crystal, 4 MHz
Diskette with source code (996020-1)
PCB may be made with the aid of the track layout.

Programmable amplifier

One way of designing a programmable amplifier is placing a HEX-coding switch in the feedback loop of an operational amplifier (op amp). A 16-position coding switch has a mother contact, COM, and four binary coded outputs. Depending on the switch position, COM is linked to the relevant output.

Op amp IC_1 is arranged as an inverting amplifier with resistors R_1 and R_4 providing the feedback. These resistors are shunted by one or two resistors, as the case may b e, by the coding switch. Each of the 16 ensuing combinations provides an amplification or attenuation factor as shown in Table 2.

The circuit as shown has a slight drawback: a changing input impedance. This means that in certain applications a buffer should precede the input. Bear in mind also that the resistor at the non-inverting input of IC_1 compensates for the input offset current and should therefore have a value that is comparable with that of the parallel combination of R_1, R_2, and R_3.

The switching may be augmented by replacing the coding switch by an analogue switch and operating this via a microcontroller.

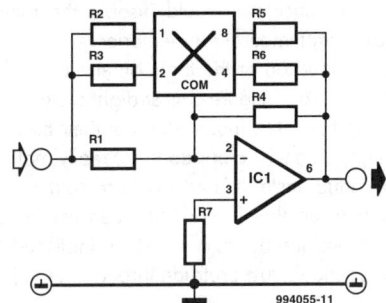

994055-11

Table 1. Resistor shunting with a hex coding switch.

Switch position	Input resistance			Feedback resistance		
0	R1			R4		
1	R1||	R2		R4		
2	R1||		R3	R4		
3	R1||	R2||	R3	R4		
4	R1			R4||	R6	
5	R1||	R2		R4||	R6	
6	R1||		R3	R4||	R6	
7	R1||	R2||	R3	R4||	R6	
8	R1			R4||		R5
9	R1||	R2		R4||		R5
A	R1||		R3	R4||		R5
B	R1||	R2||	R3	R4||		R5
C	R1			R4||	R6||	R5
D	R1||	R2		R4||	R6||	R5
E	R1||		R3	R4||	R6||	R5
F	R1||	R2||	R3	R4||	R6||	R5

Table 2. Design example
$R1, R4, R6 = 20k\Omega$,
$R2 = 20k\Omega$,
$R3, R5 = 30k\Omega$

$$A = - (R4||R5||R6)/(R1||R2||R3)$$

Switch	Amplification
C	− 0.25
4	− 0.50
E	− 0.75
0	− 1.00
6	− 1.50
5	− 2.00
2	− 3.00
1	− 4.00
3	− 6.00

Push-button dimmer switch

The dimmer does not make use of the special integrated circuit for this purpose, the Type SLB0586, but, as the diagram shows, only of standard components. It consists of a power supply, a zero-crossing detector, a comparator with triac for phase gating (phase angle control), and a serial analogue-to-digital converter (ADC).

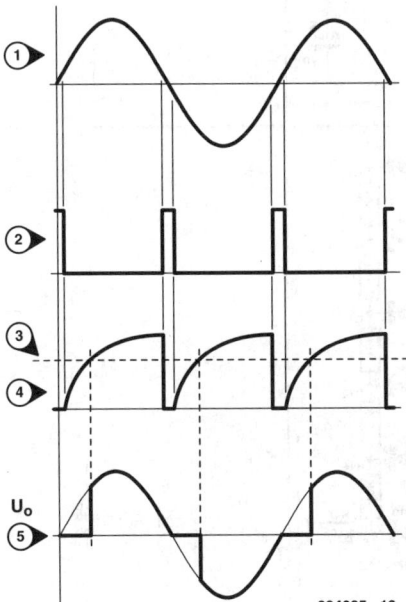

994085 - 12

The power supply section does not use a transformer, but a capacitive potential divider, C_1-R_1-D_1, rectifier D_1-D_2, voltage limiter (to 12 V) D_3, and charging cum decoupling capacitors C_2 and C_3. The 12 V direct voltage is used to power the other sections.

The zero crossings are detected by network R_2–R_5-T_1-T_2. Transistor T_1 is off when the level of the positive half wave of the attenuated mains voltage drops to about 12.6 V, that is, just before the zero crossing. Because of this, the potential at the inverting input of IC_{1a} (test point 2) becomes +12 V, but rapidly returns to its original value when the emitter potential of T_2 becomes negative with respect to earth. This results in short positive pulses around the zero crossing as shown in the timing diagram in Figure 2.

Additionally, the output pulses from IC_{1a} are inverted. Resistors R_6 and R_7 set the threshold potential to 6 V, while resistor R_8 provides some hysteresis. The inverter charges capacitor C_4 via P_1 after each and every zero crossing. During the zero crossings, the output of the inverter briefly goes low, but long enough for C_4 to become discharged rapidly via diode D_4. This results in the sawtooth-like voltage (waveform 4 in Figure 2) at the non-inverting input of comparator IC_{1b}.

The reference voltage for the comparator is provided by counter IC_2. Each time switch S_1 is pressed, the counter progresses one position. After the first time

225

the switch is pressed, Q_1 is active (high), after the second time, Q_2, after the third, Q_3, and so on, until the seventh time when the counter is reset.

Network R_{12}–R_{18}–D_5–D_{10} converts the counter position into an analogue output voltage, U_o:

$$U_o=R_{18}(U_b-U_d)/R,$$

where U_b is the supply voltage, U_d is the diode voltage, and R is the sum of all resistors interconnected by the counter output. The resistors have been selected to ensure floating transition between the various stages. For instance, at the fifth pulse,

$$U_{o(5)}=R_{18}(U_b-U_d)/(R_{16}+R_{17}+R_{18})=7.1\ V,$$

and at the sixth pulse,

$$U_{o(6)}=R_{18}(U_b-U_d)/(R_{17}+R_{18})=8.7\ V.$$

The higher the voltage, the shorter the part of the sawtooth above the threshold that turns on the triac. At $U_{o(6)}$, the output of IC_{1b} is low so that the load remains switched off. At $U_{o(0)}$ (reference voltage), there is no phase gating at all. This may be set with P_1. The phase-gated mains voltage at connector K_2 is shown in the lowest waveform in Figure 2 (U_o).

Since there is no isolation of the mains voltage either during use or test, the dimmer must be housed in a well-insulated or plastic enclosure. Switch S_1 must be suitable for carrying mains voltage.

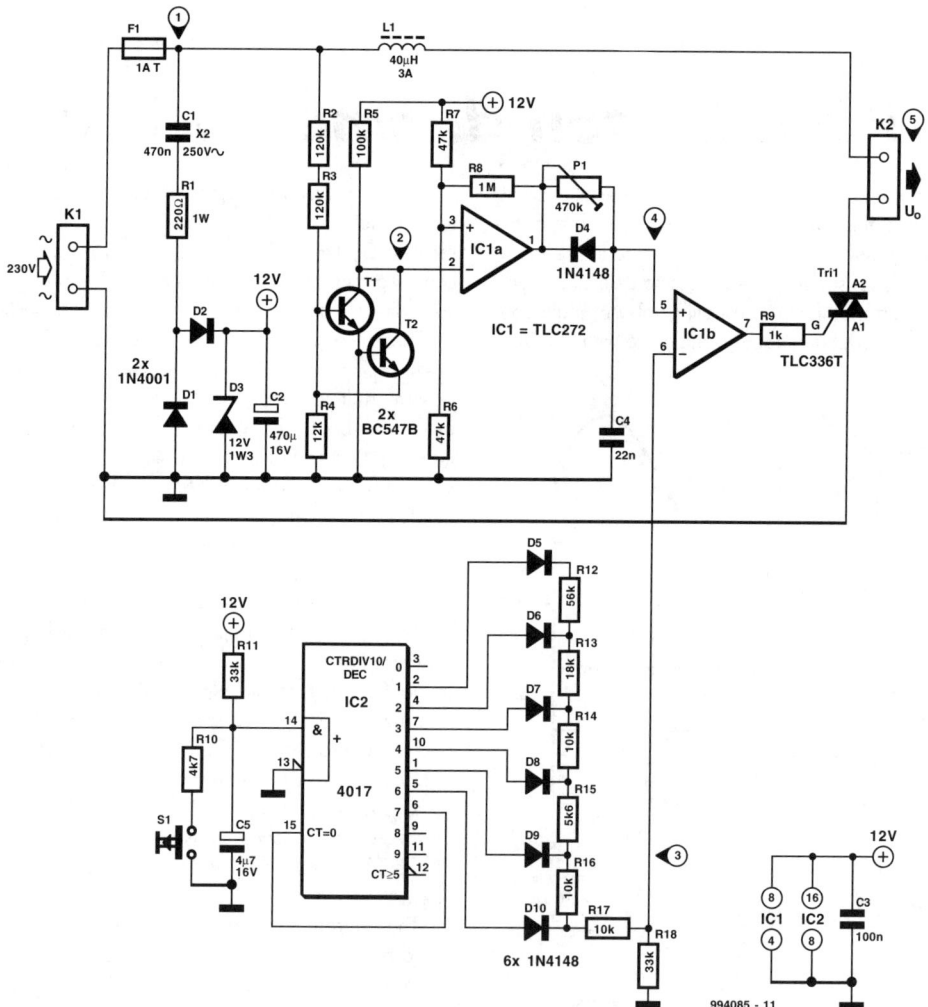

994085 - 11

Railway barrier monitor

The rails and wheels of railway carriages are liable to oxidizing no matter how carefully they are maintained. After a while this causes problems when the train moves across a level crossing since the oxidization causes unreliable contacts of the relevant switches so that correct and timely closing of the barrier (lifting boom or traditional gate) does not take place. As long as the switch contacts are sound, the barrier closes rapidly and timely. This is ensured by the circuit in the diagram, which may be used with a variety of model railway systems.

When track switch S_1 is open, supply voltage for transistor T_1 is provided via rectifiers D_1–D_4. The transistor is off, so that no current flows and the barrier is open. When an approaching train closes the track switch, capacitor C_2 is charged slowly via diode D_1. This results in the base voltage of T_1 rising from 0 V to 0.8 V, so that the transistor begins to conduct. Consequently, a gradually increasing current flows through the barrier solenoids, L_1, L_2, whereupon the barrier closes slowly.

When the switch contact is opened briefly, transistor T_1 remains on as long as the charge on, and thus the potential across, capacitor C_2 is sufficient. The barrier then remains closed. The time taken by C_2 to become discharged depends on the setting of P_1.

When the train has left the relevant section of track switch, S_1 is opened, and capacitor C_2 is discharged via R_1-P_1-R_2. The transistor is then cut off, capacitor C_1 is charged, and the barrier opens.

In the case of direct-voltage systems, diodes D_2 and D_4 are not needed.

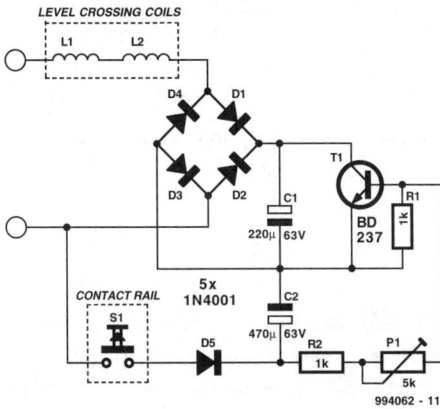

994062 - 11

Siren driver

Zetex Semiconductors have a siren driver IC Type ZSD100 available that is suitable for use in alarm systems for cars and model craft. With the addition of only a few components as shown in the diagram, the device produces an ear-splitting sound of 120 dB.

The IC contains an a.f. rectangular-wave generator that is driven by a sawtooth generator. The sawtooth sweeps the output frequency range (sweep 2:1) once every second. The frequencies of both oscillators are dependent on an internal 61.5 kΩ resistor and an external resistor, R_T=<1 MΩ, and capacitors C_{MOD} and C_{OUT}. The output driver has an inverting and a non-inverting output.

The simplest application of the ZSD100 is shown in Figure 2 in which the sound is produced by a piezo buzzer. If a dynamic 6 Ω loudspeaker is used, it is connected in an H-bridge, that is, driven symmetrically. The value of resistor R_T must then be small, or the relevant pin must be strapped to earth.

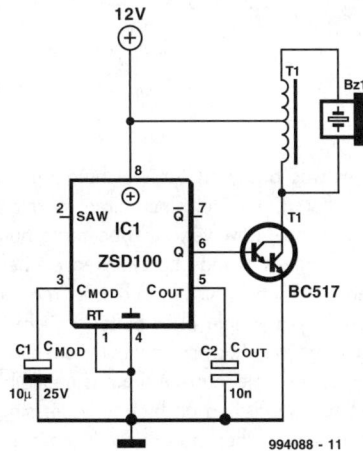

994088 - 11

The IC draws a current of about 25 mA in use, and around 1 µA in the quiescent state. It is suitable for operation over the temperature range of −40 °C to +125 °C. The output can provide 5 mA at a level of 1.4 V and sink 0.5 mA.

Provided the specified values of resistors and capacitors are maintained, the circuit is ideal for experimentation. The frequencies are given by:

$$f_{MOD}=2850/C_{MOD}(61.5+R_T)$$

and

$$f_{OUT}=1710/C_{OUT}(61.5+R_T),$$

where the capacitors are in µF and R_T is in kilohms.

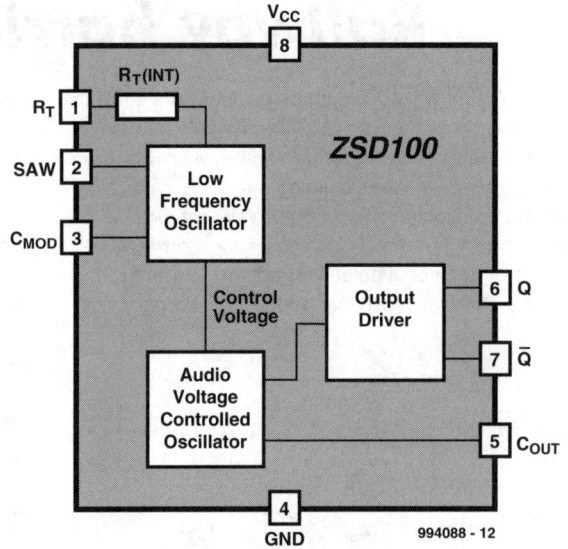

994088 - 12

Pinout of ZSD100

1	R_T	Optional resistor to earth for improved frequency control of the modulation and the output oscillator. The R_T pin may be used to disable the IC by leaving it open or connecting to V_{CC}.
2	SAW	Determines the shape of the modulating signal. When SAW remains open, the signal is triangular; if it is linked to C_{MOD}, a true sawtooth is obtained.
3	C_{MOD}	The capacitor to earth (0.1–100 µF) determines the frequency of the modulating signal.
4	GND	Earth.
5	C_{OUT}	The capacitor to earth (0.001–0.1 µF) determines the centre frequency of the output oscillator.
6	Q	Non-inverting output driver.
7	Q	Inverting output driver.
8	V_{CC}	Supply voltage, 4–18 V

Timer

The timer was designed for switching off a battery charger after a predetermined time to avoid overcharging. It may, however, be used for a number of other applications. It may be switched off before the predetermined time has elapsed and may later be retriggered by a simple push-button switch, S_1. The circuit does not need any special components or parts. In the quiescent state, current drain is negligible.

The timer is switched on by briefly pressing switch S_1. If the switch is then not touched again, the timer remains on for a period determined by time constant R_4-C_2. If during this period the switch is pressed again

briefly, the period is started anew. If, however, the switch is pressed for a time exceeding time constant R_1-C_1 during the predetermined period, the equipment being controlled is switched off. This time constant may be adapted to individual requirements by altering the value of R_1 or C_1.

The timing period (R_4-C_2) may be changed by altering the value of R_4 and/or C_2. It must be borne in mind. however, that the value of R_4 must remain greater than that of R_5 and R_6.

The slightly unorthodox configuration of the oscillator in IC_1 ensures that the polarity of capacitor C_2

does not change, so that an electrolytic type may be used here. With component values as specified, the 'on' time is six seconds per nanofarad of the value of C_2. This means that when, for example, the value of this capacitor is 10 μF, the 'on' time is more than 16 hours.

When the supply voltage is 9 V, the base current of output transistor T_1 is around 2.5 mA, which, for a voltage drop of only 0.1 V, results in an output current of up to 100 mA. If this is insufficient for a particular application, a different type of transistor may be used, or the BC327-25 followed by a relay or a MOSFET stage. If the output current is not required to be as large as stated, the value of base resistor R_3 may be increased.

The supply voltage should be not less than 4 V, since the oscillator stops operating below 3.5 V. At a supply voltage of about 5 V, the base current of T_1, and thus the output current, is rather smaller than stated. However, lowering the value of R_3 is not advisable, since the output of gate IC_{2a} is high-impedance.

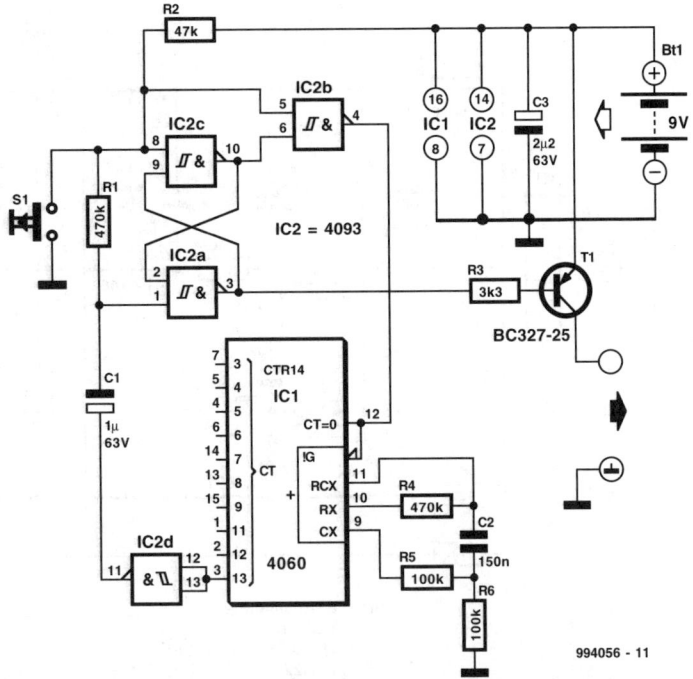

994056 - 11

Touch dimmer

The dimmer is suitable not only for normal domestic lighting, but also for dimming halogen lamps (and other inductive loads).

The dimmer is based on a Siemens Type SLB0587, a circuit specially designed for this kind of application, and to which has been added an infra-red remote control. This remote control was published in the July/August 1998 issue of this magazine.

In the circuit diagram, the series network R_2-C_4-D_1 provides a direct voltage of 5 V directly from the mains to power IC_1. The IC is also in sync with the mains voltage via resistor R_1 and capacitor C_2. The actual phase gating is effected by triac Tri_1, whose gate is linked to pin 8 (QT) of IC_1 via diode D_2. The phase-locked loop (PLL) of the IC is synchronized with the mains frequency via capacitor C_3 and resistor R_3.

The touch sensor that enables a domestic light to be dimmed consists of three 4.7 $M\Omega$ resistors, R_3–R_5.

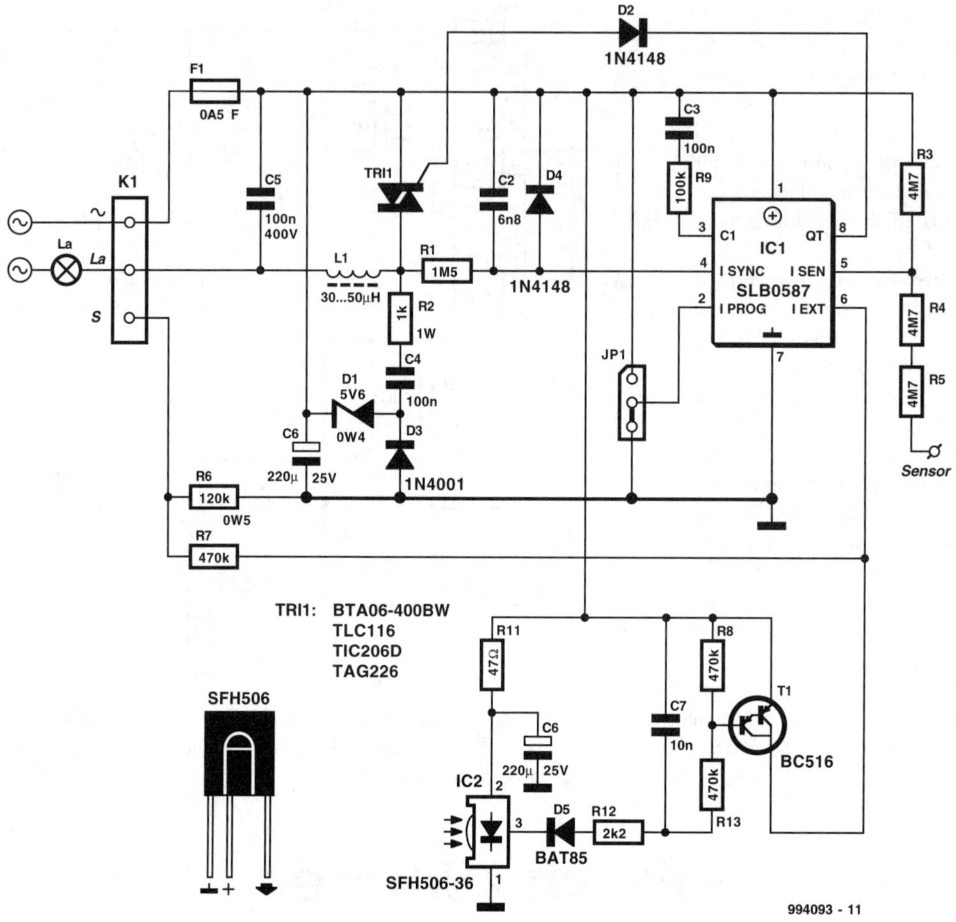

Circuit component labels (from the schematic):

D2 1N4148; F1 0A5 F; C3 100n; R3 4M7; K1; C5 100n 400V; TRI1; C2 6n8; D4; R9 100k; C1; QT; IC1 SLB0587; I SYNC; I SEN; I PROG; I EXT; R4 4M7; La; L1 30...50μH; R1 1M5; 1N4148; JP1; R5 4M7; Sensor; R2 1k 1W; C4 100n; D1 5V6; D3; C6 0W4; 1N4001; R6 120k 0W5; C6 220μ 25V; R7 470k;

TRI1: BTA06-400BW / TLC116 / TIC206D / TAG226

SFH506

R11 47Ω; IC2; C6; 220μ 25V; R8 470k; T1 BC516; C7 10n; D5; R12 2k2; BAT85; 470k; R13; SFH506-36

994093 - 11

These high resistance values make touching the sensor harmless. The output of the sensor is applied to pin 5 (I_{SEN}) of IC_1. Since in some cases it is handy to control the circuit with a push-button switch, this may be connected between S and La on connector K_1.

The IC also has an input I_{EXT}, which is for use with the optional infra-red remote control that is based on IC_2, a Type SFH506-36. This circuit picks up the signal from the sender and converts it into a control signal for IC_1 via a network that comprises among others, T_1 and R_7. The operation of the dimmer is determined by jumper JP_1. When this is left open, the dimmer will always start with the brightness with which it was switched off. Each time the dimmer is started, the direction in which it is controlled is reversed.

When the jumper is placed to connect pin 2 of IC_1 to C_3, the dimmer always starts at maximum brightness. When the jumper is placed in the position shown in the circuit diagram, the dimmer always starts at minimum brightness.

The dimmer is best built on the printed-circuit board shown, which may be made with the aid of the track layout given in the Appendix.

It cannot be emphasized too strongly to bear in mind that the circuit carries potentially lethal mains voltages. Only connect the dimmer to the mains when it has been completed and built into an insulated or plastic case. If the infra-red remote control is not used, IC_2 and T_1 and associated components, except R_7, may be omitted.

Parts list

Resistors:
$R_1 = 1.5\ \text{M}\Omega$
$R_2 = 1\ \text{k}\Omega,\ 1\ \text{W}$
$R_3-R_5 = 4.7\ \text{M}\Omega$
$R_6 = 120\ \text{k}\Omega,\ 0.5\ \text{W}$
$R_7,\ R_8,\ R_{13} = 470\ \text{k}\Omega$
$R_9 = 100\ \text{k}\Omega$
$R_{11} = 47\ \Omega$
$R_{12} = 2.2\ \text{k}\Omega$

Capacitors:
$C_1 = 47\ \mu\text{F},\ 16\ \text{V, radial}$
$C_2 = 0.0068\ \mu\text{F}$
$C_3,\ C_4 = 0.1\ \mu\text{F},\ 630\ \text{V}$
$C_5 = 0.1\ \mu\text{F},\ 400\ \text{V}$
$C_6 = 220\ \mu\text{F},\ 25\ \text{V, radial}$
$C_7 = 0.01\ \mu\text{F}$

Inductors:
$L_1 = 30-50\ \text{mH},\ 3\ \text{A}$

Semiconductors:
$D_1 = \text{zener } 5.6\ \text{V},\ 400\ \text{mW}$
$D_2,\ D_4 = 1\text{N}4148$
$D_3 = 1\text{N}4001$
$D_5 = \text{BAT85}$
$T_1 = \text{BC516}$

Integrated circuits:
$IC_1 = \text{SLB0587 (Siemens)}$
$IC_2 = \text{SFH506-36}$

Miscellaneous:
$JP_1 = 3\text{-way jumper}$
$K_1 = 3\text{-way connecting strip for board mounting, pitch 7.5 mm}$
$F_1 = \text{fuse holder with 0.5 A (F) fuse}$
$PC_1 = \text{PCB pin}$
$Tri_1 = \text{BTA06-400BW or TIC206D or TAG226}$
PCB may be made with the aid of the track layout.

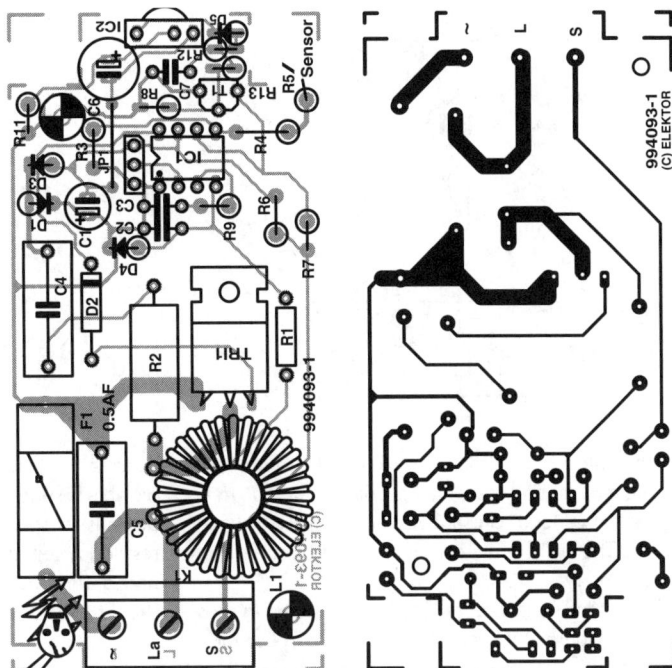

Transistor bistable

The diagram in Figure 1 shows a well-known bistable circuit that is often used (without the switches) in protection circuits in various equipment. In the quiescent state, the bistable draws no current and is set only when the U_{BE} of T_2 is exceeded. At what level of current this happens is determined by the value of resistor R_1. Both transistors are then on, so that the collector potential of T_1 goes to U_b, and that of T_2 to earth. The collector of $T2$ then connects the control input of a device at output terminals PC_3, PC_4 to earth and thus in fact switches off the direct voltage.

Normally, it is assumed that the load current circuit must be opened before the bistable can be reset after an overcurrent. This is usually effected by a relevant

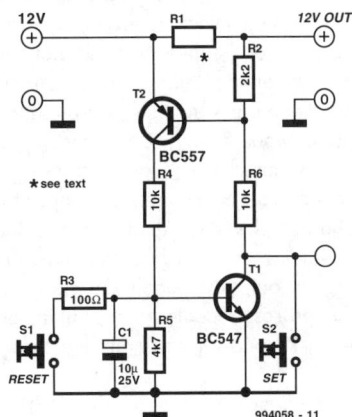

994058 - 11

2

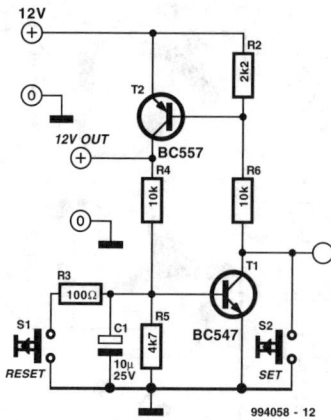

994058 - 12

3

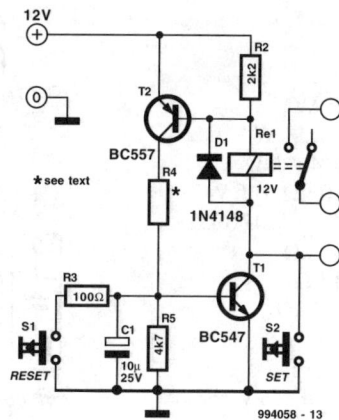

994058 - 13

switch contact or electronic device. It is, however, fairly simple to set and reset the bistable without interfering with the load circuit by placing the switch as shown. The switch only carries the control current of the bistable.

When the circuit is reduced to its essentials as shown in Figure 2, it becomes a general-purpose multivibrator that can provide a much greater current than a standard logic IC, provided suitable transistors are used. The multivibrator can work from a wide range of operating voltages. Capacitor C_1 ensures that the correct pulse/pause ratio is maintained at switch-on.

If R_3 is replaced by a relay solenoid, the circuit functions as a bistable relay which, after the circuit has

been set with switch S_2, remains stable until the bistable is switched off with switch S_1. With other components as specified, the relay should have a high-resistance coil: 900–1000 Ω in the case of 12 V types, and around 3.5 kΩ for 24 V types. The value of R_2 should be of the same order, but it is not particularly critical.

If only a power relay with low-resistance solenoid is available, the transistors and resistors R_1, R_2, and R_4, must be adapted to the current demand of the relay. Freewheeling diode D_1 may be a Type 1N4148 in case of a small relay, but when the relay current is greater than about 100 mA, it is advisable to use a Type 1N4001.

Mains/fuse failure indicator

The indicator shows when the mains is present at its output by a continuous glow of a neon bulb, La_1, and when the fuse is blown by flashing of the neon bulb.

When the fuse is intact, capacitor C_2 acts as the series resistance for the neon bulb, so that this glows continuously. When the fuse has blow, the mains voltage across diode D_1 is applied as a pulsating direct voltage to network R_1-C_1. Capacitor C_1 charges slowly and when the voltage across it reaches 80–100 V, the neon bulb comes on. Capacitor C_1 is then discharged slowly via diode D_2 and the bulb. When the voltage across it has dropped sufficiently, the bulb goes out, whereupon C_1 slowly charges again. This process repeats itself, so that, provided the values of R_1 and C_1 are right, the bulb flashes visibly.

The potential across capacitor C_2 is a ramp with

a peak value of 30 V (which is, of course, applied to the load).

Note that the neon bulb used for this purpose must not be a type that has a built-in series resistor.

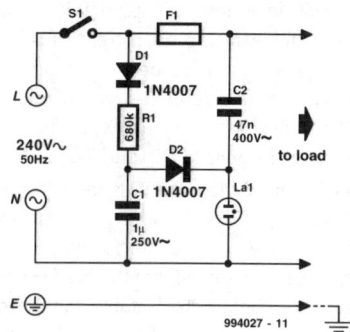

994027 - 11

5
Power Supplies
& Battery
Chargers

13.8 V power supply for mobile rigs

The power supply is based on a Type 723 voltage regulator, which is still popular among radio amateurs because it is reliable, widely available and far cheaper than many of the latest high-power (>1.5-A) three-terminal voltage regulators. The 723 comes in two

is set with preset P_1. The input voltage for the supply is obtained from a 22–25 V, 15 A transformer, a 25 A bridge rectifier and a 10 000 μF smoothing capacitor.

The current booster consists of darlington transistor T_1 and three parallel connected power transistors,

904075 - 11

packages: a 14-pin DIL case or a 10-lead metal can. The pin numbers shown in this section refer to the 14-pin DIL case.

This supply is intended for use with mobile ham radio transceivers. As many of today's combined 2m/70cm FM mobile rigs are capable of supplying r.f. powers of more than 50 watts, a heavy-duty PSU is required if such a set is not run off a vehicle battery. The present supply is capable of providing output currents of up to 12 A at 13.8 V.

The regulator, IC1, is conventionally wired, drives a power transistor array and monitors the output current by measuring the voltage drop across series resistor R_{SC}. The nominal supply output voltage of 13.8 V

T_2–T_4. Resistors, R_A, R_B and R_C are emitter current distribution elements.

The current sense resistor for the short-circuit protection, R_{SC}, has a value of 0.05 Ω which results in a protection onset level of about 0.6/0.05=12 A. This resistor is either made from resistance wire or from two parallel-connected 0.1 Ω, 5 W resistors.

A 12 A or 16 A fuse in the positive supply rail provides additional protection against output short-circuits. If the fuse blows, transistor T_5 briefly actuates an active buzzer and a flashing LED. It does so by draining the charge built up in C_9.

The 15 V overvoltage protection at the output of the supply is a crowbar circuit. If the supply is set to an output voltage other than 13.8 V, the zener diode, thyristor and associated resistor have to be omitted, or disconnected by breaking the wire link indicated in the circuit diagram. If used, the thyristor should have a current rating of about 25 A: a type 2N6506 is recommended.

The darlington and power transistors have to mounted on a large heat-sink with appropriate insulating washers. The prototype has a small 12 V fan to assist in the cooling of the heat sink. This fan is powered by the supply: two or three diodes in series are used to drop the operating voltage to about 12 V.

As the choice of components used in the supply is not critical, the circuit is easily modified for smaller output currents and/or different output voltages. The use of a 10 A transformer and two power transistors is fine for an 8 A supply. Similarly, a 5 A transformer and one power transistor are sufficient for a 4 A version of the supply.

AC-DC converter

The converter is intended to transform a sinusoidal alternating voltage into a direct voltage equal to the r.m.s. value of the input voltage with an accuracy better than 2%. What makes the circuit special is that the conversion, that is, rectification, averaging and buffering, is effected by only two op amps. The usual circuits for this kind of conversion are generally more complex.

Op amp IC_{1a} provides half-wave rectification of the applied alternating voltage. The values of R_1 and R_3 give unity amplification.

Op amp IC_{1b} amplifies the output of IC_{1a} ×2 and adds this to its input via R_2. This results in a potential that is equal to the absolute value of the input voltage. This potential is amplified ×2.22, that is, the form factor ×2 or $2\pi/2\sqrt{2}$, at a delay time R_5-C_1 (= 2.22 s).

Op amp IC_{1b} also functions as a buffer for the whole circuit.

Since the operation of this type of converter depends on the absence of any offset voltage, this is nullified by network R_6-R_7-P_1. Setting is simple: apply a 1 kHz sinusoidal signal at a level of 50 mV r.m.s. to the input of the converter and adjust

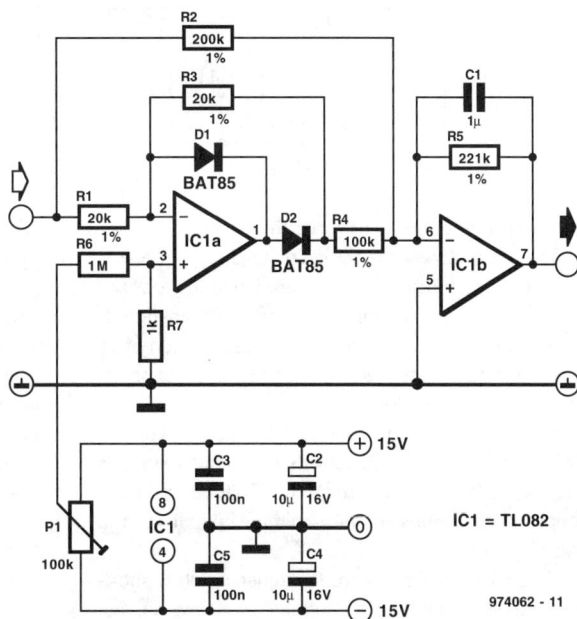

P_1 until the output is a direct voltage of 50 mV. The input voltage may vary from 50 mV to 7 V. The input frequency may vary from 10 Hz to 10 kHz.

The circuit draws a current of about 3 mA.

Auto on/off switch for power supplies

In the testing of circuits, there is sometimes a need for the supply voltage to be switched on at a given time or in steps.

The auto on/off switch presented is based on IC_1, an oscillator/binary counter. The frequency of the oscillator is determined by R_9-R_{10}-P_1-C_5. The IC has ten outputs that become high sequentially and which may be used to drive identical transistor stages (three of which are shown in the diagram). Each of these stages consists of a potential divider and a BC548B

235

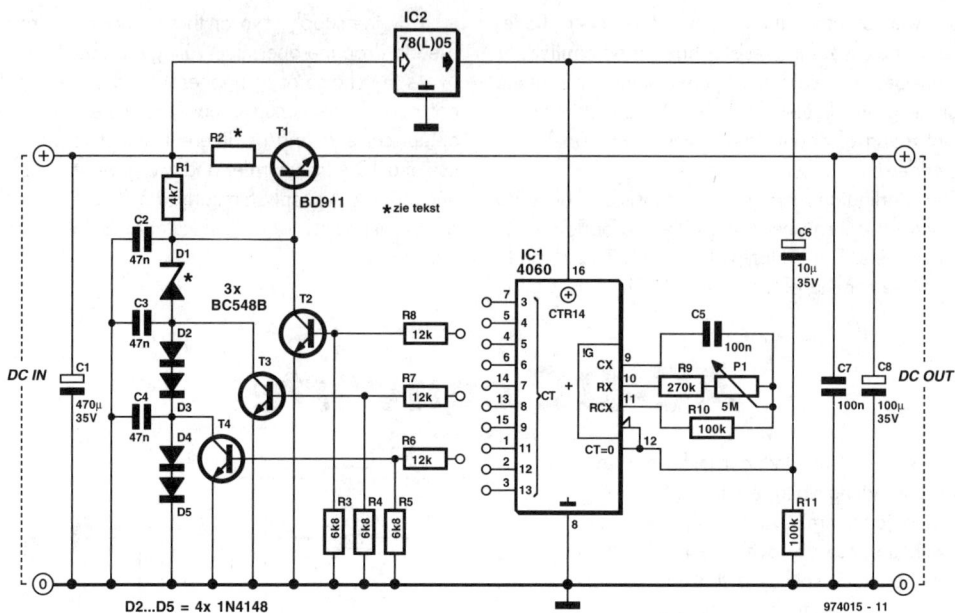

D2...D5 = 4x 1N4148   974015 - 11

transistor that functions as a switch.

The actual power supply, in conjunction with power transistor T_1, functions as a parallel regulator. The output voltage is determined by the zener voltage of D_1 less the drop across the base-emitter junction of T_1 plus the drop across those diodes that are not short-circuited by a transistor.

For example, if the rating of the zener diode is 12 V, and T_3 and T_4 are cut off, the output voltage is $12+4\times0.7-0.7 =14.1$V. The zener voltage is discretionary, but depends on the minimum output voltage plus U_{be} of T_1.

When T_2 is switched on, the zener diode is short-circuited, so that the power supply is switched off.

Resistor R_1 limits the current through the series of diodes.

Although the IC can operate from a wide range of supply voltages, it is recommended to use a 5 V regulator, IC_2. If the load current is in excess of 100 mA, T_1 must be mounted on a suitable heat sink. The transistor must not get so hot that it cannot be touched. If it does, the load current must be reduced, a higher rated heat sink used, or R_2 added. The value of this resistor in ohms is the numerical difference between the input and output voltage in volts divided by the numerical value of the peak load current in amperes. Its rating is the product of these quantities.

Finally, note that the on/off switch is not short-circuit-proof.

LM2574 switch-mode power supply

When a small, reliable and inexpensive switch-mode power supply is needed, National Semiconductor's LM257x series of switch-mode power supply controller ICs has several advantages over competitive products like the LT1070 (powerful but alas more expensive) and the TL497 (obsolescent). The LM257x family is second-sourced by Motorola.

The circuit shown here largely follows the NS rec-

ommended application configuration. The only time-consuming part is inductor L_1, a triac suppressor coil whose self-inductance depends on the output voltage and maximum anticipated output current. The required inductance may be found in the graph. Most triac suppressor coils have an inductance of about 100 mH. Since the self-inductance is (roughly) proportional to the square of the number of turns, a

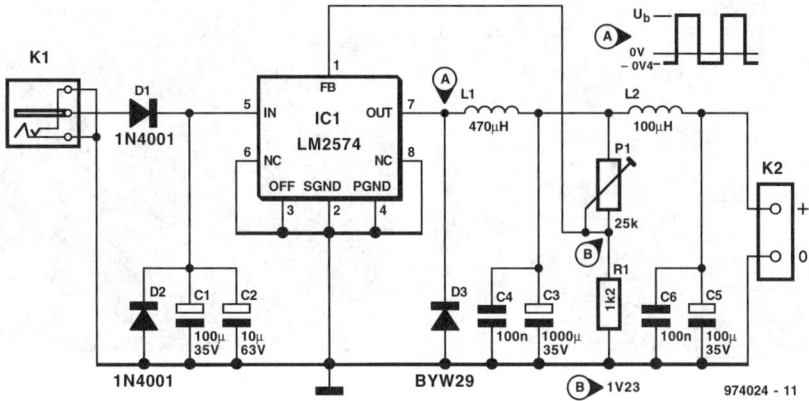

100 mH coil may be modified into a 470-mH version by multiplying the number of turns with $\sqrt{(470/100)}$ ≈ 2.17. The new winding may be of thinner enamelled copper wire to ensure it fits on the original former.

The PCB is designed for an LM2574 in an 8-pin dual in-line package (DIP). Solder the IC directly on to the board: this will aid in its cooling. For the same reason, pins 6 and 8 of the regulator, though not connected internally, must be soldered to the copper ground plane. The output capacity of the supply should not be pushed to its limit (0.5 A) regularly, as the LM2574 then works near maximum permissible performance.

The second LC circuit behind the regulator, L_2-C_5, provides additional ripple suppression. If a ripple voltage of a few millivolts is acceptable, L_2 may be replaced by a wire link. If fitted, L_2 should be a stan-dard triac suppressor coil with a self-inductance of 50–100 mH.

The output voltage is set with preset P_1, where

$$P_1 = R_1[(U_o/1.23)-1]$$

Diode D_1 at the input of the circuit provides polarity reversal protection. If D_1 is fitted, D_2 is not required. However, if the voltage across D_1 becomes unacceptably large, replace the diode by a wire link and fit D_2 which will present a virtual short-circuit to reverse input voltages.

Finally, a computer program doing all configuration calculations for these SMPSU ICs may be downloaded from

http://www.national.com/ design/index.html

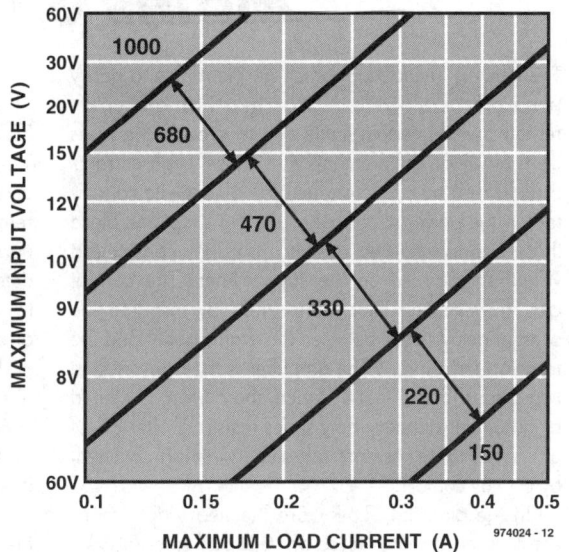

237

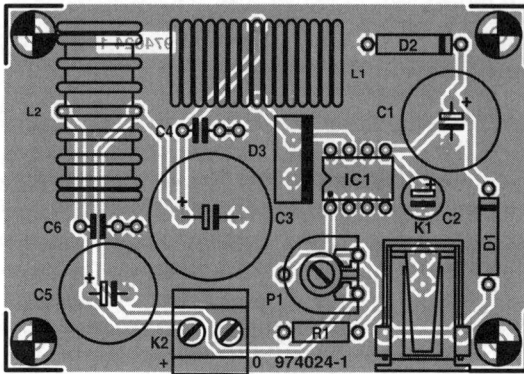

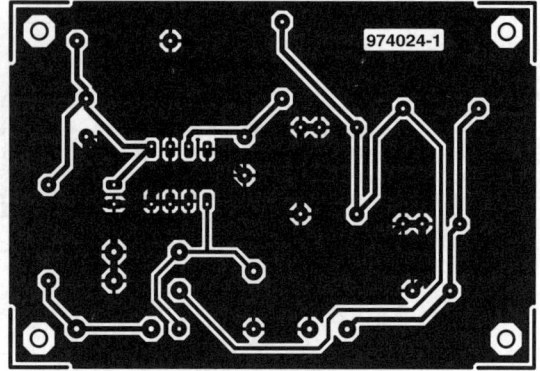

Parts list

Resistors:
$R_1 = 1.2$ kΩ
$P_1 = 25$ kΩ, preset, horizontal

Capacitors:
$C_1, C_5 = 100$ μF, 35 V, radial
$C_2 = 10$ μF, 63 V, radial
$C_3 = 1000$ μF, 35 V, radial
$C_4, C_6 = 0.1$ μF

Inductors:
$L_1 = 470$ mH triac suppressor coil (see text)
$L_2 = 100$ mH triac suppressor coil

Semiconductors:
$D_1, D_2 = $ 1N4001
$D_3 = $ BYW29 or similar fast Schottky diode

Integrated circuits:
$IC_1 = $ LM2574N (National Semiconductor)

Miscellaneous:
$K_1 = $ mains adaptor socket, PCB mount
$K_2 = $ 2-way PCB terminal block, pitch 5 mm
PCB may be made with the aid of the track layout.

Mains on delay

The delay is intended to switch on the mains to heavy loads gradually to ensure that the switch-on current remains within certain limits and to prevent the fuses from blowing. The elements that cause high currents at switch-on are, for instance, the electrolytic capacitors in the power supply of an output amplifier. Since these are not charged at switch-on, they constitute a virtual short-circuit on the supply lines. The current can, however, be kept within limits by inserting the present delay circuit between the mains outlet and the transformer primary. The amplifier is then powered in two stages: in the first instance, the current is limited by a number of heavy-duty series resistors; after about a second, these resistors are shunted (short-circuited) by a relay contact.

In the diagram, R_4–R_7 are the heavy-duty series resistors, each with a value of 10 Ω and rated at 5 W.

They limit the switch-on current to about 5.5 A.

The relay is a type whose contact is rated at 2000 VA, which will be sufficient in most cases. Its supply is derived directly from the mains via potential divider R_3-C_1-B_1-relay coil. Resistor R_3 limits the current at switch-on, after which C_1 limits the current in normal operation to about 20 mA. The delay time, determined by capacitors C_2 and C_3 in parallel with the relay, may be altered by changing the value of one or both of these capacitors.

For safety's sake, the board also has provision for a mains fuse, F_1. The rating of this depends, of course, on the load current.

It should be noted that in the case of a double-block stereo output amplifier (with separate power supplies), a mains-on delay must be fitted to each of the mono amplifiers.

As mentioned earlier, the values of R_4–R_7 refer to a switch-on current of about 5.5 A. If the power rating of the load is lower than 200 VA, it is advisable to use resistors with a slightly higher value.

Note that C_1 is a metallized paper type designed specially for mains voltage applications to Class I.

Finally, at all times bear in mind that the circuit is connected to the mains, so do not touch anything inside the unit during operation and make sure that all wiring is safe and secure.

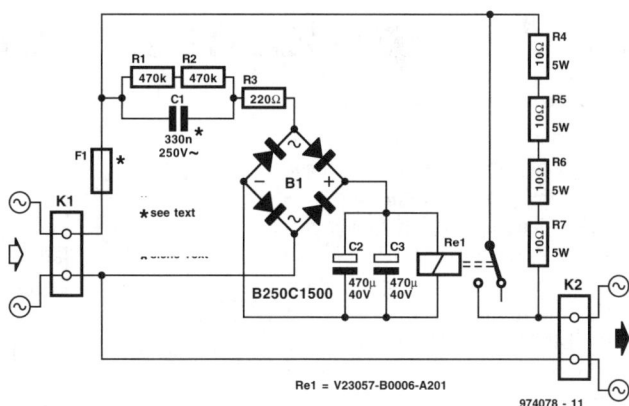

Re1 = V23057-B0006-A201

974078 - 11

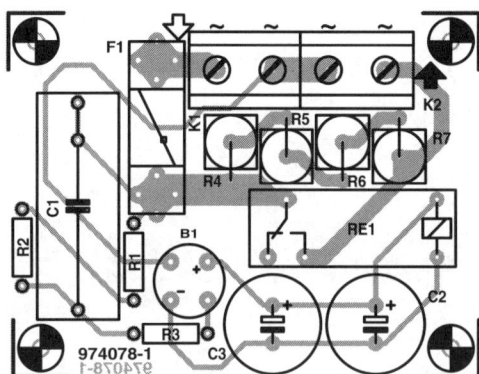

974078-1

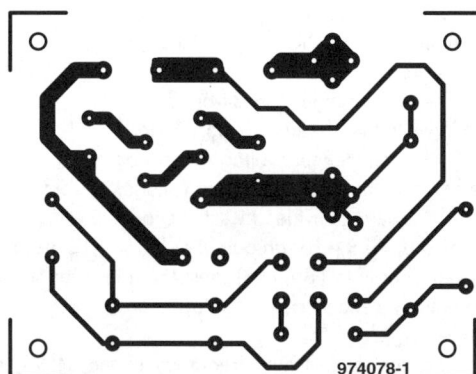

974078-1

Parts list

Resistors:
R_1, R_2 = 470 kΩ
R_3 = 220 Ω
R_4–R_7 = 10 Ω, 5 W

Capacitors:
C_1 = 0.33 μF, 250 VAC, metallized paper
C_2, C_3 = 470 μF, 40 V

Miscellaneous:
K_1, K_2 = 2-way terminal block, pitch 7.5 mm
B_1 = B250C1500, round
Re_1 = contact rating 250 V, 8 A, coil 24 V, 1200 Ω
F_1 = see text
PCB may be made with the aid of the track layout.

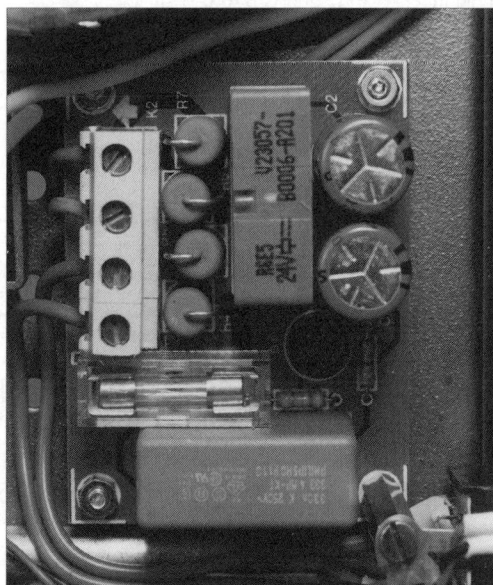

1.5 A step-down switching regulator

Step-down regulators are frequently used to derive a low voltage from a higher one without incurring losses. Owing to the steadily increasing switching rates, the suppression of interference and noise that are unavoidable by-products of the switching has become much simpler.

The L4971 from SGS-Thomson is a step-down monolithic power switching regulator delivering 1.5 A at a voltage between +3.3 V and +40 V (selected by a simple external divider). Designed in BCD mixed technology, the regulator is housed in a DIL8 or SO16-SMD enclosure.

The input voltage ranges from 8 V to 55 V, while the efficiency is 85%, rising to 95% when the input voltage is only a few volts higher than the output voltage.

The external divider to set the output voltage is formed by R_3-R_4. For an output of 3.3 V, V_{out} may be linked directly to FB (pin 8). Resistor values for some commonly needed output voltages are given in the table.

The normal switching frequency of the L4971 is 100 kHz, but this may be increased to 500 kHz by making $R_1 = 12$ kΩ and $C_2 = 0.001$ μF.

The soft start/inhibit input (SS_INH pin 2) may be used with an open-collector output to inhibit the controller or capacitor C_5 for the soft start function.

A suitable core for L_1 is the Type T-94-26 from Amidon. The specified inductance is obtained with 65 turns close-wound 0.5 mm dia. enamelled copper wire.

Other features include pulse by pulse current limit, hiccup mode for short-circuit protection, voltage feed forward regulation, protection against feedback loop disconnection, inhibit for zero current drain and thermal shutdown.

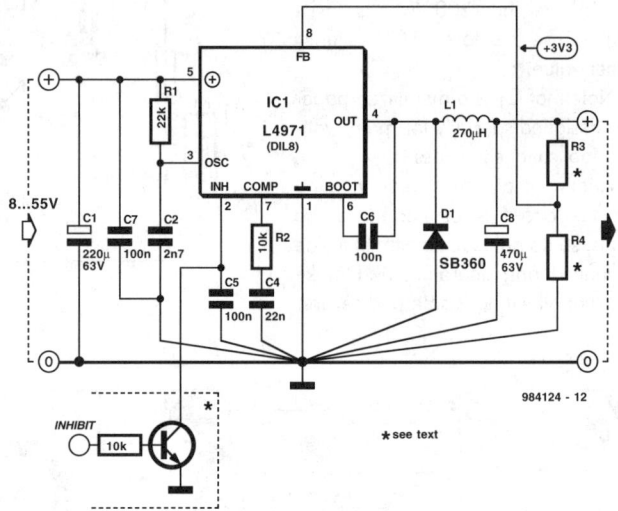

984124 - 12

V_{out} (V)	R_3 (kΩ)	R_4 (kΩ)
3.3	0	•
5.1	2.7	4.4
9.0	8.2	4.7
12	12	4.7
15	16	4.7
18	20	4.7
24	30	4.7

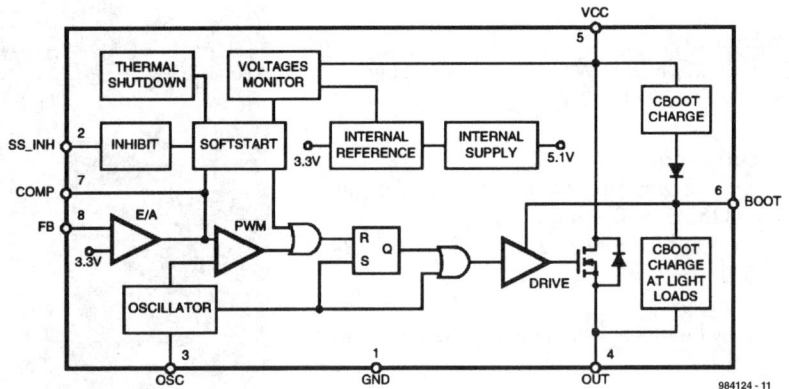

984124 - 11

Battery-charging indicator for mains adaptor

A NiCd battery with an output voltage and/or current that cannot be handled by an available charger, may be charged via a variable mains adaptor. This can only be done safelyif an indicator as described here is to hand.

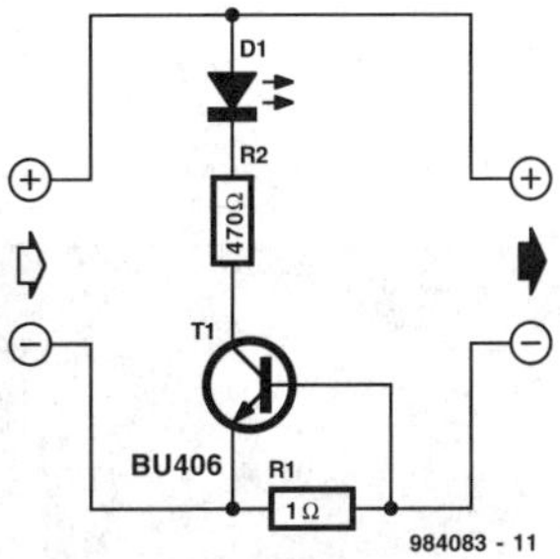

984083 - 11

In the circuit in the diagram, diode D_1 lights when the base-emitter potential of the transistor exceeds about 0.2 V. With a resistor (R_1) of 1 Ω as specified, this occurs at a current through transistor T_1 of about 200 mA, or about 40 mA when $R_1 = 4.7$ Ω.

The voltage drop across the indicator can not exceed the base-emitter voltage (U_{BE}) of the transistor (about 0.7 V). Even if the current through R_1 continues to increase beyond the level at which $U_{BE} = 0.7$ V, the base of the transistor absorbs the excess current. The BU406 transistor specified is capable of handling base currents up to 4 A.

As long as U_{BE} remains below about 0.6 V, the voltage across R_1 is a faithful indication of the charging current. Alternatively, an ammeter meter inserted in one of the charging leads will show the exact current. Charge briefly at the 5-hour rate (C/5) and then, again briefly, at the 10-hour rate (C/10). (C is the battery capacity in [milli-] ampere-hours, which is usually printed on the battery). In general, the lower the charging current, the smaller the risk of damage to the battery.

In some cases it will be possible to incorporate the indicator in the mains adaptor. This may be risky, however, owing to the presence of the mains voltage in the adaptor housing. A safer alternative is to install it in a small dedicated box.

The circuit is not protected against reversal of the battery polarity. If such protection is required, a fuse or circuit breaker should be added.

dc-dc converter

The converter, based on the CMOS version of the 555 timer (TLC555), changes a positive potential of 9 V into a negative one at the same level and is ideal for use where a single battery is to power a circuit requiring a symmetrical supply.

If a TLC555 is not available, a Type 7555 may be used instead. The device is arranged as an astable multivibrator (AMV) with R_2, R_3 and C_1 determining the operating frequency, which is about 20 kHz. The square wave produced by the AMV is fed to cascade rectifier C_3-D_1-D_2-C_4. Schottky diodes are used because of their lower forward voltage drop of about 0.4 V compared with 0.7 V in the case of silicon diodes. The rectified voltage is smoothed by C_4, while C_5 serves to bypass high-frequency noise.

The input supply voltage to IC_1 is decoupled by R_1, C_6 and C_7. The current drawn by the converter

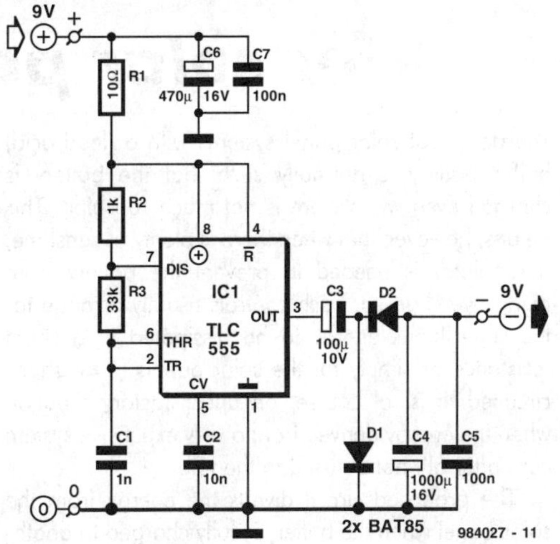

984027 - 11

241

depends largely on the load connected to the –9 V output. As evinced by the measurement data in the table, the output current may rise to about 10 mA before the output voltage collapses.

Parts list

Resistors:
R_1 = 10 Ω
R_2 = 1 kΩ
R_3 = 33 kΩ

Integrated circuits:
IC_1 = TLC555 or 7555

Capacitors:
C_1 = 0.001 μF
C_2 = 0.01 μF
C_3 = 100 μF, 10 V, radial
C_4 = 1000 μF, 16 V, radial
C_5, C_7 = 0.1 μF
C_6 = 470 μF, 16 V, radial

Semiconductors:
D_1, D_2 = BAT85

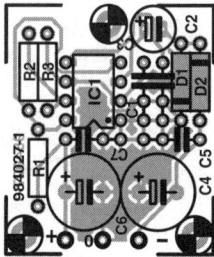

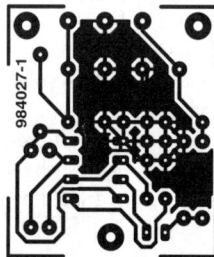

For easy incorporation in equipment in which space is at a premium, the converter is best built on the printed circuit board shown on the next page. This board may be made with the aid of the track layout.

Battery voltage: 9.1 V			
R_L	I_S	V–	Efficiency
inf.	4.8 mA	–8.89 V	0%
6.8 kΩ	6.0 mA	–8.3 V	18%
1.5 kΩ	9.55 mA	–7.2 V	40%
680 Ω	13.43 mA	–5.93 V	42%

Lead-acid-battery regulator for solar panel systems

The design of solar panel systems with a (lead-acid) buffer battery is normally such that the battery is charged even when there is not much sunshine. This means, however, that when there is plenty of sunshine, a regulator is needed to prevent the battery from being overcharged. Such controls usually arrange for the superfluous energy to be dissipated in a shunt resistance or simply for the solar panels to be short-circuited. It is, of course, an unsatisfactory situation when the energy derived from a very expensive system can, after all, not be used to the full.

The proposed circuit diverts the energy from the solar panel when the battery is fully charged to anoth-er user, for instance, a 12 V ice box with Peltier elements, a pump for drawing water from a rain butt, or a 12 V ventilator. It is, of course, also possible to arrange for a second battery to be charged by the superfluous energy. In this case, however, care must be taken to ensure that when the second battery is also fully charged, there is another control to divert the superfluous energy.

The shunt resistance needed to dissipate the super-fluous energy must be capable of absorbing the total power of the panel, that is, in case of a 100 W panel, its rating must be also 100 W. This means a current of 6–8 A when the operating voltage is 12 V. When, owing to reduced sunshine, the voltage drops below the maximum charging voltage of 14.4 V, the shunt resistance is disconnected by an n-channel power field effect transistor (FET), T_1. The disconnect point is not affected by large temperature fluctuations because of a reference voltage provided by IC_1. The requisite comparator is IC_2, which owing to R_9 has a small hysteresis voltage of 0.5 V. Capacitor C_5 ensures a relatively slow switching process, although the FET is already reacting slowly owing to C_4. The gradual switching prevents spurious radiation caused by steep edges of the switched voltage and also limits the starting current of a motor (of a possible ventilator). Finally, it prevents switching losses in the FET that might reach 25 W, which would make a heat sink unavoidable.

Setting up of the circuit is fairly simple. Start by turning P_1 so that its wiper is connected to R_5. When the battery reaches the voltage at which it will be switched off, that is, 13.8–14.4 V, adjust P_1 slowly until the output of comparator IC_2 changes from low to high, which causes the load across T_1 to be switched in.

Potentiometer P_1 is best a 10-turn model. When the control is switched on for the first time, it takes about 2 seconds for the electrolytic capacitors to be charged. During this time, the output of the compara-

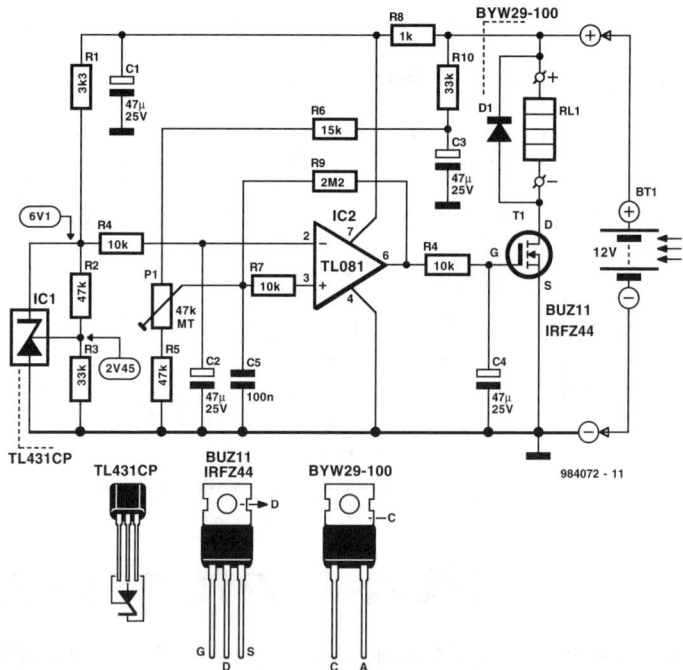

tor is high, so that the load across T_1 is briefly switched in.

In case T_1 has to switch in low-resistance loads, the BUZ11 may be replaced by an IRF44, which can handle twice as much power (150 W) and has an on-resistance of only 24 mΩ.

Because of the very high currents resulting from a short-circuited battery, it is advisable to insert a suitable fuse in the line to the regulator.

The circuit draws a current of only 2 mA in the quiescent state and not more than 10 mA when T_1 is on.

Mains filter revisited

The mains filter described in 'PC Topics' (*Elektor Electronics*, April 1998) could do with two useful additions: a fuse monitor and a voltage indicator. The filter proper consists of capacitors C_3–C_7 and inductor L_1. Its operation is described fully in the April 1998 article. The present article deals only with the additional features.

The fuse monitor indicates by the lighting of an LED that the fuse has blown. For this purpose, two capacitive potential dividers have been added: C_1-R_3-D_1 and C_2-R_4-D_3. When the fuse is intact, the

potential at the base of T_1 is 3.9 V higher than that of the (N)eutral line. The transistor is on and short-circuits D_2. When the fuse blows, T_1 is off. Its collector potential, owing to D_1, is 2.7 V, so that D_2 lights. Note that this action is only possible when the (L)ive line is positive w.r.t. the neutral line; this means that the LED flashes in rhythm with the mains frequency (since this is 50 Hz, it cannot be discerned by the human eye).

A second potential divider, C_6-R_4-D_4-D_5, at the output of the filter ensures that the LED (D_4) lights when the mains voltage is present at the output.

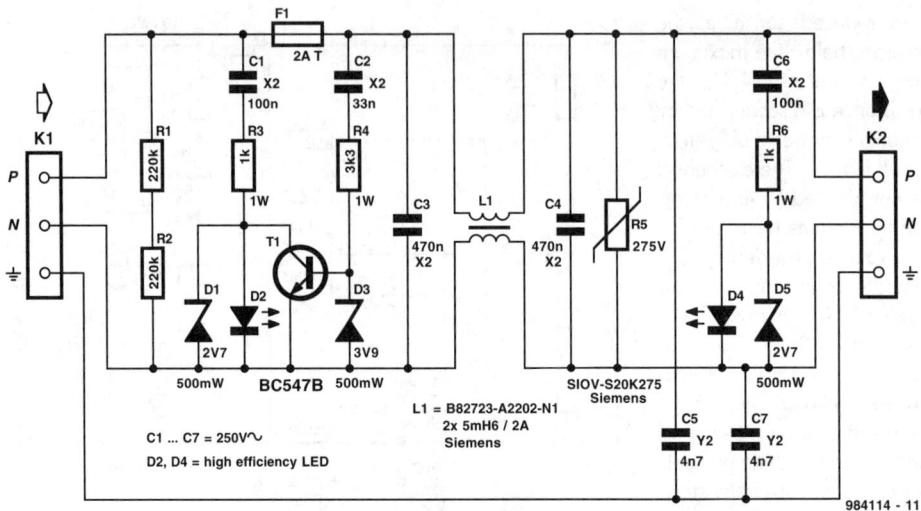

984114 - 11

NiCd battery charger

The design of the charger is similar to that of many commercially available chargers. The charger consists of a mains adaptor, two resistors and a light-emitting diode (LED). In practical use, this kind of charger is perfectly all right.

Resistor R_1 serves two functions: it establishes the correct charging current and it drops sufficient voltage to light the diode. This means that the LED lights only when a charging current flows into the battery. The charging current is about 1/4 of the battery capacity, which allows a slight overcharging, and yet the charging cycle is not too long (4–5 hours).

The value of the resistors may be calculated as follows, for which the nominal e.m.f. and the capacity of the battery must be known.

Adjust the output of the mains adaptor to 1.17 times the nominal battery voltage plus 3.3 V, which is the potential across R_1. Note that the adaptor must be capable of supplying a current of not less than half the battery capacity.

The value of R_1 in ohms is equal to 3.3 divided by

1/4 of the battery capacity. The value of the resistors for various battery voltages is given in the Table. The battery capacity is taken as 1 Ah. The rating of R_1 should be 5 W. If the battery to be charged has a different capacity, the theoretical value of R_1 in the table must be divided by the battery capacity. Its actual value is the nearest one in the E12 series. For instance, if a 6 V battery with a nominal capacity of 600 mAh is to be charged, the value of R_1 must be $20/0.6 = 33\ \Omega$.

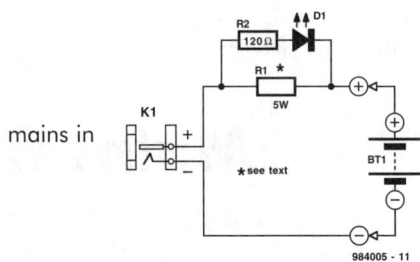

984005 - 11

Battery voltage (V)	1.2	2.4	3.6	4.8	6.0	7.2
Minimum voltage (V)	4.7	6.1	7.5	8.9	10.3	11.7
Adaptor voltage (V)	4.5	6.0	7.5	9.0	12	12
Value of R1 (theoretical) (Ω)	12.4	12.8	13.2	13.6	20	14.4
E12 value of R1 (Ω)	15	15	15	15	22	15
Value of R2 (Ω)	120	120	120	120	240	120

Balanced power supply

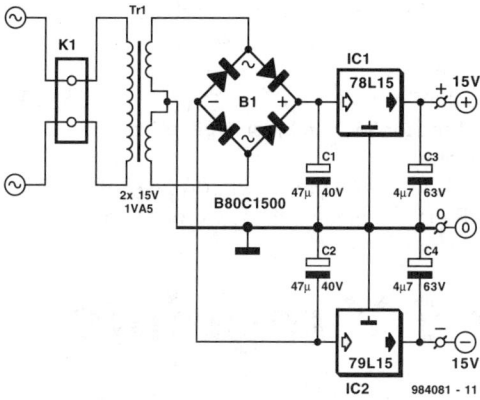

This design is useful in those cases where balanced power lines providing a relatively small output current are needed. The diagram shows a ±15 V supply that can provide a continuous output current of about 25 mA, or 100 mA peak. With other transformers and/or voltage regulators, the supply can be adapted for output voltages of ±5 V, ±9 V, ±12 V, ±15 V, ±18V and ±24 V. For the latter two voltages, however, the negative-voltage regulator may be hard to obtain. Thanks to its small size, the supply is easily incorporated into existing equipment.

A disadvantage of small (low-VA) mains transformers as used for this supply is that they often supply relatively high no-load secondary voltages. Under no-load conditions, the specified transformer, for example, supplies as much as 32 V to the regulator inputs (measured at a mains voltage of 230 V). In some cases, the no-load secondary voltage may exceed the maximum permissible input voltage of the low-power voltage regulator. Typically this will be 30 V for 5-V regulators, 35 V for 12-V and 15-V types, and 40 V for 18-V and 24-V types. When the no-load voltage is

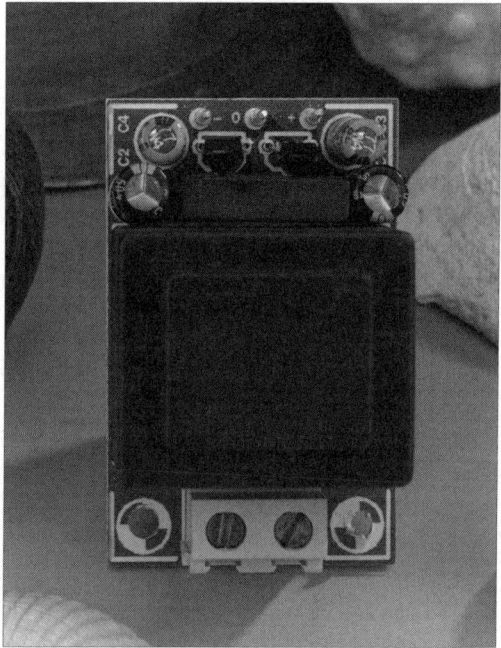

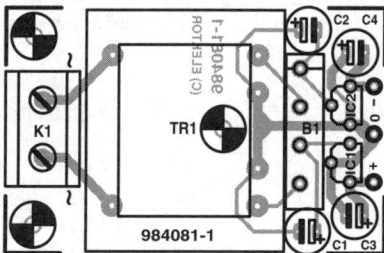

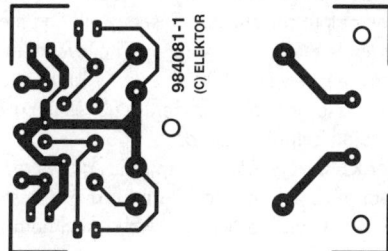

likely to approach the absolute maximum level specified for the voltage regulator, shunt resistors (bleeders) shoud be used across the transformer secondaries. The value of these resistors should be as high as possible to avoid unnecessary dissipation. In most cases, a bleeder current of a few mA is sufficient to drop the regulator input-voltage to a safe level.

Although the specified transformers have the same footprint, the 3.2 VA type is taller. If this particular transformer is used, the continuous output current capacity of the supply rises to about 55 mA, provided C_1 and C_2 are changed to, say, 100 μF, 25 V types. Note, however, that the no-load secondary voltage may have to be reduced as described above.

Parts list

Capacitors:
C_1, C_2 = 47 µF, 40 V, radial
C_3, C_4 = 4.7 µF, 63 V, radial

Semiconductors:
B1 = B80C1500

Integrated circuits:
IC_1 = 78L15 (see text)
IC_2 = 79L15 (see text)

Miscellaneous:
K_1 = 2-way PCB terminal block, raster 7.5mm
Tr_1 = mains transformer, see text.
Examples :
2x15V   1.5VA: type   VTR1215
(Monacor/Monarch) or type BV EI 302 2028
(Hahn)
2x15V 3.2VA: type BV EI 306 2078 (Hahn)
PCB may be made with the aid of the track layout.

Ultra-low-power 5 V regulator

The current drain of the regulator is minute compared with that of, say, a 78L05: at an input voltage of 9 V and open-circuit output, it is just under 50 µA.

The circuit consists of a bandgap reference based on T_1 and IC_1, followed by an amplifier formed by IC_2 and T_1.

The reference voltage is about 1.22 V, which is raised by IC_2 to 5 V. The output voltage can be set to exactly 5 V with P_1. The input voltage may lie between 6.5 V and 30 V. The maximum output current with the present configuration and component values as specified is about 10 mA.

For optimum performance, T_{1a} and T_{1b} need to be identical, which is why a dual transtor Type MAT02 is used. Other types that may be used are the MAT01, SSM2210 or LM394. In principle, two standard BC transistors may be used, provided they are selected for identical threshold voltage.

Circuits IC_1 and IC_2 are programmable op amps Type OP22. In the case of IC_2 this has the benefit that the peak output current can be set by altering the supply current to the op amp with R_9. The level of the current may be between 500 nA and 400 µA. Bear in mind that a larger output current requires the use of a higher rated output transistor.

Filter R_6-C_1 prevents any spurious pulses reaching the input of IC_2. Capacitor C_2 im-proves the stability of the regulator, particularly with maximum pulse loading.

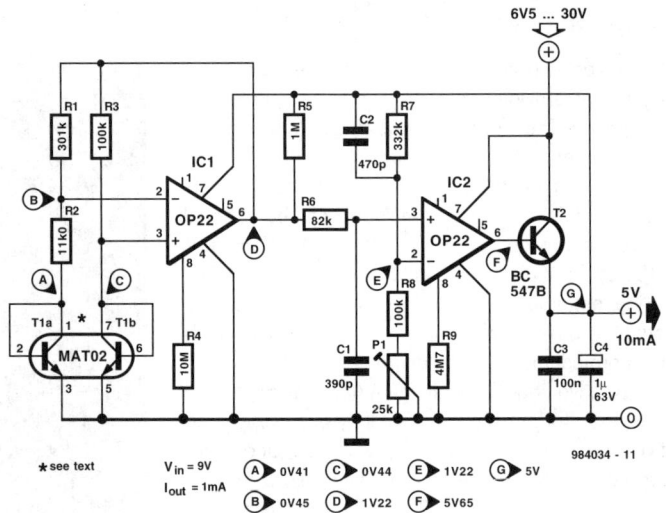

Note that the circuit has a high resistance, so that it is advisable to house it in a screened enclosure to prevent spurious magnetic and electromagnetic interference being coupled to the circuit.

The regulator was tested with a direct load current of 1 mA on which was superimposed a square-wave current of 10 mA. The test results are summarized in the table, in which I_g represents the current drawn by the circuit.

V_{in} (V)	I_g (µA)	V_{out} (V)	V_{ripple} (mV$_{rms}$)
7	43	5.002	110
9	49	5.002	100
15	68	5.002	70
30	123	5.002	50

Thrifty voltage regulator

One of the drawbacks of a three-pin voltage regulator is that the input voltage needs to be 2.5–3 V higher than the output voltage. This makes these integrated regulators unsuitable for battery power supplies. If, for instance, the output voltage is 5 V, a 9 V battery could be discharged to 7.5 V or thereabouts only. On top of this, most of these regulators draw a current of about 2 mA. Special low-drop versions sometimes offer a solution, but they are not ideal either.

The regulator described here is rather thriftier: it draws a current of only 300 μA and the difference between its input and output is only 100–200 mV.

In the circuit diagram, T_1 is arranged as a series regulator, which means that the difference between input voltage and output voltage is limited to the transistor's saturation poten-

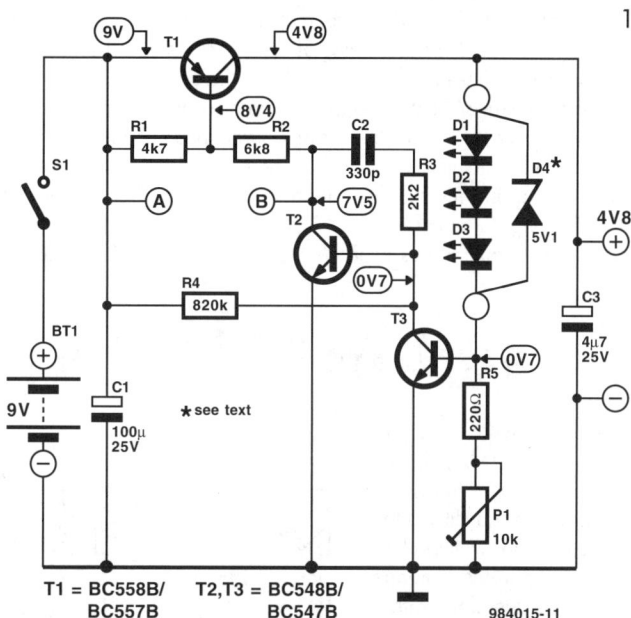

T1 = BC558B/   T2,T3 = BC548B/
BC557B   BC547B   984015-11

tial. Therefore, a 9 V battery can be discharged to about 5 V, which is quite an improvement on the situation with an integrated regulator.

Diodes D_1-D_2-D_3, or a suitable zener diode (D_4), in conjunction with R_5 and P_1, form a variable reference voltage source, which is used as the (output-dependent) base potential of T_3. If the output voltage drops below a desired level, the base potential of T_3 also drops. The transistor then conducts less hard and its collector voltage rises. The base voltage of T_2 also rises, so that T_1 is driven harder. This results in the near-instantaneous restoration of the output voltage.

The design of the reference voltage source is clearly of paramount importance. The current through the LEDs or the zener diode is of the order of only 100 μA. This means that the drop across a 5.1 V zener diode is only 4.3 V and across each LED, only about 1.43 V. For a wanted output voltage of 4.8 V, the three LEDs proved very effective, whereas the zener did not. It may well be necessary, if a zener diode is used, to try one rated at 4.7 V. If, however, an output voltage of 5 V is wanted, it will be necessary to carefully select a zener diode.

When the battery voltage has dropped to a level where it is only marginally higher than the wanted output voltage, T_1 and T_2 conduct hard. A further drop in the battery voltage will cause the collector potential of T_2 to drop rapidly to 0 V, since T_2 tries to make T_1 conduct hard.

The large drop in the collector potential of T_2 may be used to drive a BAT-LOW indicator. This may be done in three ways as shown in Figure 2.

When network a is connected between terminals A and B (Figure 1), transistor T_4 will normally be held cut

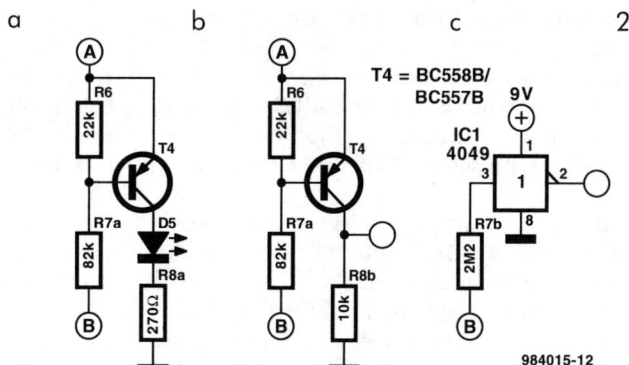

984015-12

247

off by divider R_6-R_{7a}. If then the voltage at B drops suddenly, T_4 conducts, whereupon D_5 indicates that the battery is nearly flat.

The network in b is similar to that in a, but is intended for a liquid-crystal display of BAT LOW. The collector of T_4 is linked to the IC that drives the decimal point and the BAT-LOW segment of the display.

Network c may be used if there is an unused inverter or gate in the circuit to be powered. The high value of resistor R_{7b} prevents the internal protection diodes of the IC being damaged.

When the regulator has been built, connect it to a variable power supply via a multimeter set to the mA range and set P_1 roughly at its mid-position. Turn P_1 slowly until the desired output voltage is obtained.

If with an output voltage of 4.8 V the regulator draws a current of more than 250–300 μA, the three LEDs or zener diode must be replaced.

The regulator can provide a current of up to about 25 mA. With a fresh 9 V battery, the dissipation of T_1 does not exceed 100 mW. If the input voltage is higher, it may be necessary to mount the transistor on a suitable heat sink or replace it by a power transistor, for instance, a Type BD138.

Lead-acid battery protector

Lead-acid batteries must not be discharged deeply. In the best case, this leads to an irreversible loss of capacity; in the worst case, the likelihood of serious, irreparable damage is great. It is, therefore, rather odd that there are several circuits aimed at protecting these batteries from being overcharged, but hardly any to protect them against being discharged too deeply.

This situation is put right by the protector in Figure 1. It is intended to be placed between the battery and its load, and decouples the load from the battery when this is nearly discharged.

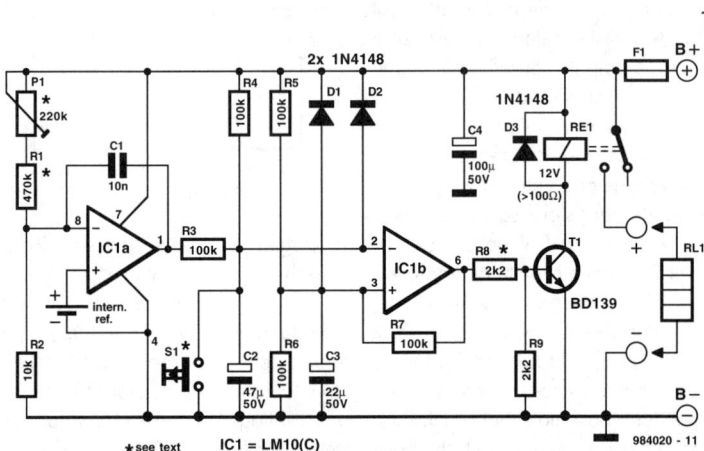

The circuit draws a current not exceeding 1 mA and is readily adapted for use with 6 V, 12 V, or 24 V lead-acid batteries. It may be equipped with an automatic or a manual reset function to override the decoupling of the load from the battery.

The circuit is based on a Type LM10C that contains an op amp and a reference amplifier. The pinout and internal circuit diagram of the device are shown in Figure 2.

The IC has an integral voltage reference source of 200 mV which is internally linked to the non-inverting (+ve) input of a reference amplifier.

The integral op amp has an output stage that can be driven almost up to the level of the supply voltage. This means that it can provide a current of $\pm$20 mA at a saturation voltage of only 400 mV.

The supply voltage range of the IC is 1.1–40 V and the device draws a current of only 270 μA. Note that the Type LM10CL cannot be used in this application since it is suitable for supply voltages of up to 7 V only.

In the circuit diagram in Figure 1, IC_{1a} is the reference amplifier, which functions as monitor of the battery voltage. The battery voltage is lowered by voltage divider R_1-R_2-P_1 to the level of the reference potential of 200 mV. As long as the battery voltage remains above the cell level of 1.83–1.85 V, the output of IC_{1a} (pin 1) is low. This is of no consequence, since the inverting (–ve) input of op amp IC_{1b} is held at half the supply voltage by R_4.

Op amp IC_{1b} is arranged as a Schmitt trigger with switching levels at 1/3 and 2/3 of the supply voltage level via R_5, R_6 and R_7. The output of the op amp, pin

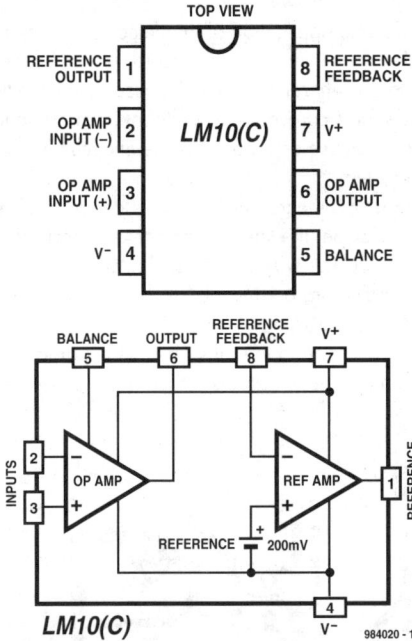

TOP VIEW

REFERENCE OUTPUT 1   8 REFERENCE FEEDBACK

OP AMP INPUT (–) 2   LM10(C)   7 V+

OP AMP INPUT (+) 3   6 OP AMP OUTPUT

V– 4   5 BALANCE

LM10(C)

984020 - 12

6, goes high immediately the supply is switched on and remains so as long as the battery is not discharged. This is arranged by C_2, C_3, D_1 and D_2.

Since the value of C_3 is only half that of C_2, the potential at pin 2 rises more slowly than that at pin 3. The output of IC_{1b} thus goes high and remains so.

Diodes D_1 and D_2 ensure that C_2 and C_3 are discharged rapidly after the supply is switched off. This enables the circuit to be reactuated quickly when necessary.

The high level at the output of IC_{1b} causes T_1 to conduct, so that the relay is energized. The impedance of the relay coil should not exceed 100 Ω.

When the battery voltage drops below the critical level, the output of IC_{1a} changes from low to high, whereupon the level at the inverting input of IC_{1b} (pin 2) rises above the level of 2/3 the supply voltage. The output of IC_{1b} goes low, T_1 is cut off, and the relay is deenergized, whereupon the load is decoupled from the battery.

Resistor R_4 limits any voltage variations at pin 2 of IC_{1b} to the upper half of the supply voltage. This means that when the output of the op amp goes low, it stays so until a reset. This arrangement ensures that when the battery voltage recovers after the load has been decoupled, the protector is not reset automatically. Such a reset would almost always be undesired, since as soon as the load would be reconnected, it would be uncoupled again. The result would be an oscillatory process.

It is clear that a manual reset is better and this is effected by short-circuiting C_2 briefly by operating push-button switch S_1. Note that a reset also occurs when the supply is switched off and then on again. If, nevertheless, an automatic reset is desired, this is easily arranged by the removal of R_4. Be warned, however, that this makes sense only if the charging of the battery is started when the auto reset occurs.

Calibrating the circuit is straightforward. Connect a variable power supply in place of the battery with a multimeter set to the 20 V direct voltage range across its terminals. Adjust P_1 until the relay opens at a voltage of 5.5 V (6 V battery); 11 V (12 V battery) or 22 V (24 V battery).

The component values in the circuit diagram are for a 12 V battery version. In case of a 6 V battery, the value of R_1 must be lowered to 220 kΩ, that of R_8 to 1.2 kΩ, and that of P_1 to 100 kΩ. For a 24 V battery, the value of R_1 must be increased to 1 MΩ and that of R_8 to 4.7 kΩ.

Finally, the voltage rating of the relay coil must, of course, be the same as the nominal battery voltage.

Mains voltage detector

The detector is intended to sense and signal to another circuit that an appliance is linked to the mains voltage. For this purpose, an optoisolator, IC_1 in the circuit, is used. The light-emitting diode in this device is connected across the mains voltage rectified by bridge B_1. The mains voltage is applied to this bridge via potential divider R_1-C_1-R_2. When the capacitor has a value as specified in the diagram, the current through the diode is 700 μA (for a mains voltage of 230 V). This results in sufficient light to make the phototransistor conduct. The drop across the LED is about 1 V.

The detector draws a current only when the monitored equipment is switched on. It is intended to be built into the appliance whose mains connection is to

994059 - 11

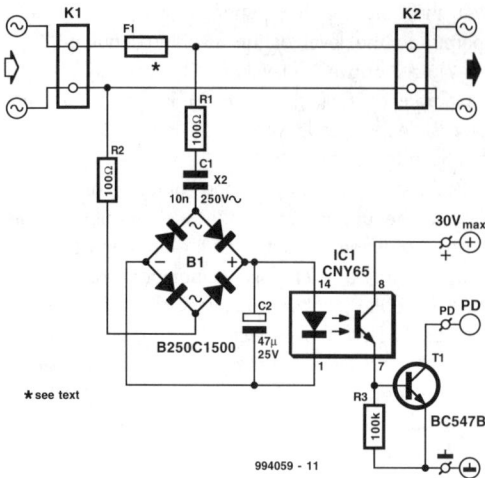

be monitored and must, of course, be connected after the mains on/off switch.

A possible application of the detector is in the preamplifier to sense when a record player is being switched on, whereupon it can be used to link the line-in of a sound card automatically to the preamplifier. Another is its use as a power-on reset circuit in a protection system.

Transistor T_1 can switch currents of up to 10 mA; in the prototype, the knee voltage of the transistor at a current of 20 mA was around 200 mV.

The maximum permissible switching voltage of the optoisolator is 30 V.

Fuse F_1 is added to make possible the omission of a fuse on the monitored appliance.

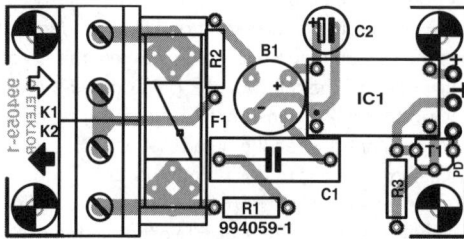

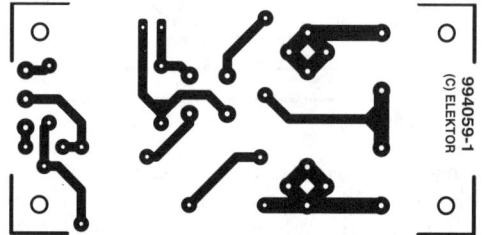

994059-1

Parts list

Resistors:
R_1, R_2 = 100 Ω
R_3 = 100 kΩ

Capacitors:
C_1 = 0.01 µF, 250 VAC
C_2 = 47 µF, 25 V, radial

Semiconductors:
B_1 = B250C1500 rectifier
T_1 = BC547B

Integrated circuits:
IC_1 = CNY65

Miscellaneous:
K_1, K_2 = 2-way terminal strip, pitch 7.5 mm
F_1 = fuse holder with fuse (rated as relevant)
PCB may be made with the aid of the track layout.

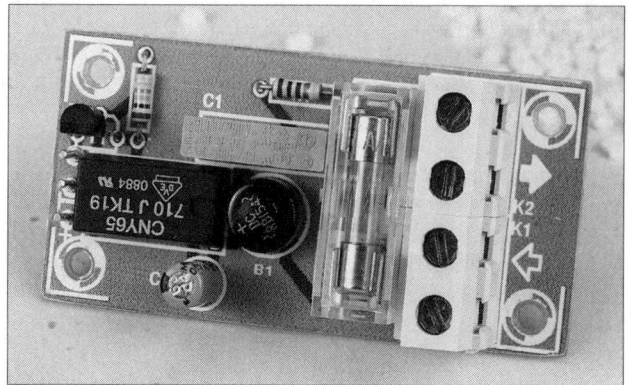

General-purpose NiCd battery charger

There is a wide variety of NiCd (nickel-cadmium) battery chargers on the market, but there are not many that can work from an in-car 12 V cigar lighter. Such a charger would, for instance, be of interest to campers and caravanners who do not have a 230 V a.c. mains supply available. To satisfy the needs of these users, a charger could be designed for operation from the cigar lighter, but it is, of course, of far greater interest if it could also work from the domestic mains supply. Furthermore, it would also be very useful if a number of cells, say, 1 to 4, of different format could be charged simultaneously. Lastly, another benefit would be if the charger would automatically switch off once the battery or cells have been charged fully.

The charger described in this article does all that: it accommodates batteries or cells Type R6 and R14. Switching off after a period of 2 h 30 m, 5 h, or 10 h is arranged by 3-way switch S_1. The 2 h 30 m period is for charging Type R6 batteries (1/2 charge), the 5 h period for fully charging Type R6 batteries or half charging Type R14 batteries, and the 10 h period for fully charging Type R14 batteries. Light-emitting diode D_1 lights when charging is taking place. Charging after the set period has elapsed can be continued, if so desired, only by switching the supply off and then on again.

The time periods are determined by counters IC_1 and IC_2, Type 4060 and 4020 respectively. The 4060 has an integral oscillator, whose frequency is set to 932 Hz with preset P_1 and the aid of a frequency meter.

For various reasons, such as the values of the components used and parasitic elements, the oscillator itself operates at a slightly higher frequency – of the order of 1 kHz. The frequency of the signal at the wiper of P_1 is divided by 2^{14}, so that the frequency of the signal at Q13 of IC_1 is 0.056 Hz, equivalent to a pulse every 17.6 s. The signal at Q13 is applied to the input, pin 10, of IC_2. When switch S_1 is in position 2 h 5 m (output Q10 of IC_2), the divisor should be 2^{10} (1024). However, contrary to what these figures

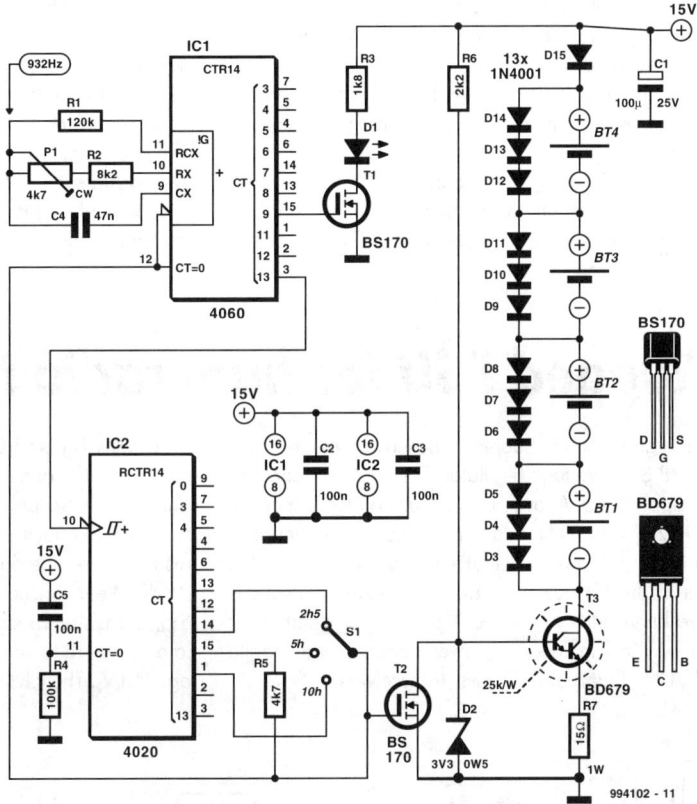

indicate, the time period stops at half that at output Q10. To obtain a charging period of 2 h 30 m, that is, 9,000 seconds, which should correspond to half a period at output Q9 of IC_2, the oscillator period must be $9000 \times 2/16.7 \times 10^6 = 1.073$ ms, which corresponds to a frequency of 932 Hz as mentioned earlier.

On power-on, only counter IC_2 is reset, since an error of a few seconds that may arise in IC_1 is of no significance. This arrangement simplifies the design. When the time set has elapsed, that is, charging is finished, diode D_1 goes out.

The charging current is fixed by darlington transistor T_3, which is a classical design of a current source with negative feedback. The transistor tends to hold its emitter potential at 1.3 V, but this requires the aid of a zener diode, D_2. In this type of design, the thermal stability is, in fact, totally acceptable, because the temperature of the zener diode, in view of the small current this draws and its consequent low temperature rise, hardly affects the charging current

Transistor T_1 is either on or off and serves to power the on/off indicator LED. It is needed to prevent an overload on the output of counter IC_1 if this would be required to absorb the total current (about 7 mA) drawn by the diode.

Transistor T_2 discontinues the charging when the time set by S_1 has elapsed by earthing the base of darlington T_3.

Diodes D_3–D_{14} are connected in threesomes across the terminals of the batteries to be charged: D_3–D_5 across those of battery Bt_1, D_6–D_8 across those of Bt_2, and so on.

Diode D_{15} prevents the batteries to be charged from being discharged when the supply fails.

When the charger is used in a vehicle, additional precautions should be taken to ensure that any spurious surges on the vehicle power lines do not adversely affect the charger's operation.

The battery holder should be one that can accommodate four size R6 (AM3; MN1500; SP/HP7; mignon) or R14 AM2; MN1400; SP/HP11; baby) batteries. The length of these batteries, but not their diameter, is the same (about 45 mm)

When the charger is used at home, it may be powered via a suitable 15 V mains adaptor. It draws a current of about 150 mA.

A final word of warning: it is possible for batteries to be connected to the charger with incorrect polarity. This may result in a very large discharge current and even destruction of the battery. It is, therefore, imperative to verify the correct polarity of the battery before inserting it into the holder.

Rugged PSU for ham radio transceivers

This rugged power supply is based on the popular LM338 3-pin voltage regulator. The LM338 is capable of supplying 5 A over an output voltage range of 1.2–32 V with all standard protections like overload, thermal shutdown, overcurrent, internal limit, etc., built in. In this power supply, some extra protections have been added to make it particularly suitable for use with low to medium-power portable and mobile VHF/UHF (ham) and 27 MHz transceivers.

Diodes D_4 and D_5 provide a discharge path for capacitors C_1 and C_2. Diode D_8 protects the supply against reverse polarity being applied to the output terminals. Capacitor C_1 assists in RF decoupling and also increases the ripple rejection from 60 dB to about 86 dB. When junction R_1–R_2 is not grounded by switch S_{1A}, transistor T_2 starts to conduct, causing the regulator to switch to zener diode D_7 for its reference voltage (13 V). The PSU output voltage will then be 12.3 V.

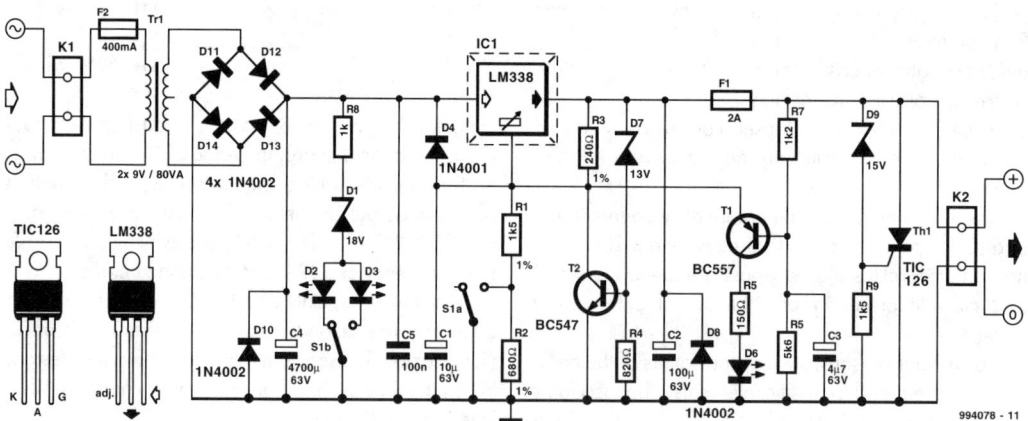

994078 - 11

Normally, T_2 will be off, however, and the PSU output voltage is then about 8.8 V. The high/low switch is useful to control the RF power level of modern VHF/UHF handhelds.

Transistor T_1, a p-n-p type BC557, acts as a blown-fuse sensor. When fuse F_1 blows, T_1 starts to conduct, causing D_6 to light. If, for whatever reason, the PSU output voltage exceeds about 15 V, thyristor THR_1 is triggered (typically in less than a microsecond). Such a high-speed 'crowbar' may look like a drastic measure, but bear in mind that this kind of protection is required by digital ICs that will not stand

much overvoltage. The crowbar, when actuated, will faithfully destroy fuse F_1 rather than allow the PSU to destroy expensive ICs.

The two LEDs on the S_{1B} contacts not only act as 'high/low' indicators but also as power-on indicators which are turned off when the mains voltage drops below about 160 V.

If heavy-duty use of the PSU is needed, voltage regulator IC_1 should be mounted on as large a heatsink as possible. The minimum would be a Fischer SK129 heatsink from Dau Components.

±20 A current monitor

The Type UCC3926 current sensor IC from Unitrode is ideally suited for use as a current monitor. It contains a 1.3 mΩ current shunt and can handle currents up to ±20 A. The common-mode voltage for the shunt is GND ±75 mV or V_{DD} ±75 mV, so that the current can be monitored either in the positive supply rail or in the negative supply line of a load. The supply voltage, V_{DD}, may lie between 4.8 V and 14 V.

The potential across the shunt is applied to an internal chopper-stabilized transimpedance amplifier, which converts it into a differential voltage at pins 5 and 12 at a level of 500 mV when the current is 15 A. The differential voltage is applied via a low-pass filter to an operational amplifier, which has unity gain and converts the voltage into a unipolar potential.

An offset voltage is superimposed on to the unipolar potential via a 1 kΩ multi-turn potentiometer to provide a preset voltage at the OUT pin. This means that the voltage level at the OUT pin is typically 500 mV plus the offset direct voltage when the current through the shunt is 15 A.

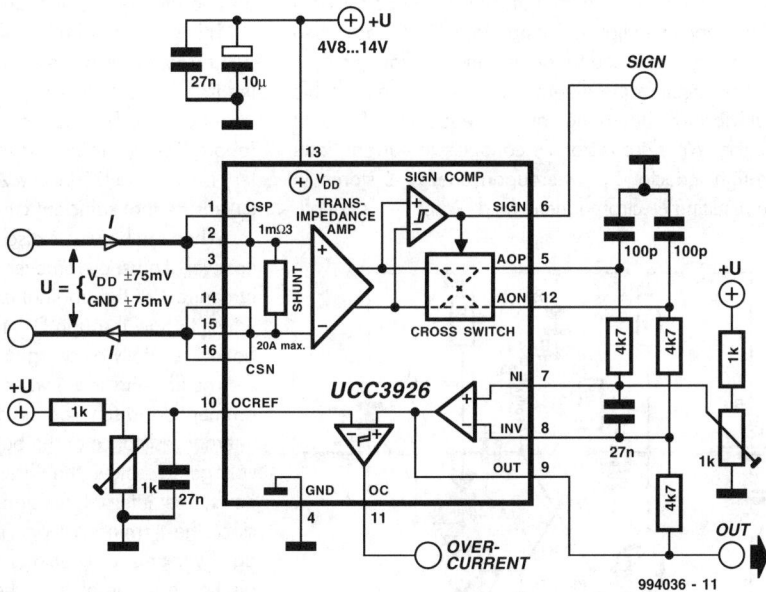

994036 - 11

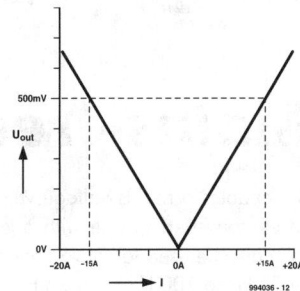

994036 - 12

The polarity of the output voltage is determined by the SIGN comparator, which sets the internal cross switch to such a position that the differential voltage between pins 5 and 12 is always positive.

The polarity may be checked at pin 6: if the level at that pin is high, the polarity is correct, that is, the current flows through the shunt from pins 1, 2, 3 to pins 14, 15, 16.

A signal to show up an overvoltage may be generated with the aid of an internal comparator. For this purpose, an overcurrent reference may be applied to pin 10 via the 1 kΩ, 10-turn potentiometer. The digital signal at pin 11 is high when there is an overcurrent.

Further information from http://www.unitrode.com

Battery discharger

The battery discharger published in the June 1998 issue of this magazine may be improved by adding a Schottky diode (D_3). This ensures that a NiCd cell is discharged not to 0.6–0.7 V, but to just under 1 V as recommended by the manufacturers. An additional effect is then that light-emitting diode D_2 flashes when the battery connected to the terminals is flat.

The circuit in the diagram is based on an astable multivibrator operating at a frequency of about 25 kHz. When transistor T_2 conducts, a current flows through inductor L_1, whereupon energy is stored in the resulting electromagnetic field. When T_2 is cut off,

the field collapses, whereupon a counter-emf is produced at a level that exceeds the forward voltage (about 1.6 V) of D_2. A current then flows through the diode so that this lights. Diode D_1 prevents the current flowing through R_4 and C_2.

This process is halted only when the battery voltage no longer provides a sufficient base potential for the transistors. In the original circuit, this happened at about 0.65 V. The addition of the forward bias of D_3 (about 0.3 V), the final discharge voltage of the battery is raised to 0.9–1.0 V. Additional resistors R_5 and R_6 ensure that sufficient current flows through D_3.

When the battery is discharged to the recommended level, it must be removed from the discharger since, in contrast to the original circuit, a small current continues to flow through D_3, R_2-R_3, and R_5-R_6 until the battery is totally discharged

The flashing of D_2 when the battery is nearing recommended discharge is caused by the increasing internal resistance of the battery lowering the terminal voltage to below the threshold level. If no current flows, the internal resistance is of no consequence since the terminal voltage rises to the threshold voltage by taking some energy from the battery. When the discharge is complete to the recommended level, the LED goes out. It should therefore be noted that the battery is discharged sufficiently when the LED begins to flash.

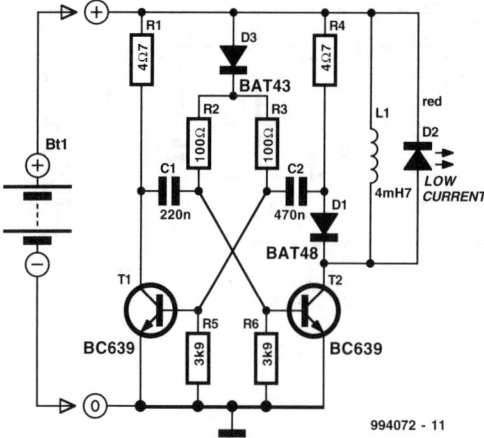

994072 - 11

Mains-frequency converter

The converter enables a mains voltage with unstable frequency to be converted to one with quartz-crystal stability. It may also be used to convert the mains frequency from 50 Hz to 100 Hz or to 25 Hz. It enables

low-power equipment that needs an accurate mains voltage frequency to be supplied from an inexpensive 230 V inverter (driven, say, by a 12 V car battery).

The time base, IC_1, of the converter is a crystal

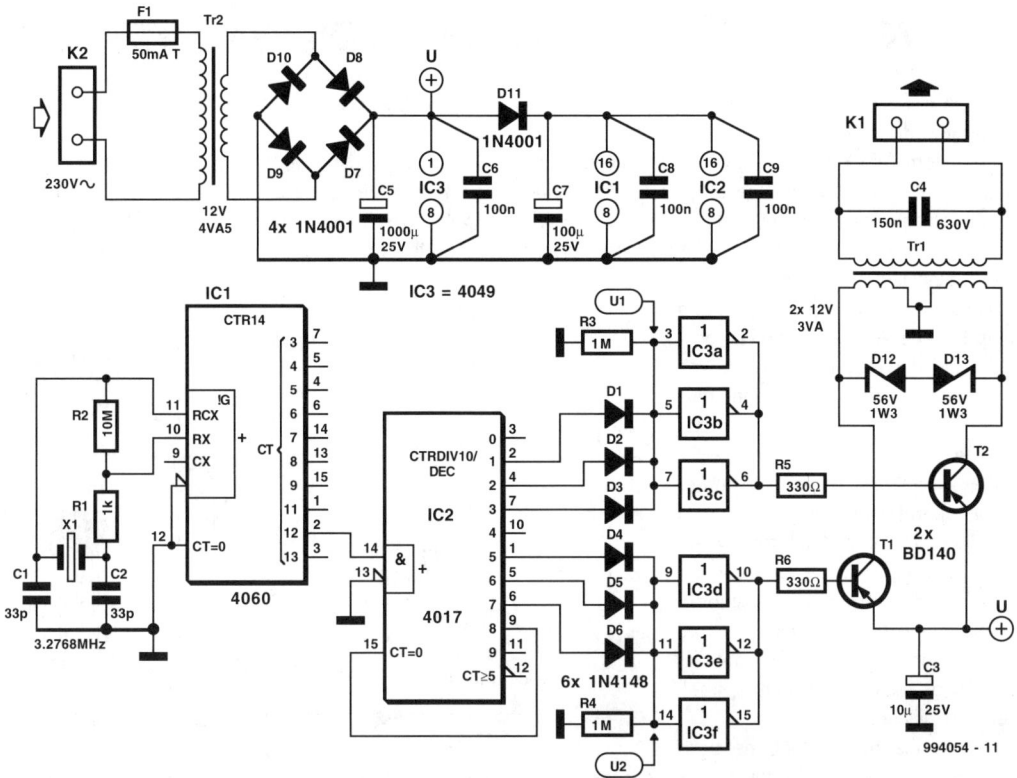

oscillator operating with a 3.2768 Mhz quartz crystal. This frequency is divided internally by 2^{13} to give an output frequency of 400 Hz. The 400 Hz signal is applied to counter/divider IC2. When the reset of this IC is linked to Q8 (pin 9), the circuit operates as an :8 divider so that its output is exactly 50 Hz.

The output signals of IC2 are bundled by wired-OR gates D1–D3 and D4–D6 into U1 and U2 respectively. These signals (shown at the top of Figure 1) are used to control drivers T1 and T2 via parallel-connected inverters contained in IC3, a Type 4049 CMOS circuit. The drivers alternately switch a 12 V transformer winding into circuit, which, because of the constantly changing magnetic flux, results in a stable 230 V alternating voltage across the primary of the transformer.

The drivers are standard Type BD140 p-n-p transistors, so that the inverters in IC3 draw current when the the relevant driver is on. It is a well-known fact that CMOS elements are far better in drawing current than providing current. The drivers are protected against high-voltage spikes by zener diodes D12 and D13.

Capacitor C4, in conjunction with the stray inductance of the transformer, forms a bypass for high-frequency components in the primary winding of the transformer.

The frequency of the output voltage may be changed to 100 Hz or 25 Hz by linking Q11 (pin 1) or Q13 (pin 3) respectively to the clock input of IC2.

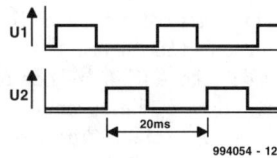

Note that in the supply circuit at the bottom of the diagram, diode D11 and capacitor C7 prevent excessive voltage fluctuations caused by the heavy load presented by the drive circuit from adversely affecting the operation of IC1 and IC2.

Power diode for solar power systems

Solar power systems cannot work without a reflow protection diode between the solar panel and the energy store. When current flows into the store, there is a potential drop across the diode which must be written off as a loss in energy. In the case of a Schottky diode, this is not less than 0.28 V at nominal current levels, but will rise with higher ones. It is clear that it is advantageous to keep the energy loss as small as possible and this may be achieved with external circuitry as shown in the diagram.

The circuit is essentially an electronic switch consisting of a high precision operational amplifier, IC_{1a}, a Type OP295 from Analog Devices, and a MOSFET, T_1. This arrangement has the advantages over a Schottky diode that it has a lower threshold voltage and the lost energy is not dissipated as heat so that only a small heat sink is needed.

When the potential at the non-inverting input of the op amp, which is configured as a comparator, rises above that at the inverting input, the output switches to the operating voltage. The transistor then comes on, whereupon light-emitting diode LD_1 lights. Diode D_3 clamps the inputs of IC_{1a} so that the peak input voltage cannot be greater than half the threshold voltage, provided the values of R_3 and R_4 are equal.

The op amp provides very high small-signal amplification, a small offset voltage, and consequent fast switching. The MOSFET changes from on to off state and vice versa at drain -source voltages in the microvolt range. In the quiescent state, when U_{DS} is 0 V, the transistor is on, so that LD_1 lights.

The operating voltage (C–A) may be between 5 V (the minimum supply for the op amp and the input control potential, U_{GS}, of the transistor) and 36 V (twice the zener voltage of D_1). Zener diode D_1 protects the MOSFET against excessive voltages (greater than ± 20 V). Diode D_3 and resistors R_3 and R_4 halve the potential across the inputs of the op amp. This ensures that operation with reversed or open terminals is harmless.

The substrate diode of the MOSFET is of no consequence since it does not become forward biased as long as the forward voltage, U_{SD}, of the transistor is held very low.

The on -resistance, $R_{SD(on)}$, of the transistor is only 8 mΩ and the transistor can handle currents of up to 75 A. When the nominal current is 10 A, the drop across the on-resistance is 80 mV, resulting in an energy loss of 0.8 W. This is low enough for a SUB type with a TO263-SMD case to be used without heat sink. When the current is 50 A, however, it is advisable to use a SUP type with a TO220 case and a heat sink since the transistor is then dissipating 12.5 W. Even then, the voltage drop, $U_{SD} = 0.32$ V is significantly lower than that across a Schottky diode in the same circumstances.

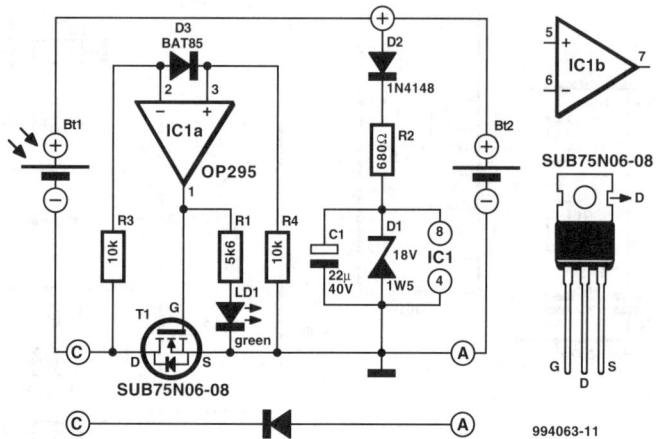

Moreover, owing to the high precision of IC_{1a}, a number of transistors may be used in parallel.

The circuit proper draws a current of 150 μA when only one of the op amps in the OP295 is used. An even lower current is drawn by the alternative Type MAX478 from Maxim. However, the differences between these two types are only relevant in the low current and voltage ranges. Both have rail-to-rail outputs that set the control voltage accurately even at very low operating voltages. This is important since the switch-on resistance of MOSFETs is not constant: it drops significantly with increasing gate potentials and decreasing temperature.

A experimental circuit may use an LM358 op amp and a Type BUZ10 transistors, but these components do not give the excellent results just described.

Switching regulator

Potentials of up to more than 100 V may be generated by a switching regulator IC complemented by a cascode circuit of diodes and reservoir capacitors. The regulator pumps the voltage up in two stages to the requisite output level determined by potential divider R_2-R_3 according to the equation

$$U_{out} = 1.245(R_2 + R_3)/R_3.$$

With resistor values specified in the diagram, the output voltage is 84 V. Resistor R_3 may be replaced by a fixed resistor and a 10 kΩ preset.

Regulator IC_1 applies the direct voltage at the input to inductor L_1. Internally, pin 3 (SW1) is periodically short-circuited to pin 4 (SW2=earth). When SW1 is opened, a counter-e.m.f. pulse is generated across L_1, which charges capacitor C_3 via diode D_3. The voltage pulse is also applied to C_2. The voltage across C_3 is applied to C_4 via D_2 together with that across C_2 via D_1. The diodes used must be fast types with a high reverse voltage rating. In the prototype, Motorola Type MURS120T3 diodes were used.

The current through switches SW1 and SW2 is determined by resistor R_1. With a resistance of 100 Ω as specified, it is about 700 mA. The peak current must not exceed 1.5 A.

CAUTION The generated voltages may be lethal, so care must be taken in handling the circuit. Also, after the input voltage has been switched off, capacitor C_4 may retain a lethal charge for some time.

Detailed information on Linear Technology's regulator IC Type LT1108 from: http://www.linear-tech.com

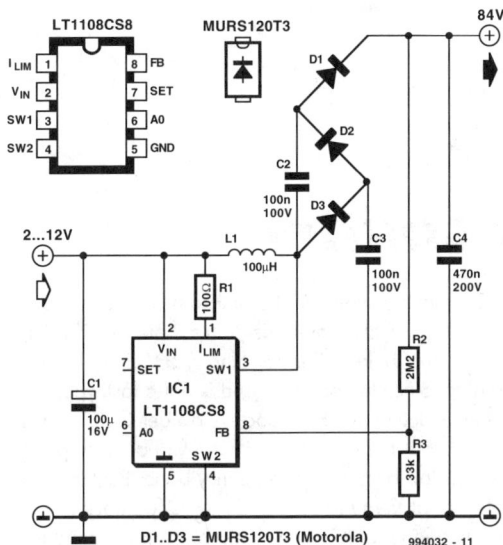

D1..D3 = MURS120T3 (Motorola)   994032 - 11

13 V/2 A PSU for handheld rigs

This compact 13-V/2-A power supply for ham radio rigs and other VHF/UHF portable PMRs is based on the STR2012/13 voltage regulator IC from Sanken Electric Co. Many power supplies for handheld amateur radio rigs are based on the LM317, LM350 or even the good old LM723. Unfortunately, these regulators are invariably associated with a fair number of external components, while we should also consider design factors like total power dissipation and input voltage range.

The STR is a hybrid power IC containing a switch-mode power supply. It supplies a fixed output voltage and accepts relatively high input voltages. Another advantage is its relatively high power dissipation rating. The 5-pin STR is available for 5.1 V, 12 V, 13 V,

15 V and 24 V at an output current rating of 2 A. Here, the STR2012 and STR2013 are suggested for output voltages of 12 V or 13 V respectively. The normal operating voltage of most handhelds being between 12.6 V and 13.8 V, the STR1303 will be the preferred device in most cases.

A high-speed crowbar circuit is added to the regulator output. Thyristor Th_1 (a TIC106 or 2N4442) is triggered when the output voltage rises above the zener voltage of D_2, that is, 15 volts (approximately). When this happens, the thyristor short-circuits the supply output, protecting the radio against overvoltage and blowing fuse F_1. Diode D_1 acts as a reverse polarity protection, also in combination with fuse F_1.

To allow for its dissipated heat, the STR regulator

should be mounted on a heat sink. Efficiency is around 80%, with ripple rejection at a comfortable 45 dB. The raw input voltage to the regulator should be in the range 18–35 V.

Inductor L_1 may be selected from the range produced by Newport. The type 1430430 is preferred, but if this proves difficult to obtain, an ordinary triac suppressor type may be used instead. Note, however, that the inductance of these coils is usually just 100 mH, so you have to count the number of turns and add another 0.7 times that number to arrive at about 300 mH.

Finally, keep the wire between pin 3 of the STR and ground as short as possible, and connect at least the negative terminals of C1 and C3 to this point to give a 'star' type ground connection.

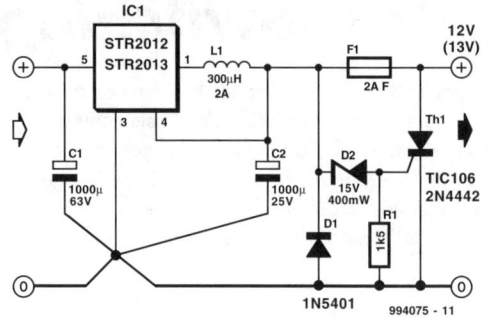

3 V supply splitter

Many modern circuits tend to work from a single supply voltage of 3 V. But often they need a virtual earth at half the supply voltage for efficient operation.

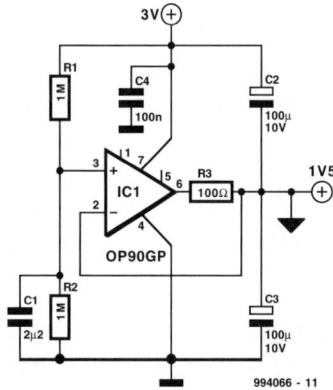

The splitter shown in the diagram bisects the supply voltage with a high-resistance potential divider, R_1-R_2, and buffers the resulting 1.5 V line with an op amp. Since the op amp used is not a fast type, the output is decoupled by capacitive divider C_2-C_3. This ensures that the impedance of the virtual earth point remains low over a wide frequency band. Because the potential at the junction C_2-C_3-R_3 is fed back to the inverting input of IC_1, the circuit becomes a standard voltage follower.

Resistor R_3 ensures that the regulation remains stable. The circuit can regulate ±2 mA without any difficulties. Because of the low current drawn by IC_1, and the high resistance of R_1 and R_2, the overall current drain is low. In the absence of a load, it was 13 μA in the prototype, of which 1.5 μA flows through R_1-R_2.

Finally, since IC_1 can operate from a voltage as low as 1.6 V, the splitter will remain fully operational when the battery nears the end of its charge or life.

Discrete voltage regulator

The title of this article may prompt readers to ask why the already generous selection of fully integrated voltage regulators needs to be extended with a version made from discrete components. In other words, what does this circuit offer that the integrated types do not have?

To start with, the present circuit is refreshingly simple for a discrete version. Three semiconductors, three resistors, a capacitor and a diode are all it needs. Of course, that's still more components than an integrated regulator, so what exactly are the advantages of this circuit? These are to be found in three areas: voltage range, bandwidth and current rating. The last of these is the primary strength of this circuit, since the maximum current depends only on the specifications of the output transistor. With the BD680, as used here, a current of 4 A can be delivered at a collector-emitter voltage of 10 V with adequate cooling (R_{th} = 3.12 K W^{-1}). The peak current is 6 A, which is not easily matched with an integrated voltage regulator.

When the illustrated version of the circuit (U_{DSmax} of T_1) is used, the peak input voltage is 30 V, but this can easily be increased by the use of high-voltage transistors. The same applies to the bandwidth, which can be extended as desired, without any modifications to the circuit, by the use of high-speed transistors. Generally speaking, a large bandwidth is not one of the strong points of integrated voltage regulators.

As mentioned earlier, the circuit is basically very simple. Zener diode D_1, which is supplied with a constant current of around 1 mA by JFET current source T_1, provides the reference potential. Capacitor C_1 is connected in parallel with D_1 to provide a soft start. This capacitor also provides additional buffering and decouples noise and other disturbances. The startup time is around three seconds.

The only additional item that is needed for the voltage regulator is an output buffer for the reference potential. This takes the form of a sort of super-Darlington consisting of T_2 and T_3. This works very well, but has the disadvantage that the output voltage

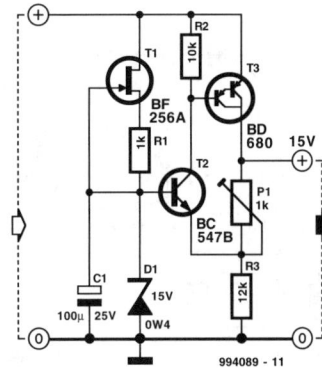

994089 - 11

is one diode drop lower than the zener voltage. Preset P_1 may be added to correct this at the cost of a reduced regulation of the circuit. If the voltage difference is not important, it is therefore better to replace P_1 with a wire bridge. The main specifications of the voltage regulator are listed below.

Main specifications

	with P1	without P1
Output voltage	15 V	14.5 V
Ripple suppression	58 dB	64 dB (I_{out}=100mA)
	46 dB	54 dB (I_{out} = 1 A)
$U_{dropout}$	1.6 V	1 V (I_{out} = 100 mA)
I_{noload}	2.1 mA	ditto
Max. input voltage	30 V	ditto

Switch-mode lithium-ion battery charger

Lithium-ion batteries are being used more and more frequently in all kinds of appliance. These batteries require a charger for which Maxim's MAX745 is ideal. It provides all the functions necessary for charging such batteries. It provides a regulated charging current of up to 4 A without getting hot, and a regulated voltage with a total error at the battery terminals of only ±0.75%. It uses low-cost, 1% resistors to set the output voltage, and a low-cost n-channel MOSFET as the power switch.

The MAX745 regulates the voltage set point and charging current using two loops that work together to transition smoothly between voltage and current regulation. The per-cell battery voltage regulation limit is set between 4.0 V and 4.4 V using standard 1% resistors, and then the number of cells is set from 1 to 4 by pin-strapping. The total output voltage error is less

than ±0.75%.

The charger is available as an evaluation kit, which is an assembled and tested printed-circuit board that implements a step-down, switching power supply designed for charging lithium-ion (Li-ion) batteries. The output voltage can be set for one to four cells. The cell voltage can be set between 4.0 V and 4.4 V.

The Li-ion battery pack is connected between BATT and GND (BATT is positive, GND is negative). The battery may be connected with the charger off without causing damage, or it can be connected after power is applied.

The charging voltage is determined by the potential at junction R_3-R_9. Replacing these resistors by a multiturn potentiometer enables the voltage to be set very accurately.

The charging current is selected with jumper JP_3.

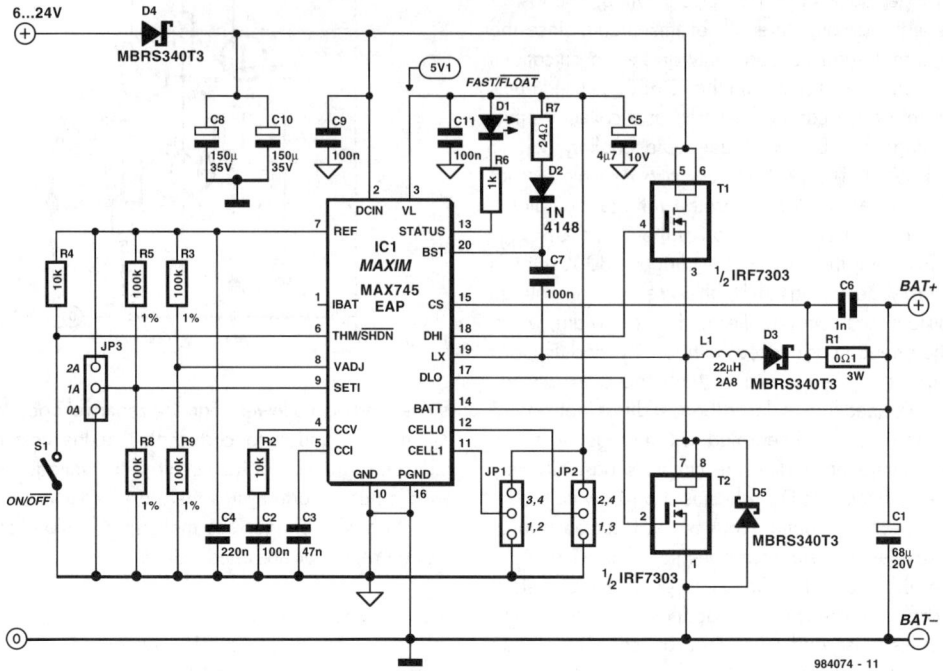

984074 - 11

Here also, a multiturn potentiometer to replace R_5 and R_8 enables a more accurate setting.

The number of cells, and thus the charging voltage, is set with jumpers JP_1 and JP_2: both to ground for one cell, only JP_2 to VL for two cells, only JP_1 to VL for three cells, both to VL for four cells.

Switch S_1 may be replaced by a resistor with negative temperature coefficient (NTC). When the voltage at pin THM drops below 2.1 V, the circuit is switched off automatically; when the voltage reaches 2.3 V again, the circuit is switched on anew.

Transistor T_1 is an n-channel FET whose auxiliary gate voltage is derived from capacitor C_7.

Diode D_1 is a freewheeling diode in case T_1 is cut off. When this happens, the diode is shunted by T_2 (which is on) to improve the efficiency. This is because the drop across the diode is 0.3–0.4 V, whereas that across a conducting transistor is only 0.1 V.

The three Schottky diodes are fast 3 A, 40 V types from Motorola. The FETs may be part of a dual FET from International Rectifier. If discrete ones are used, in view of the switching frequency of 300 kHz, types with a high input capacity must not be used: there is a current of only about 20 mA available for driving the gates. The IRF7303 has parameters: 30 V, 5 A, 0.05 Ω, and 520 pF.

Power supply regulator with sense lines

There are applications in which it is important for the supply voltage to be largely independent of the level of the output current, which is, of course, particularly so in the case of variable loads.

When the load is linked to the power supply by relatively short wires, a good variable power supply maintains the output voltage at a virtually constant level. Unfortunately, in practice, these wires can be fairly long, and since they have resistance, there is a voltage drop across them. This interferes with good regulation; the only way of avoiding this problem is to link the control part of the power supply to the load via separate sense lines.

Unfortunately, this cannot be done readily in every power supply without some tedious work, but as the diagram shows, in the case of the L200 it presents no problems.

In the diagram, A and D are the usual output terminals, while B and C are the sense input terminals. The output voltage, U_O, is

$$U_O = 2.77(1 + R_p/R_1),$$

where U_O is in volts and R_p is the effective resistance of P_1. Resistor R_2 in series with terminal A provides current limiting. The peak level of the output current, I_O, in amperes is

$$I_O = 0.45/R_2$$

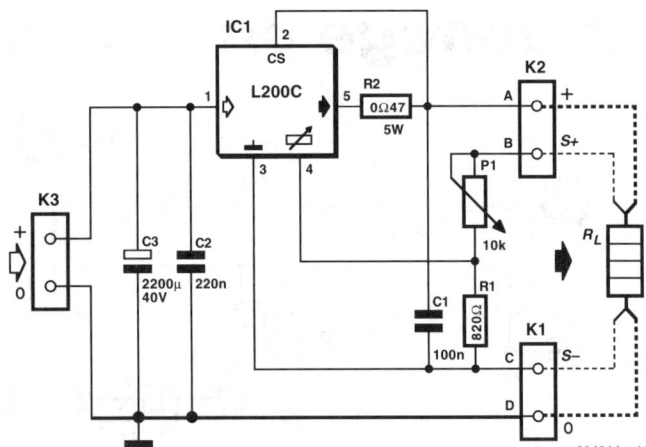

The maximum input voltage to the regulator is 40 V, and the peak output current is 2 A.

The regulator has on-board thermal protection, but this does not mean, of course, that it should not be mounted on a suitable heat sink when the dissipa-

tion is high.

The regulator is best built on the printed-circuit board which may be made with thei aid of the track layout.

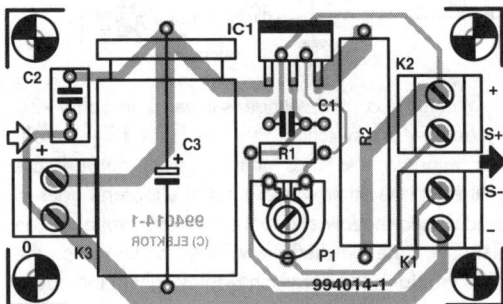

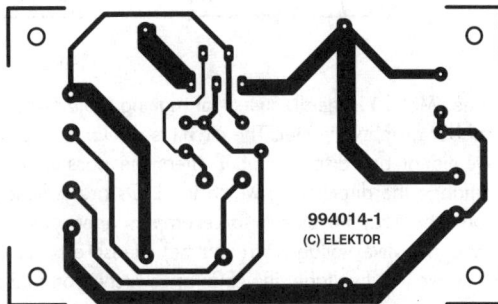

Parts list
Resistors:
R_1 = 820 Ω
R_2 = 0.47 Ω, 5 W
P_1 = 10 kΩ, preset

Capacitors:
C_1 = 0.1 μF
C_2 = 0.22 μF
C_3 = 2200 μF, 40 V
Integrated circuits:
IC_1 = L200 (ST Microelectronics)

Miscellaneous:
K_1–K_3 = terminal block for board mounting, pitch 5 mm
Heat sink for IC_1

± voltage on bargraph display

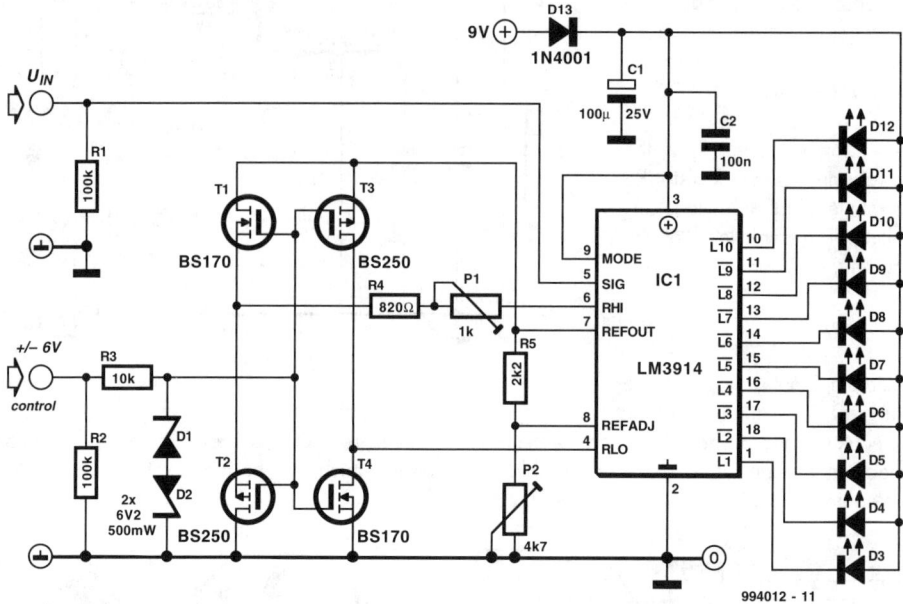

994012 - 11

The LM3914 is ideally suited for making a bidirectional bar-graph voltmeter. The circuit is similar to a conventional bar display, but it offers the possibility to change the direction in which the LEDs are switched on. This may be useful, for example, when positive and negative voltages are (to be) measured. For a positive input voltage, the LEDs are switched on in the usual manner, that is, from D_3 to D_{12}, while for negative voltages, they are switched on in the opposite direction, that is, from D_{12} to D_3. Obviously, the negative voltage must be inverted before the measurement can take place. A suitable circuit for this purpose is the 'Absolute-value meter with polarity detector' in the chapter 'Test & measurement.

A set of transistor switches (MOSFETs) controls the direction in which the LEDs light. When the control voltage is high (+6V, according to the schematics, but any voltage that is at least 3V higher than reference voltage will do), T_1 and T_4 are switched on, while the other two MOSFETs are off. In this way, the LM3194 is configured in the usual manner with the top end of the resistor network connected to the internal voltage reference and the low end connected to ground. As the input voltage rises, the comparators inside the LM3914 will cause the indicator LEDs to be switched on one by one, starting with D_3.

When the control voltage is lower than about –3V, T_2 and T_3 are switched on while T_1 and T_4 are off. Consequently, the ends of the resistor network are connected the other way around: the top end goes to ground and the low end to the reference voltage. The first LED to be switched on will then be D_{12}; i.e., the LEDs that form the bargraph display light in the opposite direction. Although not documented by the manufacturer of the LM3914, this option works well, but only in bar mode (in dot mode, internal logic disables any lower-numbered LEDs when a higher-numbered LED is on, which obviously conflicts with our purposes).

To achieve good symmetry, an adjustable resistor is added to the voltage divider in the LM3914. Using a DVM, adjust the preset until the voltage across $P_1 + R_4$ equals ⅒th part of U_{refout}.

Sensitivity is determined by the ratio of resistors R_5 and P_2. If, for example, the reference voltage is set to 2.2 V with P_2, there will be a voltage drop of 200 mV per resistor in the ladder network (including R_4–P_1). So, the first LED will switch on when the input voltage exceeds 200 mV, the second at 400 mV, and so on, and the whole display will be on at 2 V.

The circuit draws about 100 mA when all LEDs are switched on.

6
Radio, Television & Video

Active short-wave antenna

The active antenna illustrates the fact that in spite of all kinds of new component and technology, it is still possible to design useful, and interesting, circuits.

The design is based on two well-established transistors, a Type BF256C and a BF494. In conjunction with the requisite resistors and capacitors, these form a well-working antenna amplifier. Note that they are direct coupled.

Transistor T_1 is the input amplifier cum buffer, while the BF494, in a common-ground configuration, provides the necessary amplification.

The amplifier is designed for operation at frequencies between 10 MHz and 30 MHz, which is the larger part of the short-wave range, and has a gain of 20 dB.

Inductor L_1 is wound on an Amidon core Type T-37-6. The primary consists of 2 turns, and the secondary of 12 turns 0.3 mm dia. enamelled copper wire. The number of turns may be experimented with for other frequency ranges.

The input circuit is tuned to the wanted station with

capacitor C_1. The response of the tuned circuit is fairly broad, so that correct tuning is easy.

The circuit is powered by a well-decoupled mains supply converter that has an output of 9–12 V and draws a current of about 5 mA.

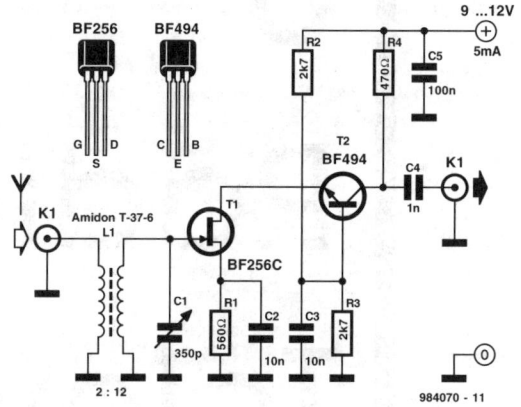

ATU for 27 MHz CB

This antenna tuning unit (ATU) enables half-wave-length ($\lambda/2$) or longer wire antennas to be matched to the 50 Ω antenna input of 27 MHz Citizens' Band (CB) rigs. The ATU is useful in those cases where a wire antenna is less obtrusive than a roof-mounted 'vertical' or ground-plane. It is also great for 'improvised' antennas used by active CB users on camping sites

and the like because it allows a length of wire to be used as a fairly effective antenna hung between, say, a tree branch at one side and a tent post, at the other. Obviously, the wire ends then have to be isolated with, say, short lengths of nylon wire. It is even possible to use the ATU to tune a length of barbed wire to 27 MHz.

The coil in the circuit consists of 11 turns of silver-plated copper wire with a diameter of about 1 mm (SWG20). The internal diameter of the coil is 15 mm, and it is stretched to a length of about 4 cm. The tap for the antenna cable to the CB radio is made at about 2 turns from the cold (ground) side. Two trimmer capacitors are available for tuning the ATU. The smaller one, C_1, for fine tuning, and the larger one, C_2, for coarse tuning.

The trimmers are adjusted with the aid of an in-line SWR (standing-wave ratio) meter which most CB enthusiasts will have or should be able to borrow. Select channel 20 on the CB rig and set C_1 and C_3 to

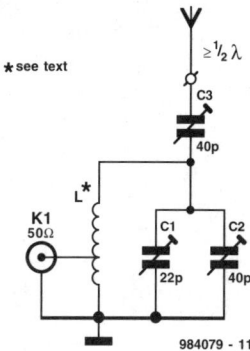

mid-travel. Press the PTT button and adjust C_2 for the best (that is, lowest) SWR reading. Next, alternately adjust C_3 and C_2 until you get as close as possible to a 1:1 SWR reading. Finally, adjust C_1 for an even better value. No need to re-adjust the ATU until another antenna is used. In case the length of the wire antenna is exactly $\lambda/2$ (5.5 metres), C_3 is set to maximum capacitance.

Although the ATU is designed for half-wavelength or longer antennas, it may also be used for physically shorter antennas. For example, if an antenna has a physical length of only 3 metres, the remaining 2.5 metres have to be wound on a length of PVC tubing. This creates a so-called BLC (base-loaded coil) electrically shortened antenna. In practice, the added coil can be made somewhat shorter than the theoretical value, so the actual length is best determined by trial and error. The ATU has to be built in an all-metal case to prevent unwanted radiation. The trimmers are accessed through small holes. The connection to the CB radio is best made via an SO239 ('UHF') or BNC socket on the ATU box and a short 50 Ω coax cable with matching plugs.

Loopstick VLF antenna

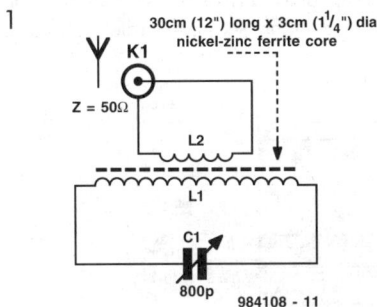

With its 30 cm long, 3.2 cm diameter composite ferrite core, the loopstick antenna covers much interesting activity between 50 kHz and 195 kHz. It is ideal for home, portable, holiday and overseas travel, but also covers the new 136 kHz band with excellent results, but for reception only.

An increasing number of radio amateurs in the UK and several European mainland counties are now actively using a new VLF (Very Low Frequency) band at 136 kHz (135.7 – 137.8 kHz). The big challenge at these extremely long wavelengths is to make transmitter antennas with reasonable efficiency. In fact, anything greater than 1 per cent is considered a feat! As

to receiver antennas, the emphasis is on noise elimination. Distances of almost 2,000 km have been covered by amateurs using CW and modest transmitter powers (EI0CF – OH1TN, 2-way CW QSO on 137.2 kHz). To amateurs in the UK, the station DA0LF in Germany is a good 'DX target'. Lots of useful information on VLF Dx-ing may be found in *RadCom*, the magazine published by the Radio Society of Great Britain (RSGB). In the US, a group calling themselves 'Lowfers' has been active for many years collecting valuable information on the quirks of 'their' 1,750-metre band.

The 'feel' of the radio spectrum below 150 kHz or so (approx. 2,000 metres) is totally different from any-

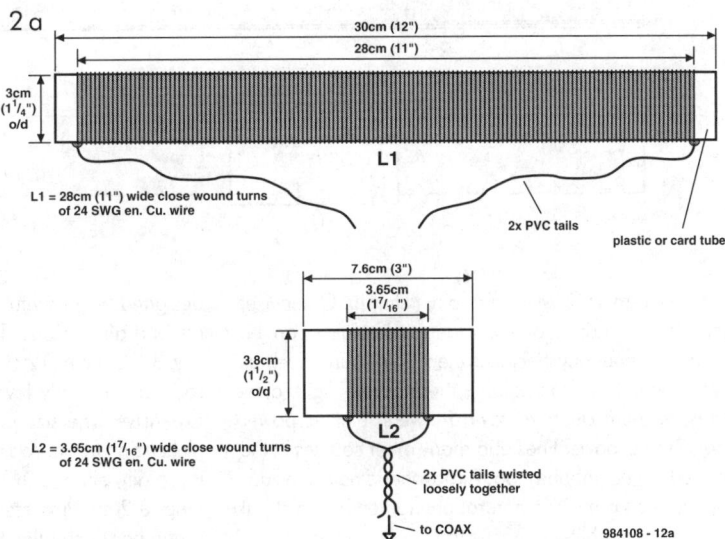

thing experienced on higher frequencies. Although there is a noticeable lack of AM broadcast stations, the most prominent feature is the high noise level which can, on occasion, have a devastating effect on reception.

For reception, a specially-designed type of receiving antenna will be found desirable, especially in urban areas where levels of man-made noise can be diabolical, especially when using a long-wire antenna. Noise experienced below about 150 kHz is either internal or external. Internal noise is either generated in the receiver, or enters via the AC mains power wiring. Remedial action can be taken. External noise is a different kettle of fish. Entering via the antenna, it is either man-made or atmospheric. Man-made noise, in its

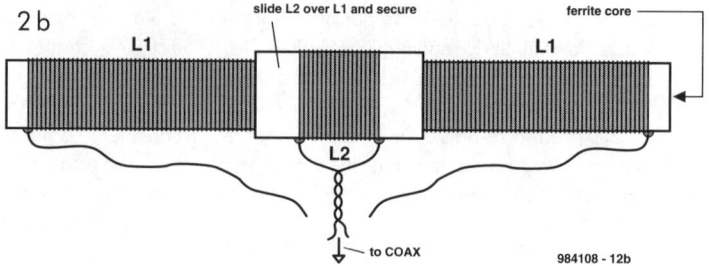

2 b

slide L2 over L1 and secure

ferrite core

L1

L1

L2

to COAX

984108 - 12b

Space to many amateurs is at a premium, and a multi-turn tuned frame loop antenna of, say, 1.2×1.2 m will be the absolute size limit. Despite all its obvious advantages, including sensitivity and good selectivity, such a directional antenna can be a cumbersome brute, and inconvenient to store when not in use.

An alternative is a ferrite-rod loop (loopstick), which in its original form is less sensitive than a frame loop. However a highly sensitive ferrite loop can be

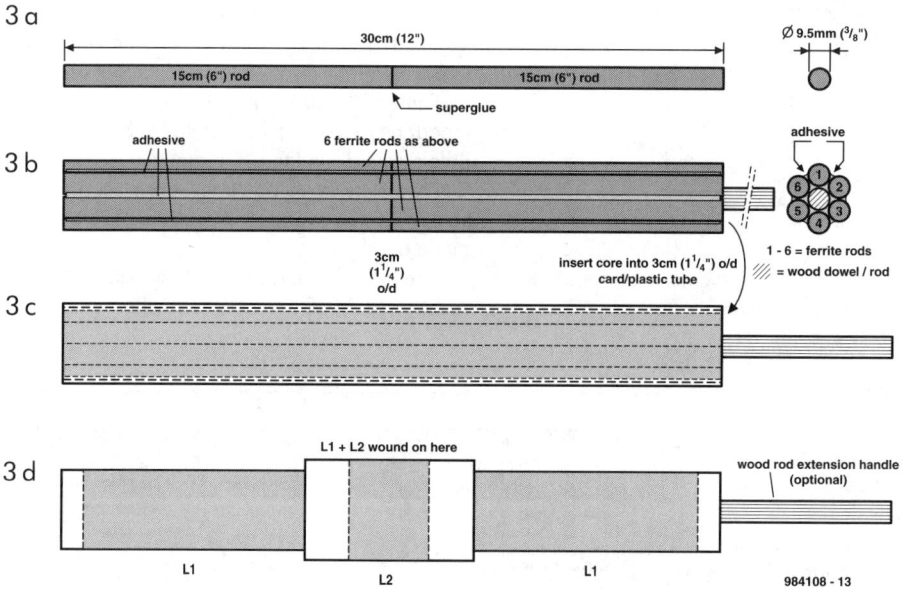

3 a

30cm (12")

15cm (6") rod

15cm (6") rod

Ø 9.5mm (³/₈")

superglue

3 b

adhesive

6 ferrite rods as above

adhesive

3cm
(1¹/₄")
o/d

insert core into 3cm (1¹/₄") o/d
card/plastic tube

1 - 6 = ferrite rods
= wood dowel / rod

3 c

3 d

L1 + L2 wound on here

wood rod extension handle
(optional)

L1

L2

L1

984108 - 13

simplest form, is QRM from another station. Otherwise, especially in urban areas, man-made noise can be from just about any electrical/electronic source such as TVs, computers, calculators, thermostats, light dimmers, vacuum cleaners, lawn mowers, electric power tools, traffic, power lines and many other sources in the immediate neighbourhood. Atmospheric noise, including electric storms, is a natural phenomenon, which, at its worst, can obliterate reception.

designed with careful selection of ferrite core materials and dimensions. The ferrite loop has to be large, say, 30×2.5 cm (12×1 inch). Unfortunately, such ferrite rods are not only few and far between but also very expensive. The size of affordable manganese-zinc or nickel-zinc ferrite rods stocked by radio parts retailers is usually either 20×1 cm or 20×1.25 cm. A 30 cm long, 3.2 cm diameter rod may be made by bundling a number of smaller zinc-nickel rods.

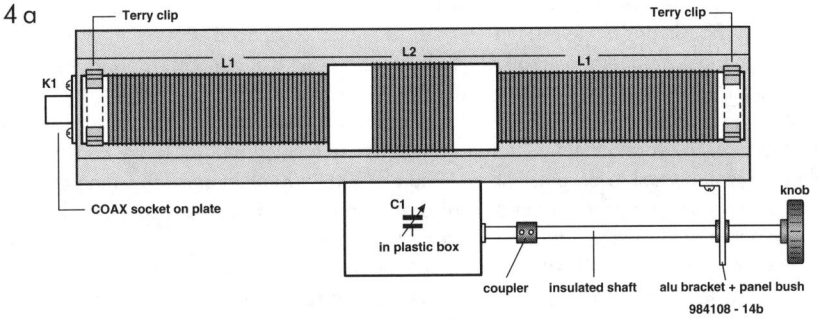

4 a

Terry clip — Terry clip —

L1   L2   L1

K1

COAX socket on plate

C1

in plastic box

coupler   insulated shaft   alu bracket + panel bush

knob

984108 - 14b

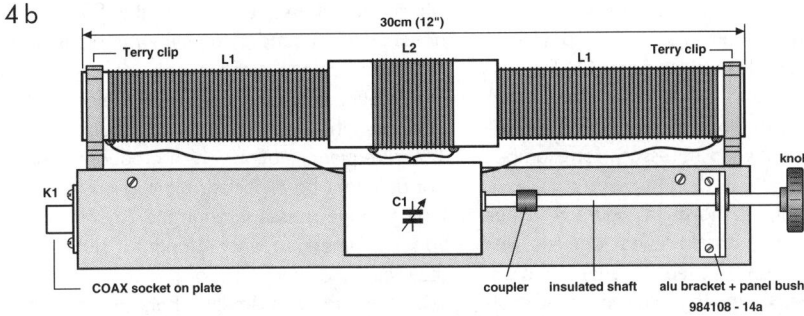

4 b

30cm (12")

Terry clip   L1   L2   L1   Terry clip

knob

K1

C1

COAX socket on plate

coupler   insulated shaft   alu bracket + panel bush

984108 - 14a

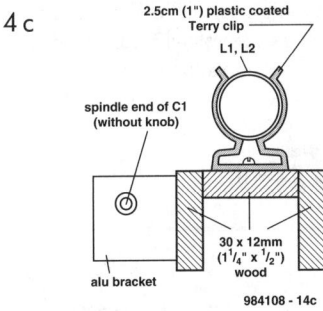

4 c

2.5cm (1") plastic coated
Terry clip

L1, L2

spindle end of C1
(without knob)

30 x 12mm
($1^1/_4$" x $^1/_2$")
wood

alu bracket

984108 - 14c

In the basic loopstick circuit shown in Figure 1, coil L_1 is brought to resonance by variable capacitor C_1. The core of L_1 is the earlier mentioned bundled loopstick. Coupling coil L_2, which provides the coaxial connection to the receiver or VLF up-converter, is wound over the centre of L_1. The coil assembly (Figure 2) is wound on a 30 cm long by 3.2 cm diameter cardboard tube (clingfilm tubing!). Coil L_1 has a width of 28 cm, and consists of 466 close-wound turns of 24 SWG (0.6 mm) enamelled copper wire (Figure 2a). The winding is terminated with lead-outs of PVC covered hook-up wire. In practice the winding ends may be held in place with a spot of superglue, with other spots every 2.5 cm or so along the winding.

This is necessary as winding the coil is a lengthy process. The coupling coil, L_2, is wound on a 7.6 cm (3 inch) length of cardboard tube of a diameter which just slips over L_1. It is a 3.6-cm (1/16 inch) wide close-wound winding of 24 SWG enamelled copper wire, terminated with hook-up wire leads lightly twisted together. The whole coil assembly is shown in Figure 2b.

The next step is to build the bundled ferrite core — this is illustrated in Figure 3. It consists of six 30 cm long, 1 cm diameter ferrite rods glued together to form one solid 30 cm by 3.2 cm (approx.) core. MMG F14 grade nickel-zinc material was used; an alternative is the US type 61 material. At these low frequencies, the difference in performance between the F14 and 61 materials is small. Each 30-cm rod is made from two 15-cm rods, secured end-to-end with superglue. The rod ends should first be lightly rubbed down with a fine glass paper in order to remove any grease, etc. (see Figure 3a). This technique effectively produces one long rod from two shorter ones. Other combinations of length could be used such as 20 + 10 cm, the 10-cm section being cut from a 20-cm rod using a hacksaw. In this way, three 20-cm rods would make two 30-cm rods.

The solid 30-cm long, 3.2 cm diameter ferrite rod consists of six 30 cm rods wrapped around a wood

267

dowel, and temporarily held in position with a couple of elastic bands, see Figures 3b, 3c and 3d. Next, the rods and dowel are adhered together to form one solid core, by cementation with a 15-minute setting adhesive such as Uhu. The adhesive is run along between all mating rod and dowel surfaces, by easing them gently apart with a thin blade. It is important to ensure that the surfaces have the adhesive between them. Several strong elastic bands are put around the rods, making sure that the circular rod formation is maintained. The assembly should be left in a warm place for at least 24 hours to make sure that the adhesive is thoroughly cured. The elastic bands are then cut away.

The core is then inserted into L_1, with any slight looseness eliminated with masking tape. On the prototype, the wood dowel is made a few centimetres longer than the rods, so that the core can be extracted from the coil if and when necessary, for example, for experiments.

The final assembly is shown in Figure 4. Three identical strips of wood are fastened together with wood screws to form an inverted 'U' shape chassis. Coil L_1-L_2 is mounted on the top with two plastic coated Terry clips fastened at the chassis ends. The twisted ends of L_2 are dropped through a hole drilled in the chassis top, and taken to the coaxial socket mounted on a piece of thin board, screwed to the chassis end.

The 800 pF tuning capacitor, C_1, is mounted on the chassis side as shown in Figure 4. On the prototype, a rigid air-spaced 800-P tuning capacitor (392+11+392+11 pF AM/FM type wired in parallel) is attached to the chassis side. It is fitted with a shaft coupler and an insulated shaft passing through a panel bush in a small bracket to the control knob. A 1000 pF (500+500 pF in parallel) tuning capacitor is also a good choice. The tuning capacitor is enclosed in a plastic dust-cover box. The ends of L_1 are taken to the tuning capacitor connections.

The tuning range of the prototype is 50–195 kHz. The 'Ultima' is used with a Palomar VLF up-converter whose output frequency is in the 80 m (3.5 MHz) amateur radio band. A simple turntable is an advantage to be able to turn the loop, which is directional. The frequency range was carefully selected. At the LF end is the 60 kHz MSF Rugby Standard Time/Frequency station, which produces a mighty signal as might be expected. Moving up frequency, the tuning range passes through the 73 kHz band where various Time/Frequency Standard stations can be received, and much else of interest. Next comes the 2,000-m European Amateur band around 136 kHz.

Compared with a traditional 20×1.25 cm diameter loopstick, the present design gives appreciably increased signal strength, with atmospheric and man-made noise being either eliminated, or reduced to an acceptable level by simple loop rotation.

Video amplifier

The video amplifier is a well-known design. Simple, yet very useful, were it not for the ease with which the transistors can be damaged if the potentiometers (black level and signal amplitude) are in their extreme position. Fortunately, this can be obviated by the addition of two resistors.

If in the diagram R_3 and R_4 were direct connections, as in the original design, and P_1 were fully clockwise and P_2 fully anticlockwise, such a large base current would flow through T_1 that this transistor would give up the ghost. Moreover, with the wiper of P_2 at earth level, the base current of T_2 would be dangerously high. Resistors R_3 and R_4 are sufficient protection against such mishaps, since they limit the base currents to a level of not more than 5 mA.

Shunt capacitor C_4 prevents R_4 having an adverse effect on the amplification.

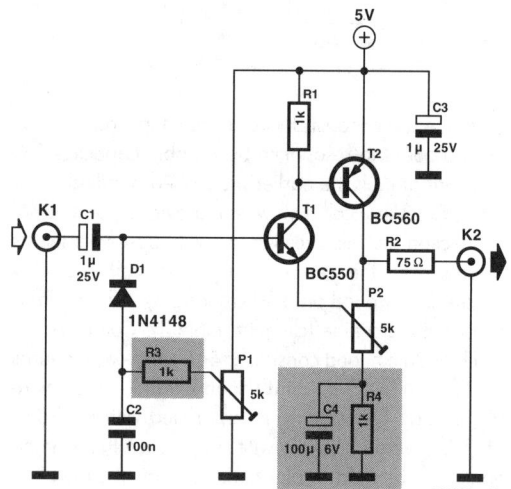

984066 - 11

Dark-level clamp

In video processing, it is often desirable to hold the dark level of the video signal at a defined voltage level. In the clamp, this is earth level. The circuit complements the video contrast expander elsewhere in this issue. The expander cannot easily handle the situation when the sync and dark levels are dissimilar.

The clamp may also be used to advantage with a video mixer or video fader. Merely adding a potentiometer and a video switch with correct timing is sufficient to obtain the desired effects. The output signals of the sync separator used in the clamp may often prove very useful.

After it has been decoupled by C_1 and C_2, the video signal is applied to buffer/ amplifier IC_2. This is an op amp with a slew rate of 300 V μs^{-1} and a unity-gain bandwidth of 50 MHz.

The voltage level between the colour burst and a given video line is sampled via T_1 and C_3. This is the back porch which usually provides a reference level of 5.8 μs. The back porch appears at the output of IC_2 and is then at earth level.

Since the output of IC_1 forms part of the feedback loop of IC_2, it provides high amplification, which means that the value of R_6 can be fairly high. This lessens the effect on the video signal.

The normal video signal is amplified ×2, and the potential at C_3 ×–2. The amplification factor ensures the correct signal across the terminating impedance of 75 Ω.

Analysing the video signal is effected by sync separator IC_4. Since the pulse at the burst output is too wide (typically 4 μs) for the present application, IC_{3a} (which provides the control pulse for T_1) cannot be triggered at the trailing edge of the signal. The sample pulse would then arrive too late and might be taken from the video signal. Therefore, IC_{3b} is triggered at the leading edge of the burst signal (3.6 μs).

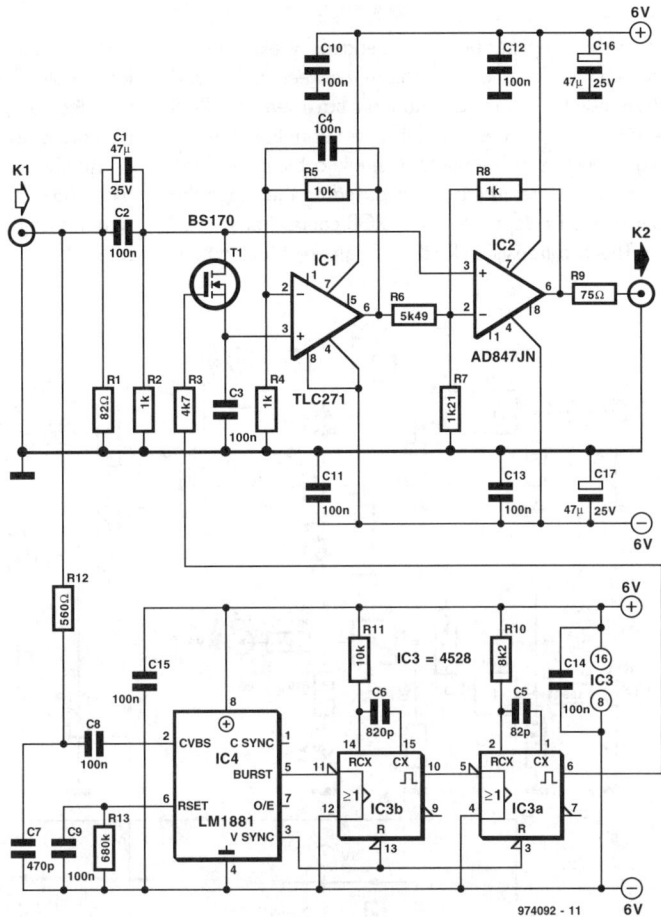

The output of IC_{3b} is used to start IC_{3a} (0.6 μs), whereupon it is possible to determine where the samples are taken by giving C_5 and C_6 appropriate values. In the present design this is 4 μs after the frame-sync pulse.

Since the burst output of IC_4 also generates pulses during the frame synchronization, which may occur at an awkward moment and thus create contradictory situations, sampling during the vertical synchronization is temporarily stopped by resetting IC_3 with the signal from the vertical sync output.

Resistor R_3 prevents any glitches caused by the switching of T_1 from penetrating the video signal.

The clamps draws a current of about 20 mA.

974092 - 11

RGB video amplifier

The video amplifier board is intended for experimenting with RGB video connections between a PC and VGA monitor. Many VGA monitors have separate RGB V/Hsync inputs besides the more familiar 15-way high-density sub-D input for a single cable connection to the VGA card. The circuit is based on the use of the better quality: separate (coax) RGB connections.

The two-transistor RGB (red, green, blue) ampli-

fiers are identical, each containing adjustment points for the black (reference) level and the signal level. In the R(ed) amplifier, for example, the respective controls are presets P_8 and P_7.

Another, similar, amplifier, T_8-T_7, supplies a combined (G+CSYNC) signal. The CSYNC portion of this signal is adjusted to individual requirements with preset P_2.

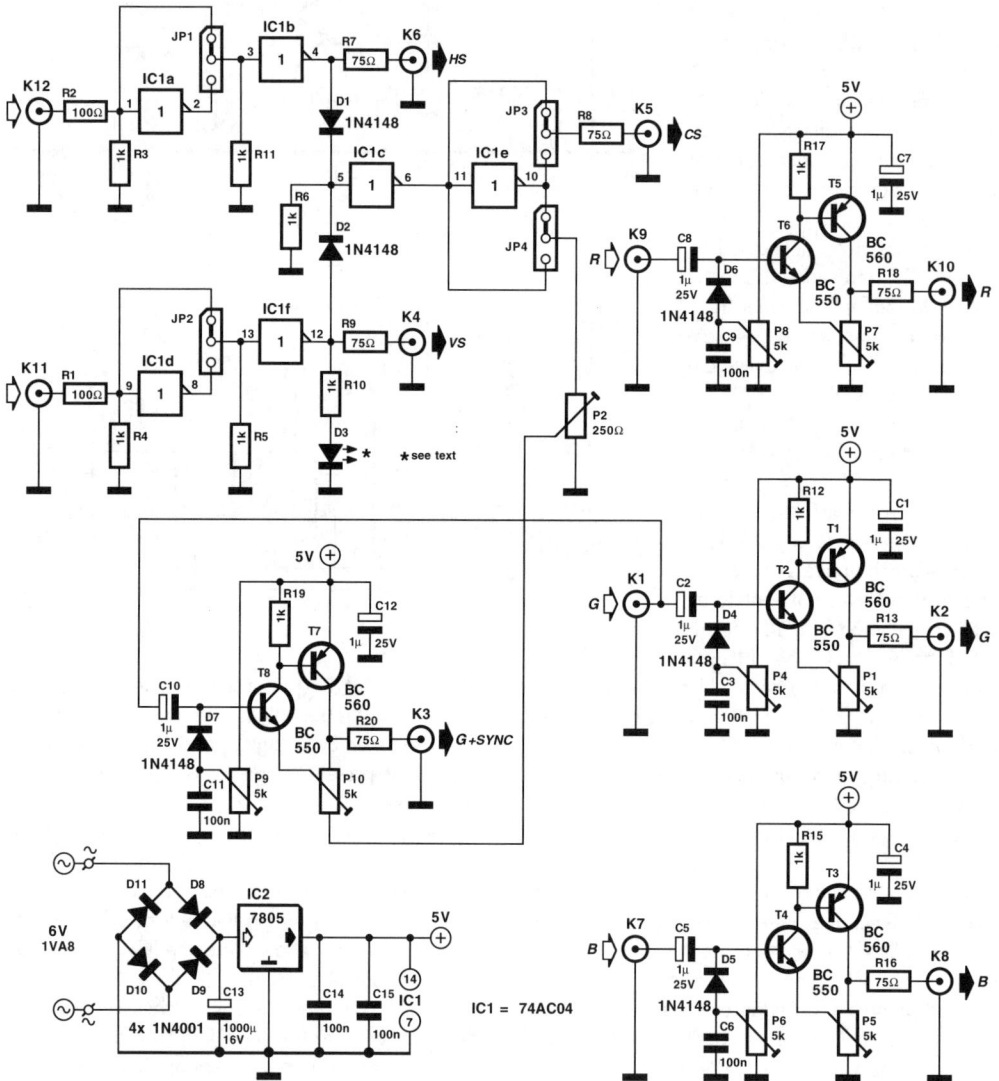

974042 - 11

The RGB and (G+CSYNC) amplifiers have 75 Ω output resistors to ensure a good match to coax cable. Their drive capacity is such that relatively long (up to 3 metres) coax cables may be used without causing bandwidth reduction problems.

Jumpers JP_1 and JP_2 enable the HS (horizontal sync) and VS (vertical sync) signals to be output in inverted or true form as required by the monitor (RTFM). The VS and HS signals are also combined by diodes D_1 and D_2 to create a composite sync (CS) signal. This, too, is available in true and inverted form on socket K_5, the polarity selection being made with jumper JP_3. The output impedance of the CSYNC output is 75W. The brightness of D_3 indicates the polarity of the VS signal: bright = negative VS; weak = positive VS. Jumper JP_4, selects between true or inverted CSYNC for use in the (G+CSYNC) adder, T_7-T_8.

The amplifier board has its own power supply consisting of rectifier diodes D_8-D_{11}, smoothing capacitor C_{13} and regulator IC_2. The board may be powered by a small 6-volt mains transformer.

Parts list

Resistors:
R_1, R_2 = 100 Ω
R_3–R_6, R_{10}, R_{11}, R_{12}, R_{15}, R_{17}, R_{19} = 1 kΩ
R_7, R_8, R_9, R_{13}, R_{16}, R_{18}, R_{20} = 75 Ω
P_1, P_4–P_9 = 5 kΩ, multiturn, vertical
P_2 = 250 Ω. multiturn, vertical
P_{10} = 500 Ω, multiturn, vertical

Capacitors:
C_1, C_2, C_4, C_5, C_7, C_8, C_{10}, C_{12} = 1 µF, 25 V, radial
C_3, C_6, C_9, C_{11}, C_{14}, C_{15} = 0.1 µF
C_{13} = 1000 µF, 16 V, radial

Semiconductors:
D_1, D_2, D_4–D_7 = 1N4148

D_3 = LED
D_8–D_{11} = 1N4001
T_1, T_3, T_5, T_7 = BC560C
T_2, T_4, T_6, T_8 = BC550C

Integrated circuits:
IC_1 = 74AC04
IC_2 = 7805

Miscellaneous:
JP_1–JP_4 = 3-way, 2.54 mm pin strip and pin jumper
K_1–K_{12} = audio socket, PCB mount
PCB may be made with the aid of the track layout.

974042-1

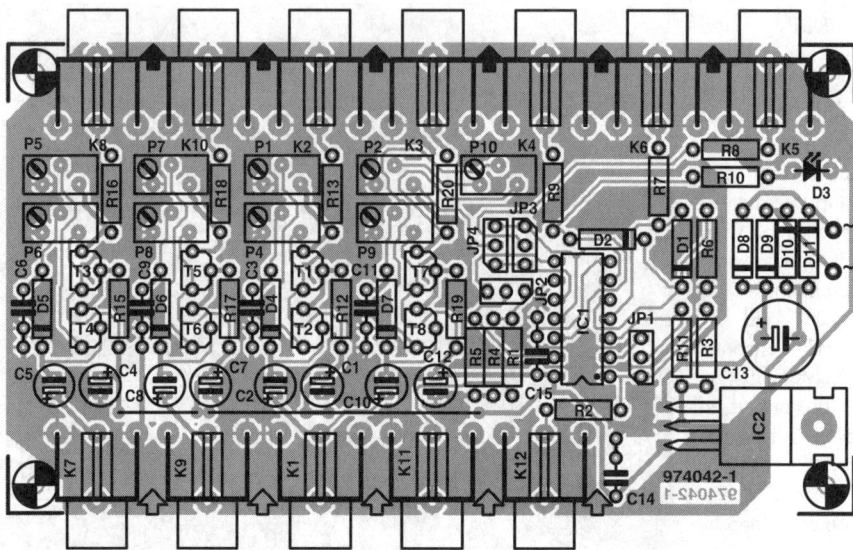

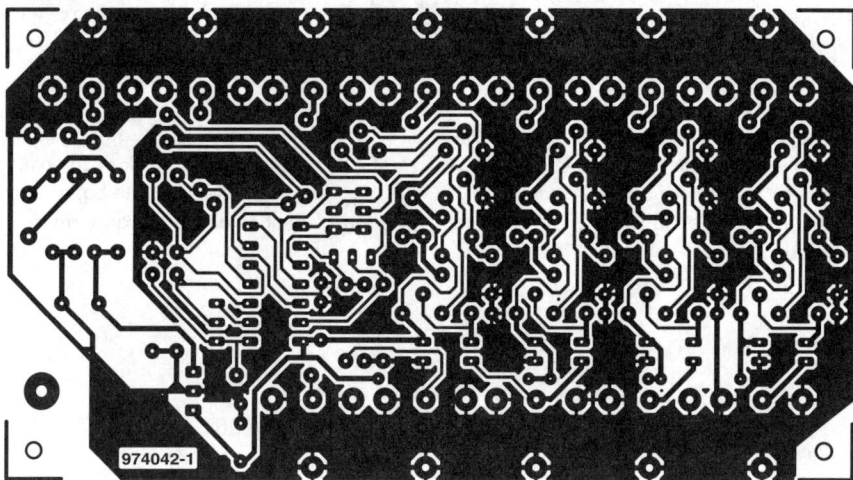

Video-contrast expander

It may happen that a video recording is a little too dark so that certain nuances disappear and the picture is no longer clear. The expander may rectify this to some extent by increasing the contrast in the dark passages. Provided that the circuit is set up correctly, the nominal black and white levels are not affected.

The circuit has four calibration points, which make the use of an oscilloscope a must. It is, of course, important that the existing black and white levels are retained and that the synchronization remains fully functional.

The circuit has a few drawbacks: (a) owing to the added amplification, the level of the colour burst changes which requires the saturation to be readjusted; (b) the contrast in bright images diminishes; and (c) there is a risk that when the dark levels are amplified too much noise becomes visible.

The input signal at K_1 is decoupled by C_1, C_2 and R_2 and then amplified by IC_{1a}. Diode D_1, in conjunction with R_4 and P_1, ensures that the earth level is used as reference for the black level. The output level is set with P_2.

Diode D_3 in series with P_3 in the feedback loop of IC_{1b} holds the white level at 100%. This ensures that small signal levels (dark levels) are amplified in accordance with the setting of P_4, while larger signals are also affected by the setting of P_3.

Diode D_2 limits the level of the sync signals which, owing to the chosen amplification, may become too high.

Experimenters may replace D_3 by one or two Type BAT85 diodes or a simple germanium diode, which, of course, changes the operating characteristic of the circuit.

Note that the signal input must give a level of $1 V_{pp}$ across 75Ω – no more, no less. Remember that 30% of the available space must be reserved for the sync signals.

The circuit draws a current of ± 15 mA.

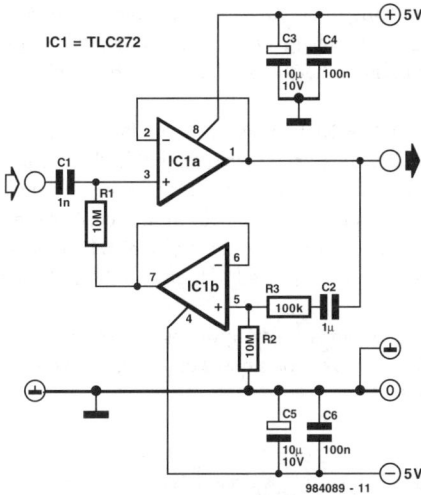

Wideband VHF preamplifier

The VHF preamplifier uses the BF324 (TO92 case) p-n-p transistor in a grounded-base configuration. The circuit may be used as a signal booster with VHF receivers whose front end suffers from low sensitivity (such as many valved and army surplus types). The frequency range of the preamplifier is roughly 75–150MHz.

The two inductors in the circuit are home made: L_1 consists of 10turns of 24 SWG enamelled copper wire; the internal diameter is 3mm, air core. Inductor L_2 has 13 turns of the same wire, and an internal diameter of 5mm; air core also. A construction tip: close-wind the inductors around 3 mm and 5mm drill bits respectively as temporary formers.

The prototype of the preamplifier is used successfully with an 88-108 MHz FM broadcast receiver and a 2 metre VHF ham receiver.

The preamplifier draws a current of about 2.5mA from a 5 V supply.

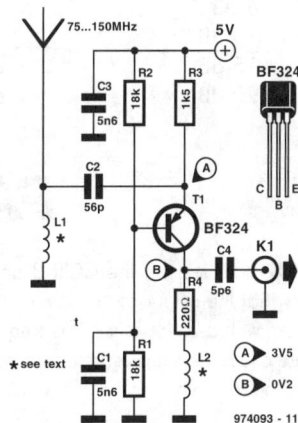

273

Matching attenuator

When r.f. signals are (to be) attenuated, it is essential that the requisite network retains correct matching to

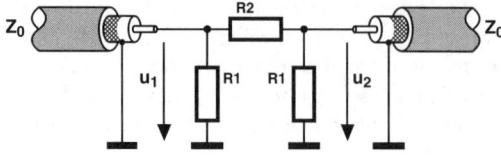

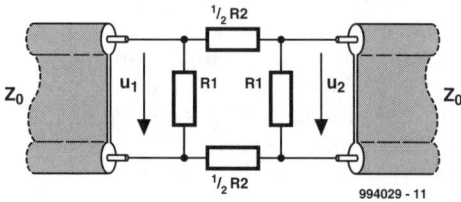

the relevant (coaxial) cable(s). If this is not so, reflected signal waves will ensue along the cable(s), which may attenuate or magnify the forward signal waves. In other words, there will be points along the cable(s) where the resulting signal is much smaller than the original and others where the resulting signal is much larger.

With the attenuators show in the diagram, the cables are correctly terminated, that is, there is proper matching. If the link is via balanced cables, the network must also be balanced. In that case, there is a series resistor in both signal lines that is half the value of the series resistor used with an unbalanced connection.

The formulas below give resistor values for both 50 Ω and 75 Ω cables and wanted attenuation as listed in the table.

$$R1 = Z_0 \cdot \frac{A+1}{A-1} \qquad R2 = Z_0 \cdot \frac{A^2-1}{2A} \qquad A = \frac{U_1}{U_2} = 10^{\left(\frac{-a}{20\,\text{dB}}\right)}$$

	50 Ω		75 Ω	
Attenuation (A)	R_1 (Ω)	R_2 (Ω)	R_1 (Ω)	R_2 (Ω)
1 dB	909	5.62	1300	4.32
2 dB	475	10	619	18.2
3 dB	274	18.2	432	27.4
6 dB	150	35.7	221	56.2
10 dB	100	68.1	150	100
15 dB	68.1	150	110	200
20 dB	61.9	243	90.9	392

PAL timing

In the PAL television system, the CCIR B and G standards specify that the colour carrier is directly coupled to the line rate, with a 25 Hz offset. The frequency ratio and offset are chosen to suppress interference patterns, according to the formula

$$f_{colour} = 283.75\, f_{line} + 25\ \text{Hz}$$

At a line rate of 15,625 Hz, this means that the PAL colour carrier frequency is 4.43361875 MHz. Single-sideband modulation is frequently used to obtain the

correct relationship with the line rate. For example, the frequency of a crystal oscillator can be offset by 25 Hz, divided by 1135 and then multiplied by 8 to obtain twice the actual line rate. This is a rather complicated procedure, which we think could be made a lot simpler.

There is a fixed ratio between the 25 Hz frame rate and four times the colour carrier frequency. This is calculated as follows: four times the colour carrier frequency is exactly equal to 709 379 times the frame rate! An obvious approach is to use a crystal oscillator running at four times the colour carrier frequency and

IC1 = 74HCU04
IC2 = 74HC74
IC6 = 74HC32

R1 1M
IC1a 1 — IC1b 1
X1 15p
C4
C1 22p C2 27p C3 47p
X1 = 17.734475MHz

IC1d 8 1 9
IC1e 10 1 11
IC1f 12 1 13
IC6c 8 ≥1 9 10
IC6d 11 ≥1 12 13
5V

IC2a D C R S
IC2b D C R S 4.43361875MHz
5V

IC3 CTR8 1,2,4– 2C3 CT=255 C3/G4 G2 EN1 3CT 1CT=0 74HC40103
IC4 CTR8 1,2,4– 2C3 CT=255 C3/G4 G2 EN1 3CT 1CT=0 74HC40103
IC5 CTR8 1,2,4– 2C3 CT=255 C3/G4 G2 EN1 3CT 1CT=0 74HC40103

60 ns
IC6a ≥1 25Hz

+5V
C5 (14) IC1 (7) 100n
C6 (14) IC2 (7) 100n
C7 (16) IC3 (8) 100n
C8 (16) IC4 (8) 100n
C9 (16) IC5 (8) 100n
C10 (14) IC6 (7) 100n

IC6b 6 ≥1 4 5
IC1c 6 1 5

994086 - 11

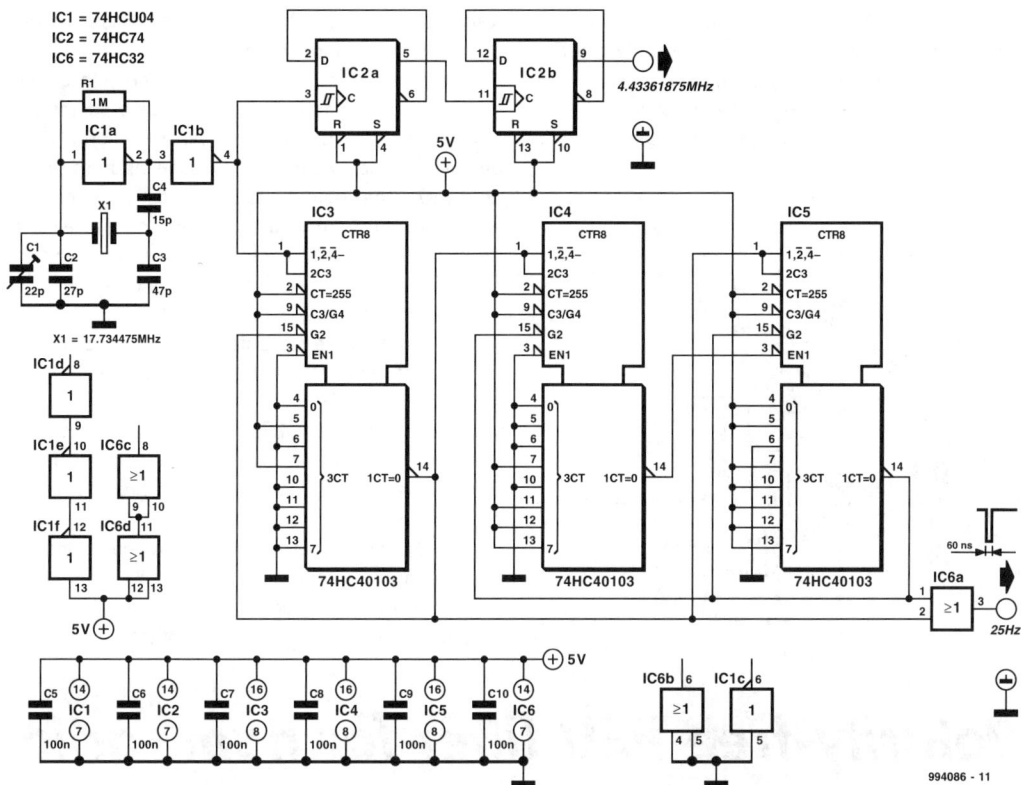

divide its output by 709 379 to obtain the frame rate. The line rate can then be derived from the frame rate with the help of a PLL circuit (see next article).

The crystal oscillator is a standard Pierce configuration with a trimmer capacitor, built around a 74HCU04 (IC1). The values of C_2 and C_3 must be chosen carefully to obtain the specified load capacitance for the crystal. An incorrect value of C_{load} can make it impossible to adjust the oscillator to the exact frequency. Two D-type bistables arranged as binary scalers are used to obtain the colour carrier frequency.

Four ICs are used for the division needed to obtain the frame rate signal. Circuits IC_3, IC_4 and IC_5 are presettable synchronous down-counters (type 74HC40103) that are very well suited for timing and frequency-division applications. The requisite divisor is split into two factors, namely 11 and 64 489. The 74HC40103 works as an (1+N) divider, so a value of 10 is applied to the preset inputs of IC_3 for the first factor. The second factor is obtained by wiring IC_4 and IC_5 as synchronous 16-bit dividers, with the output of IC_5 fed back to both synchronous preset inputs. The preset value is again 1 less than the division factor.

A disadvantage of the 74HC40103 is that glitches can occur owing to differences in internal delay times. These glitches are eliminated by clocking OR gate IC_{6a} at the divider output with the divider input signal. The 25 Hz output signal has an active low pulse approximately 60 ns long, which is roughly equal to one period of the crystal oscillator.

The current consumption of the circuit is slightly higher than 12 mA, mainly due to IC_1.

Active rod antenna

In r.f. reception, the length of a rod antenna does not affect the signal-to-noise ration, provided that the noise of the receiver does not exceed the received noise. It is, therefore, possible to obtain good reception results with a telescopic rod antenna over a frequency range of 10 kHz to 30 MHz.

However, the antenna must, of course, be matched correctly to the receiver. The shorter the antenna, the larger its reactance. This reactance may be as high as some kilohms. Since the sum of the radiation resistance and reactance must be about equal to the input resistance of 50–75 Ω, measures to ensure correct matching are invariably needed.

The present circuit is intended to provide correct matching. Transistor T_1 is arranged as a source follower, which has a high input resistance and a low output resistance. Resistor R_1 fixes the input resistance at 1 MΩ. The output resistance depends on the transconductance of T_1 and the value of source resistor R_2. Inductor L_1 increases the source impedance of T_1 at high frequencies. Diodes D_1 and D_2 limit the signal voltage to +12.6 V and –0.6 V.

The length of the rod antenna may be 0.5–1 m.

The reception range extends from 10 kHz to 100 MHz.

The current drain is relatively high (20–30 mA), so that power must be provided by a mains converter or drawn from the receiver. The use of batteries is not recommended.

The connection between the rod antenna and the gat of T_1 must be kept as short as possible. The unbalanced 50 Ω or 75 Ω coaxial lead between the antenna amplifier and receiver may be rather longer.

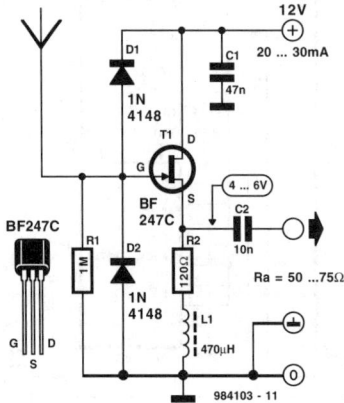

Polarity-free *PSU* filter for ham radio

Many radio amateurs are well aware of the chaotic situations that may occur during field days or contests, when several radios have to be connected in a hurry and under circumstances less comfortable than those in the shack at home. For instance, many operators may be busy at the same time connecting up the power leads to equipment which is not theirs. In such situations, supply polarity errors may readily occur, with disastrous results.

Many currently available handhelds from Sony, Yaesu, Standard, Kenwood, Alinco and other makes may be powered from an external vehicle battery. However, the supply polarity on the external power connector may not always be known or easily found out when a hectic situation arises (traditionally, few hams have the Owners Manual available...).

The filter was developed to allow handheld rigs to be connected to an external 12V vehicle battery without paying attention to polarity. This function is due to a bridge rectifier, D_1–D_4, at the input of the circuit. Irrespective of the battery polarity, the radio will always receive the correct supply voltage.

Additional functions of the circuit include an effective noise filter L_1-L_2-C_4-C_5, high-voltage DC protection (zener diode D_7), and blown fuse/power indicators D_6 and D_5 respectively. Inductors L_1 and L_2 consist of 8 turns of 24 SWG (0.6 mm) enamelled copper wire on a large ferrite toroidal core from Amidon's T series (check coil saturation specification!). Alternatively, use EMI suppression beads Type 4330 020 3326 from Philips Components. The LEDs should be high-efficiency types.

The filter as shown may be used with any modern handheld that draws less than 2 A at supply voltages between 4.5 V and 14 V. In fact, most of these rigs will draw 1.3-1.5 A at 13.8 V for 5 watts of RF output power.

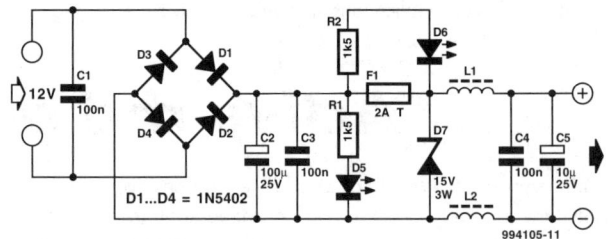

7

Test & Measurement

12/24/48 V d.c. tester

The tester is intended primarily for testing the 24 V electrical circuits found on most pleasure craft. However, if the resistors are given different values, the circuit may, of course, be used for other voltage ranges. For 12 V, the value of the resistors should be 1.2 kΩ, and for 48 V, 4.7 kΩ.

The tester should be connected to the +ve and −ve voltage rails with test clips or crocodile clips, whereupon the test probe is placed on the point to be tested. When the potential at the point is positive, the red LED lights; if it is negative, the green one does.

If the supply is not connected to earth, the tester may be used as ground-leak tester. In this situation, one of the LEDs lights when the test probe touches a point at earth potential and there is a leakage.

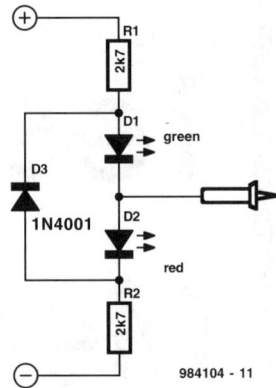

FET probe

Reliable measurement of a signal is possible only if the circuit in which the measurement is carried out is not loaded by the measurement instrument. The higher the input impedance of the instrument, the closer the ideal measurement is approached. The proposed probe may be used to increase the input impedance to about 10 MΩ.

A field-effect transistor (FET) is used to design a high-impedance voltage follower. In this circuit, R_1 determines the input impedance. The resistor is shunted by a parasitic capacitance of 3 pF. The output impedance depends on T_1 and R_4: with values as specified in the diagram, it is about 65 Ω.

The potential at the second gate (U_{g2s}) is set with P_1 such that the d.c. offset at the output is 0 V.

Unfortunately, a small price has to be paid for the simplicity of the circuit: since the overall amplification is ×0.8, the value measured by the oscilloscope must be corrected as appropriate.

The bandwidth of the probe is ≥ 15 MHz.

The probe draws a current of about 10 mA.

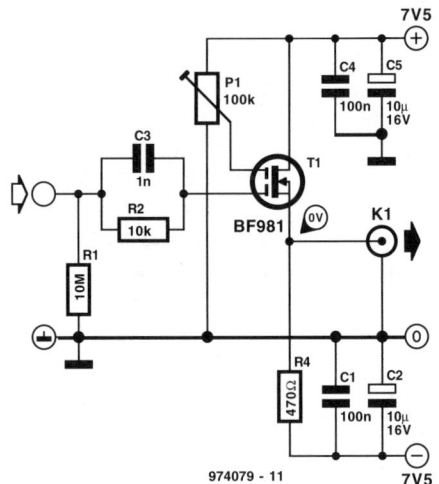

Hygrometer

A hygrometer is a device for measuring, or giving an output signal proportional to, ambient humidity. It is, therefore, very suitable for switching on a ventilator or dehumidifier in spaces such as a bathroom or kitchen where the humidity at certain times can reach uncomfortable or unacceptable levels.

Normally, hygrometers use a hygristor as sensor, but the present circuit uses a capacitor whose capacitance is dependent on the degree of humidity.

With values as specified in the diagram, the frequency of the output signal of oscillator IC_2 varies from 30 kHz in dry conditions to 25 kHz when the ambient humidity is 100%.

The output of the oscillator is applied to retriggerable monostable multivibrator (MMV) IC_{1b}, whose Q output remains high as long as the oscillator frequency is high.

When the humidity rises, the oscillator frequency drops and short pulses appear at the Q output of IC_{1b}. These trigger MMV IC_{1a}, whose output then goes high, whereupon the electro-optical relay trips. Resistor R_4 provides some hysteresis to prevent relay clatter.

The humidity at which the relay should trip is set with P_1. The desired hysteresis is set with P_2.

The mono time of IC_{1a} is set to 30 ms, which is more than ample to switch on T_1.

The electro-optical relay is a Type S201S02 from Sharp, which can switch loads up to 1 A.

The circuit draws a current of about 25 mA.

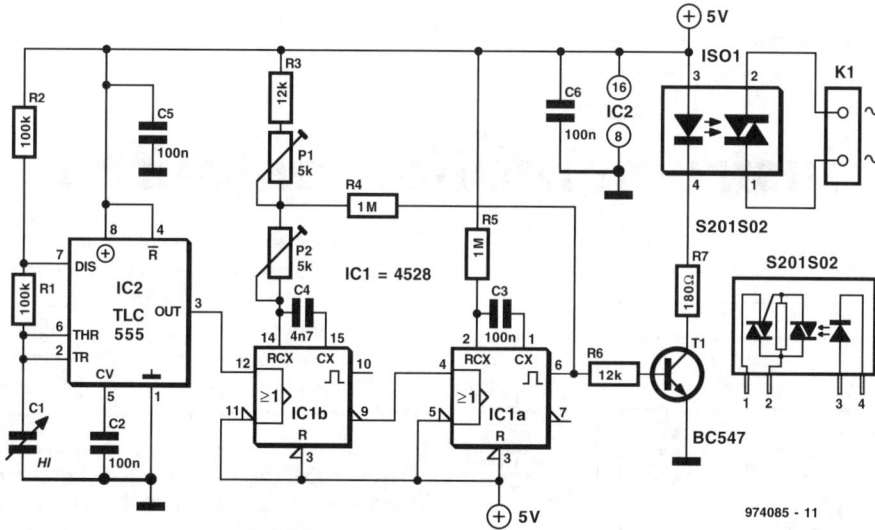

974085 - 11

Infra-red-illuminance meter

When a photodiode is illuminated, it produces a significant photocurrent whose value depends on the level of illuminance*.

When the drop across a photodiode is measured, it appears that this is normally ≤ 500 mV. This voltage is not, or hardly, dependent on the photocurrent. If therefore a low-value resistor is placed in parallel with the diode, the drop across the parallel combination remains below the diode (forward) voltage. The potential across the resistor is then directly proportional to the illuminance.

A high-impedance digital ammeter set to its lowest d.c. µA range will show that the diode voltage hardly varies with the light intensity. If a suitable d.c. µA range is not available on the meter, connect a resistor of a few kΩ across the meter input.

In the diagram, the shunt resistance of the diodes is formed by a moving-coil ammeter with 30 µA full-scale deflection. Its internal resistance is 6.5 kΩ.

Calibration of the circuit and gradation of the scale must be carried out with the aid of a light source of well-defined luminous intensity. Even without calibration the meter may be used to compare the emission of an infra-red headphone transmitter with that of incident daylight. It may also be used to verify whether a remote controller is still working properly.

The prototype is built with photodiodes Type BP104, which appear to have the best sensitivity of a number of different types. With five photodiodes in parallel as in the diagram, the infra-red light from a remote controller at a distance of 100 mm from the meter gives a reasonably clear deflection of the pointer.

If light sources emitting light of different frequency are to be checked, appropriate photodiodes must, of course, be substituted for the BP104s.

Note that the meter does not need a power supply.

*Illumination = illuminance = the quantity of light or luminous flux falling on unit area of a surface.

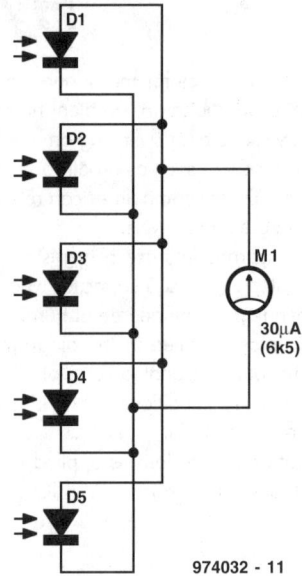

974032 - 11

Instrumentation amplifier

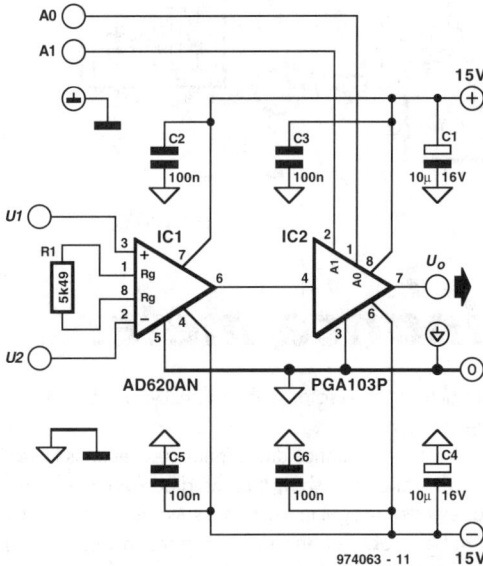

974063 - 11

The broad-band instrumentation amplifier may be used for a number of applications. It has balanced inputs and a variable amplification factor.

The design is based on two op amps, of which IC_1 is the actual instrumentation amplifier. Its amplification factor is determined by R_1 and is here ×10.

Op amp IC_2 is a programmable gain amplifier whose amplification is ×1, ×10, or ×100. The amplification depends on the level at pins A_0 and A_1. When both pins are strapped to earth, the amplification is unity. When A_0 is linked to +15 V and A_1 to earth, the amplification is ×10. When A_1 is connected to +15 V and A_0 to earth, the amplification is ×100. The overall amplification may thus be set to ×10, ×100 or ×1000.

Building the amplifier is simple and is best done on a piece of prototyping board. A good earthed link between pins 5 of IC_1 and 3 of IC_2 is pivotal.

Power requirements are ±15 V at a current of 10 mA.

The bandwidth of the amplifier is 250 kHz.

The common-mode rejection is 95 dB up to 1 kHz.

Li-ion battery capacity meter

The meter exploits the fact that the remaining capacity of a Lithium-ion (Li-ion) rechargeable battery is virtually proportional to the battery voltage.

When discharged at constant current, a single Li-ion cell is marked by an almost linear voltage decrease from about 4.1 V (fully charged) down to about 3.5 V (10% charge left). So, all that is needed to measure the battery capacity is an accurately defined, small, voltage window with a span of 3.5–4.1 V, and a calibrated voltmeter or equivalent circuit providing a percentage readout.

Li-ion batteries usually come in one of three shapes: single-cell (4.1 V), dual-cell (8.2 V) and three-cell (12.3 V). The indicated voltages apply to fully loaded batteries. The 12.3 V block is a particularly popular type as it is often used in camcorders. The BT-L1 block as used in Sharp camcorders is probably the best known. The present tester is suitable for all three Li-ion battery types mentioned above.

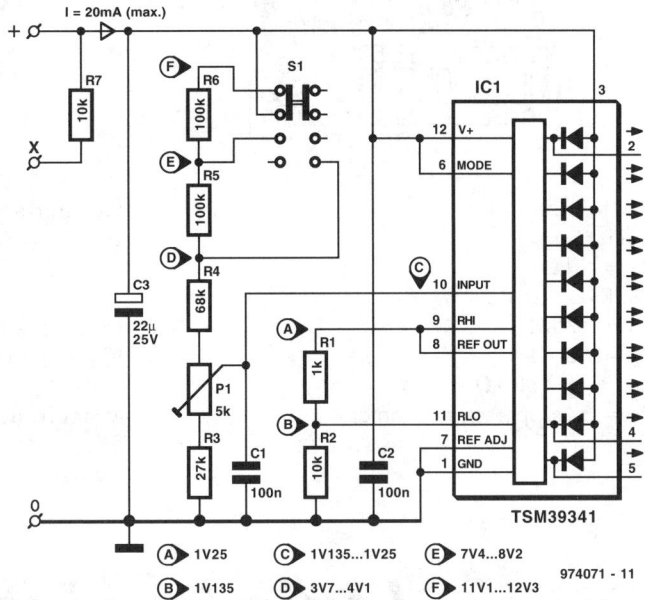

The number of cells of the battery to be tested is set with slide switch S_1. A conventional resistor ladder is used to reduce the battery voltage to a level which is suitable for applying to the input of IC_1, an integrated analogue-to-digital converter (ADC) with direct LED drive outputs. Although the operation of the TSM39341 is similar to that of the familiar LM3914, a major difference is the LED array which is internal to the chip.

The TSM39341 is wired to drive the ten LEDs in 'bar' mode (as opposed to 'dot' mode). The LED drive current is set to about 1.3 mA per LED by R_1 and R_2.

Resistor R_7 is necessary to flag 'all-clear' to the output protection circuit built in the Sharp BT-L1 battery.

The circuit is simple to adjust: connect a fully charged Li-ion battery, set the slide switch to the appropriate range (4.1 V, 8.2 V or 12.3 V), and adjust P_1 until the '100%' LED just lights.

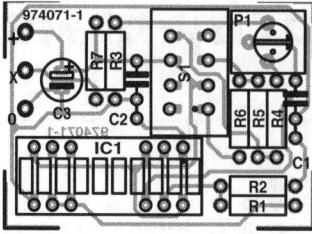

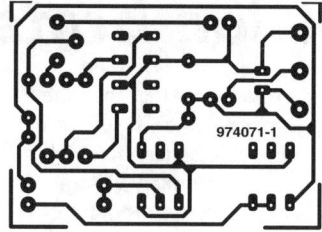

Parts list

Resistors:
$R_1 = 1$ kΩ
$R_2, R_7 = 10$ kΩ
$R_3 = 27$ kΩ
$R_4 = 68$ kΩ
$R_5, R_6 = 100$ kΩ
$P_1 = 5$ kΩ, preset, horizontal

Capacitors:
$C_1, C_2 = 0.1$ μF
$C_3 = 22$ μF, 25 V, radial

Integrated circuits:
$IC_1 = $ TSM39341

Miscellaneous:
$S_1 = $ slide switch, PCB mount, 3 positions, 2 rows of 4 contacts.

Logic-level tester

The tester provides 16 LEDs arranged on a printed-circuit board in the shape of a 16-pin dual-in-line (DIL) package. Each of the LEDs indicates the logic level present at the corresponding pin of the logic IC whose function is to be examined. This is achieved by placing a 16-way test clamp on to the IC on test, and feeding the logic levels to the present tester via a 20-way flatcable. Obviously, as the (corner) supply pins of the IC on test are also included, the relevant indicator LED will show a permanent logic 0 for the V_{ss} or GND pin,.7 or 8, and a logic 1 for the V_{dd}/V_{cc} pin, 14 or 16.

The circuit consists of two clocked octal bistables/latches type 74HCT574. Each latch drives a LED via a current limiting resistor (R_{19}–R_{34}). The set of logic levels at the latch inputs is refreshed with the aid of a com-

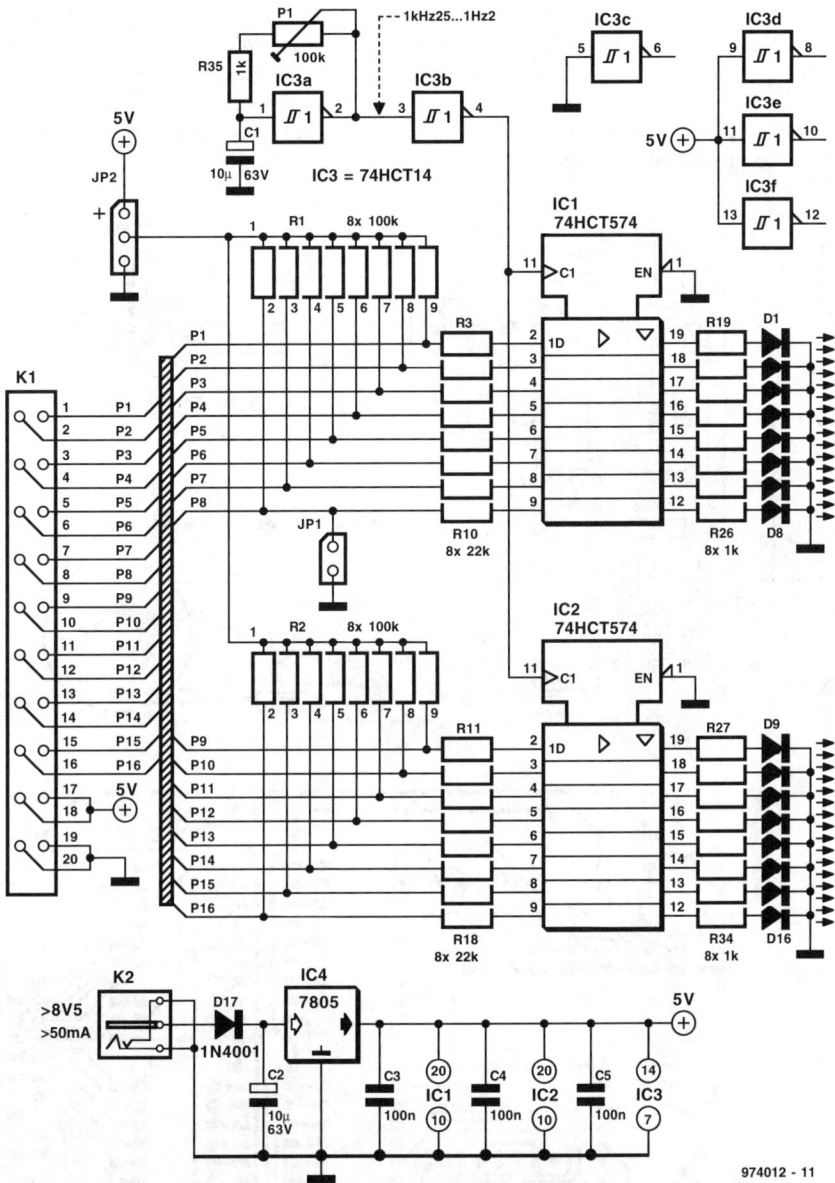

```
974012 - 11
```

mon clock signal supplied by a two-gate free-running *RC* oscillator built around IC$_{3a}$ and IC$_{3b}$. Potentiometer P$_1$ allows the refresh frequency to be adjusted to any value between about 1.2 Hz and 1250 Hz. The 'display refresh' gives the user an idea about the functioning of a suspect IC, and allows him/her to draw up elementary truth tables.

Apart from the 16 logic levels, the flatcable between the tester and the circuit on test also car-ries the +5 V supply voltage (pins 17, 18) and ground (pins 19, 20). The relevant four wires are separated from those connected to the 16-way DIL clamp. Normally, a flexible ground wire is taken from wires 19/20 and connected to ground of the circuit on test, while the +5 V wires are not used. Alternatively, the tester may power the circuit on test via the +5 V wires, but only if it is capable of sourcing the requisite current.

283

As pin 8 of the IC on test will usually be at ground potential, the corresponding input line may be tied to ground permanently with the aid of jumper JP1.

The common junctions of resistor arrays R_1 and R_2 may be tied to ground or +5 V, depending on the type of logic circuit being tested: For TTL, set JP_2 to +5 V; for CMOS, set JP_2 to $\perp$.

The tester draws a current $\geq$ 50 mA with all LEDs on. The direct voltage applied to power socket K_2 must be 8.5–12 V from a small mains adaptor.

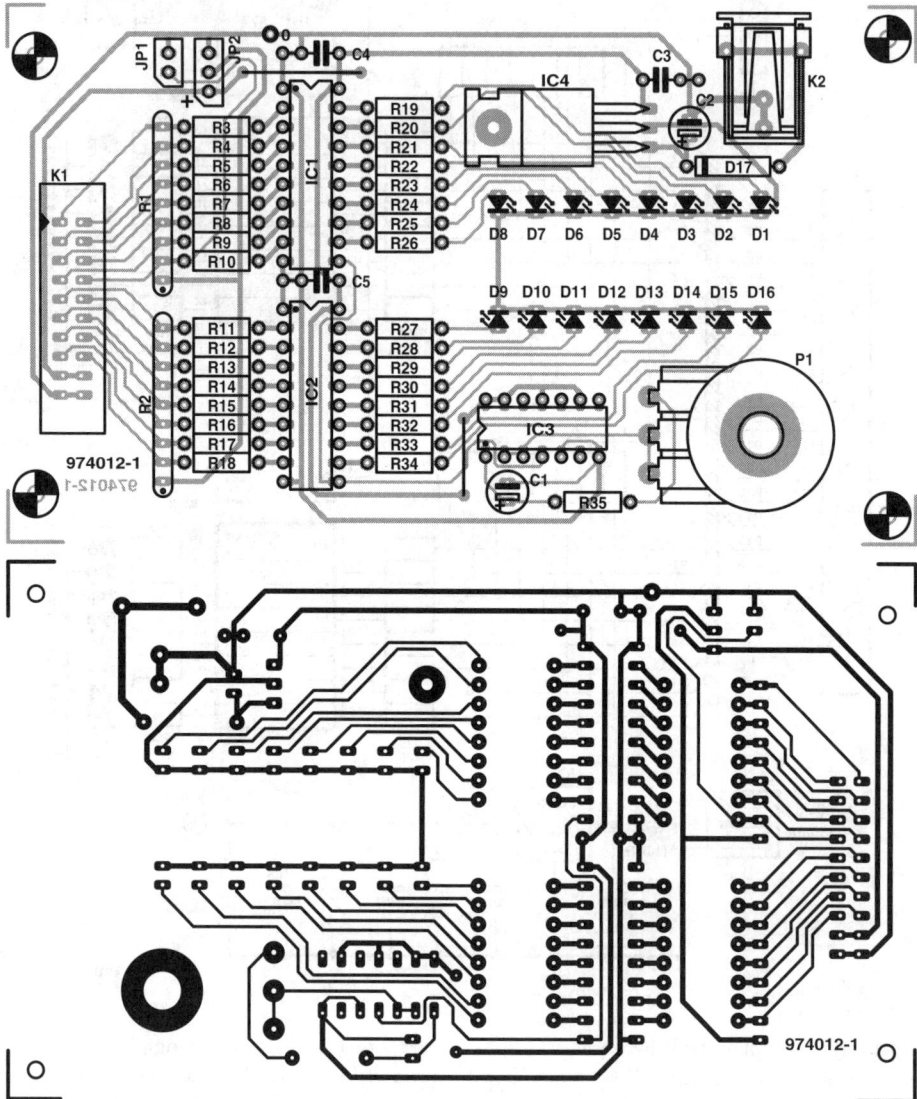

Parts list

Resistors:
R_1, R_2 = 100 kΩ, 8-way SIL array
R_3–R_{18} = 22 kΩ
R_{19}–R_{35} = 1 kΩ

P_1 = 100 kΩ linear potentiometer

Capacitors:
C_1, C_2 = 10 μF, 63 V, radial
C_3, C_4, C_5 = 0.1 μF

Semiconductors:
D_1–D_{16} = LED, high efficiency
D_{17} = 1N4001

Integrated circuits:
IC_1, IC_2 = 74HCT574
IC_3 = 74HCT14
IC_4 = 7805

Miscellaneous:
JP_1 = 2-way 2.54 mm pinheader and pin jumper
JP_2 = 3-way 2.54 mm pinheaderand pin jumper
K_2 = mains adaptor socket, PCB mount
K_1 = 20-way boxheader or pinheader

Mains-THD meter

The circuit described may be used in conjunction with a digital voltmeter (DVM) to measure the total harmonic distortion (THD) of the mains supply voltage. A knowledge of this may be useful when the effect of switch-mode supplies or dimmers on the mains supply is to be determined.

The 230 V mains voltage is divided symmetrically :230 by R_1–R_5, so that the potential difference, pd, across R_5 is 1 V. The divider is symmetrical to prevent potentially dangerous voltages at the output terminals.

There follows a notch filter with a centre frequency of 50 Hz. Provided the filter is set up

properly, the mains frequency is attenuated by 70 dB. This means that the only frequencies that can appear at the output of the circuit are harmonics of the mains frequency.

The design ensures that every 1 mV r.m.s. measured by the DVM at the output corresponds to 0.1% THD.

Start the calibration by setting P_1, P_2 and P_3 to the centre of their travel and, *taking great care*, applying the mains voltage to the input terminals. Adjust P_1 for the lowest possible reading on the DVM. Note the position of this preset and then set it exactly between this and the cen-

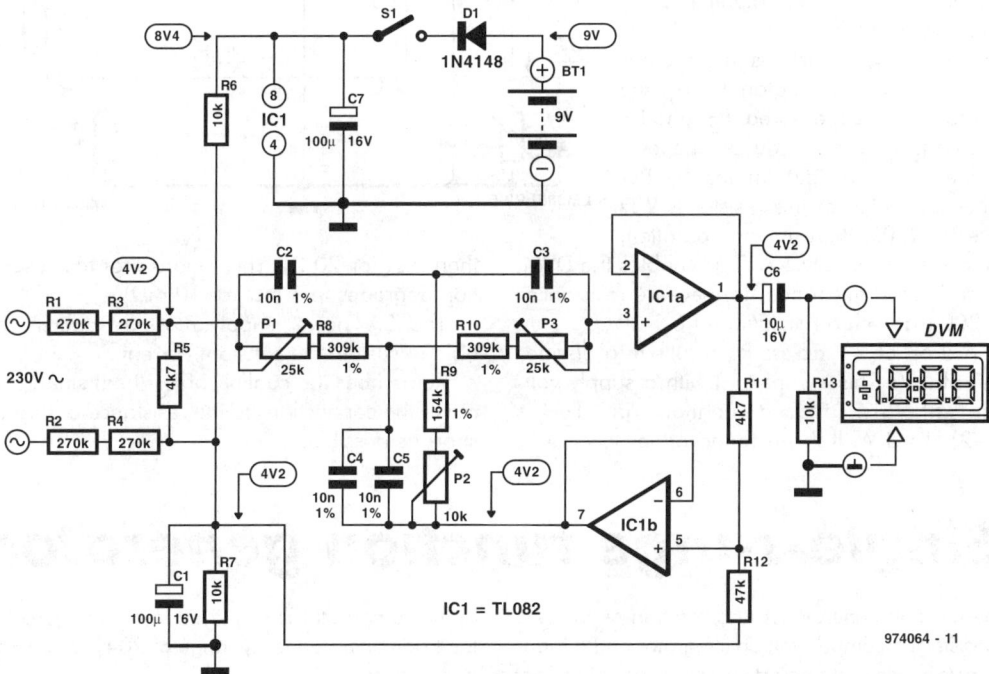

974064 - 11

285

tre position. Adjust P_3 for the lowest possible reading on the DVM. Then, alternately adjust P_1 and P_3 for a minimum reading on the DVM. When this has been established, adjust P_2 for the lowest possible reading on the DVM.

Since the circuit draws a current of only 5 mA, power may be obtained from a 9 V alkaline or rechargeable battery.

In the construction of the meter, make absolutely certain that all mains-carrying part are well insulated.

Milliohm adaptor for DVM

Most DVMs (digital voltmeters) offer a resolution of only 0.1 Ω in the lowest resistance range. This does not allow the measurement of low-value resistors or the transfer resistance of plug-and-socket connections. The adaptor overcomes this deficiency.

The adaptor sends a constant current through the device or connection on test, whereupon the resulting potential drop across it is measured with the DVM.

The circuit is based on a 2.5 V voltage reference source, IC_1. Part of the reference voltage, 2.0 V, is applied to the non-inverting (+ve) input of the op amp via P_1. The op amp will attempt to hold the potential at its inverting (–ve) input also at 2.0 V. Consequently, it functions in combination with super darlington T_1–T_3, the resistance to be measured, R_2, and R_3 as a constant-current source. The level of the current is determined by R_3. When the value of this resistor is 2 Ω (five 10 Ω, 0.5 W, resistors in parallel), the current is exactly 1 A. This enables the DVM, set to its 200 mV range, to measure resistances of 200 mΩ with a resolution of 0.1 mΩ.

A drop of 2 V across R_3 results in a dissipation of 2 W. This means that with a supply voltage of 5 V, the dissipation in T_3 is $(5–2)\times1=3$ W. If this is found rather too much,

the value of R_3 may be increased to 20 Ω (or 200 Ω). The level of the constant current then drops to 100 mA (or 10 mA). The resistance range with the DVM set to the 200 mV range is

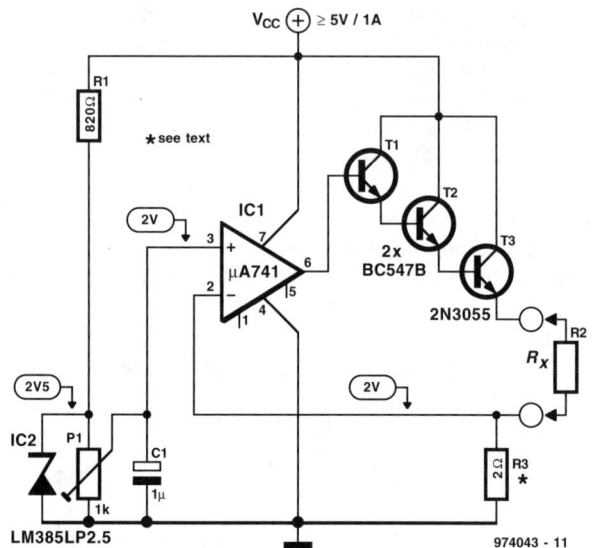

then 2 Ω (or 20 Ω). This means that the resolution degrades to 1 mΩ (or 10 mΩ).

The 5 V power supply must, of course, be able to sustain the constant current.

Remember the cooling of T_3 (heat sink!) and make the connections to the resistance on test as short as possible.

Single-range function generator

The function generator is traditional in as far as it consists of a comparator, an integrator and a triangular/sine wave shaper. However, a special variant

of the comparator is employed to be able to cover the traditional frequency range of 20 Hz to 25 kHz in one sweep.

The circuit is based on integrator IC_3, with R_{10}-C_4 forming the integration network. Unusually, the +input of the integrator is not connected to ground, so that the output signal is not just determined by the instantaneous level of the rectangular input voltage (and, of course, the RC network). The main function of comparator IC_1 is to control electronic switch IC_{2a}. This switch and IC_{2b} shift the integrator input (R_{10}) between ground and a positive potential which is set with frequency control potentiometer P_2. This arrangement ensures a positive-only rectangular voltage. However, R_{11} and R_{12} also hold the +input of IC_3 at half the potential on the CMOS switches. The fact that the output signal of the integrator is determined by the voltages at both op amp inputs allows a single capacitor, C_4, to cover well over three frequency decades. Resistors R_6 and R_7 determine the extreme frequencies that may be set on the generator.

Assuming that IC_{2b} is closed, the ramp voltage at the integrator output drops until the zener voltage of D_1 or D_2 is reached. When one of the zener diodes starts to conduct, the comparator toggles and its output becomes negative. Schottky diode D_3 and resistor R_9 prevent a negative voltage at the control inputs of IC_2, which is powered by the positive supply rail only. IC_{2b} opens, IC_{2a} is closed via inverter IC_{2c}, and the ramp voltage at the integrator output starts to rise until the zener voltage is reached again. Thereupon, the comparator toggles and the oscillation cycle starts again, creating a triangular and a rectangular signal at the output of IC_3 and

IC_1 respectively. Because the triangular-to-sine wave converter requires a virtually constant drive signal, the reference level created with the zener diodes can be adjusted with preset P_1. Capacitor C_1 eliminates a small rise of the triangular signal at higher frequencies, depending on component tolerances and construction.

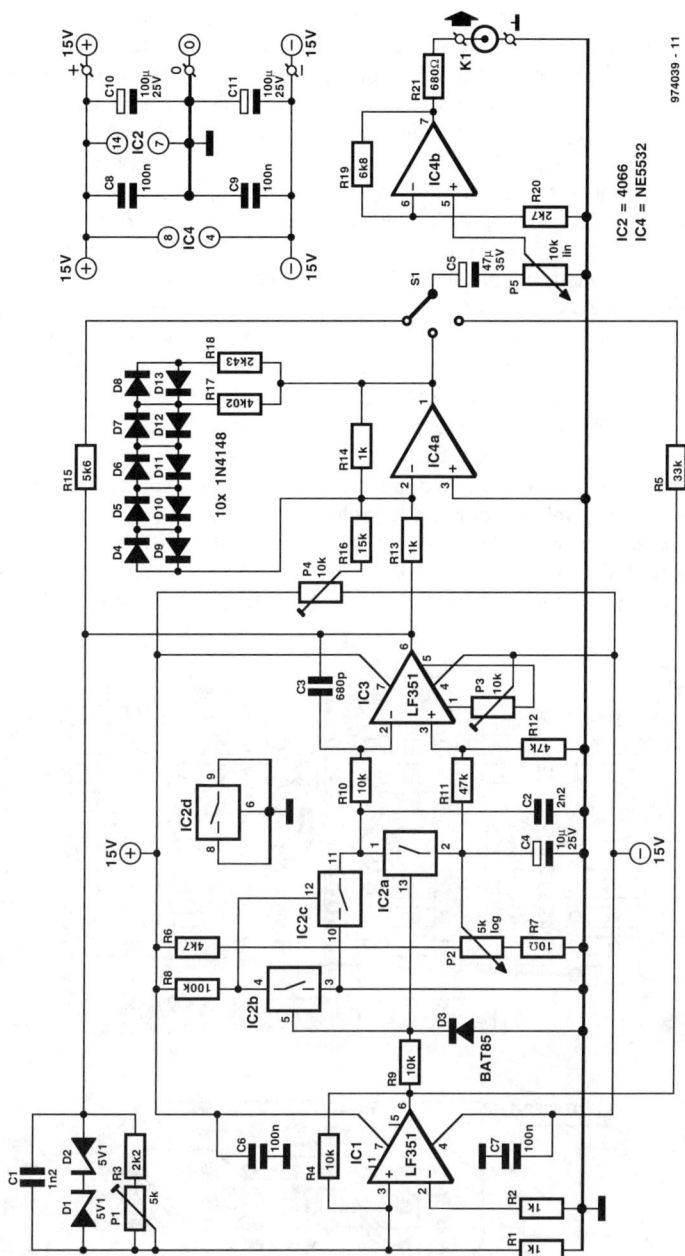

The values of resistors R_5 and R_{15} ensure roughly equal peak-to-peak values of the generator output voltage at all three positions of waveform selector S_1. The generator output impedance is about 600 Ω, the maximum (no-load) output voltage, about 20 V_{pp}.

The generator is powered by a symmetrical, regulated, 15-volt supply. Current drain is about 22 mA on each voltage rail.

The only critical parts in the circuit are high-frequency compensation capacitor C_1 and frequency control potentiometer P_2. The optimum value of C_1 may have to be established empirically, while a good-quality logarithmic potentiometer has to be used for P_2. If at all possible, a gear and dial assembly should be used, because the full frequency range is compressed into a span of 270 degrees. As regards level control P_5, a logarithmic pot may be preferred over a linear one if setting set small output levels accurately is needed.

The generator is adjusted with the aid of a dual-beam oscilloscope and presets P_1, P_3 and P_4. Initially, set P_2 and P_5 to mid-position, and connect one scope channel to the output of IC_3. Turn down the frequency with P_2, and carefully adjust P_3 for optimum symmetry of the triangular wave. Move the probe to the output of IC_{4a}, and set the generator to a frequency of about 1 kHz. Adjust P_1 and P_4 for the best possible sine wave shape. For these adjustments, it is convenient to have the other scope channel display the triangular signal (at the same sensitivity), and move the trace onto the sine wave. In this way, any asymmetry of the sine wave is easily detected and eliminated.

Monitor the output of IC_3 again, this time for stability of the output level across the full frequency range. If necessary change the exact (equivalent) value of C_1 until the level is virtually constant.

Finally, check the upper and lower frequency extremes, which should be a little over 25 kHz and a little under 20 Hz respectively. If necessary, alter the value of R_6 and/or R_7.

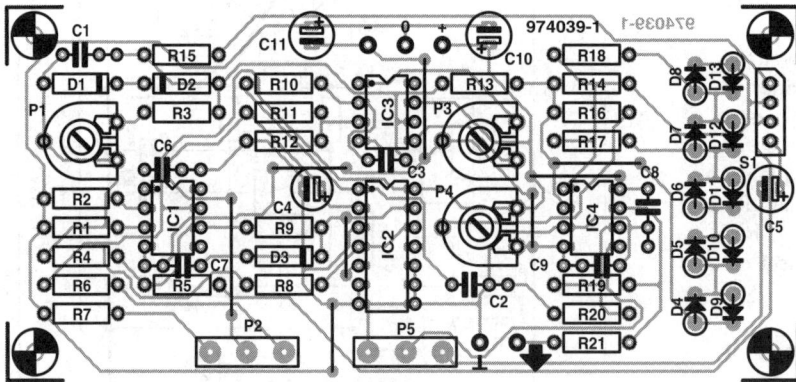

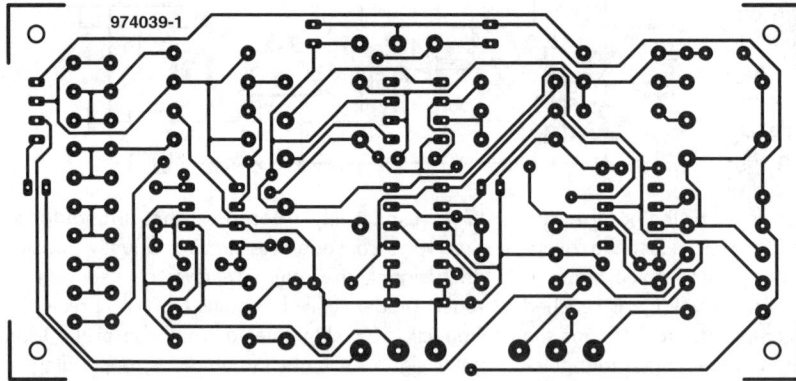

Parts list

Resistors:
$R_1, R_2, R_{13}, R_{14} = 1 k\Omega$
$R_3 = 2.2 k\Omega$
$R_4, R_9, R_{10} = 10 k\Omega$
$R_5 = 33 k\Omega$
$R_6 = 4.7 k\Omega$
$R_7 = 10 \Omega$
$R_8 = 100 k\Omega$
$R_{11}, R_{12} = 47 k\Omega$
$R_{15} = 5.6 k\Omega$
$R_{16} = 15 k\Omega$
$R_{17} = 4.02 k\Omega$
$R_{18} = 2.43 k\Omega$
$R_{19} = 6.8 k\Omega$
$R_{20} = 2.7 k\Omega$
$R_{21} = 680 \Omega$
$P_1 = 5 k\Omega$, preset, horizontal
$P_2 = 5 k\Omega$, logarithmic potentiometer
$P_3, P_4 = 10 k\Omega$, preset, horizontal
$P_5 = 10 k\Omega$, linear potentiometer

Capacitors:
$C_1 = 0.0012 \mu F$ (see text)
$C_2 = 0.0022 \mu F$
$C_3 = 680 pF$
$C_4 = 10 \mu F$, 25 V, radial
$C_5 = 47 \mu F$, 35 V, radial
C_6–$C_9 = 0.1 \mu F$
$C_{10}, C_{11} = 100 \mu F$, 25 V, radial

Semiconductors:
$D_1, D_2 = $ zener diode 5.1 V, 400 mW
$D_3 = $ BAT85
D_4–$D_{13} = $ 1N4148 (matched pairs)

Integrated circuits:
$IC_1, IC_3 = $ LF351
$IC_2 = $ 4066
$IC_4 = $ NE5532

Miscellaneous:
$K_1 = $ BNC socket
$S_1 = $ rotary switch, 1 pole, 3 positions

State-of-charge tester

The tester is intended to determine the state of charge of a NiCd or NiMH battery precisely. Also, the counter module linked to it it enables the capacity to be read. It is meant for a battery containing up to 12 cells.

When the start knob is pressed, the battery on test is discharged at a predetermined current of, say, 50 mA. At the same time, a pulse generator is enabled which produces a number of pulses per hour corresponding to the discharge current in mA (i.e., 50). When the battery has been discharged to about 75% of its nominal e.m.f., discharging is terminated and the pulse generator is disabled. The number of pulses generated during the discharge period is shown on the display of the counter.

The current source instrumental in the discharging of the battery is based on IC_{1a} and T_1. The op amp corrects the drive to T_1 until the potential across R_1 corresponds to the reference voltage set with P_1. The reference is provided by zener diode D_1, the current through which is held constant by T_3. The values specified in the diagram refer to a discharge current of 125 mA.

The values given in brackets refer to a discharge current of 50 mA. Which is to be used depends on the capacity of the battery: generally, a discharge current of 1/10 of the capacity of the battery gives best results.

To determine whether the battery is discharged, its terminal voltage is compared with a second reference potential derived from the supply lines for the digital part of the circuit via R_{10}-R_{11}-P_1. Comparator IC_{1b} changes state when the battery voltage drops below 11.25 V (set with P_2). Thereupon T_2 comes on, so that the wiper of P_1 is linked to earth and the current source is disabled. The hysteresis provided by R_8-R_9-D_1 prevents the comparator changing state again when the battery voltage rises slightly owing to the ceasing of the discharge current. Pressing S_1 resets the comparator to its original state.

The digital section of the circuit consists of pulse generator IC_2-IC_3-IC_4-T_5-T_6 and a ready-made counter module (e.g. Voltcraft Type 195650). The generator provides 50 or 125 pulses per hour, depending on the level of the

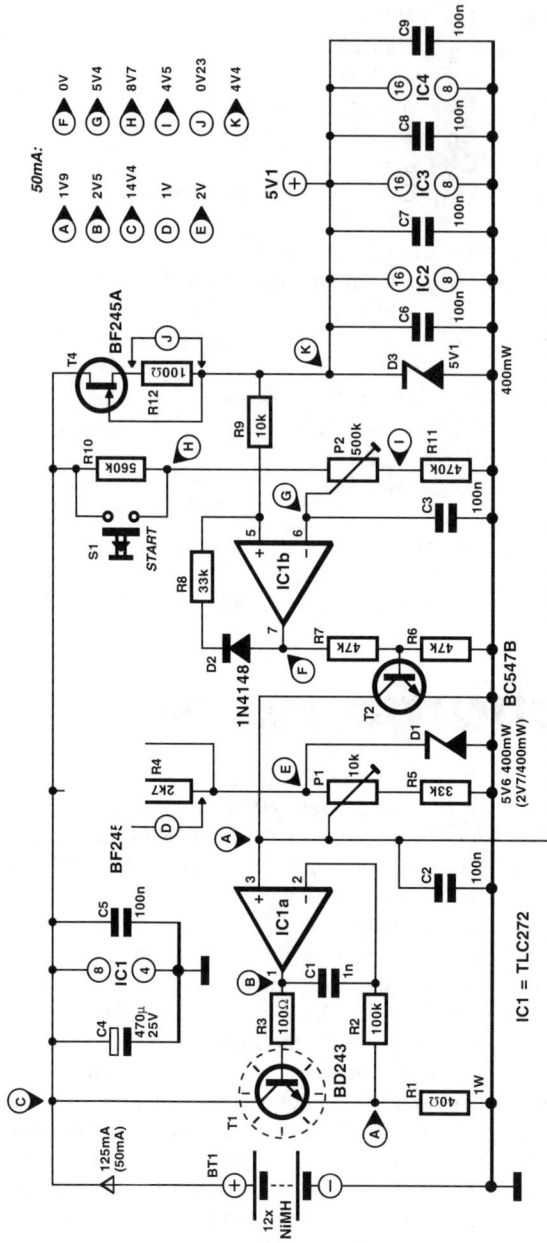

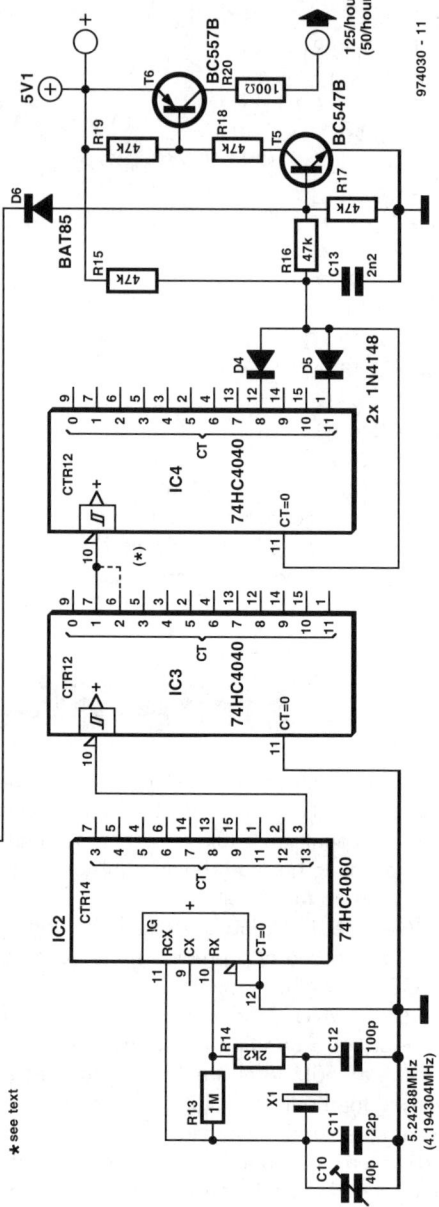

discharge current. In the first case, the oscillator in IC_2 runs at 4.194304 MHz, which is lowered to 0.0139 Hz (50 pulses per hour) by a three-stage divider. When the discharge current is 125 mA, the divisor of IC_3 is changed from 2^3 to 2^2, and the crystal frequency to 5.24288 MHz. These two alterations result in a

raising of the pulse rate by a factor 2.5 (= 125 pulses).

The counter module at the output is powered by an AA battery. Counting is commenced when the counter module is short-circuited briefly ($\leq 100\,\mu s$) to the positive supply line. Capacitor C_{13} ensures that the short-circuit pulses are of the correct width, while T_5 and T_6 arrange correct matching between the output of the divider chain and the input of the counter module.

The +ve terminal of the counter module is linked to the +5.1 V supply to the pulse generator, so that T_6 comes on for every output pulse, whereupon the counter increases by 1. When the battery is fully discharged, so that T_2 begins to conduct, the output of the pulse generator is blocked because T_5 is disabled via D_6.

If a counter module different from the one specified is used, it may be necessary to alter output stage T_5-T_6 slightly. To what extent changes are necessary will be clear from the data of the module.

Transistor matcher

In balanced preamplifiers and output amplifiers, it is highly advisable, if not imperative, to use truly complementary transistors. This means that the base-emitter voltage and the current amplification of the p-n-p and n-p-n transistors must be equal or very nearly so. The absolute values of these parameters are not that important. The present circuit is intended compare these two

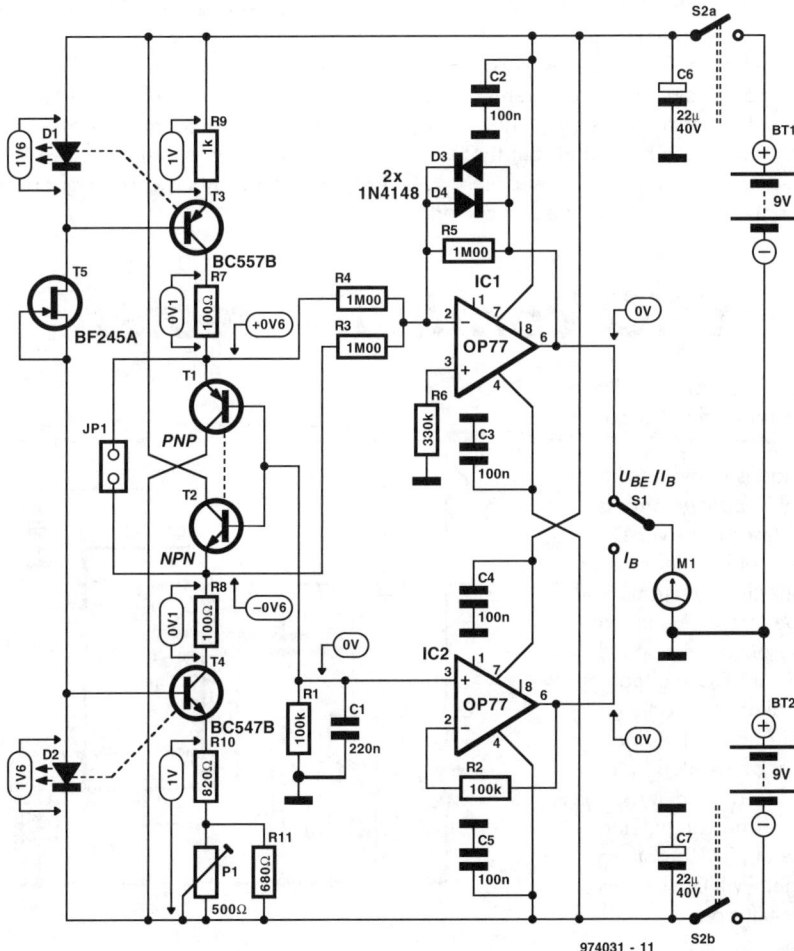

974031 - 11

parameters of a pair of transistors in one operation.

The collector current of the transistors to be paired, T_1 and T_2 respectively, is set accurately to 1 mA with the aid of current sources T_3 and T_4. Accuracy is vital and T_3 and T_4 are therefore thermally coupled to reference diodes D_1 and D_2 respectively. The current through these LEDs is held stable by current source T_5. It is imperative that the currents through T_3 and T_4 are not only stable, but also equal, and this is achieved by R_7, R_8 and P_1. The preset is adjusted so that the potential across R_7 is equal to that across R_8.

Circuit IC_1 functions as an adder. When the base-emitter voltages of T_1 and T_2 are equal, the output potential of IC_1 is equal to the base voltages of T_1 and T_2, but inverted w.r.t. them. This can be accomplished only if the amplification factors of the two transistors are equal. So, in case of a truly complementary pair, moving-coil meter M_1 will read 0 or very nearly so.

Even if a pair looks truly complementary, there is still a theoretical possibility that their base-emitter voltages are not equal, but that the inequality is compensated by a difference in h_{FE}. Circuit IC_2 enables this to be verified. It buffers

the voltage at the base terminals of T_1 and T_2 and these can be compared on the meter by briefly changing over switch S_1.

Any unwanted output offset may be obviated by linking the fixed terminals of a 25 kΩ preset potentiometer to pins 1 and 8 of IC_1, and its wiper to the +9 V line. Short-circuit JP_1 temporarily and adjust the preset until the meter reads 0 V. This procedure may also have to be carried out with IC_2, but in this case R_1 should be short-circuited temporarily.

Transistors T_3 and T_4 are to be thermally coupled to the relevant diode, which is most conveniently accomplished if the diodes are flat (rectangular) types. In that case, the component pairs are easily held together with a cable tie.

It ius advisable tio use sockets for T_1 and T_2: these need not be transistor sockets; IC sockets do nicely as well. Clamp the two transistors firmly together with a clothes peg or crocodile clip and allow them a little time to reach the same temperature. Bear in mind that temperature differences exert a great influence.

The matcher is powered by two 9 V batteries, from which it draws a current of about 7 mA.

VCO continuity tester

Although the tester could hardly be simpler, it has a fixed place in the tool box of the designer. The test voltage is derived from a standard 9 V battery and is applied to a test probe via D_1 and S_1. This probe is linked to one end of the line whose continuity is to be tested. A second probe, whose output is applied to the input of a VCO (voltage-controlled oscillator) is connected to the other end of the line.

The range of oscillation of the VCO is determined by C_2 and R_1 (top of the range) and the resistance at pin 12 (bottom of the range), which in this application is left open.

In the absence of a voltage

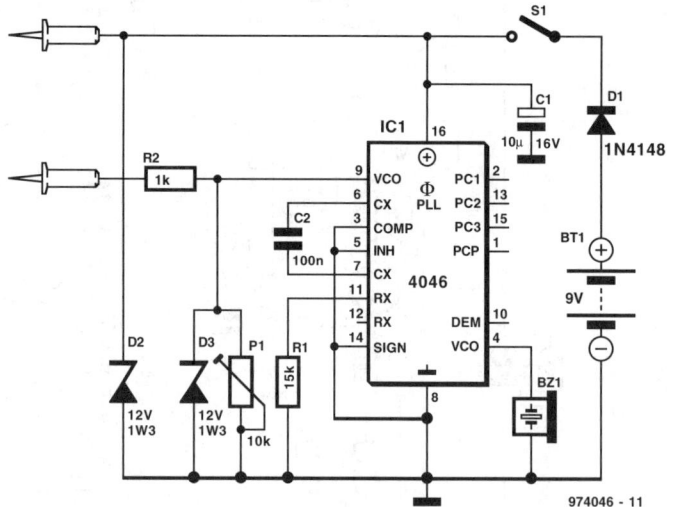

974046 - 11

at the input of the VCO, the oscillator is disabled. When the test voltage reaches pin 9 (line is all right), the VCO oscillates at the maximum frequency of 1.2 kHz. This signal is made audible via a piezobuzzer, Bz_1.

Because of its discrete supply, the tester can also be used for tests on active circuits. Zener diodes D_2 and D_3 prevent damaging voltages from reaching the input of the VCO.

The level of the test current is set with P_1, which is useful when the connection to be tested is a high-impedance one.

The tester draws a current of around 3 mA.

Function generator I

Simple triangle-wave generators have a weakness in that the waveform of their output signal normally cannot be modified. The circuit presented here makes it possible to smoothly alter the waveform of a linearly rising and steeply trailing sawtooth signal through a symmetrical triangular-wave to a slowly trailing, steeply rising linear sawtooth. The wanted waveform may be selected independently of the frequency, which can also be varied uniformly from 0.2 Hz to 8 kHz. At the same time, a rectangular signal with variable duty factor (also independent of frequency) is available at the rectangular-signal output of the circuit.

The circuit consists of integrator IC_{1b}, whose output is applied to comparator IC_{1c}. The output of the comparator is a rectangular signal

The output of IC_{1b} is raised by amplifier IC_{1d} to a level that allows the full output voltage range of the operational amplifier to be used.

Op amp IC_{1a} provides a stable virtual earth, whose level is set to half the supply voltage with P_1.

Smooth setting of the frequency is made possible by feedback of part of the output of the comparator to the input of the integrator via P_2. This preset is usually not provided in standard triangular-wave generators.

Network D_1-R_1-D_2-R_2-P_3 makes it possible to give integrator capacitor C_3 different charging

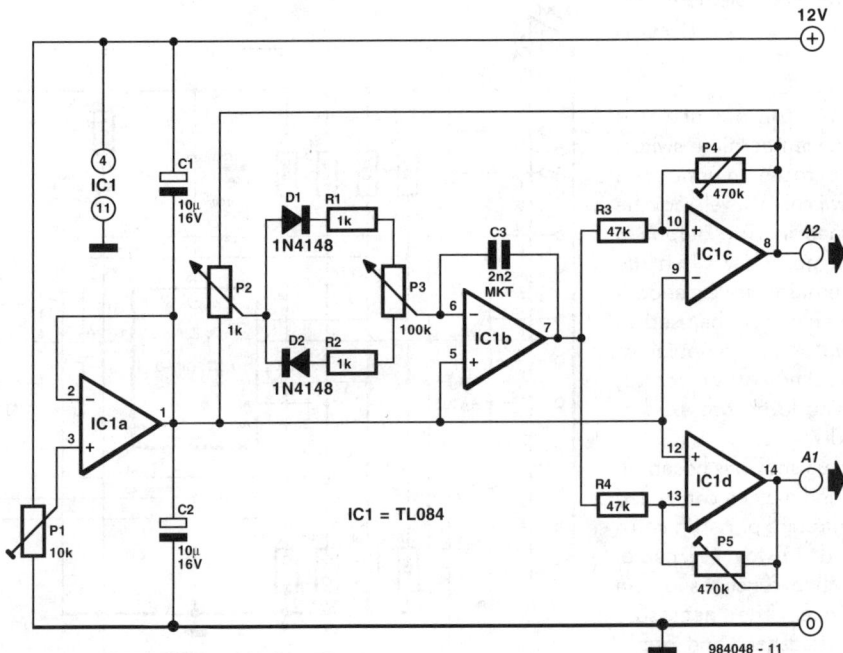

984048 - 11

293

and discharge times. This arrangement enables the output signal at A_1 and the duty factor of the rectangular wave signal at A_2 to be varied.

Varying the amplification factor with P_5 has no effect on the frequency set with P_2.

The slope of the signal edges, the transient responses, and the output voltage range (rail-to-rail or with some voltage drop) depend on the type of op amp used. The TL084 used in the prototype offers a good compromise between price and meeting the wanted parameters.

The circuit is best built on a small piece of prototyping board.

The circuit draws a current of not more than 12 mA.

Summary of preset actions:
P_1 – sets virtual earth to a level equal to $U_{cc}/2$;
P_2 – sets the frequency;
P_3– sets the waveform;
P_4 – sets the hysteresis of the comparator (frequency and amplitude of the triangle-wave signal)
P_5 – sets the amplification of the triangle-wave and sawtooth signals.

Applications
• Test and measurement
• Pulse-width control

I²C™ *temperature sensor*

The LM75 from National Semiconductor is a temperature sensor, Delta-Sigma analogue-to-digital converter (ADC), and digital over-temperature detector with I²C interface. It is manufactured in surface-mount technology (SMT) for operation from 5 V or 3.3 V. The temperature may be read in half degrees in the range –55 °C to +125 °C. It provides a 9-bit output in twos complement (that is, 0FAH is +125 °C; 192H is –55 °C; 001H is +0.5 °C; 1FFH is –0.5 °C).

The LM75 can operate as a stand-alone temperature switch, for which purpose an upper and a lower switching level may be programmed in. The output of the device goes low when the set temperature is exceeded. This output may also be used as an interrupt for a computer or microcontroller. At power-up, the switching levels are fixed at 80 °C and 75 °C

The circuit shown is based on an LM75 and may be connected to the Centronics port of a computer via a 25-way 1:1 cable. The port then functions as an I²C interface. The necessary software, datasheet and appli-

cation note may be downloaded from www.national.com/pf/LM/ LM75.html.

Operation of the software is simple: at the top left is a button which when set to 'off' renders the Centronics port voltage-less. Connect the board and select the relevant Centronics address and an I²C address. This means that the highest

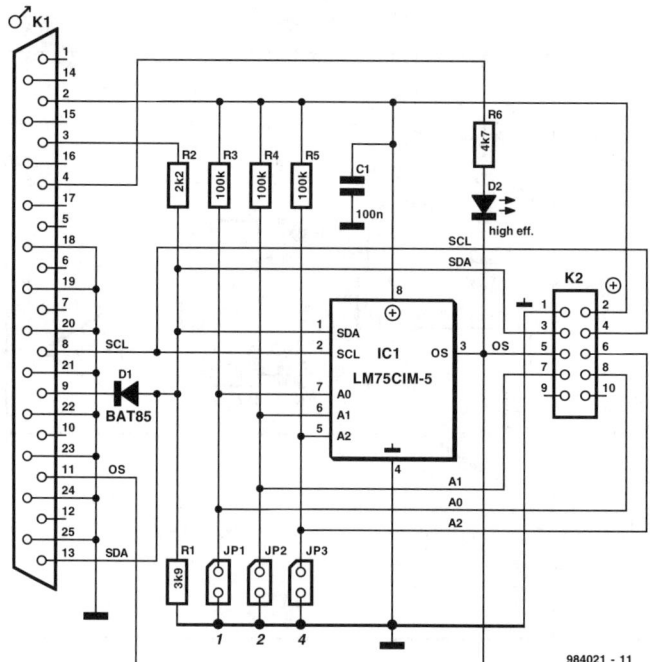

984021 - 11

address on the board (lowest in the list) must be selected without the use of jumpers. Set the button to 'on' and temperature monitoring starts.

Since the circuit draws current from the Centronics port, a dedicated power supply is not required. However, readers who worry about the additional load placed on their PC, may note that the LM75 draws a current not exceeding 250 μA.

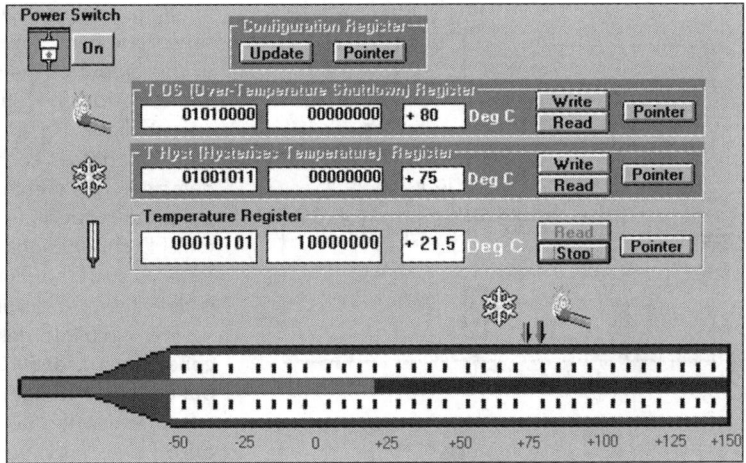

I^2C is a trademark of the Philips Corporation.

Parts list
Resistors:
R_1 = 3.9 kΩ
R_2 = 2.2 kΩ
R_3–R_5 = 100 kΩ
R_6 = 4.7 kΩ

Capacitors:
C_1 = 0.1 μF

Semiconductors:
D_1 = BAT85
D_2 = LED, high efficiency

Integrated circuits:
IC_1 = LM75CIM-5

Miscellaneous:
K_1 = DB25 connector, male, right-angled, for board mounting
K_2 = 10-way box header for board mounting
JP_1–JP_3 = 2-way pin strip header with jumper link

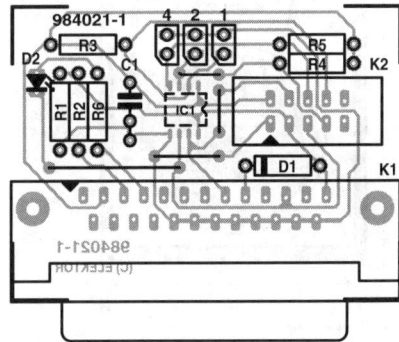

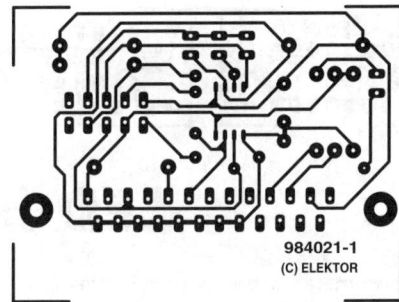

Infra-red remote control tester

The battery-operated tester is invaluable for quick go/no-go checking of almost any remote control transmitting infra-red (IR) light. It may be fitted in a small case.

Schmitt trigger gate IC_{1f} is used as a quasi-analogue amplifier with, unusually, an infra-red

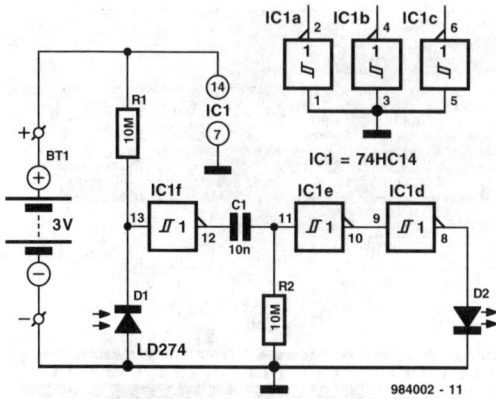

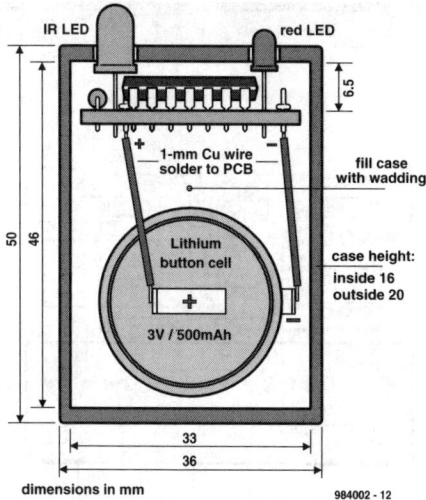

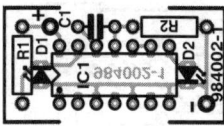

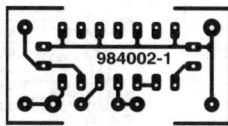

emitting diode (IRED) type LD274 acting as the sensor element. An RC network, C_1-R_2, is used at the output of the gate because all IR remote controls transmit pulse bursts, and to prevent output diode D_2 lighting constantly when daylight or another continuous source of IR light is detected. This creates a useful quick test facility: point the tester at direct daylight, and the indicator LED should light briefly. The sensitivity of the tester is such that IR light from a remote control is detected at a distance of up to 50 cm.

The circuit is designed for very low power consumption, drawing less than 1 mA from the battery when IR light is detected, and practically no current when no light is detected. Hence no on/off switch is required. The construction drawing shows how the tester may be fitted in a small ABS case.

Parts list
Resistors:
R_1, R_2 = 10 MΩ

Capacitors:
C_1 = 0.01 μF

Semiconductors:
D_1 = LD274
D_2 = LED, 3 mm, low-current

Integrated circuits:
IC_1 = 74HC14

Miscellaneous:
Bt_1 = 3V Li-ion battery with solder tags (560 mAh)
Case, 50×30×13 mm (approx.)

LED barometer

The barometer is based on air pressure transducer Type MPXS4100A from Motorola, IC_1, and two LM3914 bargraph drivers, IC_3 and IC_4, each of which generates a reference voltage of 1.25 V. The reference of IC_3 is with respect to ground. When the the RLO and REFADJ inputs of IC_4 are linked to the reference voltage from IC_3, the REFOUT pin of IC_4 is at 2.5 V with respect to ground. In this way, the LED drivers are cascad-ed to give a scale of twenty LEDs, each representing an air pressure increase of 5 mbar (sometimes, particularly by scientists, called hecto Pascal).

Because the output voltage of the pressure sensor follows any change in the supply voltage, a stable 5 V supply is is provided by opamp IC_{2a} which doubles the 2.5 V reference voltage from IC4. The sensor output voltage, U_s, is expressed

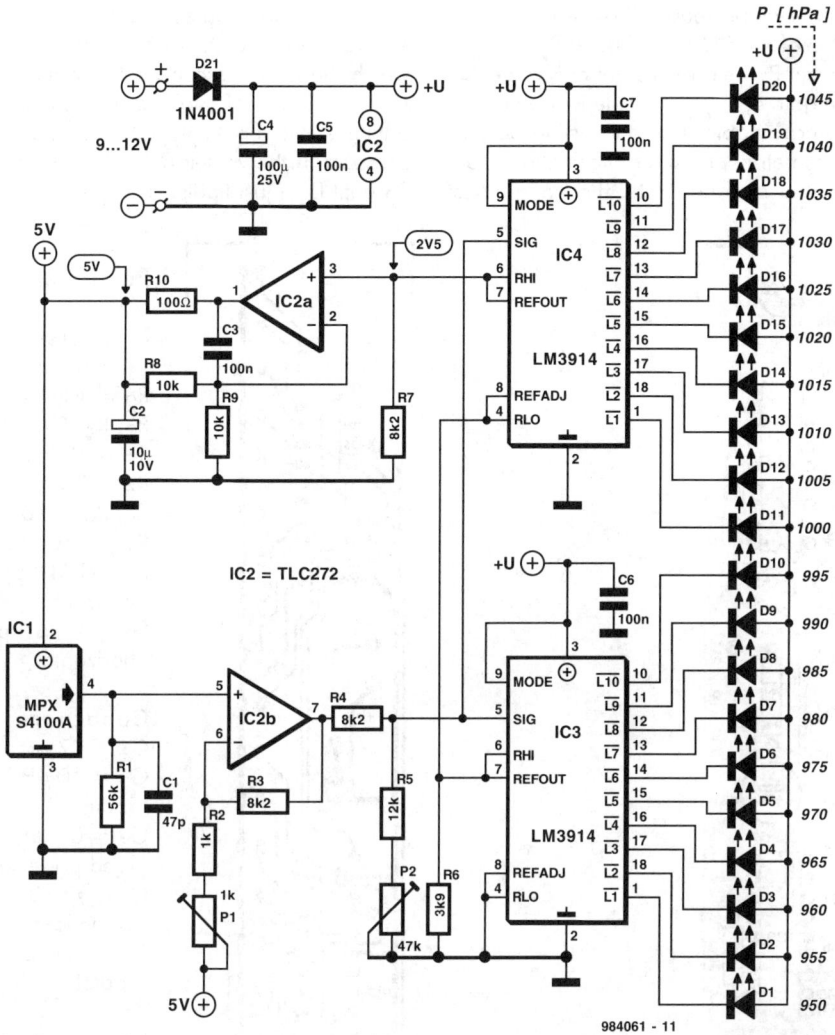

by the equation

$$U_s = 5(0.001059P - 0.1518) \quad \text{[V]}$$

where P is the barometric pressure in mbar. In most of the world, a useful barometer range is 945–1045 mbar (lowerst ever recorded is 870 mbar in a typhoon; highest ever recorded is 1084 mbar in Siberia). If the lower reading is represented by all LEDs off, and the higher by all LEDs on, sensor outputs are needed as follows

$$U_{s(low)} = 5(0.001059P - 0.1518) = 5(1.000755 - 0.1518) = 4.245 \text{ V}$$

$$U_{s(high)} = 5(0.001059P - 0.1518) = 5(1.106655 - o.1518) = 4.774 \text{ V}$$

The required amplification, A, between the sensor output and the input of the readout is:

$$A = U_{ref}/(U_{s(high)} - U_{s(low)}) = 2.5/(4.774 - 4.245) = 4.726$$

In addition, a negative offset of 4.245 V is needed, so that the output voltage is 0 V at an air pressure of 945 mbar. Components IC_{2b}, P_1, P_2, R_2–R_5 provide the gain and offset compensation. The 5 V reference voltage, IC_{2b}, P_1, R_2 and R_3 cancel the offset and provide an amplification of

297

×6.65, which may be reduced to the earlier calculated value of ×4.726 by adjusting P_2.

Since preset P_1 determines not only the offset but also the gain of IC_{2b}, multiple two-point calibration is unavoidable. Use an existing barometer or barometric pressure information from your national or regional Met Office, and adjust the circuit several times at different air pressures. Alternatively, if a pressure vessel is to hand in which the pressure can be accurately controlled to 945 mbar, set P_2 to mid-travel, and adjust P_1 until the output of IC_{2b} is 0 V. Then, increase the pressure in the vessel to 1045 mbar, and adjust P_2 until D_{20} just lights.

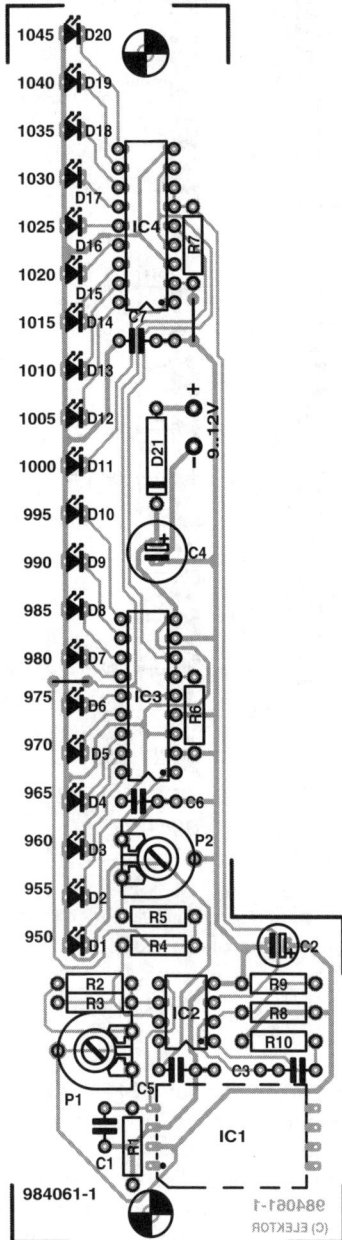

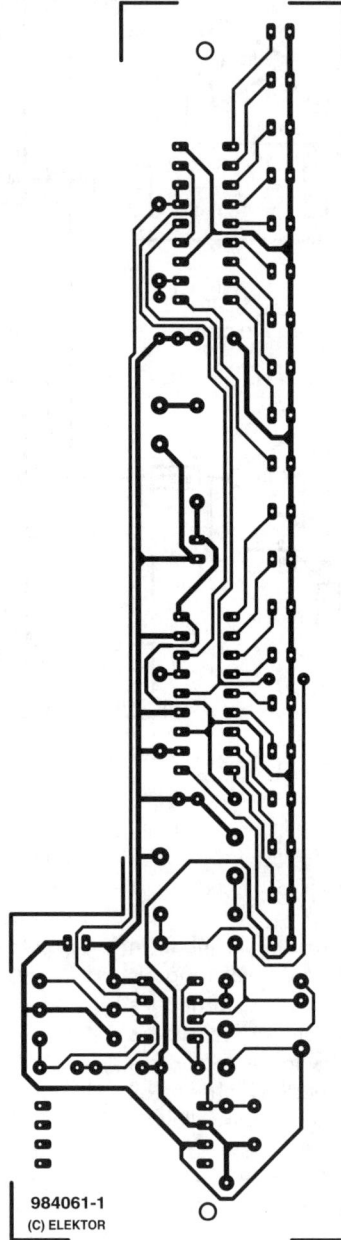

Parts list

Resistors:
R_1 = 56 kΩ
R_2 = 1 kΩ
R_3, R_4, R_7 = 8.2 kΩ
R_5 = 12 kΩ
R_6 = 3.9 kΩ
R_8, R_9 = 10 kΩ
R_{10} = 100 Ω
P_1 = 1 kΩ, preset, horizontal
P_2 = 47 kΩ, preset, horizontal

Capacitors:
C_1 = 47 pF ceramic
C_2 = 10 μF, 10 V, radial
C_3 = 0.1 μF, metallized polyester (MKT)
C_5–C_7 = 0.1 μF ceramic
C_4 = 100 μF, 25 V, radial

Semiconductors:
D_1–D_7 = LED, red, 3 mm, high efficiency
D_8–D_{13} = LED, yellow, 3 mm, high efficiency
D_{14}–D_{20} = LED, green, 3 mm, high efficiency
D_{21} = 1N4001

Integrated circuits:
IC_1 = MPXS4100A (Motorola)
IC_2 = TLC272CP
IC_3, IC_4 = LM3914N

984061-1

984061-1
(C) ELEKTOR

Liquid-crystal display (LCD) tester

Liquid-crystal displays come in all sorts and sizes, and this applies also to their pinouts. In fact, many of these displays cannot be used properly without the manufacturers' documentation. But, of course, this can never be found when it is needed, and a small tester to unravel the terminals may, therefore, be found very handy.

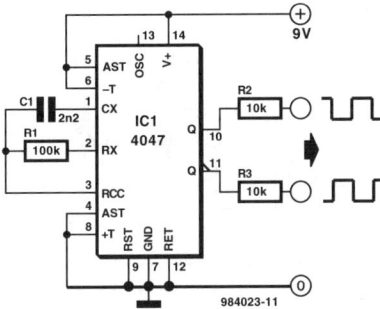

A liquid-crystal display consists of two thin sheets of glass, the facing surfaces of which have been given thin conducting tracks. When the glass is looked through at right or near-right angles, these tracks cannot be seen. At certain viewing angles, they become visible, however.

The space between the sheets of glass is filled with a liquid that, stimulated by an electric voltage, alters the polarization of the incident light. In this way, segments may appear light or dark and give rise to the display of lines or shapes.

A segment may be tested by applying an alternating voltage of a few volts across it. Note that the application of a direct voltage will dam-age the display irreversibly: the resulting current will remove the tracks. The alternating voltage should contain not even a tiny direct voltage component. An alternating current also removes part of the tracks when the current flows in one direction, but restores it when the current flows in the opposite direction.

The tester described here consists of a square-wave generator that produces an absolutely symmetrical alternating voltage without any d.c. component. Most logic oscillators are incapable of producing a square-wave signal: they generate rectangular waveforms whose duty factor is about 50%. The 4047 used in the tester has a binary scaler at its output that guarantees symmetry.

The oscillator frequency is about 1 kHz. It may be powered from a 3–9 V source. Normally, this will be a battery, but a variable power supply has advantages. It shows at which voltage the display works satisfactorily and also that there is a clear relationship between the level of the voltage and the angle at which the display is clearly legible. The tester draws a current not exceeding 1 mA.

The test voltage must at all times be connected between the common terminal, that is, the back plane, and one of the segments. If it is not known which of the terminals is the back plane, connect one probe of the tester to a segment and the other successively to all the other terminals until the segment becomes visible. Note, however, that there are LCDs with more than one back plane. Therefore, if a segment does not become visible, investigate whether the display has a second back plane terminal.

Function generator II

The generator supplies sine wave, rectangular and triangular waveforms within the frequency range 1–15 kHz. The output level is adjustable between 0 and about 10 V_{pp}. Though modest, these specifications make the generator a useful piece of test equipment for audio design, experimentation and repair purposes. Since only cstandard components are used, the generator can be built for a modest outlay.

Inverter gates IC_{1a} and IC_{1b} are connected to resistors R_2 and R_3 to form a buffer with some hysteresis. Gate IC_{1f}, in conjunction with R_1, P_1 and C_1, acts as an integrator. Potentiometer P_1 defines the integrator's time-constant. A buffer acting as a comparator with hysteresis, together with the integrating effect provided by IC_{1f}

299

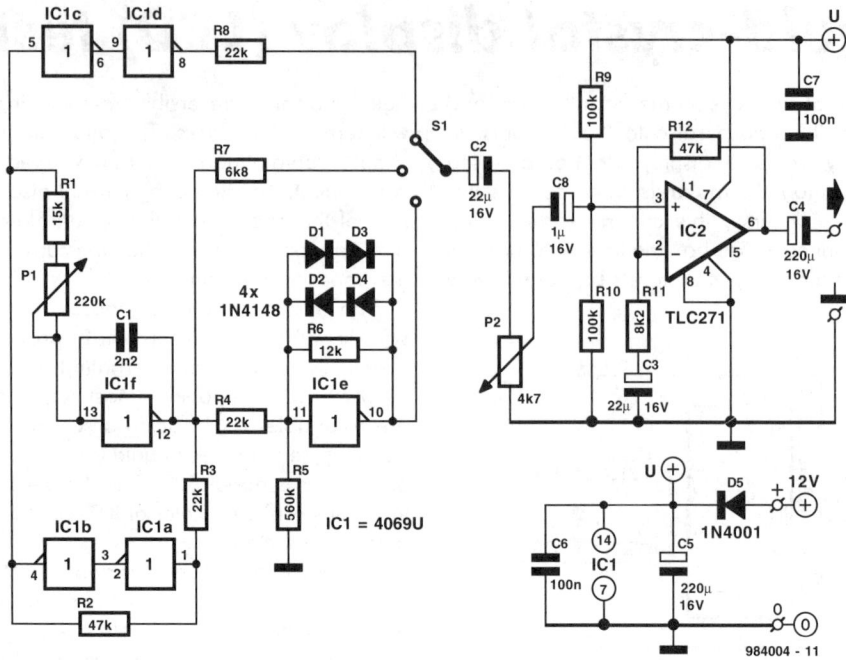

984004 - 11

results in an oscillator whose output frequency is controlled by potentiometer P_1. The buffer supplies a rectangular output signal; the integrator, a triangular one. The rectangular signal is further shaped and buffered by gates IC_{1c} and IC_{1d} before it is applied, via R_8, to one of the contacts of waveform selector S_1. The triangular signal is also applied to the switch: via R_7. The triangular signal supplied by IC_{1f} is fed to a sine wave shaper consisting of IC_{1e}, R_4, R_6, R_5 and diodes D_1–D_4. The output signal is applied directly to the waveform selector.

Because the three waveforms have different individual levels, that of the sine wave being the smallest, they have to be made roughly equal before they can be applied to output amplifier IC_2. This levelling is achieved with the aid of resistors R_7 and R_8 for the triangular and rectangular wave respectively, in combination with output level control P_2. The

TLC271 op amp is wired for a gain of ×6.7 to achieve a maximum (no-load) output level of about 10 V_{pp}. The minimum load impedance is about 600 ohms.

The generator is powered by a regulated 12 V supply, from which it draws a current of about 20 mA, depending, of course, on the load connected to the output.

The printed circuit board is designed to contain all the controls, i.e., the frequency control pot, the waveform selection switch and the output level control pot, so that no tedious wiring is required.

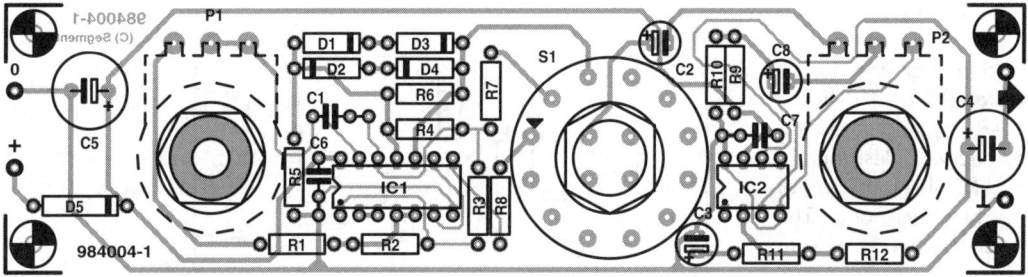

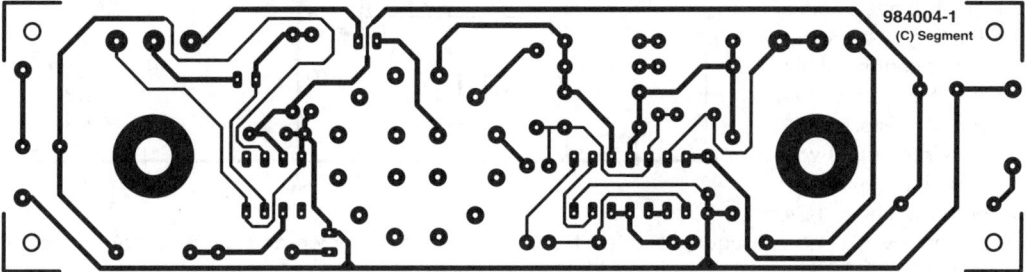

Parts list

Resistors:
R_1 = 15 kΩ
R_2, R_{12} = 47 kΩ
R_3, R_4, R_8 = 22 kΩ
R_5 = 560 kΩ
R_6 = 12 kΩ
R_7 = 6.8 kΩ
R_9, R_{10} = 100 kΩ
R_{11} = 8.2 kΩ
P_1 = 220 kΩ, linear potentiometer
P_2 = 4.7 kΩ, linear potentiometer

Capacitors:
C_1 = 0.0022 µF metallized polyester (MKT)
C_2, C_3 = 22 µF, 16 V, radial

C_4, C_5 = 220 µF, 16 V, radial
C_6, C_7 = 0.1 µF, miniature ceramic, high stability
C_8 = 1 µF, 16 V, radial

Semiconductors:
D_1–D_4 = 1N4148
D_5 = 1N4001

Integrated circuits:
IC_1 = 4069U (U = unbuffered version!)
IC_2 = TLC271CP

Miscellaneous:
S_1 = 3-way rotary switch, 4 poles, PCB mount

Audio signal generator

A small audio test generator is very useful for quickly tracing a signal through an audio unit. Its main purpose is speed rather than refinement. A single sine-wave signal of about 1 kHz is normally all that is needed: distortion is not terribly important. It is, however, important that the unit does not draw too high a current.

The generator described meets these modest requirements. It uses standard components, produces a signal of 899 Hz at an output level of 1 V r.m.s. and draws a current of only 20 µA. In theory, the low current drain would give a 9 V battery a life of 25,000 hours.

The circuit is a traditional Wien bridge oscillator based on a Type TLC271 op amp. The frequency determining bridge is formed by C_1, C_2

and R_1–R_4. The two inputs of the op amp are held at half the supply voltage by dividers R_3-R_4 and R_5-R_6 respectively. Resistors R_5 and R_6 also form part of the feedback loop. The amplification is set to about ×3 with P_1.

Diodes D_1 and D_2 are peak limiters. Since the limiting is based on the non-linearity of the diodes, there is a certain amount of distortion. At the nominal output voltage of 1 V r.m.s., the distortion is about 10%. This is, however, of no consequence in fast tests. Nevertheless, if 10% is considered too high, it may be improved by linking pin 8 of IC_1 to ground. This increases the current drain of the circuit to 640 μA, but the distortion is down to 0.7%, provided the circuit is adjusted properly. If a distortion meter or similar is not available, simply adjust the output to 1 V r.m.s.

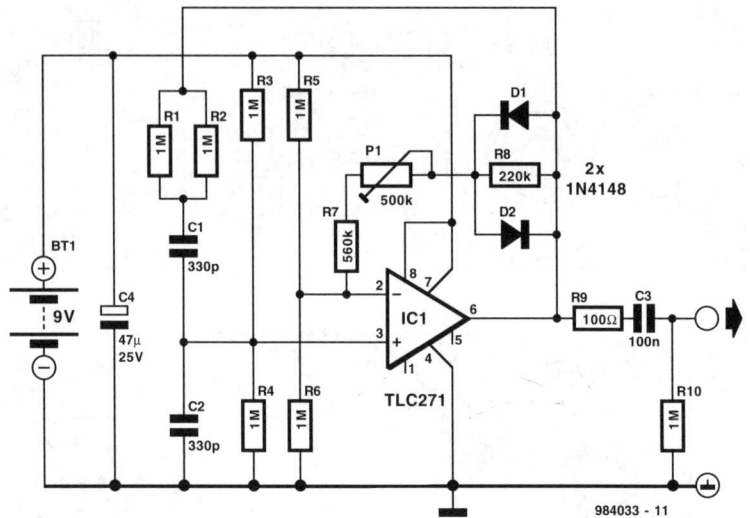

Since the distortion of the unit is not measured in hundredths of a per cent, C_1 and C_2 may be ceramic types without much detriment.

Oscillator monitor

The monitor was originally designed for checking an oscillator, but it can also be used as a general-purpose level indicator for a.c. signals. It is based on a quadruple IC containing four NAND gates. Only three of the gates are used, making the fourth free for other purposes. All the gates have a Schmitt trigger input.

When a 5 V supply is used, the Type 74HC132 is recommended; for higher voltage, a Type 4093. Note, however, that these two ICs have different pinouts. In the diagram, the differing pins of a 4093 are shown in brackets.

The signal to be monitored is applied to the input of the first gate via capacitor C_1. Resistor R_2, in conjunction with the protection diode in the IC, guards the input against high voltages.

In the absence of a signal, resistor R_1 holds the input high so that the output of the gate is low.

When a signal of sufficient strength is received, the input of the gate goes low during the negative half cycle of the signal, so that the output of the gate goes high in rhythm with the input signal. However, the Schmitt trigger converts sinusoidal signals into rectangular ones, which charge capacitor C_3 via diode D_1. When the potential across C_3 exceeds the threshold at the input of the second gate, this gate also toggles. The output of the second gate is then low, which disables the third gate, which functions as an oscillator.

When the level of the input signal drops, C_3 is discharged via R_3. The potential across the capacitor then no longer exceeds the threshold at the input of IC_{1b}, whereupon IC_{1c} is enabled and the LED flashes.

The LED may be connected as shown or as indicated by the dashed line. As shown, the diode remains off when there is an input signal of sufficient strength and begins to flash when the signal fails or its level drops. When the diode is linked to earth, it is on continuously when there is an input signal, and begins to flash when the input drops.

When a 5 V power supply is used, $R_5 = 1\ k\Omega$, and the circuit draws a current, including that of the LED, of 3 mA. The frequency of the input signal may lie between 10 Hz and 10 MHz.

When a 9–12 V supply is used, the value of R_5 must be altered as necessary. Owing to the 4093 being slower than the 74HC132, the upper frequency of the input signal is then limited to 3 MHz.

When the wiper of P_1 is at the level of the supply voltage, the response threshold, U_{SS}, lies between 3.5 V (when $U_b = 5\ V$) and 7 V (when $U_b = 12\ V$). When the wiper is moved away from the positive supply line, U_{SS} (max) is 1.5 V (when $U_b = 5\ V$).

The response threshold is quite precise: a drop in the input signal level of 50–100 mV is sufficient to disable the input.

When the input level is too high, a preset across the input terminals enables the level to be reduced to a value that lies in the desired range above the response threshold.

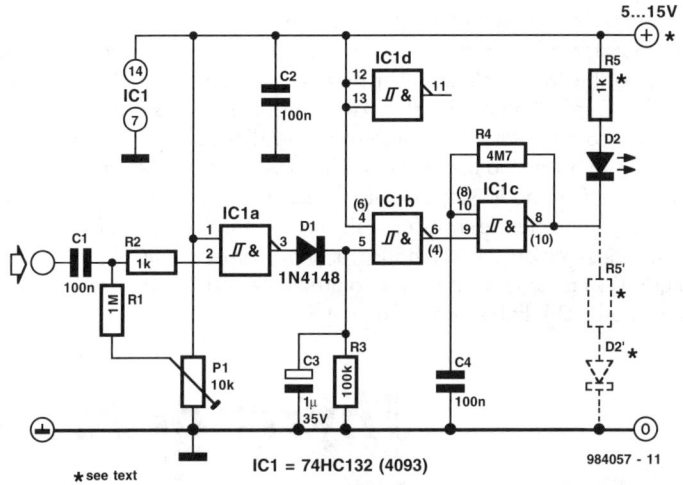

IC1 = 74HC132 (4093)

★ see text

984057 - 11

Three-state continuity tester

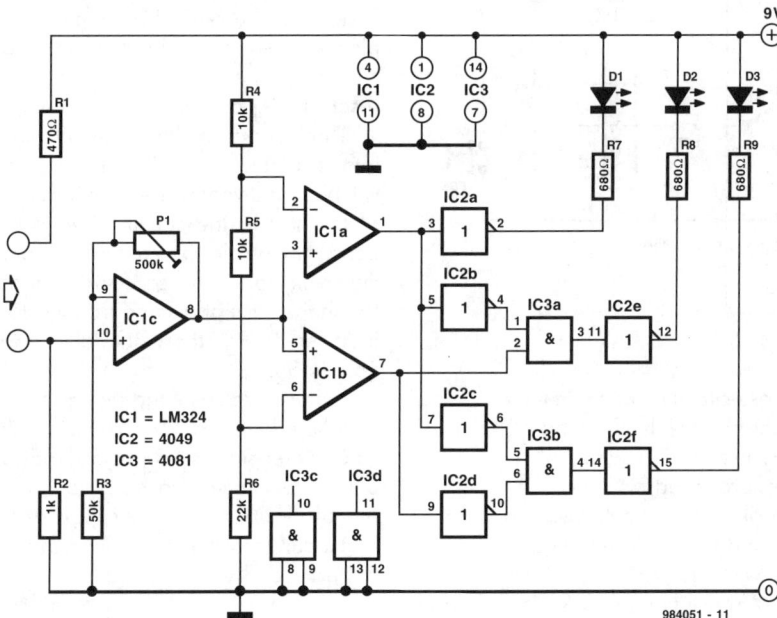

IC1 = LM324
IC2 = 4049
IC3 = 4081

984051 - 11

The continuity tester can distinguish between high-, medium-, and low-resistance connections. When there is a conductance between the inputs, which are linked to small probes, a current flows from the +9 V line to earth via R_1 and R_2. The consequent potential difference, p.d., across R_2 is used to determine the transfer resistance.

Operational amplifier IC_{1c} amplifies the p.d. across R_2 to a level that is set with P_1. A window comparator, IC_{1a} and IC_{1b}, likens the output of IC_{1c} to the two levels set with potential divider R_4–R_6. Depending on the state of the outputs of the two comparators, three light-emitting diodes (LEDs) are driven via the gates and inverters contained in IC_3 and IC_2 respectively in such a way that they indicate the transfer resistance in three categories. When the resistance is high, green diode D_3 lights; when it is of medium value, yellow diode D_2 lights, and when it is low, red diode D_1 lights. The levels at which the diodes light is set with P_1, but note that in any case the minimum value depends on the p.d. across R_2.

It is possible to reduce the value of the p.d. to enable lower transfer resistances to be detected, but this would mean an increase in the test current through R_2.

With values as specified, the circuit in its quiescent state draws a current of about 17 mA, but in operation each LED adds about 10 mA to this.

The LM324 (IC_1) may be operated from a single supply line: R_1 prevents the voltage at the input from reaching the level of the supply line (which is not permissible).

The supply voltage may be 5–18 V. The LEDs are driven directly by the inverters in the 4049 (IC_2), which can switch currents of up to 20 mA to earth.

Thyristor tester

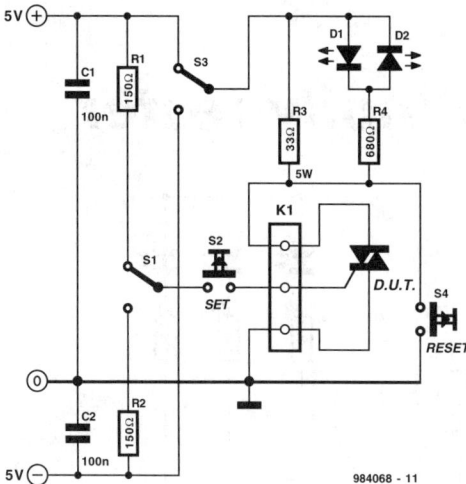

984068 - 11

The tester, whose circuit is shown in the diagram, is a handy tool for rapidly checking all kinds of thyristor (SCR, triac, ...). In case of a triac, all four quadrants are tested, which is done with S_3, while in case of a standard thyristor, a positive power supply and trigger current need to be set, which is done with S_1.

The value of resistors R_1 and R_2 is chosen to obtain a current of about 28 mA, which is more than sufficient for most thyristors. The hold current is determined by R_3, and is 125 mA, which is more than adequate to keep the thyristor in conduction after it has been triggered.

Since D_1 is a red, low-current LED, and D_2 a green, low-current LED, it can be seen at a glance in which quadrant the thyristor conducts.

Testing is started with S_2, and the circuit is reset with S_4 after the test has been concluded.

Three short lengths of circuit wire terminated into insulated crocodile clips on connector K_1 will be found very convenient for linking any kind of thyristor to the circuit. Mind correct connections, though: in the case of a triac, MT_1/A_1 is linked to earth, the gate to S_2 and MT_2/A_2 to R_3; in the case of a standard thyristor, the anode is linked to R_3, the cathode to earth, and the gate to S_2.

If, in a rare case, the trigger current needs to be altered, this can be done by changing the value of resistors R_1–R_3 as appropriate. The trigger current may also be made variable by the use of a variable power supply. If that is done, make sure that the dissipation in the resistors is not exceeded.

Mains pulser

The pulser is intended to switch the mains voltage on and off at intervals between just under a second and up to 10 minutes. This is useful, for instance, when a mains-operated equipment is to be tested for long periods, or for periodic switching of machinery.

In the diagram on the next page, transformer Tr_1, the bridge rectifier, and regulator IC_1 provide a stable 12 V supply rail for IC_2 and the

pins 2 and 6 of the timer IC, in conjunction with the relevant resistors, determines the time. The value of this capacitor may be slightly lower than shown.

The two preset potentiometers enable the on and off periods to be set. The 1 kΩ resistor in series with one of the presets determines the minimum discharge time.

The timer IC switches a relay whose double-

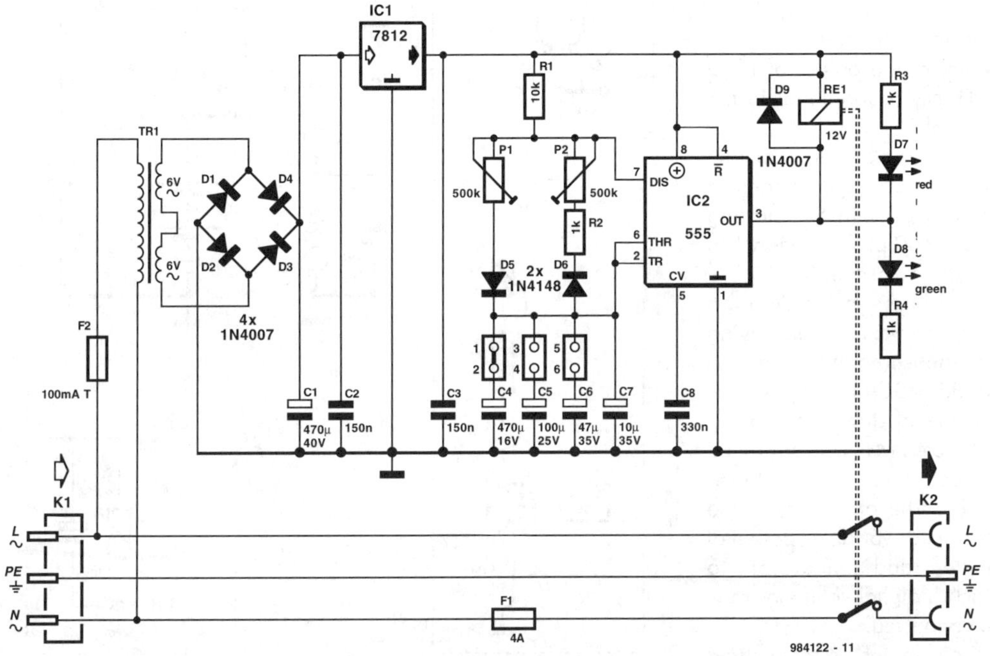

984122 - 11

relay. The timer is arranged so that the period-determining capacitor can be charged and discharged independently. Four time ranges can be chosen by selecting capacitors with the aid of jumpers. Short-circuiting positions 1 and 2 gives the longest time, and short-circuiting none, the shortest. In the latter case, the 10 μF capacitor at

pole contacts switch the mains voltage.

The LEDs indicate whether the mains voltage is switched through (red) or not (green).

The 100 mA slow fuse protects the mains transformer and low-voltage circuit. The 4 A medium slow fuse protects the relay against overload.

Capacitance meter

The capacitance meter, which is attractive and easy to build, has five measurement ranges and allows the test result to be displayed on an ana-

logue or a digital meter.

Gates IC_{1a} and IC_{1b} are arranged as an astable multivibrator (AVM), whose frequency is determined by capacitor C_2 and the resistor, R_1, R_2, or R_3, selected with switch S_1.

The output of the multivibrator is applied to a monostable multivibrator (MMV) formed by IC_{1c} and IC_{1d}. The pulses generated by the AMV appear in integrated form at the output of the MMV for a period of time determined by the capacitor on test, C_x, and the resistor, R_5 or R_6, selected with S_1. It follows that the pulse duration (width) at the output of IC_{1d} is directly proportional to the value of C_x.

The values of resistors R_1–R_6 enable five ranges to be selected with S_1, starting with 0–100 pF in position 1 to 0–1 μF in position 5.

The boxes in the diagram show two ways of displaying the measurement result. A standard BC547 transistor is capable of driving a moving coil meter, M_1, which may be calibrated with the aid of preset P_1. The other way is the use of a variable potential divider and integrator to which a digital voltmeter may be connected.

The meter is calibrated with the aid of a number of capacitors of accurately known value for each of the ranges. Set switch S_1 to range 1, connect one of the test capacitors (of relevant value!) across the test terminals and adjust P_1 or P_2, as the case may be, until the meter reading coincides with the value of the capacitor.

The accuracy of the meter is, of course, dependent on the tolerance of the resistors, which should therefore be 1% in all cases.

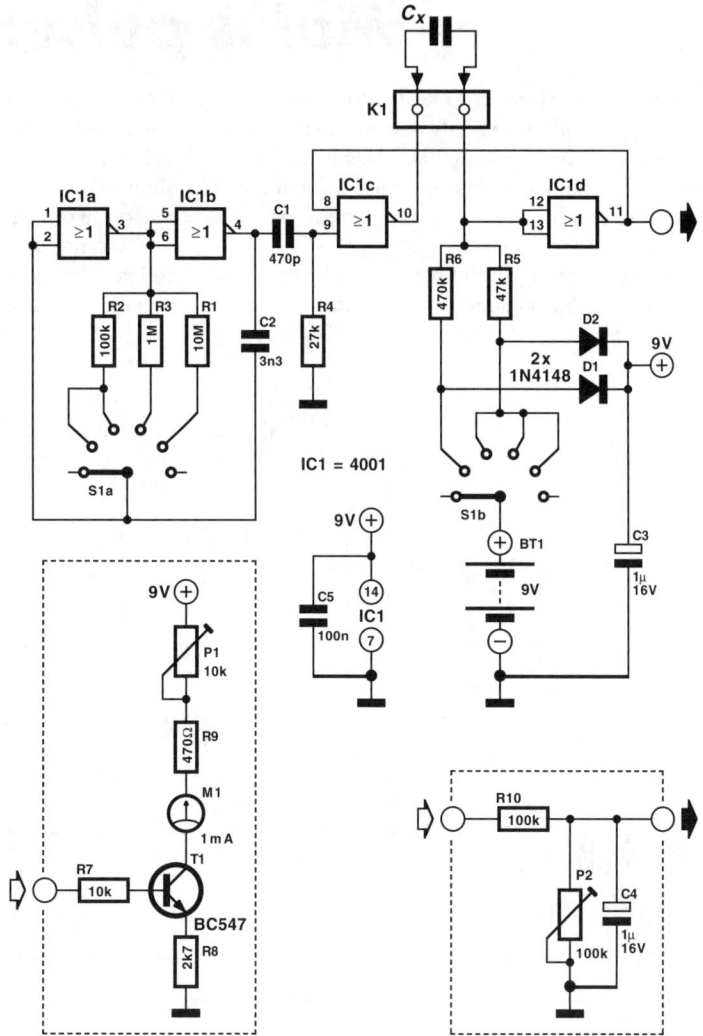

12-bit ADC with I²C™ interface

Until not so long ago, only 8-bit analogue-to-digital converters (ADC) offered an I²C™ interface. Unfortunately, a resolution of 256 samples is inadequate for a number of measurements. The 12-bit ADC described in this short article has a resolution ×16 or even ×64 as high. This makes it a lot more convenient to read, say, the usual temperature sensors, such as the LM335, which out-

R1 1k C1 100n R2 10k R4 10k C2 100n R3 1k
K1

5V (+) L1 100μH
C13 10μ 63V C9 100n
D2 1N4148

K6
9 10
7 8
P3.4 5 6
3 4 P3.5
P3.6 1 2
P3.7
D1
R18 1k5
S1

R5 1k C3 100n R6 10k R8 10k C4 100n R7 1k
K2

IC1 MAX128
11 SHDN
SCL 5
SDA 7
13 CH0
14 CH1
15 CH2   A0 6
16 CH3   A1 10
17 CH4   A2 8
18 CH5
19 CH6   VREF 23
20 CH7   REFADJ 21
A 12   D 4

1 (+) 2 (+)

SCL
SDA

A2 1 0   A1 1 0   A0 1 0

5V (+)
K7
4 3
5 2
6 1

R9 1k C5 100n R10 10k R12 10k C6 100n R11 1k
K3

C12 10n C11 4μ7 63V

R13 1k C7 100n R14 10k R16 10k C8 100n R15 1k
K4

R17 100Ω (+) 5V
K5
+R
0
D3 5V6 1W3
C10 100μ 6V

994018 - 11

put 10 mV °C^{-1}. With a 12-bit ADC and an internal reference voltage of 4.096 V, the resolution is 1 mV, which means that the sensor reading can be read in 0.1 °C. With an 8-bit converter, the resolution would be only 1.6 °C.

The program given below is written for the Matchbox computer a book about which, *Matchbox BASIC Computer* (ISBN 0 905705 53 X) is available from Elektor Electronics (Publishing).

When START is enabled, the program arranges for an interrupt per second to be produced, which means that a sample is taken every second. At the same time, the liquid-crystal display (LCD) is initialized. The final line sends the result of the print instruction to the LCD; if this and the penultimate line are ignored, the results are automatically passed to the monitor screen.

After initialization under START, the program is in an endless loop waiting for interrupts. When an interrupt arrives, a byte is sent to the ADC, whereupon the result (2 bytes) is read. Note that this means that the result must previously be declared as an array of two bytes. Subsequently, the cursor of the display is set to the correct position and the result is written to the LCD. The interrupt routine is ended with the instruction IRETURN.

There are two types of ADC: the MAX 127 and the MAX128. In the MAX127, the software allows an input range of 0–10, 0–5, ±10, or ±5 V, to be selected. In the MAX128, these ranges are 0–V$_{REF}$, 0–V$_{REF}$/2, ±V$_{REF}$, and ±V$_{REF}$/2.

The analogue-to-digital conversion is started

307

by sending a byte to the converter: base address 50_H. Bit 7 must be 1, bits 6–4 determine which of the eight inputs is selected; bits 3, 2 determine which input range is chosen, and bits 1, 0 differentiate between active mode (current drain about
10 mA) and power down (current drain 700 or 120 μA respectively).

After the first byte has been sent, the converter may be read by reading two bytes via the I^{2}C. The first received byte is the most significant bit (MSB), and the four highest bits in the second byte contain the lowest significant bit (LSB). The four lowest bits are zero.

Datasheets for the MAX127 and MAX128 may be downloaded via the Internet: http://www.maxim-ic.com

The interface is best built on the printed-circuit board, which may made with the aid of the track layout.

A final note: if the inputs are used for voltages only, resistors R_2, R_4, R_6, R_8, R_{10}, R_{12}, R_{14}, and R_{16}, may be omitted.

· I^{2}C is a trademark of the Philips Corporation.

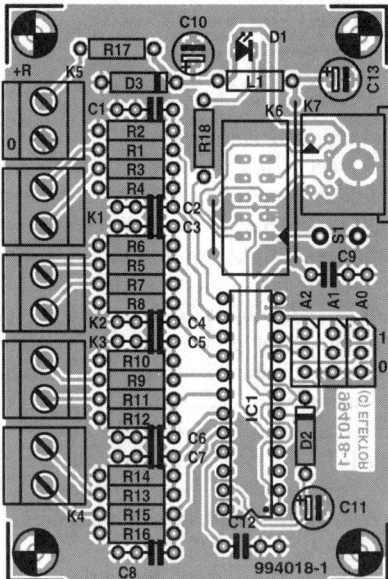

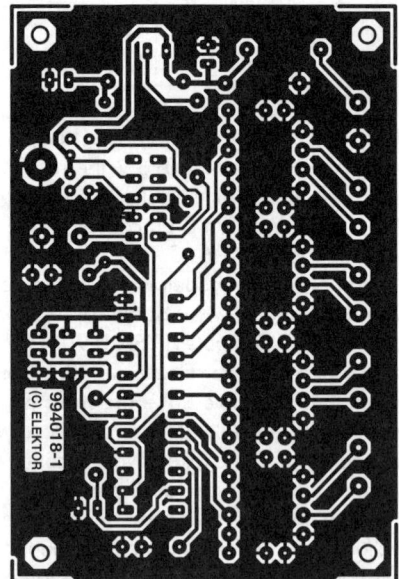

Parts list

Resistors:
$R_1, R_3, R_5, R_7, R_9, R_{11}, R_{13}, R_{15} = 1\ k\Omega$
$R_2, R_4, R_6, R_8, R_{10}, R_{12}, R_{14}, R_{16} = 10\ k\Omega$
  but see text
$R_{17} = 100\ \Omega$
$R_{18} = 1.5\ k\Omega$

Capacitors:
C_1–$C_9 = 0.1\ \mu F$
$C_{10} = 100\ \mu F$, 6 V, radial
$C_{11} = 4.7\ \mu F$, 63 V, radial
$C_{12} = 10\ nF$
$C_{13} = 10\ \mu F$, 63 V, radial

Inductors:
$L_1 = 100\ \mu F$

Semiconductors:
$D_1 = $ LED, high efficiency
$D_2 = $ 1N4148
$D_3 = $ zener, 5.6 V, 1.3 W

Integrated circuits:
$IC_1 = $ MAX128BNCG (Maxim)

Miscellaneous:
JP_1–$JP_3 = $ 3-way jumper
K_1–$K_5 = $ 2-way terminal block for board
  mounting
$K_7 = $ 6-pin mini DIN socket for board mounting
$S_1 = $ push-button switch, 1 make contact

Listing

```
; MAX128.MBL
; MAXIM 128 12 BIT A/D TEST
; 08/04/99 BY W

RESOURCE   IIC-EEPROM     0100H   BYTES
@05000H
RESOURCE 8051-IRAM   10H   BYTES @070H

BYTE RESULT[2]              ; Array for I2C
BYTE CNTRL

START:
  ON INT GOSUB CONVERSION
  TIMER(0,0)               ; Stop Timer
  TIMER(192,4800)          ; Start Timer 1s inter-
val
  SETBITS(INTena,TIMena)   ; Enable interrupts
and Timer interrupt
  LCDSET                   ; Init LCD
  FORMAT(LCD D U LENGTH=5 Z I) ; Output to
LCD, decimal, no sign, 5 digits

LOOP:                      ; Endless loop
GOTO LOOP

CONVERSION:                ; Interrupt routine
every 1s
  CNTRL:=10001000B         ; Start A/D con-
version, input 0, 0..Vref
  IICWR(01010000B,1,CNTRL) ; Write to A/D
  IICRD(01010000B,2,RESULT) ; Read two
bytes (msb & lsb)
  LCDCOM(128)              ; Position cursor
LCD
  PRINT(RESULT[0]*16+RESULT[1]/16) ;

  CLEARBITS(TIMint)        ; Reset timer inter-
rupt flag.
  IRETURN                  ; Return from timer
interrupt

END
```

General scale enhancement

The set frequency of inexpensive frequency or function generators is often indicated by a simple scale on the tuning knob, which is not normally very accurate. It is possible to improve this by adding a counter discriminator circuit consisting of a small number of standard components as

UB V	IB mA	R3 kΩ	R2 min Ω
5	9	15	56
9	15	39	100
12	19 (12)	56	120
15	22 (14)	68	150

shown in the diagram.

The input should be a square wave at TTL level to drive transistor T_1 which functions as a switch. The RC network in the base circuit improves the switching operation at high frequencies.

One of capacitors C_3–C_6, depending on the position of switch S_1, is charged and discharged via T_1 and R_2 in synchrony with the input signal. Resistor R_2 plays an important role in achieving the wanted accuracy. If its value is high, the relevant capacitor cannot be charged fast enough at high frequencies, with the result that the signal at the collector of T_1 is no longer a true square wave. On the other hand, when T_1 is on, the current through the resistor, and thus the total current drain, is small. If, however, the value of R_2 is small, the relevant capacitor is charged at the correct speed, which improves the accuracy, but the current drain is higher. Clearly, the value of the resistor specified is a compromise between these requirements.

The charging and discharge currents of the relevant capacitor flow through the base-emitter junction of T_2 and diode D_1. Owing to the reactance of the capacitor, the currents vary linearly with the frequency. The relatively high currents would result in relatively high pulses at the collector of T_2 were it not for capacitor C_7 which integrates the pulses into a smooth direct voltage, whose level also depends on the frequency. The direct voltage is applied to a milliammeter, M_1, via preset P_1 and resistor R_3.

The circuit may be used for voltages in the range 5–15 V: the current drain, I_B, at several voltages, U_B, and the requisite value of R_3 for a 100 μA meter are shown in the table. The current level in brackets refers to R_2 having a value of

1 kΩ, but this results in the measurement error, which is normally about 2%, rising somewhat. If current drain is of no consequence, the value of R_2 may be lowered until the current through T_1 just does not exceed the maximum permissible level of 100 mA. The minimum values of R_2 for various conditions are shown in the table.

The level of the supply voltage is not critical, but the voltage should be well regulated. If a digital display instead of an analogue meter is used, replace the meter by a 1 kΩ resistor and measure the voltage across this in the 200 mV range. When a liquid-crystal display (LCD) voltmeter is to be used, bear in mind that this needs a discrete supply. To use the total display range of an LCD voltmeter, the ranges may be set to 200 Hz, 2 kHz, 20 kHz, and 200 kHz. It is then necessary to use an r.f. transistor, such as the BSX20, in the T_1 position.

For good accuracy, capacitors C_3–C_6 must have the same tolerance and their values should be in a ratio of 1:10. Note that the circuit capacitance may have an effect when the switch is in the C_3 position.

To calibrate the circuit, apply in each range a

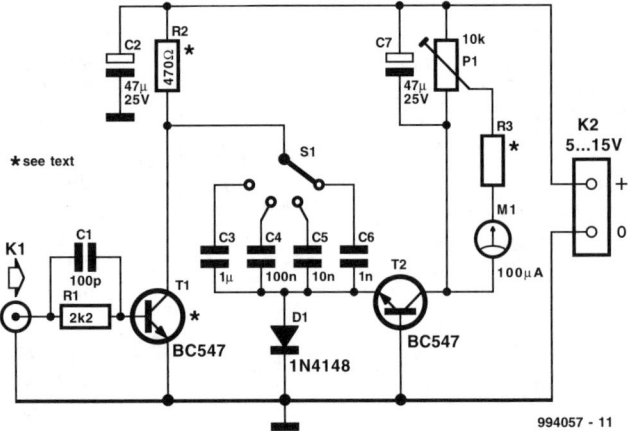

994057 - 11

square-wave input signal at a frequency of about 2/3 of the maximum in that range (660 Hz in the 1 kHz range) and adjust preset P_1 to obtain a voltage reading of about 2/3 of the maximum for that range. Without altering the setting of P_1, find the range in which the highest voltage reading is obtained. Adjust P_1 to exactly the input frequency. In the three other ranges, the readings are then slightly too small, but this is rectified by placing small value capacitors in parallel with the range capacitors until the correct frequency id displayed.

If all this is too cumbersome, replace P_1 by a 10 kΩ resistor and solder four parallel-connected presets in series with R_3. These presets must be switched by a second wafer on S_1. This arrangement allows each range to be calibrated independently of the others. If, furthermore, a three-wafer switch is used, the third wafer enables the decimal point of an LCD to be switched in.

High-resolution AC/DC voltmeter with LED display

Voltmeters with LED display are very useful instruments for repair and maintenance work on electronic equipment. Unfortunately, their resolution is normally poor: of the order of ten per cent at the top end of the range. The instrument described in this article has a resolution at least twice as good and also measures audio frequency signals up to 10 kHz with small error.

Almost all such voltmeters use the well-known Type LM3914 LED driver which can control up to ten light-emitting diodes. This means that with a total measuring range of 0–2 V, the voltage difference between two adjacent LEDs is 200 mV. When a rectangular voltage at a level of half this voltage difference is added to the measurand, and the resulting signal exceeds 200 mV, the next higher LED flashes. This arrangement ensures a resolution of about five per cent at the top end of the range.

The diagram shows the circuit of the voltmeter arranged for measuring ranges of 2 V and 20 V. The divider at the input lowers the analogue signal to about 1/20 when the 20 V position of switch S_1 is selected. Capacitors C_1–C_3 provide frequency compensation up to about 10 kHz. Diodes D_1 and D_2 provide protection against overvoltages. Operational amplifier IC_{1a} functions as an impedance inverter.

Alternating voltages are measured with the aid of active rectifier IC_{1b}, which has excellent linearity/ During a positive half-wave of the signal the op amp operates as an inverting unity-gain amplifier ($R_5=R_6$). Its output is negative, so that diode D_4 conducts, whereupon capacitor C_4 is charged to the peak level of the signal applied pin 6. During a negative half-wave, the output is positive, D_4 is reverse-biased, but the feedback is retained since diode D_3 then conducts. In the case of a direct-voltage input, the

Brief technical data	
Measurement range	100 mV–2 V, resolution 100 mV (AC/DC)
	1–20 V, resolution 1 V (AC/DC)
Frequency range	10 Hz – >10 kHz ±2 dB
	about –2 dB at 100 kHz
Input impedance	about 4.6 MΩ//20 pF
Power supply	9 V dry or rechargeable battery
Current drain	5 mA (stand-by)
	≤ 6 mA (dot mode)
	≤ 20 mA (bar mode)

rectifier is bypassed by switch S_2.

Circuit IC_{1d} is a summing amplifier that combines the measurand and the 100 mV rectangular voltage. Since this stage inverts, whereas the LED driver requires a positive signal, the signals applied to the inputs of IC_{1d} must be negative with respect to earth. This is the reason for the somewhat unusual polarity of the rectifiers and the input terminals.

The rectangular signal is generated by IC_{1c} which, with component values as specified, oscillates at 1.6 Hz. This frequency may be altered within a wide range by changing the value of resistor R_{14} to individual requirements.

The output of the generator drives transistor

switch T_1, which applies the −2.5 V signal periodically to potential divider R_{17}-P_3. The preset enables the exact level of the rectangular voltage to be set, whereupon this is applied via R_8 to pin 13 of IC_1 together with the measurand.

The LED driver, IC_2 is configured in a conventional manner. Its input pin 5 is preceded by filter R_{10}-C_6. The display mode is set with jumper JP_1: when this is left open, a bar display ensues; when it is closed, a dot display results.

The internal reference voltage of 1.25 V, which may be varied with preset P_2, is available at pin 7 (REF_{OUT}) and is used to set the brightness of the display. The current flowing from pin 7 via R_{11} and P_2 determines the LED current (about 1.7 mA). It is, of course, advantageous to use high-efficiency LEDs.

In the power supply section reference diode D_6 ensures that even with fluctuating currents and falling battery voltage the negative supply line remains a stable −2.5 V with respect to earth. This ensures that the voltage across P_3 is independent of the state of the battery.

Diode D_5 provides protection against polarity reversal and is a Schottky type to minimize voltage losses. The battery voltage may drop to 5.5 V before the accuracy of the display begins to suffer.

So as to save energy, an on/off indicator is not provided, but where this is wanted, an LED with a 2.2 kΩ series resistor may be added between the output of IC_{1d} and the +6 V line.

Before the voltmeter can be used, it must, of course, be calibrated. This is begun by checking the potential divider at the input by connecting a digital voltmeter (DVM) at the output of IC_{1a} and applying a 2 V direct voltage (+ to earth) across the input terminals With S_1 in position 2 V, the DVM should indicate about 1 V. When S_1 is then set to position 20 V, the DVM should indicate about 0.1 V.

Next, set S_2 to position = and short-circuit P_3. Apply a voltage of exactly 2 V or 20 V (set S_1 accordingly) to the input terminals and adjust P_2 until D_{16} just lights. Remove the short-circuit from P_3 and apply a voltage of exactly 10 V (set S_1 to 20 V) across the input terminals. In the dot mode, D_{11} should light; in the bar mode, D_7–D_{11} should light. If D_{12} shows a tendency to flash, turn P_3 slightly towards earth until D_{12} is out.

Increase the input voltage to 11 V and adjust P_3 until D_{12} just begins to flash. When the input voltage is raised to 12 V, D_{12} should be fully on. In the dot mode, only one LED should light, but in practice it will be found that when the input is raised from 10 V to 11 V, D_{10} lights constantly and D_{11} begins to flash. When the input is raised slightly more, both diodes light until D_{10} goes out. A clearer indication is obtained in the bar mode, but this adds to the current drain from the battery.

The AC range is calibrated by setting trimmer C_1 to its centre position (use an insulated small screwdriver) and applying a 100 Hz sinusoidal signal to junction R_1-R_2 (earth to K_2). Set S_1 to 20 V. Connect a DVM (10 V d.c. range) across C_4, set S_2 to <> and adjust the generator output until the DVM reads ≤ 1.4 V. Keep the generator output constant, change the frequency to 10 kHz, and adjust trimmer C_2 (insulated screwdriver) until the DVM reading is the same as with 100 Hz.

Next, set S_1 to 2 V and repeat the procedure just described at the same frequencies and adjust C_1 as before. Since the settings of the trimmers influence one another, the calibration of the AC ranges must repeated a couple of times, setting C_1 with S_1 in the 2 V position and C_2 with S_1 in the 20 V position. A properly calibrated divider is virtually linear up to about 10 kHz.

Finally, to set the sensitivity of the instrument, apply a 100 Hz sinusoidal signal at an accurately known level to the input terminals and adjust P_1 until the LED display shows the r.m.s. value of the input signal.

Hold adaptor for voltmeters

Modern, good-quality multimeters have a hold function that enables a measurand to be read even after the test prods have been removed. The present adaptor is intended to add this facility to multimeters and voltmeters that are not so equipped.

In meters with a hold function, the measurand is quantized by an analogue-to-digital converter (ADC) and held in memory. This type of hold function is fairly sophisticated, but for can be attained by a rather less expensive and simpler analogue circuit.

A hold function is invariably based on a capacitor that is charged to the full value of the measurand. The resulting voltage across the capacitor is used to drive the display. In the case of analogue memories a variety of

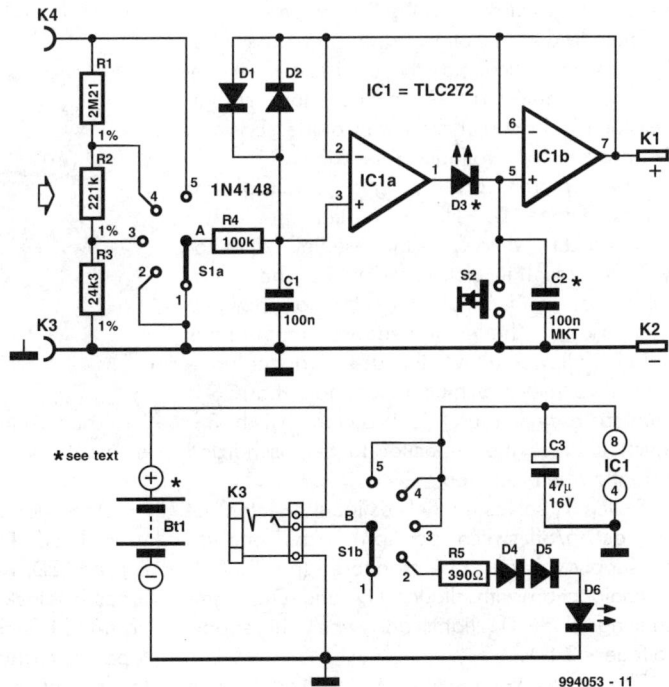

313

phenomena may affect the reading and lead to errors. Also, the charge on the capacitor does not remain constant owing to self-discharge, leakage currents on the board, the input current to a measurement amplifier, and so on. These deficiencies can, however, be negated by the use of a capacitor with a very high insulation resistance, a customized board layout, a current operational amplifier (op amp) with an input impedance in the region of teraohms.

The discharge current of a capacitor cannot be controlled properly by a silicon diode: it is far better to use a light-emitting diode (LED) that is totally screened, so that no light can escape. The photo current is then reduced from a few nA to some pA. This property of LEDs has been known for may years, but is hardly ever used for practical purposes. As an illustration, in the prototype the output voltage had dropped by only 1%, from 1.000 V to 0.990 V, thirty minutes after the original measurement was taken.

The circuit is simplicity itself as is evident from the diagram. In this type of circuit, the potential divider at the input is unavoidable. Here, the divider is arranged to measure input voltages of 2 V, 20 V, and 200 V: the voltmeter connected to the output remains set to the 2 V range.

The divider is followed by low-pass filter R_4-C_1 which decouples the op amps from any noise and interference on the input signal. Resistor R_4, in conjunction with diodes D_1 and D_2, provides overvoltage protection.

The two op amps are arranged as voltage followers. Diode D_3 is the earlier mentioned screened LED, while C_2 is the reservoir capacitor with a very high insulation resistance. The potential across C_2 is applied to the non-inverting input of IC_{2b}. The low-impedance output of this voltage follower allows the use of a analogue pointer voltmeter or multimeter. Note that C_2 is shunted by miniature push-button switch S_2, which enables the capacitor to be discharged instantly when required.

Switch S_1 serves not only as input selector, but also as on/off switch (position 1), and enables the supply voltage to be monitored (position 2). In conjunction with diodes D_4 and D_5, light-emitting diode D_6 lights only when the supply voltage <2.8 V.

The supply voltage may be derived from a 3.6 V Li-ion battery, a 9 V size PP3 dry or rechargeable battery, or a mains adaptor. The current drawn by the hold adaptor is small: in the quiescent mode <1 mA, and in the absence of a hold voltage about 0.2 mA. The specified value of R_5 is right for supply voltages up to 5 V; when a 9 V battery or mains adaptor is used, the value should be increased to 1.2 kΩ.

Pinout of K_1		
K_1	LPT	COM
1	STROBE	
2	AUTOFEED	
3	DATA 0	TxD
4	ERROR	
5	DATA 1	RxD
6	RESET	
7	DATA 2	RTS
8	SELECT IN	
9	DATA 3	CTS
10	GND	
11	DATA 4	DSR
12		
13	DATA 5	GND
14		DTR
15	DATA 6	DCD
16		
17	DATA 7	
18		
19	ACKNOWLEDGE	
20		
21	BUSY	
22		
23	PAPER OUT	
24		
25	ONLINE	
26	n.c.	

When set to positions 3–5, switch S_1 functions as input selector. Diode D_6 is then not in operation.

Only good-quality components should be used. Diode D_3 is a standard (not high-efficiency) red LED, which has been dipped a couple of times in black lacquer, but it can also be made light-tight with good-quality black insulating tape. Its terminals should also be well insulated to prevent leakage currents. The IC should be placed in a good-quality suitable socket. If a board is used, make sure that it is very clean (use pure alcohol) and free of any smudges such as fingerprints or handprints. When the construction work has been completed, it is beneficial to spray the track side with a good insulating lacquer. This will reduce the risk of leakage current sand also keep the tracks clean and free from corrosion.

The adaptor need not be calibrated. After it has been switched on and after every measurement, press S_2 for about a second to make sure that capacitor C_2 is thoroughly discharged. Even then, an offset voltage of 2–3 mV may be measured. However, this is of no consequence, as long as the measurand is higher than this potential.

In principle, the circuit may also be used for measuring alternating voltages: the value of C_1 must then be reduced to about 0.001 μF, which gives an upper limit of 1000 Hz. Owing to the high input resistance and to prevent stray voltages from interfering with the measurement, the input sockets should be replaced by a BNC socket. Owing to the asymmetric supply voltage, the IC, in conjunction with D_3, forms a peak-voltage rectifier. This means that with a sinusoidal input signal, the output of the hold adaptor is ×1.414 higher than the r.m.s. value of the input signal.

LPT/COM *tester*

Any computer owner/enthusiast who likes to experiment should really have an interface tester. This need not be an expensive, proprietary instrument: in most cases a simple Go/NoGo tester as shown in Figure 1 is perfectly adequate. The circuit may be used for testing the level on serial interface lines (COM) as well as on parallel ones (LPT). As a reminder: parallel interfaces normally work with TTL levels (0–5 V), whereas serial interfaces usually work with levels of ±12 V to ±15 V (although interfaces used with laptop and palmtop computers invariably use TTL levels). For both cases a common 26-pin box header with protective shoulder and polarity and positioning lug. The link to the computer is via 25-core flatcable with pressed-on sub-D connectors. Note that pin 26 of the box header is not connected. If the COM interface on the computer has a 9-pin connector, a suitable adaptor must be used. The interconnections between the box header and the interface connectors are shown in the table.

LPT display
The display for the parallel interface consists of a red light-emitting diode (LED) and series resistor

for each wire linked to earth. If the level on the wire is +5 V, the diode lights, except in the case of the ACKNOWLEDGE and STROBE lines. On these lines, the LED is connected via a inverter, T_2 and T_3 respectively, so that it lights when the line is active low. The supply voltage is derived from active low lines AUTOFEED, ERROR, RESET, SELECT IN, and ONLINE, via 'OR gates' D_{25}–D_{29}. Network

R_{18}-C_1 (R_{20}-C_2) stretches the display time, which can be altered to individual requirements by changing the component values within certain limits.

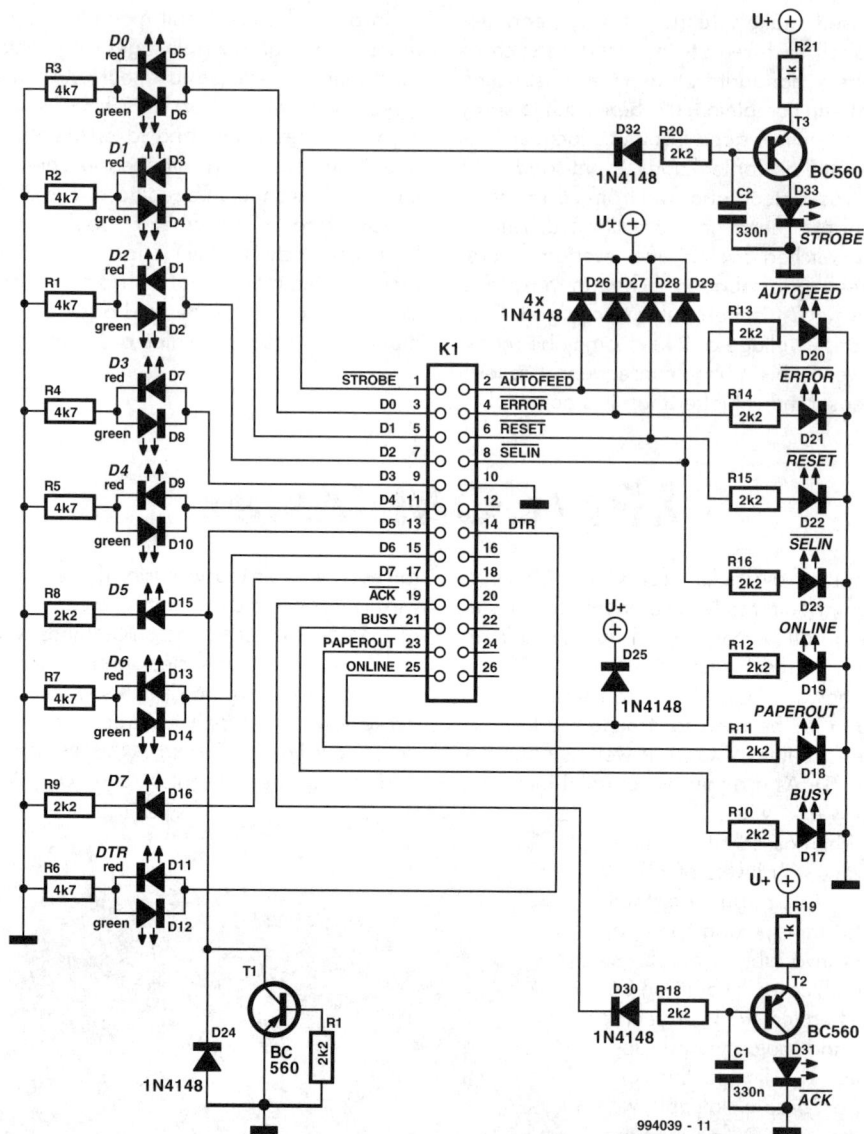

994039 - 11

COM display

The level display of the serial interface uses the same LEDs as the LPT display, but, since negative voltages occur on certain lines, green LEDs are connected in antiparallel with the red LEDs where relevant.

Additionally, there is a DTR line and transistor T_1, which is linked to the (LPT) DATA5 line. In the case of a serial interface, this line is linked to the interface earth, which is passed on as a reference potential via D_{24} when the levels are posi-

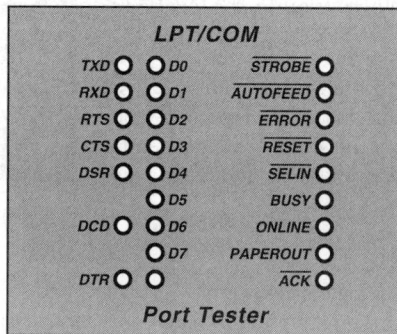

LPT/COM		
TXD ○	○ D0	STROBE ○
RXD ○	○ D1	AUTOFEED ○
RTS ○	○ D2	ERROR ○
CTS ○	○ D3	RESET ○
DSR ○	○ D4	SELIN ○
	○ D5	BUSY ○
DCD ○	○ D6	ONLINE ○
	○ D7	PAPEROUT ○
DTR ○	○	ACK ○

Port Tester

994039 - 12

316

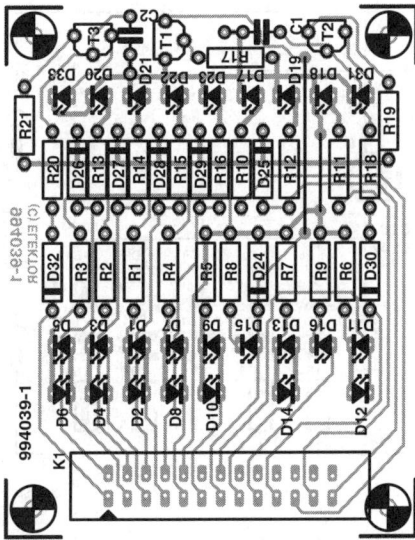

994039-1

(C) ELEKTOR

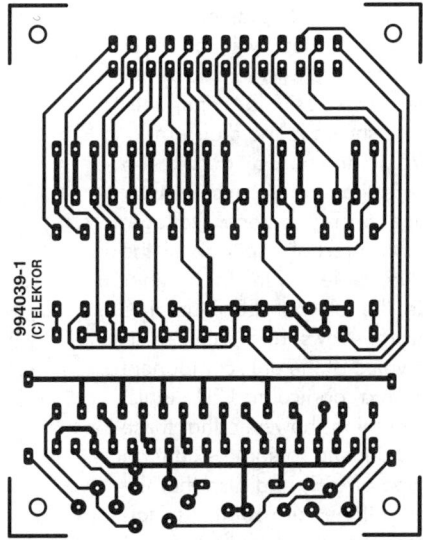

994039-1
(C) ELEKTOR

tive, and via T_1 when the levels are negative. Note that the transistor operates with reduced current amplification since otherwise the brightness of D_{15} in the LPT mode would be insufficient.

Construction

The tester is best built on the printed-circuit board shown, which may be made with the aid of the track layout. The construction is very simple, but do not forget the two wire bridges and make sure that all components, where relevant, are wired in with correct polarity. The cathodes of the LEDs are identified by a short terminal, whereas those of the other diodes are marked by a ring. To ensure that all LEDs are at the same height above the board, first fix the board to the ready drilled lod of the enclosure and only then solder the LEDs in place.

At the left adjacent to K_1 is the COM display (negative levels). In the centre is the common LPT/COM display for positive COM levels as well

as LPT levels of data lines DATA0 to DATA7. At the right is the display for the LPT control levels.

The design of a suggested front cover is shown above.

Parts list
Resistors:
R_1–R_7 = 4.7 kΩ
R_8–R_{18}, R_{20} = 2.2 kΩ
R_{19}, R_{21} = 1 kΩ

Semiconductors:
D_1, D_3, D_5, D_7, D_9, D_{11}, D_{13} = LED, red
D_2, D_4, D_6, D_8, D_{10}, D_{12}, D_{14} = LED, green
D_{15}–D_{23}, D_{31}, D_{33} = LED
D_{24}–D_{30}, D_{32} = 1N4148
T_1–T_3 = BC560

Miscellaneous:
K_1 = 2×13 polarized box header
25-way sub-D male connector for cable fitting
25-way sub-D female connector for cable fitting

Multiple continuity tester

The continuity tester is a handy adjunct to an ohmmeter. The unit or component whose continuity is to be checked is connected between terminals E_1 and E_2 (which may be probes or croc clips). The test current then flowing through the

unit/component on test causes a potential drop across resistor R_2, which is applied to the non-inverting input of buffer IC_2. The output of the op amp is applied to transistor T_1, in the emitter circuit of which there are a number of parallel-con-

317

nected light-emitting diodes. Each LED is in series with a zener diodes and a resistor. The zener diodes have dissimilar zener voltages as shown in the diagram.

When the drop across R_2 exceeds the sum of base-emitter voltage of T_1, a zener voltage, and the threshold voltage of the LED in series with that zener diode, the relevant LED lights. The diagram shows at which resistance value of the unit/component on test a particular LED lights. Bear in mind, however, that these values depend to some extent on the type of LED, and also that the zener voltages are subject to tolerances. Serious deviations may be corrected by the addition of a standard diode or a Schottky diode. It is also possible to add branches to individual requirements, or to use a bar display instead of LEDs.

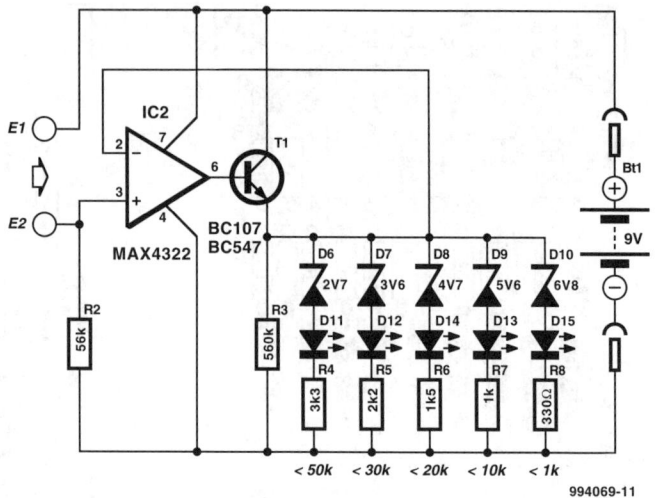

994069-11

It is important that the op amp used has a rail-to-rail output since the input voltages as well as the output may rise to the peak supply voltage. This requirement is met by the MAX4322 as used in the prototype.

AM modulator and 50Ω RF output stage

The 10 MHz Function Generator design published in *Elektor Electronics* in June 1995 has one serious deficiency: it cannot provide amplitude modulation. The standard configuration of the MAX038 IC has no provision for amplitude modulation, in contrast to frequency modulation which is easily achieved. The circuit presented here makes amplitude modulation possible, and also has the significant advantage that it replaces the somewhat exotic and quite expensive OP603AP output opamp with a standard type. Of course, this amplitude modulator can also be used with other models of function generator or for other purposes.

The gain of an NE592 video opamp can be set to 400, 100 or 10 by means of an external

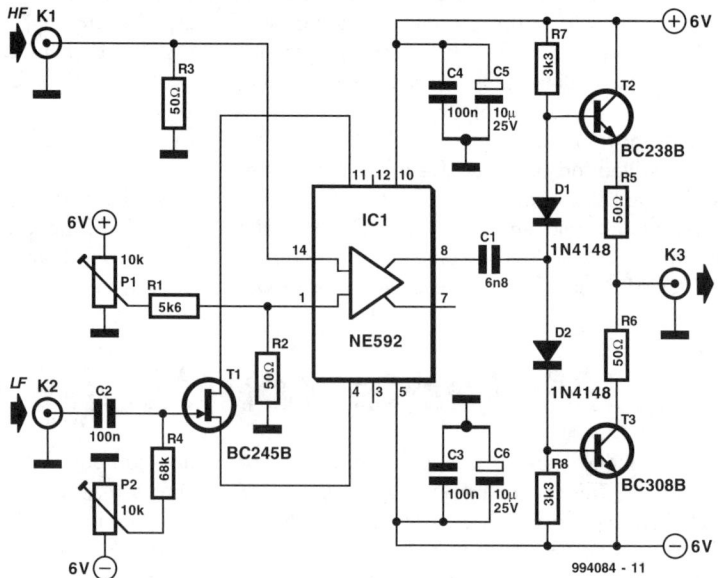

994084 - 11

jumper. Intermediate settings can be achieved by using a suitable resistance in place of the jumper. This adjustment takes place in the emitter leads of the differential amplifier, directly at the input to the opamp, where the signal amplitude is low. A BF245B FET is used as a variable resistance. With suitably low signal levels, it provides at least 50% of clean amplitude modulation for modulating signals (LF) up to 10 kHz and modulated signals (HF) up to 20 MHz. The FET can also be driven with a DC voltage to control the amplitude of the output signal over a 10:1 range with low distortion. Any slight asymmetry of the modulated signal can be corrected by applying a small correction voltage via P_1. Preset P_2 is used to bias the FET at around –2.5 V. The output stage is built from discrete transistors and guarantees a 50 Ω output impedance with low DC offset.

The complete circuit can deliver a constant amplitude output signal of up to 2.5 V_{pp} (unmodulated) for frequencies ranging to over 20 MHz. If the signal is not modulated, the maximum amplitude can be increased somewhat. Output level controls (a potentiometer and/or range switches), if used, should be placed between the NE592 output and the input of the output stage. In such cases, an emitter-follower stage with a high input impedance might be a good idea, since the opamp should operated with a load of at least 1 kΩ. Conceivably, the gate of the FET could be driven via an additional opamp, together with the demodulated signal from the output of the NE592 applied as negative feedback, to achieve higher modulation levels.

Measurement interval generator

The purpose of this circuit is to generate a pulse with a predefined duration when a button is pressed. It is especially well suited for use as a

timing window generator for a frequency counter. It uses only inexpensive standard components and can be quickly put together.

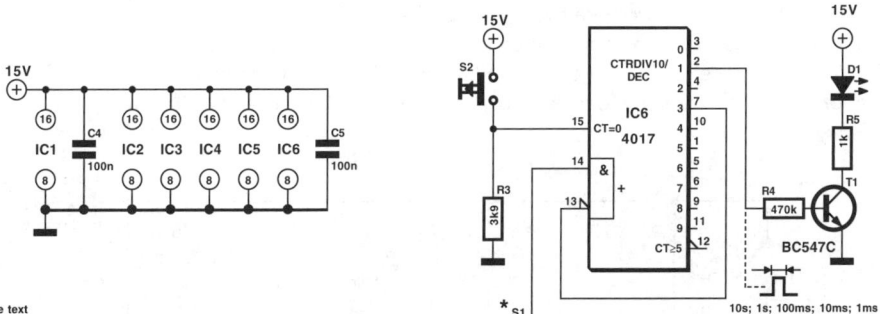

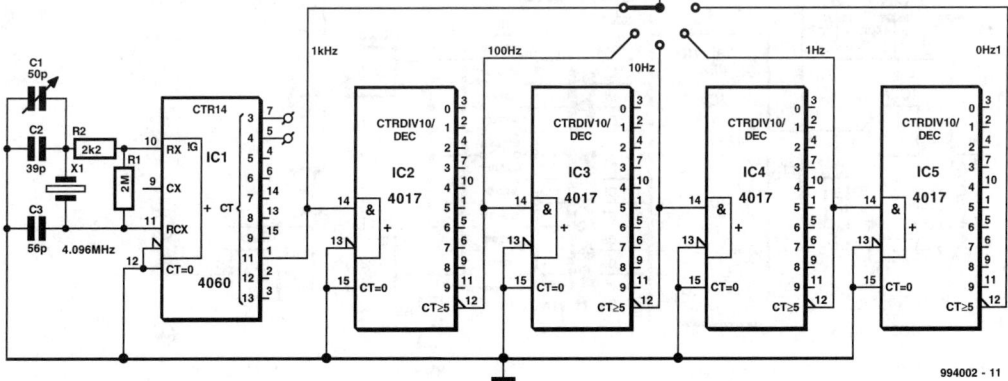

994002 - 11

In the diagram, IC_1 (a 4060) is a 14-stage binary counter with an integrated oscillator. An inexpensive 4.096 MHz crystal is used as the timing reference, which means that a 1 kHz signal appears at the output (pin 1) after division by 2^{14}. Circuit IC_1 is followed by decimal counters IC_2–IC_5, all Type 4017, that are cascaded via their Carry Out outputs (pin 12). These counters produce reference frequencies of 100 Hz, 10 Hz, 1 Hz and 0.1 Hz.

The non-shorting rotary switch S_1 selects one of the reference frequencies and applies it to the clock input of an additional 4017. In contrast to the other ICs of this type, its control inputs Reset and /Enable (pins 15 and 13) are used dynamically. When pushbutton S_2 is pressed, the count is reset to zero. When S_2 is released, the first leading edge at pin 14 clocks the counter. The input signal, divided by 2, appears at the Q1 output (pin 2). However, since Q3 (pin 7) is connected to the /Enable input, the counter is disabled after the first period of the output signal, so that only one pulse is generated. Depending on the input signal, this pulse has a length of 10 s, 1 s, 0.1 s, 0.01 s or 0.001 s.

A simple transistor buffer drives an LED that lights for the duration of the pulse. An additional, similar buffer stage at the output may well be considered. The circuit is powered by a stabilized 15 V supply, from which it draws a current of around 10 mA.

Pulse generator with selectable duty factor

Many pulse generators with adjustable duty cycle have a significant drawback, in that the pulse repetition rate also changes when the duty cycle is adjusted. The circuit shown here avoids this problem. It uses a square wave generator consisting of IC_{2a} and an RC network. Range switch S_1 selects one of three RC combinations that control the pulse repetition rate. With the specified values of C_1, C_2, C_3, P_1, P_2 and R_1, there are three frequency ranges:

I) 0.1 Hz to 10 Hz
II) 10 Hz to 1 kHz
III) 1 kHz to 100 kHz

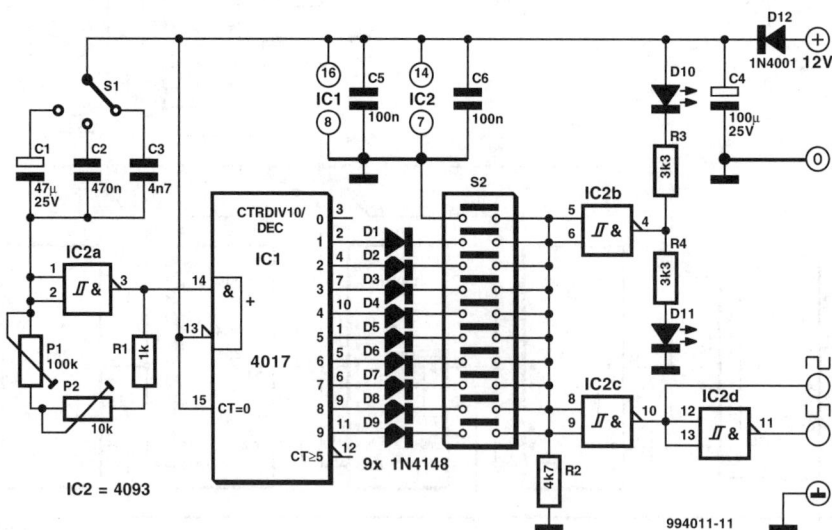

994011-11

Note that the actual frequency depends on the hysteresis of the 4093, so that it is strongly dependent on the specific manufacturer and fabrication of the IC used. For this reason, P_1 and P_2 are provided for coarse and fine frequency adjustment, respectively, to allow the pulse repetition rate to be set exactly to the necessary value. If the desired frequencies cannot be obtained in spite of the wide adjustment range of the circuit, the capacitor values should be modified.

The rectangular-wave generator clocks a synchronous decimal counter. Each time the count goes up, the associated decoded output goes high, while the remaining outputs stay low. DIP switch S_2 allows several outputs to be connected to a single lead. The diodes prevent short-circuits between outputs selected by S_2, whenever the outputs have differing signal levels. For a high/low ratio of 0.1, only the Q1 switch of S_2 should be closed; for a ratio of 0.2, Q1 and Q2 should be closed, and so on. If all outputs of the 4017 (except for Q0) are selected, the resulting duty factor is 0.9. The switch connected to earth can be closed to disable the PWM generator if all other switches are open.

The signal from the counter IC is connected to two inverters. Two LEDs are connected to IC_{2b}. The lower the duty cycle, the brighter D_{11} shines and the dimmer D_{10}. Circuit IC_{2b} buffers (and inverts) the output signal of the circuit, and IC_{2d} re-inverts the signal levels, so that two complimentary signals are available at the output. The circuit draws a current of around 4 mA.

Pulse generator with variable duty factor

The duty factor of the pulse generator in the diagram is variable in 10% steps from 10% to 90%.

With the aid of thumb wheel switch S_1, a 4-bit word, S, is added to one input of IC_1 and to IC_2. Circuit IC_1 is a full adder, while IC_2 functions as a switch. The binary equivalent of 5 is applied to the other input of IC_1. The output of IC_1 is linked to the second input of IC_2.

The output of IC_2 is applied to the programmable input of up/down counter IC_3 via bistable IC_{4a}. The terminal count output of IC_3 clocks IC_{4a} via transistor T_1. The output of IC_{4a} switches the counter from up to down and at the same time links data S or data S+5 to the programmable input of the counter and vice versa.

Imagine that the non-inverting output (pin 1) of IC_{4a} is high. Counter IC_3 then counts downward and the programmable input is linked with S, which is, say, in

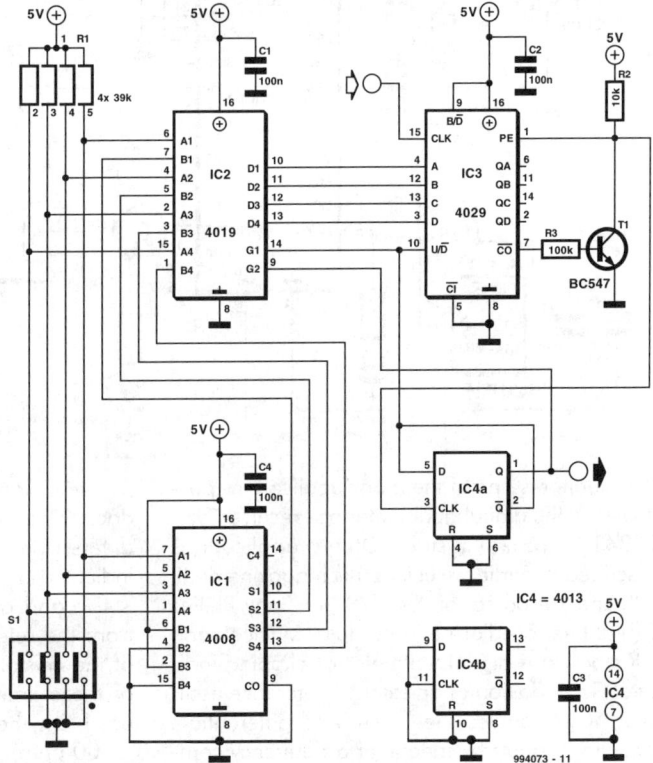

994073 - 11

position 6. The counter counts downward until it reaches 0, which is after six input pulses. The terminal count output goes low, whereupon the non-inverting output of IC_{4a} also goes low, IC3 starts counting upward and is programmed with data S+5 (here, 11), so that the terminal count output becomes high again. When the counter reaches position 15, the terminal count output goes low again, pin 1 of IC_{4a} becomes high, and the whole process starts repeating itself. In short: in position 6 of the thumbwheel switch, the non-inverting output of IC_{4a} (pin 1) is high for six and low for four of every ten input pulses.

S/PDIF monitor

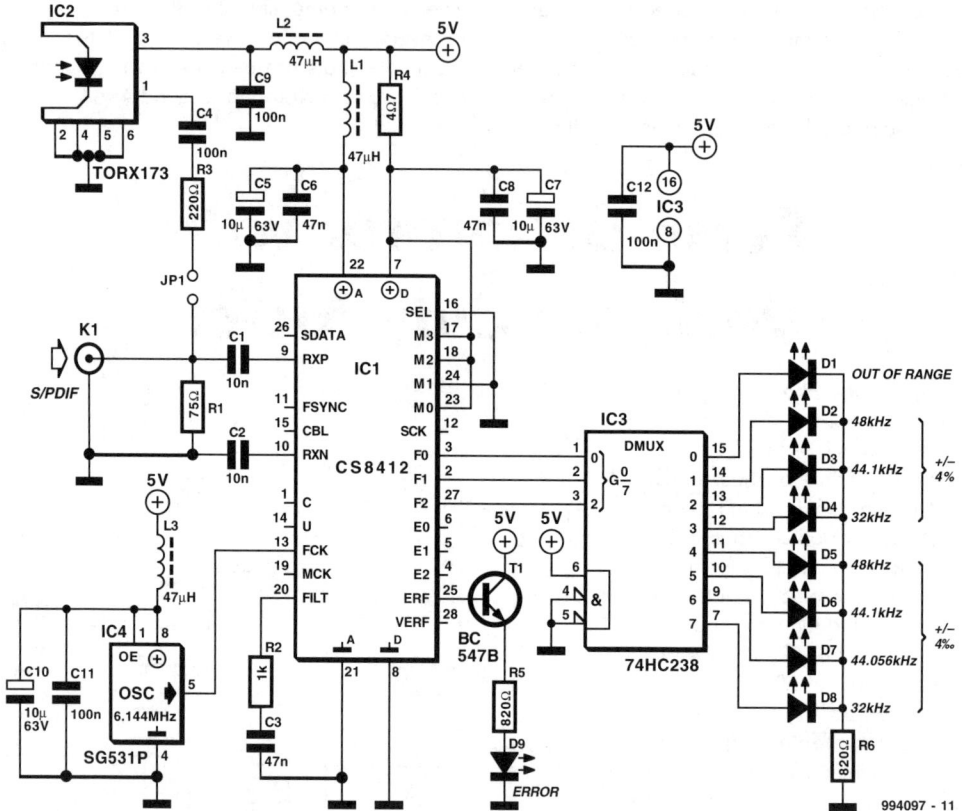

994097 - 11

The monitor is one of the many applications possible with the digital audio interface receiver Type CS8412 from Crystal. Other applications described in earlier issues of this magazine dealt with the decoding of the S/PDIF (Sony/Philips Digital Interface Format) into data, bit clock and L/R clock in a digital voltmeter or clipping indicator. The addition of an external reference oscillator, IC_4, enables the receiver to differentiate incoming signals by means of a frequency comparator – and this is what the present monitor does. When the frequency of an incoming signal differs from a reference value, the difference is indicated in one of three ways: <400 ppm; <4%; and out of range (differs more than 4% from the reference value). Clearly, the accuracy of the crystal oscillator determines the precision of these limits (the SG531P crystal from Epson used in the diagram has an accuracy of ±100 ppm).

The optical input provided by IC_2 is a useful addition. The output of this circuit is applied across R_1 via C_4, R_3 and jumper JP_1. The potential across R_1 may also be used as a digital output, in which case the value of R_3 needs to be adapted as necessary.

The circuit may also be used as a kind of relay station or as a means for reducing jitter. For these purposes, IC_1 is connected in a special mode (mode 13) when M_3 is made 1, and M_0–M_2, 1, 0, and 1, respectively. When these levels are set the received S/PDIF data, including the preamble, is transferred directly to the out-put. The bit clock, SCK, then has a value twice as high as would be the case with coded data. It is possible to connect a TOSLINK module, or a coaxial output via a buffer (such as a number of parallel-linked 74HC04 inverters), to the SDATA output.

A demultiplexer, that is, 3-to-8 line decoder IC_3, is used to decode the data at F_0–F_2 to eight separate light-emitting diodes. Diode D_9 indicates whether IC_1 receives no or poor data. The overall circuit draws a current of not more than 35 mA.

Absolute-value meter with polarity sensor

This circuit breaks an input voltage signal down into its components: (1) the absolute value and (2) the polarity or 'sign' (+ or –). It will handle direct input voltages as well as alternating voltages up to several kHz. With a supply voltage of ±9 V, the input level should remain below ±6 V.

The circuit consists of two sections, each having its own function. Operational amplifiers IC_{1a} and IC_{1b} form a full-wave rectifier, its output terminal supplying the absolute value of the input signal, while operational amplifiers IC_{1c} and IC_{1d} examine the polarity of the input voltage.

When the input voltage is negative, the output of IC_{1a} goes high. Consequently D_2 is reverse-biased so that IC_{1a} has no effect on the rest of the circuit. Circuit IC_{1b} then acts as an inverter because its amplification is $-R_5/R_3 = -1$. So, the output voltage is positive.

When the input voltage is positive, D_2 is forward-biased and the amplification of IC_{1a} is –1. The output voltage is then determined by the sum of currents that flow

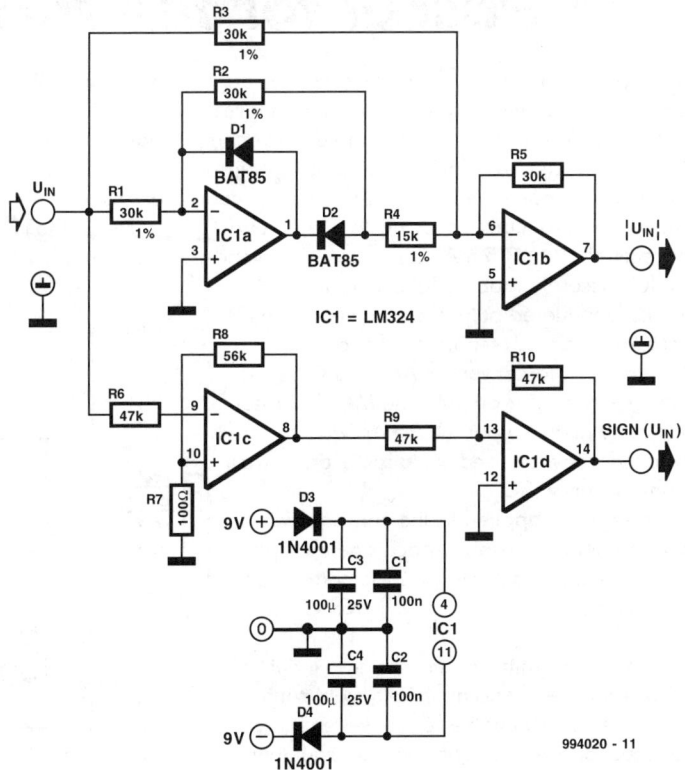

IC1 = LM324

994020 - 11

through R_3 and R_4. Taking into account the polarities and the value of all resistors, the overall amplification is

$$-R_5/R_3+(-R_5/R_4)\leftrightarrow(-R_2/R_1)=-1+2=1$$

This means that the numerical value of the voltage at the output terminal is the same as that of the input voltage, but the polarity is always positive. The accuracy of the rectification process is determined by the accuracy of resistors R_1–R_4; close-tolerance (1%) types are recommended.

At low input voltages below 20 mV, the input offset voltage of the operational amplifiers may introduce significant errors. If this is the case, discrete operational amplifiers instead of multiple ones (TL061, TLC271, AD548, ...) should be used, since these have pins for offset voltage compensation. Alternatively, use an operational amplifier with a low offset voltage like the OP07.

In the polarity sensor, IC_{1c} functions as a comparator with positive feedback provided by

R_7 and R_8. This feedback causes a hysteresis of 20 mV that prevents oscillation when the input voltage changes slowly. Circuit IC_{1d} is an ordinary inverter. With input voltages ≥ 10 mV, the SIGN output terminal swings to almost the positive supply level. When the input voltage ≤ -10 mV, the SIGN terminal drops low, almost to the negative supply voltage. With input voltages between these two thresholds, the output voltage is well defined, too, because it stays at its previous level.

The meter is a perfect complement to the '$\pm$ voltage on bargraph display' (page 262). The $|U_{in}|$ and SIGN outputs of the meter may be connected directly to U_{in} and CONTROL IN inputs of the bargraph display circuit. The ± 6 V sign indicator signal may be used as the control voltage for the $\pm$voltage display as long as the reference voltage remains ≤ 3 V. Although presented as a pair, both circuits may of course be used individually for other purposes.

Pascal for the MAX512

The MAX512 is a simple triple digital-to-analogue converter with a serial interface and a resolution of eight bits. Two of the three converters (DAC A and DAC B) provide a unipolar or bipolar buffered output voltage. Converter A can provide or sink currents of up to 5 mA, and B currents of up to 0.5 mA. Converter C is intended for accurate applications and therefore has an unbuffered output. The reference voltages are applied separately (in contrast to the diagram) to converters A/B and C. Apart from the converter outputs, the MAX512 also has a digital output (1.6 mA) which, for instance, may be used for directly driving a high-efficiency LED.

The data is applied to the converter via a 3-wire interface. The interface operates with frequencies up to 5 MHz and is compatible with standards such as SPI, QSPI, and Microwire.

The serial shift register at the input is 16 bits wide: eight data bits and eight control bits. The latter enable a converter to be selected or switched off. In the shutdown

mode, the R2R network of the relevant converter is isolated from the reference source. The DAC registers may be charged at the leading edge of CS either independently of one another or simultaneously.

The MAX512 may be operated from a single +5 V supply or a symmetrical ±2.5 V supply. It

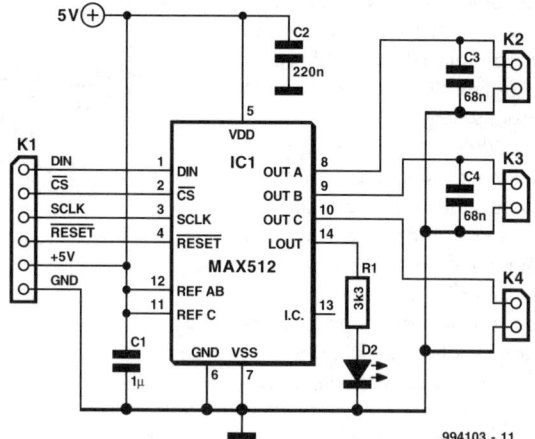

994103 - 11

draws a current of about 1 mA during normal operation and <1 µA in the shutdown mode.

The Pascal program shows clearly how one can work with the MAX512. Serial lines D_{IN}, Chip-Select, CS, clock signal SCLK, and RESET are to be connected to ports P4.0–P4.3 of an 87537 processor, but other devices may also be used for controlling the process. It may then be necessary to adapt the port addresses at the start of the program.

After the program has been started, all three converters generate a five-step staircase voltage at a frequency of about 5 Hz. The introduction of variations in the FOR loop that determines the length of the steps enables the voltage steps to be altered.

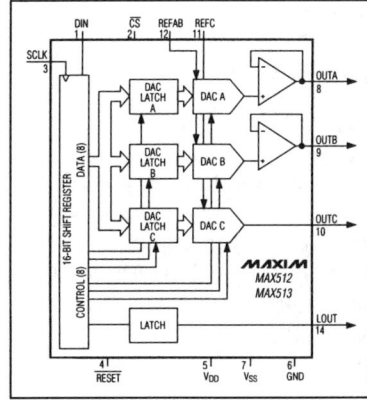

994103 - 12

```
program seri_dau;

const                          DIN = $E8;            (* Serial data line on Port P40 *)
                               CS = $E9;             (* Serial Chip Select line on Port P41 *)
                               SCLK = $EA;           (* Serial Clock line on Port P42 *)
                               RESET = $EB;          (* Reset input of DAU on Port P43 *)


procedure init_dau;                                  (* Bus line defaults *)

begin
  setbit(CS);                                        (* Off state of serial Bus: *)
  clearbit(SCLK);                            (* CS\=HIGH, SCLK=LOW, RESET\=HIGH *)
  setbit(RESET);                                     (* and level of DIN does not matter! *)
end;


procedure reset_dau;                                 (* D/A converter basic settings *)
                                                     (* clear *)

begin
  clearbit(RESET);                                   (* Pull Reset line to active state, *)
  setbit(RESET);                                     (* to load all registers with their *)
end;                                                 (* default values! *)
procedure rausbytes(control,data:byte);              (* Serial transmission of 2 Bytes, *)
                                                     (* always MSB first, LSB last *)

var PEGEL, TEILER, i : byte;


begin
  clearbit(CS);                                      (* Start condition of serial Ds, CS=LOW *)


  Teiler := 128;                                     (* Mask, with 1st iteration for MSB *)
  for i:=1 to 8 do                                   (* Loop for transmission of *)
    begin                                            (* first 8 data bits (Contr. Byte) *)
      PEGEL:=control and TEILER;                     (* Mask off other 7 Bits *)
      if (PEGEL>=1) then setbit(DIN) else clearbit(DIN);
```

```
        setbit(SCLK);                                    (* Depending on bit level, build data line *)
        clearbit(SCLK);                                  (* and apply clock pulse to clock line *)
        TEILER:=TEILER div 2                                                        (* New mask! *)
     end;

  Teiler := 128;                                         (* Mask, with 1st iteration for MSB *)
  for i:=1 to 8 do                                              (* Loop for transmission of *)
     begin                                                (* second 8 data bits (Data Byte) *)
        PEGEL:=data and TEILER;                                  (* Mask off other 7 Bits *)
        if (PEGEL>=1) then setbit(DIN) else clearbit(DIN);
        setbit(SCLK);                                    (* Depending on bit level, build data line *)
        clearbit(SCLK);                                  (* and apply clock pulse to clock line *)
        TEILER:=TEILER div 2                                                        (* New mask! *)
     end;

  setbit(CS);                                            (* Stop condition of serial Ds, CS=HIGH *)
end;

procedure treppe(kanal:byte);                            (* Generate sawtooth on channel: *)
                                               (* 1->channel A, 2->channel B, 3->channel C or *)
                                                          (* 0->on all three channels! *)
var                        kontrollbyte : byte;          (* Aux. variable for channel info *)
                           i : byte;                                     (* Loop variable *)

begin
  case kanal of
  0 : kontrollbyte:=%00000111;                           (* Load all Registers, shut down none  *)
  1 : kontrollbyte:=%00110001;                     (* Load Reg. A only, shut down channel B & C  *)
  2 : kontrollbyte:=%00101010;                     (* Load Reg. B only, shut down channel A & C  *)
  3 : kontrollbyte:=%00001100;                     (* Load Reg. C only, shut down channel A & B  *)
  end;
  for i:=0 to 4 do                                       (* Loop var. = 4 -> staircase w. 5 steps *)
     rausbytes(kontrollbyte, (i*50));
end;

begin
  reset_dau;                                             (* Reset serial D/A converter *)
  init_dau;                                              (* Initialise serial D/A converter *)
  repeat                                       (* Endless loop, create periodic staircase voltage *)
     treppe(0);                                          (* Show starircase on all three channels *)
  until false;
end.
```

Universal countdown timer

The countdown timer is good example of what can be achieved with bare-bones hardware when a powerful microcontroller like the AT89C2051 from Atmel is used. This 20-pin microcontroller has a 2-kByte flash ROM which is compatible with the 8051 Intel architecture. In this article, the AT89C2051 is loaed with a program that simply eliminates a lot of hardware.

The microcontroller is available ready-pro-grammed from the Publishers (Order no. 996511-1).

The user interface of the timer consists of two pushbuttons and three multiplexed 7-segment LED displays. The circuit diagram shows that a small number of inexpensive external parts is required to make it all work. To ensure the necessary degree of electrical isolation, a solid-state relay, IC_1, is used to control an external (mains-powered) load. This load is switched on when the timer starts, and is switched off again when the programmed time has elapsed. The maximum current that may be switched by the relay is about 2 A.

The timer has its own mains power supply consisting of mains transformer Tr_1, bridge recti-fier B_1 and voltage regulator IC_2. This section of the circuit may be separated from the rest by cut-ting the printed circuit board in two (see compo-nent overlay). Due attention should be given to electrical safety when connecting the load and all other mains wiring.

Pressing switch S_1 selects the desired digit and its value is raised by pressing S_2. The time format is [xx s], where the seconds digit indicates tens of seconds. So, the maximum time that can be set is 99.5 minutes = 99 minutes and 30 sec-onds. The resolution of the timer is 10 seconds; the accuracy is determined by quartz crystal X_1.

Once all digits are programmed, a shot peri-od must be allowed until the display stops flash-ing (time-out for entry). Next, press S_2 to switch the load on and to start the countdown process. The programming of the timer is illustrated in the state diagram.

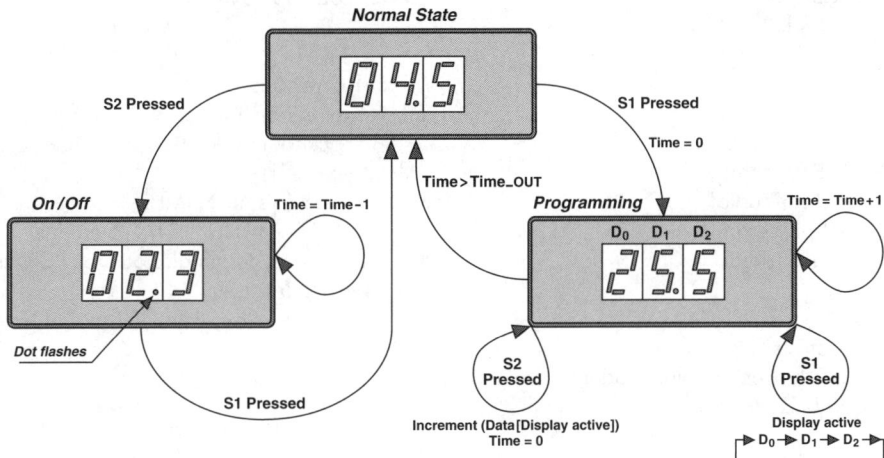

Normal State

S2 Pressed   S1 Pressed

Time = 0

Time > Time_OUT

On/Off   Time = Time-1   **Programming**   Time = Time+1

D_0 D_1 D_2

Dot flashes

S1 Pressed

S2 Pressed   S1 Pressed

Increment (Data [Display active])   Display active
Time = 0 $D_0 \rightarrow D_1 \rightarrow D_2$

994015 - 12

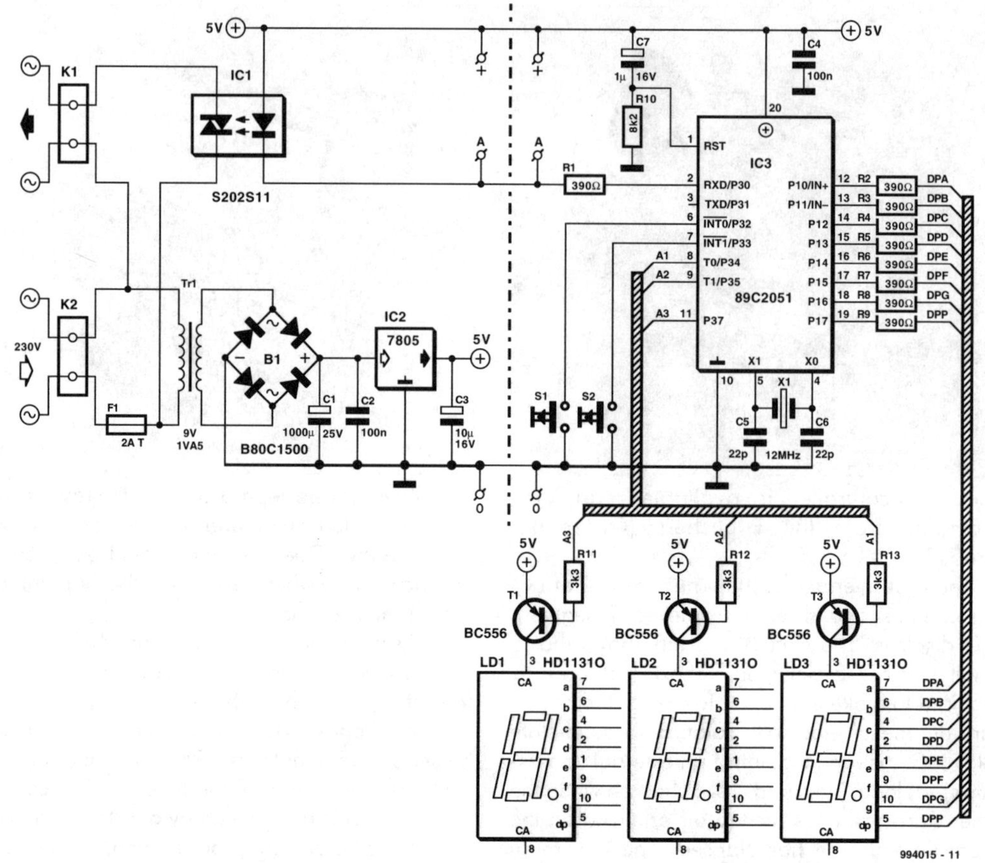

Parts list

Resistors:
R_1–R_9 = 390 Ω
R_{10} = 8.2 kΩ
R_{11}–R_{13} = 3.3 kΩ

Capacitors:
C_1 = 1000 μF, 25 V radial
C_2, C_4 = 0.1 μF, ceramic
C_3 = 10 μF, 16 V radial
C_5, C_6 = 22 pF, ceramic
C_7 = 1 μF, 16 V radial

Semiconductors:
B_1 = B80C1500 (rectangular model)
T_1, T_2, T_3 = BC556

Integrated circuits:
IC_1 = S202S11 (Sharp; distributor: Eurodis)
IC_2 = 7805
IC_3 = AT89C2051 (programmed: Publishers order code 996511-1)

Miscellaneous:
Tr1 = mains transformer, 9 V, 1.5 VA
S_1, S_2 = pushbutton for PCB mounting, e.g., MEC type 3CTL
X_1 = quartz crystal, 12 MHz
LD_1, LD_2, LD_3 = HD1131O (Siemens)
K_1, K_2 = 2-way terminal block for PCB mounting, pitch 7.5 mm
F_1 = fuse, 2 A, slow, with PCB mount holder

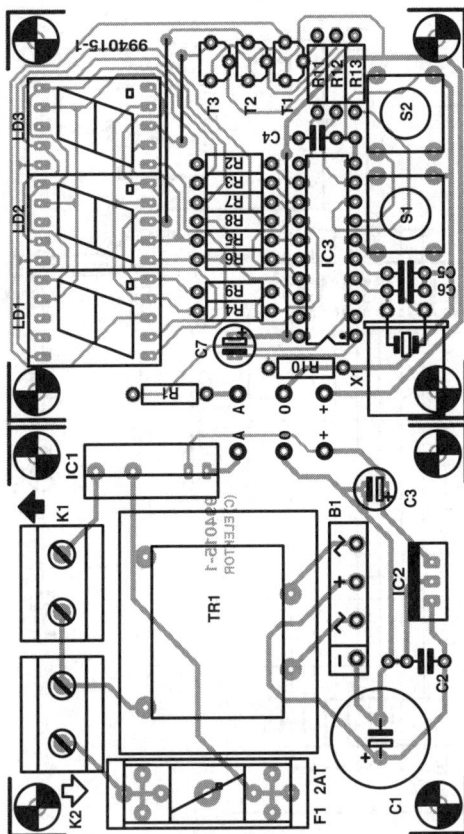

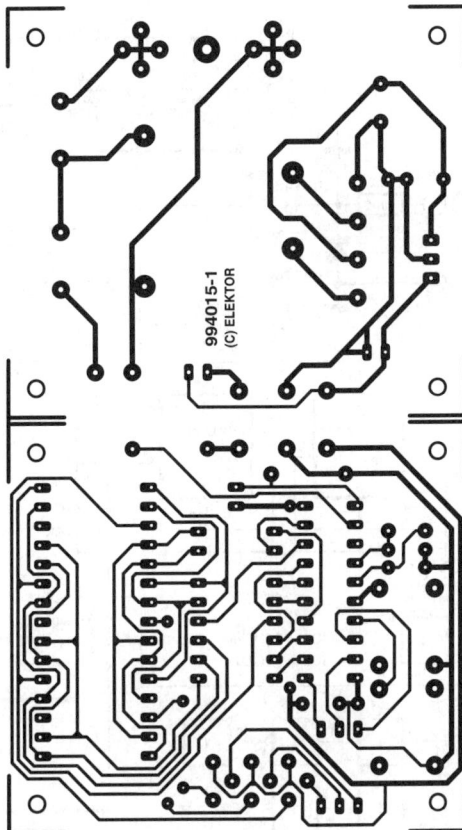

S/PDIF test generator

The generator is intended primarily for checking S/PDIF (Sony/ Philips Digital Interface Format) receivers and any associated digital-to-analogue converters (DAC) and/or output filters. The external clock – standard TTL level – enables 128 sample frequencies to be generated. The clock may also be used for generating standard frequencies with the remaining inverters serving as crystal oscillators (provided a 74HCU04 is used).

The sender is a Type CS8402A digital audio interface transmitter from Crystal. In this short article it is not possible to list all settings that may be obtained with switch S_1: the reader is referred to the data sheet of the IC or to the 'sampling rate converter' published in the October 1996 issue of this magazine. The connections to the switch are exactly as described in that article.

There is an optical (IC_4) as well as a coaxial output (K_1, K_2). Toroidal transformer Tr_1 provides electrical isolation of the coaxial sockets and also serves to prevent earth loops. Capacitors C_2 and C_3 provide the earth connections for the sockets.

The transformer is wound on a TN13/7.5/5-3E25 core with a transformation ratio of 20:2:2 since TXP and TXN (on IC_3) are differential outputs. The primary voltage is 10 V_{pp} to give a signal across the 75 Ω coaxial outputs of 0.5 V_{pp}. After a reset, both outputs are low and are not short-circuited by Tr_1. A coarse audio signal is added to prevent, for instance, muting of the outputs.

Jumper JP_1 enables either the left-hand or right-hand signal to contain a rectangular signal at peak value and half the sampling frequency. This enables, for instance, the channel separa-

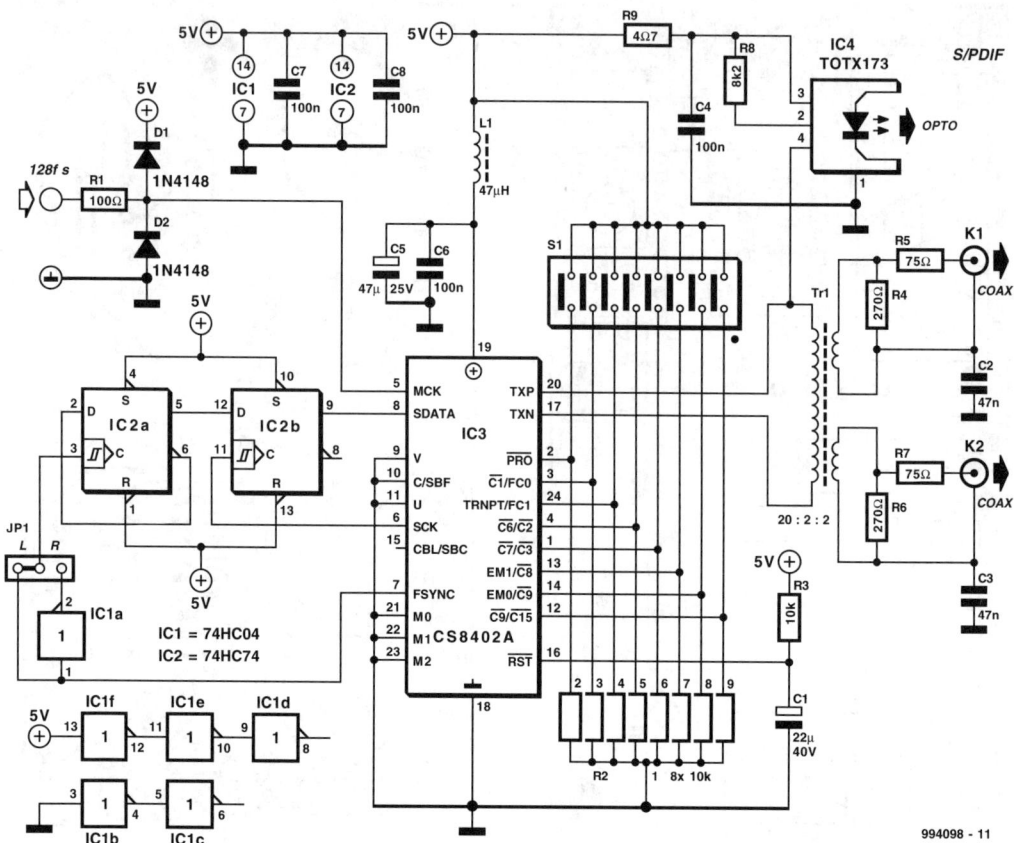

994098 - 11

tion and the combination of digital and analogue signals to be checked. In most DACs, filter action commences at half the sampling frequency. At that instance, there is hardly any attenuation by the analogue filters, so that the level of the sinusoidal signal more or less coincides with that of a 0 dB signal. At this frequency, it is also clearly discernible whether de-emphasis correction is present (S_{1-4} off: de-emphasis on) and, if so, whether this provides the requisite attenuation of 10 dB.

The CS8402A is used in mode 0 (low level at inputs M_0–M_2). This mode is really intended for interfacing with analogue-to-digital converters (ADC), but is used here since it enables the FSYNC of the L/R clock and the bit clock, SCK, to be derived internally from the MCK clock, and to be arranged as outputs. The data for half the sampling frequency are obtained by halving the L/R clock in IC_{2a}. Since the data must be the 2s complement, they are shifted by one clock period in IC_{2b}, so that, depending on the phase of the L/R clock, that is, inverter IC_{1a}, either the left-hand or the right-hand channel contains a peak-level signal. The other channel then toggles one LSB at identical frequency.

It should be noted that some DACs, particularly 1st generation 1-bit types, fail to operate correctly with 0 dB signals, which may cause difficulties with overdriven CDs (see 'Clipping and the CD' in the Readers' Letters column in the April 1999 issue of this magazine). This may be checked with the present generator. If the audio signal is not wanted, the SDATA input should be linked to earth and IC_1 and IC_2 omitted.

Resistor R_1 and diodes D_1, D_2 protect the MCK input against excessive or unbalanced clock signals.

The generator draws a current of about 30 mA.

330

Spike detector for oscilloscope

Many inexpensive digital and analogue/digital oscilloscopes are not capable of rendering spikes in a signal clearly visible. This increases the risk that an undefined voltage pulse is missed altogether, which, when a digital circuit is being tested, can be very frustrating. After all, it means that the oscilloscope is not suitable for detecting and remedying electronic faults.

Fortunately, something can be done about this, provided the oscilloscope has two channels. In the diagram, the level at or above which a spike should be detected is set with preset P_1 and switch S_1. The switch enables unity gain (1:1) or an attenuation of 1:10 to be selected (provided there is a 1:10 probe connected to the input of the oscilloscope). This means that the coarse setting is provided by the switch, whereas the preset provides the fine setting.

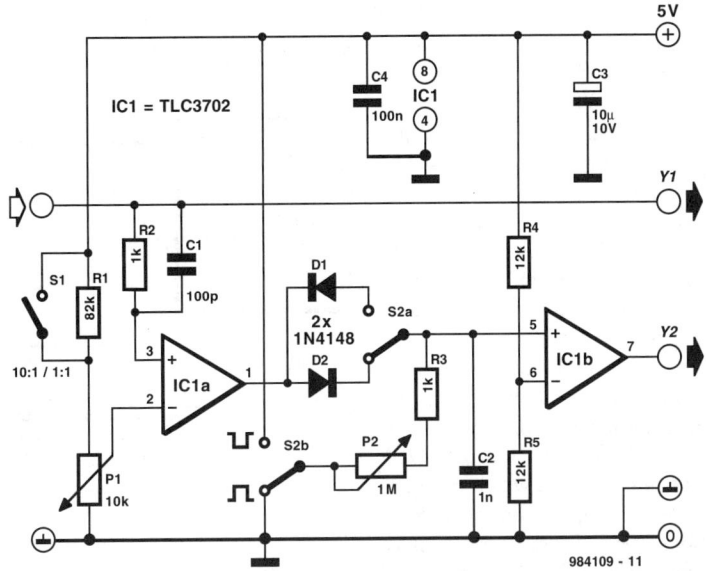

Switch S_2 is used to select whether a logic 0 or logic 1 is to be detected. Spikes as short-lived as $1\,\mu s$ can be sensed by the analogue-to-digital converter (ADC) in the oscilloscope. The duration can be set between $5\,\mu s$ and $500\,\mu s$ with preset P_2.

The detector draws a current of about 1 mA.

Load share controller

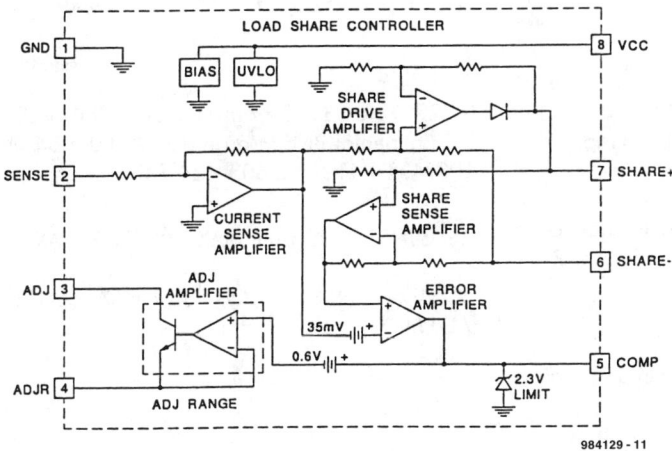

Parallel connected power supplies require a dedicated control mechanism, called load share circuitry to ensure full utilization of the system. The purpose of a load share controller, like Unitrode's UC3902, is to provide for equal distribution of the load current among the parallel connected power supplies. By equalizing the output currents, uniform thermal stress of the individual modules is also ensured which has the utmost importance for long-term reliability of electronic components.

For the UC3902 to work properly, the modular power system has to consist of power supplies that have their own feedback circuits. Furthermore, the stand-alone modules have to be equipped with true remote sense capability or with an output voltage adjustment terminal. Each module must have its own load share controller.

The operating principle of a load share mechanism is to measure the output current of each individual module and to be able to modify the output voltage of the units until all participating power supplies deliver equal output currents. It is accomplished by the UC3902 integrated circuits which are connected to the common load share bus and adjust the positive sense voltage (or the voltage of the output adjust pin) of their respective modules to provide equal load sharing.

The diagram shows an example of two coupled power supplies: it is, of course, essential that each has its own sense input. A low-value current sense resistor is inserted in each of the OUT– lines.

In each of the SENSE+ lines a adjustment resistor, R_{ADJ}, is inserted. This resistor enables the load share controller to apply a correction in the feedback loop of the relevant power supply when a current flows to ground. This forces the power supply to increase its output voltage.

Since the voltage drop across the current sense resistor reduces the range available for output voltage adjustment, this factor has to be accounted for as shown in the equation for R_{ADJ}:

$$R_{ADJ} = [V_{o(m)} - I_{o(m)}(R_s)] / I_{ADJ(m)},$$

where the suffix (m) means maximum, and s means sense.

Apart from the load share controller only a few additional components are needed.

Further information and design formulas are given in Application Note U-163 and Data Sheets Load Share Controller UC1902, UC2902, UC3902 from Unitrode Corporation, 7 Continental Blvd, Merrimack, NH 03054; Tel. 603-424-2410; fax 603-424-3460.

$$R_{SENSE} = V_{SHARE,MAX} / 40 \cdot I_{SOR,MAX}$$

$$R_{ADJ} = (\Delta V_{SOR,MAX} - I_{SOR,MAX} \cdot R_{SENSE}) / I_{ADJ,MAX} = 20...100 \ \Omega$$

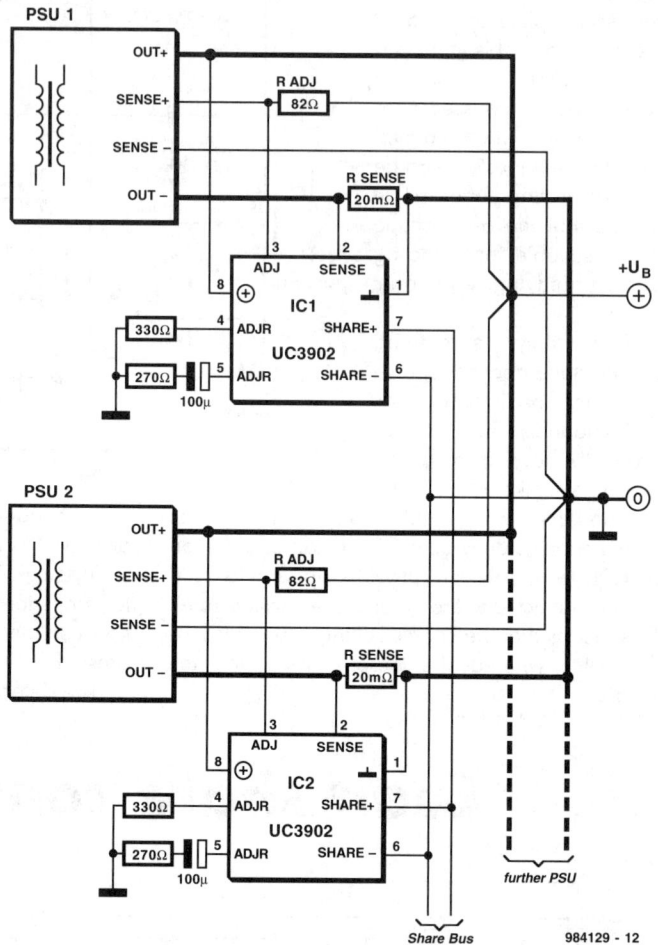

Share Bus   984129 - 12

Hidden switch

In some circumstances, it may be very useful to switch on a row of socket outlets simultaneously. The drawback of this is that any unauthorized person can operate the switch and use the equipment connected to the outlets. This can be prevented by hiding a switch in one of the outlets that breaks the live line and can be actuated only by those with a need to know.

The switch consists of a relay, a drive circuit consisting of a thyristor and two reed contacts, S_1 and S_2. When S_1 is operated by a magnet, a trigger current flows to the thyristor via R_1. The thyristor then energizes the relay and the mains live line is reconnected to the outlets. At the same time, D_2 lights to indicate that the mains voltage is available.

The thyristor remains on after the holding current has ceased, since the holding voltage is exceeded. This situation obtains until reed contact S_2 is closed and short-circuits the thyristor. Since the holding voltage is then insufficient, the thyristor is cut off, and the mains to the outlets is switched off.

In the diagram, Tr_1, Br_1 and C_1 form an unregulated power supply. Diode D_1 is the usual freewheeling diode for the relay coil.

The previous page shows the cross-section of one of the mains socket outlet boxes. The reed contacts are fixed in such a way that they can be operated by a magnet through the wall (lid) of the box.

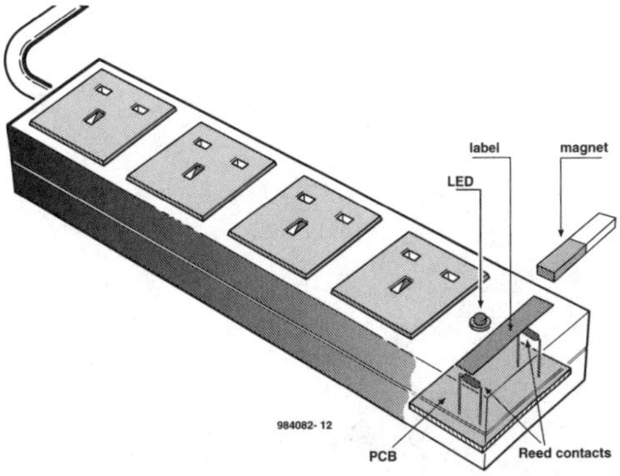

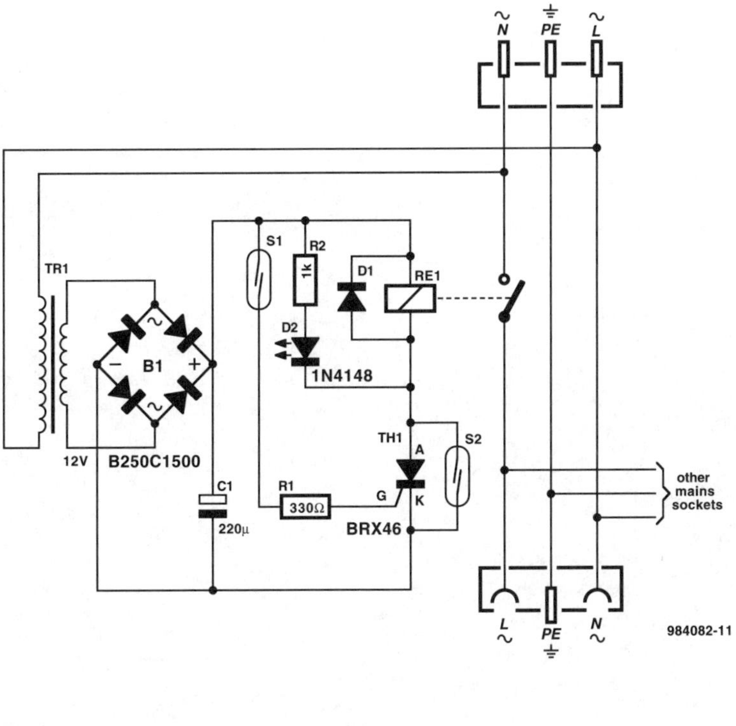

Index